Lecture Notes in Artificial Intelligence   4754

Edited by J. G. Carbonell and J. Siekmann

Subseries of Lecture Notes in Computer Science

T0241059

Marcus Hutter   Rocco A. Servedio
Eiji Takimoto (Eds.)

Algorithmic Learning Theory

18th International Conference, ALT 2007
Sendai, Japan, October 1-4, 2007
Proceedings

Springer

Series Editors

Jaime G. Carbonell, Carnegie Mellon University, Pittsburgh, PA, USA
Jörg Siekmann, University of Saarland, Saarbrücken, Germany

Volume Editors

Marcus Hutter
RSISE @ ANU and SML @ NICTA
Canberra, ACT, 0200, Australia
E-mail: marcus@hutter1.net

Rocco A. Servedio
Columbia University
New York, NY, USA
E-mail: rocco@cs.columbia.edu

Eiji Takimoto
Tohoku University
Sendai 980-8579, Japan
E-mail: t2@maruoka.ecei.tohoku.ac.jp

Library of Congress Control Number: 2007935310

CR Subject Classification (1998): I.2.6, I.2.3, F.1, F.2, F.4, I.7

LNCS Sublibrary: SL 7 – Artificial Intelligence

ISSN   0302-9743
ISBN-10 3-540-75224-2 Springer Berlin Heidelberg New York
ISBN-13 978-3-540-75224-0 Springer Berlin Heidelberg New York

Springer is a part of Springer Science+Business Media

springer.com

© Springer-Verlag Berlin Heidelberg 2007
Printed in Germany

Typesetting: Camera-ready by author, data conversion by Scientific Publishing Services, Chennai, India
Printed on acid-free paper   SPIN: 12164653   06/3180   5 4 3 2 1 0

Preface

This volume contains the papers presented at the 18th International Conference on Algorithmic Learning Theory (ALT 2007), which was held in Sendai (Japan) during October 1–4, 2007. The main objective of the conference was to provide an interdisciplinary forum for high-quality talks with a strong theoretical background and scientific interchange in areas such as query models, on-line learning, inductive inference, algorithmic forecasting, boosting, support vector machines, kernel methods, complexity and learning, reinforcement learning, unsupervised learning and grammatical inference. The conference was co-located with the Tenth International Conference on Discovery Science (DS 2007).

This volume includes 25 technical contributions that were selected from 50 submissions by the Program Committee. It also contains descriptions of the five invited talks of ALT and DS; longer versions of the DS papers are available in the proceedings of DS 2007. These invited talks were presented to the audience of both conferences in joint sessions.

- Avrim Blum (Carnegie Mellon University, Pittsburgh, USA): "A Theory of Similarity Functions for Learning and Clustering" (invited speaker for ALT 2007)
- Thomas G. Dietterich (Oregon State University, Corvallis, Oregon, USA): "Machine Learning in Ecosystem Informatics" (invited speaker for DS 2007)
- Masaru Kitsuregawa (The University of Tokyo, Tokyo, Japan): "Challenge for Info-plosion" (invited speaker for DS 2007)
- Alex Smola (National ICT Australia / ANU, Canberra, Australia): "A Hilbert Space Embedding for Distributions" (invited speaker for ALT 2007)
- Jürgen Schmidhuber (IDSIA, Lugano, Switzerland): "Simple Algorithmic Principles of Discovery, Subjective Beauty, Selective Attention, Curiosity and Creativity" (joint invited speaker for ALT 2007 and DS 2007)

Since 1999, ALT has been awarding the *E. Mark Gold* Award for the most outstanding paper by a student author. This year the award was given to Markus Maier for his paper "Cluster Identification in Nearest-Neighbor Graphs," co-authored by Matthias Hein and Ulrike von Luxburg. We thank Google for sponsoring the E.M. Gold Award.

ALT 2007 was the 18th in a series of annual conferences established in Japan in 1990. Another ancestor of ALT 2007 is the conference series Analogical and Inductive Inference, held in 1986, 1989, and 1992, which merged with the ALT conference series after a collocation in 1994. ALT subsequently became an international conference series which has kept its strong links to Japan but has also regularly been held at overseas destinations including Australia, Germany, Italy, Singapore, Spain and the USA.

Continuation of the ALT series is supervised by its Steering Committee, consisting of: Thomas Zeugmann (Hokkaido University, Japan) Chair, Steffen Lange

(FH Darmstadt, Germany) Publicity Chair, Naoki Abe (IBM Thomas J. Watson Research Center, Yorktown, USA), Shai Ben-David (University of Waterloo, Canada), Marcus Hutter (Australian National University, Canberra, Australia), Roni Khardon (Tufts University, Medford, USA), Phil Long (Google, Mountain View, USA), Akira Maruoka (Ishinomaki Senshu University, Japan), Rocco Servedio (Columbia University, New York, USA), Takeshi Shinohara (Kyushu Institute of Technology, Iizuka, Japan), Frank Stephan (National University of Singapore, Republic of Singapore), Einoshin Suzuki (Kyushu University, Fukuoka, Japan), and Osamu Watanabe (Tokyo Institute of Technology, Japan).

We would like to thank all of the individuals and institutions who contributed to the success of the conference: the authors for submitting papers, and the invited speakers for accepting our invitation and lending us their insight into recent developments in their research areas. We wish to thank the following sponsors for their generous financial support: the Air Force Office of Scientific Research (AFOSR); the Asian Office of Aerospace Research and Development (AOARD)[1]; Google for sponsoring the E.M.Gold Award; Graduate School of Information Sciences (GSIS), Tohoku University for providing secretarial assistance and equipment as well; the Research Institute of Electrical Communication (RIEC), Tohoku University; New Horizons in Computing, MEXT Grant-in-Aid for Scientific Research on Priority Areas; and the Semi-Structured Data Mining Project, MEXT Grant-in-Aid for Specially Promoted Research.

We are also grateful for the Technical Group on Computation (COMP) of the Institute of Electronics, Information and Communication Engineers (IEICE) for its technical sponsorship; the Division of Computer Science, Hokkaido University for providing the Web page and online submission system; and the Institute for Theoretical Computer Science, University of Lübeck where Frank Balbach developed a part of the online submission system.

We thank the Local Arrangements Chair Akira Ishino (Tohoku University, Japan) for his great assistance in making the conference a success in many ways. We thank Vincent Corruble for making the beautiful poster. We thank Springer for its continuous support in the preparation of this volume.

We would also like to thank all Program Committee members for their hard work in reviewing the submitted papers and participating in on-line discussions. We thank the external referees whose reviews made a substantial contribution to the process of selecting papers for ALT 2007.

We are grateful to the Discovery Science conference for its ongoing collaboration with ALT. In particular we would like to thank the Conference Chair Ayumi Shinohara (Tohoku University, Japan) and the Program Committee Chairs Vincent Corruble (UPMC, Paris, France) and Masayuki Takeda (Kyushu University, Japan) for their cooperation and support.

Finally, we would like to express special thanks to Thomas Zeugmann for his continuous support of the ALT conference series and in particular for his great

---

[1] AFOSR/AOARD support is not intended to express or imply endorsement by the U.S.Federal Government.

service in maintaining the ALT Web pages and the ALT submission system, which he programmed together with Frank Balbach and Jan Poland. Thomas Zeugmann assisted us in many ways by answering countless questions related to running the conference and preparing the proceedings.

July 2007

Marcus Hutter
Rocco A. Servedio
Eiji Takimoto

Table of Contents

Online Learning

Unsupervised Learning

Language Learning

Query Learning

Kernel-Based Learning

Other Directions

Editors' Introduction

Marcus Hutter, Rocco A. Servedio, and Eiji Takimoto

Philosophers have pondered the phenomenon of learning for millennia; scientists and psychologists have studied learning for more than a century. But the analysis of learning as a *computational* and *algorithmic* phenomenon is much more recent, going back only a few decades. Learning theory is now an active research area that incorporates ideas, problems, and techniques from a wide range of disciplines including statistics, artificial intelligence, information theory, pattern recognition, and theoretical computer science. Learning theory has many robust connections with more applied research in machine learning and has made significant contributions to the development of applied systems and to fields such as electronic commerce and computational biology.

Since learning is a complex and multi-faceted phenomenon, it should come as no surprise that a wide range of different theoretical models of learning have been developed and analyzed. This diversity in the field is well reflected in the topics addressed by the invited speakers to ALT 2007 and DS 2007, and by the range of different research topics that have been covered by the contributors to this volume in their papers. The research reported here ranges over areas such as unsupervised learning, inductive inference, complexity and learning, boosting and reinforcement learning, query learning models, grammatical inference, online learning and defensive forecasting, and kernel methods. In this introduction we give an overview first of the five invited talks of ALT 2007 and DS 2007 and then of the regular contributions in this volume. We have grouped the papers under different headings to highlight certain similarities in subject matter or approach, but many papers span more than one area and other alternative groupings are certainly possible; the taxonomy we offer is by no means absolute.

Avrim Blum works on learning theory, online algorithms, approximation algorithms, and algorithmic game theory. His interests within learning theory include similarity functions and clustering, semi-supervised learning and co-training, online learning algorithms, kernels, preference elicitation and query learning, noise-tolerant learning, and attribute-efficient learning. In his invited talk for ALT 2007, Avrim spoke about developing a theory of similarity functions for learning and clustering problems. Some of the aims of this work are to provide new insights into what makes kernel functions useful for learning, and to understand what are the minimal conditions on a similarity function that allow it to be useful for clustering.

Alexander Smola works on nonparametric methods for estimation, in particular kernel methods and exponential families. He studies estimation techniques including Support Vector Machines, Gaussian Processes and Conditional Random Fields, and uses these techniques on problems in bioinformatics, pattern recognition, text analysis, computer vision, network security, and optimization for parallel processing. In his invited lecture for ALT 2007, co-authored with Arthur Gretton, Le Song, and Bernhard Schölkopf, Alexander spoke about a

M. Hutter, R.A. Servedio, and E. Takimoto (Eds.): ALT 2007, LNAI 4754, pp. 1–8, 2007.

technique for comparing distributions without the need for density estimation as an intermediate step. The approach relies on mapping the distributions into a reproducing kernel Hilbert space, and has a range of applications that were presented in the talk.

Masaru Kitsuregawa works on data mining, high performance data warehousing, high performance disk and tape arrays, parallel database processing, data storage and the Web, and related topics. His invited lecture for DS 2007 was about "Challenges for Info-plosion."

Thomas G. Dietterich studies topics in machine learning including sequential and spatial supervised learning, transfer learning, and combining knowledge and data to learn in knowledge-rich/data-poor application problems. He works on applying machine learning to a range of problems such as ecosystem informatics, intelligent desktop assistants, and applying AI to computer games. His invited lecture for DS 2007 discussed the role that machine learning can play in ecosystem informatics; this is a field that brings together mathematical and computational tools to address fundamental scientific and application problems in the ecosystem sciences. He described two on-going research efforts in ecosystem informatics at Oregon State University: (a) the application of machine learning and computer vision for automated arthropod population counting, and (b) the application of linear Gaussian dynamic Bayesian networks for automated cleaning of data from environmental sensor networks.

Jürgen Schmidhuber has worked on a range of topics related to learning, including artificial evolution, learning agents, reinforcement learning, metalearning, universal learning algorithms, Kolmogorov complexity and algorithmic probability. This work has led to applications in areas such as finance, robotics, and optimization. In his invited lecture (joint for ALT 2007 and DS 2007), Jürgen spoke about the algorithmic nature of discovery, perceived beauty, and curiosity. Jürgen has been thinking about this topic since 1994, when he postulated that among several patterns classified as "comparable" by some subjective observer, the subjectively most beautiful is the one with the simplest (shortest) description, given the observer's particular method for encoding and memorizing it. As one example of this phenomenon, mathematicians find beauty in a simple proof with a short description in the formal language they are using.

We now turn our attention to the regular contributions contained in this volume.

Inductive Inference. Research in inductive inference follows the pioneering work of Gold, who introduced a recursion-theoretic model of "learning in the limit." In the basic inductive inference setting, a learning machine is given a sequence of (arbitrarily ordered) examples drawn from a (recursive or recursively enumerable) language L, which belongs to a known class C of possible languages. The learning machine maintains a hypothesis which may be updated after each successive element of the sequence is received; very roughly speaking, the goal is for the learning machine's hypothesis to converge to the target language after finitely many steps. Many variants of this basic scenario have been studied in inductive inference during the decades since Gold's original work.

John Case, Timo Kötzing and Todd Paddock study a setting of learning in the limit in which the time to produce the final hypothesis is derived from some ordinal which is updated step by step downwards until it reaches zero, via some "feasible" functional. Their work first proposes a definition of feasible iteration of feasible learning functionals, and then studies learning hierarchies defined in terms of these notions; both collapse results and strict hierarchies are established under suitable conditions. The paper also gives upper and lower runtime bounds for learning hierarchies related to these definitions, expressed in terms of exponential polynomials.

John Case and Samuel Moelius III study *iterative learning*. This is a variant of the Gold-style learning model described above in which each of a learner's output conjectures may depend only on the learner's current conjecture and on the current input element. Case and Moelius analyze two extensions of this iterative model which incorporate parallelism in different ways. Roughly speaking, one of their results shows that running several distinct instantiations of a single learner in parallel can actually increase the power of iterative learners. This provides an interesting contrast with many standard settings where allowing parallelism only provides an efficiency improvement. Another result deals with a "collective" learner which is composed of a collection of communicating individual learners that run in parallel.

Sanjay Jain, Frank Stephan and Nan Ye study some basic questions about how hypothesis spaces connect to the class of languages being learned in Gold-style models. Building on work by Angluin, Lange and Zeugmann, their paper introduces a comprehensive unified approach to studying learning languages in the limit relative to different hypothesis spaces. Their work distinguishes between four different types of learning as they relate to hypothesis spaces, and gives results for vacillatory and behaviorally correct learning. They further show that every behaviorally correct learnable class has a *prudent* learner, i.e., a learner using a hypothesis space such that it learns every set in the hypothesis space.

Sanjay Jain and Frank Stephan study Gold-style learning of languages in some special numberings such as Friedberg numberings, in which each set has exactly one number. They show that while explanatorily learnable classes can all be learned in some Friedberg numberings, this is not the case for either behaviorally correct learning or finite learning. They also give results on how other properties of learners, such as consistency, conservativeness, prudence, iterativeness, and non U-shaped learning, relate to Friedberg numberings and other numberings.

Complexity aspects of learning. Connections between complexity and learning have been studied from a range of different angles. Work along these lines has been done in an effort to understand the computational complexity of various learning tasks; to measure the complexity of classes of functions using parameters such as the Vapnik-Chervonenkis dimension; to study functions of interest in learning theory from a complexity-theoretic perspective; and to understand connections between Kolmogorov-style complexity and learning. All four of these aspects were explored in research presented at ALT 2007.

Vitaly Feldman, Shrenek Shah, and Neal Wadhwa analyze two previously studied variants of Angluin's exact learning model that make learning more challenging: learning from equivalence and incomplete membership queries, and learning with random persistent classification noise in membership queries. They show that under cryptographic assumptions about the computational complexity of solving various problems the former oracle is strictly stronger than the latter, by demonstrating a concept class that is polynomial-time learnable from the former oracle but is not polynomial-time learnable from the latter oracle. They also resolve an open question of Bshouty and Eiron by showing that the incomplete membership query oracle is strictly weaker than a standard perfect membership query oracle under cryptographic assumptions.

César Alonso and José Montaña study the Vapnik-Chervonenkis dimension of concept classes that are defined in terms of arithmetic operations over real numbers. Such bounds are of interest in learning theory because of the fundamental role the Vapnik-Chervonenkis dimension plays in characterizing the sample complexity required to learn concept classes. Strengthening previous results of Goldberg and Jerrum, Alonso and Montaña give upper bounds on the VC dimension of concept classes in which the membership test for whether an input belongs to a concept in the class can be performed by an arithmetic circuit of bounded depth. These new bounds are polynomial both in the depth of the circuit and in the number of parameters needed to codify the concept.

Vikraman Arvind, Johannes Köbler, and Wolfgang Lindner study the problem of properly learning k-juntas and variants of k-juntas. Their work is done from the vantage point of parameterized complexity, which is a natural setting in which to consider the junta learning problem. Among other results, they show that the consistency problem for k-juntas is $W[2]$-complete, that the class of k-juntas is fixed parameter PAC learnable given access to a $W[2]$ oracle, and that k-juntas can be fixed parameter improperly learned with equivalence queries given access to a $W[2]$ oracle. These results give considerable insight on the junta learning problem.

The goal in transfer learning is to solve new learning problems more efficiently by leveraging information that was gained in solving previous related learning problems. One challenge in this area is to clearly define the notion of "relatedness" between tasks in a rigorous yet useful way. M. M. Hassan Mahmud analyzes transfer learning from the perspective of Kolmogorov complexity. Roughly speaking, he shows that if tasks are related in a particular precise sense, then joint learning is indeed faster than separate learning. This work strengthens previous work by Bennett, Gács, Li, Vitányi and Zurek.

Online Learning. Online learning proceeds in a sequence of rounds, where in each round the learning algorithm is presented with an input x and must generate a prediction y (a bit, a real number, or something else) for the label of x. Then the learner discovers the true value of the label, and incurs some loss which depends on the prediction and the true label. The usual overall goal is to keep the total loss small, often measured relative to the optimal loss over functions from some fixed class of predictors.

Jean-Yves Audibert, Rémi Munos and Csaba Szepesvári deal with the stochastic multi-armed bandit setting. They study an Upper Confidence Bound algorithm that takes into account the empirical variance of the different arms. They give an upper bound on the expected regret of the algorithm, and also analyze the concentration of the regret; this risk analysis is of interest since it is clearly useful to know how likely the algorithm is to have regret much higher than its expected value. The risk analysis reveals some unexpected tradeoffs between logarithmic expected regret and concentration of regret.

Jussi Kujala and Tapio Elomaa also consider a multi-armed bandit setting. They show that the "Follow the Perturbed Leader" technique can be used to obtain strong regret bounds (which hold against the best choice of a fixed lever in hindsight) against adaptive adversaries in this setting. This extends previous results for FPL's performance against non-adaptive adversaries in this setting.

Vovk's Aggregating Algorithm is a method of combining hypothesis predictors from a pool of candidates. Steven Busuttil and Yuri Kalnishkan show how Vovk's Aggregating Algorithm (AA) can be applied to online linear regression in a setting where the target predictor may change with time. Previous work had only used the Aggregating Algorithm in a static setting; the paper thus sheds new light on the methods that can be used to effectively perform regression with a changing target. Busuttil and Kalnishkan also analyze a kernel version of the algorithm and prove bounds on its square loss.

Unsupervised Learning. Many of the standard problems and frameworks in learning theory fall under the category of "supervised learning" in that learning is done from labeled data. In contrast, in unsupervised learning there are no labels provided for data points; the goal, roughly speaking, is to infer some underlying structure from the unlabeled data points that are received. Typically this means clustering the unlabeled data points or learning something about a probability distribution from which the points were obtained.

Markus Maier, Matthias Hein, and Ulrike von Luxburg study a scenario in which a learning algorithm receives a sample of points from an unknown distribution which contains a number of distinct clusters. The goal in this setting is to construct a "neighborhood graph" from the sample, such that the connected component structure of the graph mirrors the cluster ancestry of the sample points. They prove bounds on the performance of the k-nearest neighbor algorithm for this problem and also give some supporting experimental results. Markus received the E. M. Gold Award for this paper, as the program committee felt that it was the most outstanding contribution to ALT 2007 which was co-authored by a student.

Kevin Chang considers an unsupervised learning scenario in which a learner is given access to a sequence of samples drawn from a mixture of uniform distributions over rectangles in d-dimensional Euclidean space. He gives a streaming algorithm which makes only a small number of passes over such a sequence, uses a small amount of memory, and constructs a high-accuracy (in terms of statistical distance) hypothesis density function for the mixture. A notable feature of the algorithm is that it can handle samples from the mixture that are presented in any arbitrary order. This result extends earlier work of Chang and Kannan which dealt with mixtures of uniform distributions over rectangles in one or two dimensions.

Language Learning. The papers in this group deal with various notions of learning languages in the limit from positive data. Ryo Yoshinaka's paper addresses the question of what precisely is meant by the notion of efficient language learning in the limit; despite the clear intuitive importance of such a notion, there is no single accepted definition. The discussion focuses particularly on learning very simple grammars and minimal simple grammars from positive data, giving both positive and negative results on efficient learnability under various notions.

François Denis and Amaury Habrard study the problem of learning stochastic tree languages, based on a sample of trees independently drawn according to an unknown stochastic language. They extend the notion of rational stochastic languages over strings to the domain of trees. Their paper introduces a canonical representation for rational stochastic languages over trees, and uses this representation to give an efficient inference algorithm that identifies the class of rational stochastic tree languages in the limit with probability 1.

Query Learning. In query learning the learning algorithm works by making queries of various types to an oracle or teacher; this is in contrast with "passive" statistical models where the learner typically only has access to random examples and cannot ask questions. The most commonly studied types of queries are membership queries (requests for the value of the target function at specified points) and equivalence queries (requests for counterexamples to a given hypothesis). Other types of queries, such as subset queries (in which the learner asks whether the current hypothesis is a subset of the target hypothesis, and if not, receives a negative counterexample) and superset queries, are studied as well.

Sanjay Jain and Efim Kinber study a query learning framework in which the queries used are variants of the standard queries described above. In their model the learner receives the *least* negative counterexample to subset queries, and is also given a "correction" in the form of a positive example which is nearest to the negative example; they also consider similarly modified membership queries. These variants are motivated in part by considerations of human language learning, in which corrected versions of incorrect utterances are often provided as part of the learning process. Their results show that "correcting" positive examples can sometimes give significant additional power to learners.

Cristina Tîrnăucă and Timo Knuutila study query learning under a different notion of correction queries, in which the prefix of a string (the query) is "corrected" by the teacher responding with the lexicographically first suffix that yields a string in the language. They give polynomial time algorithms for pattern languages and k-reversible languages using correction queries of this sort. These results go beyond what is possible for polynomial-time algorithms using membership queries alone, and thus demonstrate the power of learning from these types of correction queries.

Lev Reyzin and Nikhil Srivastava study various problems of learning and verifying properties of hidden graphs given query access to the graphs. This setting lends itself naturally to a range of query types that are somewhat different from those described above; these include edge detection, edge counting, and shortest path queries. Reyzin and Srivastava give bounds on learning and verifying general graphs, degree-bounded graphs, and trees with these types of queries. These results extend our understanding of what these types of queries can accomplish.

Rika Okada, Satoshi Matsumoto, Tomoyuki Uchida, Yusuke Suzuki and Takayoshi Shoudai study learnability of finite unions of linear graph patterns from equivalence queries and subset queries. These types of graph patterns are useful for data mining from semi-structured data. The authors show that positive results can be achieved for learning from equivalence and subset queries (with counterexamples), and give negative results for learning from restricted subset queries (in which no counterexamples are given).

Kernel-Based Learning. A kernel function is a mapping which, given two inputs, implicitly represents each input as a vector in some (possibly high-dimensional or infinite dimensional) feature space and outputs the inner product between these two vectors. Kernel methods have received much attention in recent years in part because it is often possible to compute the value of the kernel function much more efficiently than would be possible by performing an explicit representation of the input as a vector in feature space. Kernel functions play a crucial role in Support Vector Machines and have a rich theory as well as many uses in practical systems.

Developing new kernel functions, and selecting the most appropriate kernels for particular learning tasks, is an active area of research. One difficulty in constructing kernel functions is in ensuring that they obey the condition of positive semidefiniteness. Kilho Shin and Tetsuji Kuboyama give a sufficient condition under which it is ensured that new candidate kernels constructed in a particular way from known positive semidefinite kernels will themselves be positive semidefinite and hence will indeed be legitimate kernel functions. Their work gives new insights into several kernel functions that have been studied recently such as principal-angle kernels, determinant kernels, and codon-improved kernels.

Guillaume Stempfel and Liva Ralaivola study how kernels can be used to learn data separable in the feature space except for the presence of random classification noise. They describe an algorithm which combines kernel methods, random projections, and known noise tolerant approaches for learning linear separators over finite dimensional feature spaces, and give a PAC style analysis of the algorithm. Given noisy data which is such that the noise-free version would be linearly separable with a suitable margin in the implicit feature space, their approach yields an efficient algorithm for learning even if the implicit feature space has infinitely many dimensions.

Adam Kowalczyk's paper deals with analyzing hypothesis classes that consist of linear functionals superimposed with "smooth" feature maps; these are the types of hypotheses generated by many kernel methods. The paper studies continuity of two important performance metrics, namely the error rate and the area under the ROC (receiver operating characteristic curve), for hypotheses of this sort. Using tools from real analysis, specifically transversality theory, he shows that pointwise convergence of hypotheses implies convergence of these measures with probability 1 over the selection of the test sample from a suitable probability density.

Other Directions. Several papers presented at ALT do not fit neatly into the above categories, but as described below each of these deals with an active and interesting area of research in learning theory.

Hypothesis boosting is an approach to combining many weak classifiers, or "rules of thumb," each of which performs only slightly better than random guessing, to obtain a high-accuracy final hypothesis. Boosting algorithms have been intensively studied and play an important role in many practical applications. In his paper, Takafumi Kanamori studies how boosting can be applied to estimate conditional probabilities of output labels in a multiclass classification setting. He proposes loss functions for boosting algorithms that generalize the known margin-based loss function and shows how regularization can be introduced with an appropriate instantiation of the loss function.

Reinforcement learning is a widely studied approach to sequential decision problems that has achieved considerable success in practice. Dealing with the "curse of dimensionality," which arises from large state spaces in Markov decision processes, is a major challenge. One approach to dealing with this challenge is *state aggregation*, which is based on the idea that similar states can be grouped together into meta-states. In his paper Ronald Ortner studies pseudometrics for measuring similarity in state aggregation. He proves an upper bound on the loss incurred by working with aggregated states rather than original states and analyzes how online aggregation can be performed when the MDP is not known to the learner in advance.

In defensive forecasting, the problem studied is that of online prediction of the binary label associated with each instance in a sequence of instances. In this line of work no assumption is made that there exists a hidden function dictating the labels, and in contrast with other work in online learning there is no comparison class or "best expert" that is compared with. One well-studied parameter of algorithms in this setting is the calibration error, which roughly speaking measures the extent to which the forecasts are accurate on average. In his paper Vladimir V. V'yugin establishes a tradeoff between the calibration error and the "coarseness" of any prediction strategy by showing that if the coarseness is small then the calibration error can also not be too small. This negative result comes close to matching the bounds given in previous work by Kakade and Foster on a particular forecasting system.

July 2007   Marcus Hutter
                                                  Rocco A. Servedio
                                                  Eiji Takimoto

A Theory of Similarity Functions for Learning and Clustering

Avrim Blum

Department of Computer Science
Carnegie Mellon University
Pittsburgh, PA 15213
avrim@cs.cmu.edu

Abstract. Kernel methods have proven to be powerful tools in machine learning. They perform well in many applications, and there is also a well-developed theory of sufficient conditions for a kernel to be useful for a given learning problem. However, while a kernel can be thought of as just a pairwise similarity function that satisfies additional mathematical properties, this theory requires viewing kernels as implicit (and often difficult to characterize) maps into high-dimensional spaces. In this talk I will describe work on developing a theory that applies to more general similarity functions (not just legal kernels) and furthermore describes the usefulness of a given similarity function in terms of more intuitive, direct properties, without need to refer to any implicit spaces.

An interesting feature of the proposed framework is that it can also be applied to learning from purely unlabeled data, i.e., clustering. In particular, one can ask how much stronger the properties of a similarity function should be (in terms of its relation to the unknown desired clustering) so that it can be used to *cluster* well: to learn well without any label information at all. We find that if we are willing to relax the objective a bit (for example, allow the algorithm to produce a hierarchical clustering that we will call successful if some pruning is close to the correct answer), then this question leads to a number of interesting graph-theoretic and game-theoretic properties that are sufficient to cluster well. This work can be viewed as an approach to defining a PAC model for clustering.

This talk is based on work joint with Maria-Florina Balcan and Santosh Vempala.

M. Hutter, R.A. Servedio, and E. Takimoto (Eds.): ALT 2007, LNAI 4754, p. 9, 2007.
© Springer-Verlag Berlin Heidelberg 2007

Machine Learning in Ecosystem Informatics*

Thomas G. Dietterich

Oregon State University, Corvallis, Oregon, USA
tgd@eecs.oregonstate.edu
http://web.engr.oregonstate.edu/~tgd

The emerging field of Ecosystem Informatics applies methods from computer science and mathematics to address fundamental and applied problems in the ecosystem sciences. The ecosystem sciences are in the midst of a revolution driven by a combination of emerging technologies for improved sensing and the critical need for better science to help manage global climate change. This paper describes several initiatives at Oregon State University in ecosystem informatics.

In the area of sensor technologies, there are several projects at Oregon State University that are developing new ways of sensing the environment. One project seeks to develop a dense network of battery-free temperature sensors that will be deployed in a watershed in the H. J. Andrews Experimental Forest, which is operated by Oregon State. Each sensor combines a digital thermometer, a radio transmitter, a radio receiver, and a circuit that harvests energy from spread-spectrum radio frequency signals broadcast from a base station. A second project is applying computer vision techniques to automatically identify and count small arthropods with applications in water quality monitoring and biodiversity studies. The project has developed robotic devices for manipulating and photographing specimens as well as computer vision algorithms that learn to classify specimens to genus and species levels.

Once data is collected by sensors, a second challenge is to clean the data to remove anomalies due to sensor failures. A third project at Oregon State has developed dynamic Bayesian network methods to model the normal behavior of an array of 12 temperature sensors deployed at the Andrews forest. Low-probability departures from normal are identified as anomalies. The corresponding data values are marked before being made available on the Andrews web site.

Oregon State has also developed two educational programs to train people to work in interdisciplinary Ecosystem Informatics research teams. One program is a 10-week residential summer research program at the Andrews forest. The other is an interdisciplinary graduate program that brings together students in mathematics, computer science, and the ecosystem sciences to learn about each others' fields and work together on joint projects. Students in this program earn a Ph.D. in their home field, but they also earn a "minor" in Ecosystem Informatics by taking a series of courses and writing one chapter in their dissertation relating to Ecosystem Informatics.

---

* The full version of this paper is published in the Proceedings of the 10th International Conference on Discovery Science, Lecture Notes in Artificial Intelligence Vol. 4755.

M. Hutter, R.A. Servedio, and E. Takimoto (Eds.): ALT 2007, LNAI 4754, pp. 10–11, 2007.

The long term goal of this work is to transform ecology from a discipline that relies on hand-crafted analytical and computational models to a data exploration science in which models are built and tested more automatically based on massive data sets collected automatically. We encourage more people to work in this important new research area.

Challenge for Info-plosion[*]

Masaru Kitsuregawa

Institute of Industrial Science, The University of Tokyo
4-6-1 Komaba, Meguro-ku, Tokyo 153-8505, Japan
kitsure@tkl.iis.u-tokyo.ac.jp

Abstract. Information created by people has increased rapidly since the year 2000, and now we are in a time which we could call the "information-explosion era." The project "Cyber Infrastructure for the Information-explosion Era" is a six-year project from 2005 to 2010 supported by Grant-in-Aid for Scientific Research on Priority Areas from the Ministry of Education, Culture, Sports, Science and Technology (MEXT) of Japan. The project aims to establish the following fundamental technologies in this information-explosion era: novel technologies for efficient and trustable information retrieval from explosively growing and heterogeneous information resources; stable, secure, and scalable information systems for managing rapid information growth; and information utilization by harmonized human-system interaction. It also aims to design a social system that cooperates with these technologies. Moreover, it maintains the synergy of cutting-edge technologies in informatics.

---

[*] The full version of this paper is published in the Proceedings of the 10th International Conference on Discovery Science, Lecture Notes in Artificial Intelligence Vol. 4755.

M. Hutter, R.A. Servedio, and E. Takimoto (Eds.): ALT 2007, LNAI 4754, p. 12, 2007.
© Springer-Verlag Berlin Heidelberg 2007

A Hilbert Space Embedding for Distributions

Alex Smola[1], Arthur Gretton[2], Le Song[1], and Bernhard Schölkopf[2]

[1] National ICT Australia, North Road, Canberra 0200 ACT, Australia
alex.smola@nicta.com.au,lesong@it.usyd.edu.au
[2] MPI for Biological Cybernetics, Spemannstr. 38, 72076 Tübingen, Germany
{arthur,bernhard.schoelkopf}@tuebingen.mpg.de

Abstract. We describe a technique for comparing distributions without the need for density estimation as an intermediate step. Our approach relies on mapping the distributions into a reproducing kernel Hilbert space. Applications of this technique can be found in two-sample tests, which are used for determining whether two sets of observations arise from the same distribution, covariate shift correction, local learning, measures of independence, and density estimation.

Kernel methods are widely used in supervised learning [1, 2, 3, 4], however they are much less established in the areas of testing, estimation, and analysis of probability distributions, where information theoretic approaches [5, 6] have long been dominant. Recent examples include [7] in the context of construction of graphical models, [8] in the context of feature extraction, and [9] in the context of independent component analysis. These methods have by and large a common issue: to compute quantities such as the mutual information, entropy, or Kullback-Leibler divergence, we require sophisticated space partitioning and/or bias correction strategies [10, 9].

In this paper we give an overview of methods which are able to compute distances between distributions *without* the need for intermediate density estimation. Moreover, these techniques allow algorithm designers to specify which properties of a distribution are most relevant to their problems. We are optimistic that our embedding approach to distribution representation and analysis will lead to the development of algorithms which are simpler and more effective than entropy-based methods in a broad range of applications.

We begin our presentation in Section 1 with an overview of reproducing kernel Hilbert spaces (RKHSs), and a description of how probability distributions can be represented as elements in an RKHS. In Section 2, we show how these representations may be used to address a variety of problems, including homogeneity testing (Section 2.1), covariate shift correction (Section 2.2), independence measurement (Section 2.3), feature extraction (Section 2.4), and density estimation (Section 2.5).

M. Hutter, R.A. Servedio, and E. Takimoto (Eds.): ALT 2007, LNAI 4754, pp. 13–31, 2007.

1   Hilbert Space Embedding

1.1   Preliminaries

In the following we denote by $\mathcal{X}$ the domain of observations, and let $\mathbf{P}_x$ be a probability measure on $\mathcal{X}$. Whenever needed, $\mathcal{Y}$ will denote a second domain, with its own probability measure $\mathbf{P}_y$. A joint probability measure on $\mathcal{X} \times \mathcal{Y}$ will be denoted by $\mathbf{P}_{x,y}$. We will assume all measures are Borel measures, and the domains are compact.

We next introduce a reproducing kernel Hilbert space (RKHS) $\mathcal{H}$ of functions on $\mathcal{X}$ with kernel k (the analogous definitions hold for a corresponding RKHS $\mathcal{G}$ with kernel l on $\mathcal{Y}$). This is defined as follows: $\mathcal{H}$ is a Hilbert space of functions $\mathcal{X} \to \mathbb{R}$ with dot product $\langle \cdot, \cdot \rangle$, satisfying the reproducing property:

$$\langle f(\cdot), k(x, \cdot) \rangle = f(x) \tag{1a}$$

$$\text{and consequently } \langle k(x, \cdot), k(x', \cdot) \rangle = k(x, x'). \tag{1b}$$

This means we can view the linear map from a function f on $\mathcal{X}$ to its value at x as an inner product. The evaluation functional is then given by $k(x, \cdot)$, i.e. the kernel function. Popular kernel functions on $\mathbb{R}^n$ include the polynomial kernel $k(x, x') = \langle x, x' \rangle^d$, the Gaussian RBF kernel $k(x, x') = \exp\left(-\lambda \|x - x'\|^2\right)$, and the Laplace kernel $k(x, x') = \exp\left(-\lambda \|x - x'\|\right)$. Good kernel functions have been defined on texts, graphs, time series, dynamical systems, images, and structured objects. For recent reviews see [11, 12, 13].

An alternative view, which will come in handy when designing algorithms is that of a *feature map*. That is, we will consider maps $x \to \phi(x)$ such that $k(x, x') = \langle \phi(x), \phi(x') \rangle$ and likewise $f(x) = \langle w, \phi(x) \rangle$, where w is a suitably chosen "weight vector" (w can have infinite dimension, e.g. in the case of a Gaussian kernel).

Many kernels are universal in the sense of [14]. That is, their Hilbert spaces $\mathcal{H}$ are dense in the space of continuous bounded functions $C_0(\mathcal{X})$ on the compact domain $\mathcal{X}$. For instance, the Gaussian and Laplacian RBF kernels share this property. This is important since many results regarding distributions are stated with respect to $C_0(\mathcal{X})$ and we would like to translate them into results on Hilbert spaces.

1.2   Embedding

At the heart of our approach are the following two mappings:

$$\mu[\mathbf{P}_x] := \mathbf{E}_x \left[k(x, \cdot) \right] \tag{2a}$$

$$\mu[X] := \frac{1}{m} \sum_{i=1}^{m} k(x_i, \cdot). \tag{2b}$$

Here $X = \{x_1, \ldots, x_m\}$ is assumed to be drawn independently and identically distributed from $\mathbf{P}_x$. If the (sufficient) condition $\mathbf{E}_x \left[k(x, x) \right] < \infty$ is satisfied, then $\mu[\mathbf{P}_x]$ is an element of the Hilbert space (as is, in any case, $\mu[X]$). By virtue of the reproducing property of $\mathcal{H}$,

$$\langle \mu[\mathbf{P}_x], f \rangle = \mathbf{E}_x \left[f(x) \right] \text{ and } \langle \mu[X], f \rangle = \frac{1}{m} \sum_{i=1}^{m} f(x_i).$$

That is, we can compute expectations and empirical means with respect to $\mathbf{P}_x$ and X, respectively, by taking inner products with the means in the RKHS, $\mu[\mathbf{P}_x]$ and $\mu[X]$. The representations $\mu[\mathbf{P}_x]$ and $\mu[X]$ are attractive for the following reasons [15, 16]:

Theorem 1. *If the kernel k is universal, then the mean map $\mu : \mathbf{P}_x \to \mu[\mathbf{P}_x]$ is injective.*

Moreover, we have fast convergence of $\mu[X]$ to $\mu[\mathbf{P}_x]$ as shown in [17, Theorem 15]. Denote by $R_m(\mathcal{H}, \mathbf{P}_x)$ the Rademacher average [18] associated with $\mathbf{P}_x$ and $\mathcal{H}$ via

$$R_m(\mathcal{H}, \mathbf{P}_x) = \frac{1}{m} \mathbf{E}_{x_1, \dots, x_m} \mathbf{E}_{\sigma_1, \dots, \sigma_m} \left[\sup_{\|f\|_{\mathcal{H}} \leq 1} \left| \sum_{i=1}^{m} \sigma_i f(x_i) \right| \right]. \tag{3}$$

$R_m(\mathcal{H}, \mathbf{P}_x)$ can be used to measure the deviation between empirical means and expectations [17].

Theorem 2. *Assume that $\|f\|_\infty \leq R$ for all $f \in \mathcal{H}$ with $\|f\|_{\mathcal{H}} \leq 1$. Then with probability at least $1 - \delta$, $\|\mu[\mathbf{P}_x] - \mu[X]\| \leq 2R_m(\mathcal{H}, \mathbf{P}_x) + R\sqrt{-m^{-1} \log(\delta)}$*

This ensures that $\mu[X]$ is a good proxy for $\mu[\mathbf{P}_x]$, provided the Rademacher average is well behaved.

Theorem 1 tells us that $\mu[\mathbf{P}_x]$ can be used to define distances between distributions $\mathbf{P}_x$ and $\mathbf{P}_y$, simply by letting $D(\mathbf{P}_x, \mathbf{P}_y) := \|\mu[\mathbf{P}_x] - \mu[\mathbf{P}_y]\|$. Theorem 2 tells us that we do not need to have access to actual distributions in order to compute $D(\mathbf{P}_x, \mathbf{P}_y)$ approximately — as long as $R_m(\mathcal{H}, \mathbf{P}_x) = O(m^{-\frac{1}{2}})$, a finite sample from the distributions will yield error of $O(m^{-\frac{1}{2}})$. See [18] for an analysis of the behavior of $R_m(\mathcal{H}, \mathbf{P}_x)$ when $\mathcal{H}$ is an RKHS.

This allows us to use $D(\mathbf{P}_x, \mathbf{P}_y)$ as a drop-in replacement wherever information theoretic quantities would have been used instead, e.g. for the purpose of determining whether two sets of observations have been drawn from the same distribution. Note that there is a strong connection between Theorem 2 and uniform convergence results commonly used in Statistical Learning Theory [19, 16]. This is captured in the theorem below:

Theorem 3. *Let $\mathcal{F}$ be the unit ball in the reproducing kernel Hilbert space $\mathcal{H}$. Then the deviation between empirical means and expectations for any $f \in \mathcal{F}$ is bounded:*

$$\sup_{f \in \mathcal{F}} \left| \mathbf{E}_x \left[f(x) \right] - \frac{1}{m} \sum_{i=1}^{m} f(x_i) \right| = \|\mu[\mathbf{P}_x] - \mu[X]\|.$$

Bounding the probability that this deviation exceeds some threshold ϵ is one of the key problems of statistical learning theory. See [16] for details. This means that we have at our disposition a large range of tools typically used to assess the quality of estimators. The key difference is that while those bounds are typically used to bound the deviation between empirical and expected means under the assumption that the data *are* drawn from the same distribution, we will use the bounds in Section 2.1 to test whether this assumption is actually true, and in Sections 2.2 and 2.5 to motivate strategies for approximating particular distributions.

This is analogous to what is commonly done in the univariate case: the Glivenko-Cantelli lemma allows one to bound deviations between empirical and expected means for functions of bounded variation, as generalized by the work of Vapnik and Chervonenkis [20, 21]. However, the Glivenko-Cantelli lemma also leads to the Kolmogorov-Smirnov statistic comparing distributions by comparing their cumulative distribution functions. Moreover, corresponding q-q plots can be used as a diagnostic tool to identify where differences occur.

1.3   A View from the Marginal Polytope

The space of all probability distributions $\mathcal{P}$ is a convex set. Hence, the image $\mathcal{M} := \mu[\mathcal{P}]$ of $\mathcal{P}$ under the linear map μ also needs to be convex. This set is commonly referred to as the marginal polytope. Such mappings have become a standard tool in deriving efficient algorithms for approximate inference in graphical models and exponential families [22, 23].

We are interested in the properties of $\mu[\mathbf{P}]$ in the case where $\mathbf{P}$ satisfies the conditional independence relations specified by an undirected graphical model. In [24], it is shown for this case that the sufficient statistics decompose along the maximal cliques of the conditional independence graph.

More formally, denote by $\mathcal{C}$ set of maximal cliques of the graph G and let x_c be the restriction of $x \in \mathcal{X}$ to the variables on clique $c \in \mathcal{C}$. Moreover, let k_c be universal kernels in the sense of [14] acting on the restrictions of $\mathcal{X}$ on clique $c \in \mathcal{C}$. In this case [24] show that

$$k(x, x') = \sum_{c \in \mathcal{C}} k_c(x_c, x'_c) \tag{4}$$

can be used to describe all probability distributions with the above mentioned conditional independence relations using an exponential family model with k as its kernel. Since for exponential families expectations of the sufficient statistics yield injections, we have the following result:

Corollary 1. *On the class of probability distributions satisfying conditional independence properties according to a graph G with maximal clique set $\mathcal{C}$ and with full support on their domain, the operator*

$$\mu[\mathbf{P}] = \sum_{c \in \mathcal{C}} \mu_c[\mathbf{P}_c] = \sum_{c \in \mathcal{C}} \mathbf{E}_{x_c}[k_c(x_c, \cdot)] \tag{5}$$

is injective if the kernels k_c are all universal. The same decomposition holds for the empirical counterpart $\mu[X]$.

The condition of full support arises from the conditions of the Hammersley-Clifford Theorem [25, 26]: without it, not all conditionally independent random variables can be represented as the product of potential functions. Corollary 1 implies that we will be able to perform all subsequent operations on structured domains simply by dealing with mean operators on the corresponding maximal cliques.

1.4   Choosing the Hilbert Space

Identifying probability distributions with elements of Hilbert spaces is not new: see e.g. [27]. However, this leaves the obvious question of which Hilbert space to employ. We could informally choose a space with a kernel equalling the Delta distribution $k(x, x') = \delta(x, x')$, in which case the operator μ would simply be the identity map (which restricts us to probability distributions with square integrable densities).

The latter is in fact what is commonly done on finite domains (hence the L_2 integrability condition is trivially satisfied). For instance, [22] effectively use the Kronecker Delta $\delta(x_c, x'_c)$ as their feature map. The use of kernels has additional advantages: we need not deal with the issue of representation of the sufficient statistics or whether such a representation is minimal (i.e. whether the sufficient statistics actually span the space).

Whenever we have knowledge about the class of functions $\mathcal{F}$ we would like to analyze, we should be able to trade off simplicity in $\mathcal{F}$ with better approximation behavior in $\mathcal{P}$. For instance, assume that $\mathcal{F}$ contains only linear functions. In this case, μ only needs to map $\mathcal{P}$ into the space of all expectations of x. Consequently, one may expect very good constants in the convergence of $\mu[X]$ to $\mu[\mathbf{P}_x]$.

2   Applications

While the previous description may be of interest on its own, it is in application to areas of statistical estimation and artificial intelligence that its relevance becomes apparent.

2.1   Two-Sample Test

Since we know that $\mu[X] \to \mu[\mathbf{P}_x]$ with a fast rate (given appropriate behavior of $R_m(\mathcal{H}, \mathbf{P}_x)$), we may compare data drawn from two distributions $\mathbf{P}_x$ and $\mathbf{P}_y$, with associated samples X and Y, to test whether both distributions are identical; that is, whether $\mathbf{P}_x = \mathbf{P}_y$. For this purpose, recall that we defined $D(\mathbf{P}_x, \mathbf{P}_y) = \|\mu[\mathbf{P}_x] - \mu[\mathbf{P}_y]\|$. Using the reproducing property of an RKHS we may show [16] that

$$D^2(\mathbf{P}_x, \mathbf{P}_y) = \mathbf{E}_{x,x'}[k(x, x')] - 2\mathbf{E}_{x,y}[k(x, y)] + \mathbf{E}_{y,y'}[k(y, y')],$$

where x' is an independent copy of x, and y' an independent copy of y. An unbiased empirical estimator of $D^2(\mathbf{P}_x, \mathbf{P}_y)$ is a U-statistic [28],

$$\hat{D}^2(X, Y) := \tfrac{1}{m(m-1)} \sum_{i \neq j} h((x_i, y_i), (x_j, y_j)), \tag{6}$$

where

$$h((x, y), (x', y')) := k(x, x') - k(x, y') - k(y, x') + k(y, y').$$

An equivalent interpretation, also in [16], is that we find a function that maximizes the difference in expectations between probability distributions. The resulting problem may be written

$$D(\mathbf{P}_x, \mathbf{P}_y) := \sup_{f \in \mathcal{F}} |\mathbf{E}_x[f(x)] - \mathbf{E}_y[f(y)]|. \tag{7}$$

To illustrate this latter setting, we plot the witness function f in Figure 1, when $\mathbf{P}_x$ is Gaussian and $\mathbf{P}_y$ is Laplace, for a Gaussian RKHS kernel. This function is straightforward to obtain, since the solution to Eq. (7) can be written $f(x) = \langle \mu[\mathbf{P}_x] - \mu[\mathbf{P}_y], \phi(x) \rangle$.

The following two theorems give uniform convergence and asymptotic results, respectively. The first theorem is a straightforward application of [29, p. 25].

Theorem 4. *Assume that the kernel k is nonnegative and bounded by 1. Then with probability at least $1 - \delta$ the deviation $|D^2(\mathbf{P}_x, \mathbf{P}_y) - \hat{D}^2(X, Y)|$ is bounded by $4\sqrt{\log(2/\delta)/m}$.*

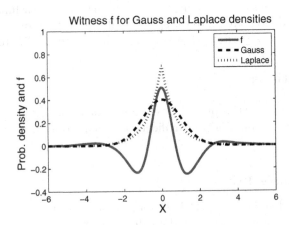

Fig. 1. Illustration of the function maximizing the mean discrepancy in the case where a Gaussian is being compared with a Laplace distribution. Both distributions have zero mean and unit variance. The function f that witnesses the difference in feature means has been scaled for plotting purposes, and was computed empirically on the basis of 2×10^4 samples, using a Gaussian kernel with $\sigma = 0.5$.

Note that an alternative uniform convergence bound is provided in [30], based on McDiarmid's inequality [31]. The second theorem appeared as [30, Theorem 8], and describes the asymptotic distribution of $\hat{D}^2(X,Y)$. When $\mathbf{P}_x \neq \mathbf{P}_y$, this distribution is given by [28, Section 5.5.1]; when $\mathbf{P}_x = \mathbf{P}_y$, it follows from [28, Section 5.5.2] and [32, Appendix].

Theorem 5. *We assume* $\mathbf{E}\left(h^2\right) < \infty$. *When* $\mathbf{P}_x \neq \mathbf{P}_y$, $\hat{D}^2(X,Y)$ *converges in distribution [33, Section 7.2] to a Gaussian according to*

$$m^{\frac{1}{2}} \left(\hat{D}^2(X,Y) - D^2(\mathbf{P}_x, \mathbf{P}_y) \right) \xrightarrow{D} \mathcal{N}\left(0, \sigma_u^2\right),$$

where $\sigma_u^2 = 4 \left(\mathbf{E}_z \left[(\mathbf{E}_{z'} h(z,z'))^2 \right] - [\mathbf{E}_{z,z'}(h(z,z'))]^2 \right)$ *and* $z := (x,y)$, *uniformly at rate* $1/\sqrt{m}$ *[28, Theorem B, p. 193]. When* $\mathbf{P}_x = \mathbf{P}_y$, *the U-statistic is degenerate, meaning* $\mathbf{E}_{z'} h(z,z') = 0$. *In this case,* $\hat{D}^2(X,Y)$ *converges in distribution according to*

$$m\hat{D}^2(X,Y) \xrightarrow{D} \sum_{l=1}^{\infty} \lambda_l \left[g_l^2 - 2 \right], \tag{8}$$

where $g_l \sim \mathcal{N}(0,2)$ *i.i.d.,* λ_i *are the solutions to the eigenvalue equation*

$$\int_X \tilde{k}(x,x')\psi_i(x)dp(x) = \lambda_i \psi_i(x'),$$

and $\tilde{k}(x_i, x_j) := k(x_i, x_j) - \mathbf{E}_x k(x_i, x) - \mathbf{E}_x k(x, x_j) + \mathbf{E}_{x,x'} k(x, x')$ *is the centered RKHS kernel.*

We illustrate the MMD density by approximating it empirically for both $\mathbf{P}_x = \mathbf{P}_y$ (also called the null hypothesis, or H_0) and $\mathbf{P}_x \neq \mathbf{P}_y$ (the alternative hypothesis, or H_1). Results are plotted in Figure 2. We may use this theorem directly to test whether two distributions are identical, given an appropriate finite sample approximation to the $(1 - \alpha)$th quantile of (8). In [16], this was achieved via two strategies: by using the bootstrap [34], and by fitting Pearson curves using the first four moments [35, Section 18.8].

While uniform convergence bounds have the theoretical appeal of making no assumptions on the distributions, they produce very weak tests. We find the test arising from Theorem 5 performs considerably better in practice. In addition, [36] demonstrate that this test performs very well in circumstances of high dimension and low sample size (i.e. when comparing microarray data), as well as being the only test currently applicable for structured data such as distributions on graphs. Moreover, the test can be used to determine whether records in databases may be matched based on their statistical properties. Finally, one may also apply it to extract features with the aim of *maximizing* discrepancy between sets of observations (see Section 2.4).

2.2 Covariate Shift Correction and Local Learning

A second application of the mean operator arises in situations of supervised learning where the training and test sets are drawn from different distributions,

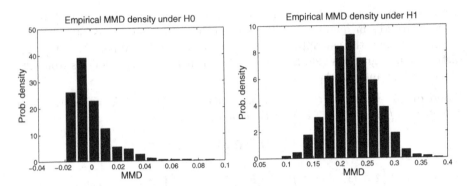

Fig. 2. Left: Empirical distribution of the MMD under H_0, with $\mathbf{P}_x$ and $\mathbf{P}_y$ both Gaussians with unit standard deviation, using 50 samples from each. **Right:** Empirical distribution of the MMD under H_1, with $\mathbf{P}_x$ a Laplace distribution with unit standard deviation, and $\mathbf{P}_y$ a Laplace distribution with standard deviation $3\sqrt{2}$, using 100 samples from each. In both cases, the histograms were obtained by computing 2000 independent instances of the MMD.

i.e. $X = \{x_1, \ldots, x_m\}$ is drawn from $\mathbf{P}_x$ and $X' = \{x'_1, \ldots, x'_{m'}\}$ is drawn from $\mathbf{P}_{x'}$. We assume, however, that the labels y are drawn from the same *conditional* distribution $\mathbf{P}_{y|x}$ on both the training and test sets.

The goal in this case is to find a weighting of the training set such that minimizing a reweighted empirical error on the training set will come close to minimizing the expected loss on the test set. That is, we would like to find weights $\{\beta_1, \ldots, \beta_m\}$ for X with $\sum_i \beta_i = 1$.

Obviously, if $\mathbf{P}_{y|x}$ is a rapidly changing function of x, or if the loss measuring the discrepancy between y and its estimate is highly non-smooth, this problem is difficult to solve. However, under regularity conditions spelled out in [37], one may show that by minimizing

$$\Delta := \left\| \sum_{i=1}^{m} \beta_i k(x_i, \cdot) - \mu[X'] \right\|$$

subject to $\beta_i \geq 0$ and $\sum_i \beta_i = 1$, we will obtain weights which achieve this task. The idea here is that the expected loss with the expectation taken over $y|x$ should not change too quickly as a function of x. In this case we can use points x_i "nearby" to estimate the loss at location x'_j on the test set. Hence we are re-weighting the empirical distribution on the training set X such that the distribution behaves more like the empirical distribution on X'.

Note that by re-weighting X we will assign some observations a higher weight than $\frac{1}{m}$. This means that the statistical guarantees can no longer be stated in terms of the sample size m. One may show [37], however, that $\|\beta\|_2^{-2}$ now behaves like the effective sample size. Instead of minimizing Δ, it pays to minimize $\Delta^2 + \lambda \|\beta\|_2^2$ subject to the above constraints. It is easy to show using the reproducing property of $\mathcal{H}$ that this corresponds to the following quadratic program:

$$\underset{\beta}{\text{minimize}} \ \frac{1}{2}\beta^{\top} (K + \lambda \mathbf{1}) \beta - \beta^{\top} l \tag{9a}$$

$$\text{subject to } \beta_i \geq 0 \text{ and } \sum_i \beta_i = 1. \tag{9b}$$

Here $K_{ij} := k(x_i, x_j)$ denotes the kernel matrix and $l_i := \frac{1}{m} \sum_{j=1}^{m'} k(x_i, x_j')$ is the expected value of $k(x_i, \cdot)$ on the test set X', i.e. $l_i = \langle k(x_i, \cdot), \mu[X'] \rangle$.

Experiments show that solving (9) leads to sample weights which perform very well in covariate shift. Remarkably, the approach can even outperform "importance sampler" weights, i.e. weights β_i obtained by computing the ratio $\beta_i = \mathbf{P}_{x'}(x_i)/\mathbf{P}_x(x_i)$. This is surprising, since the latter provide unbiased estimates of the expected error on X'. A point to bear in mind is that the kernels employed in the classification/regression learning algorithms of [37] are somewhat large, suggesting that the feature mean matching procedure is helpful when the learning algorithm returns relatively smooth classification/regression functions (we observe the same situation in the example of [38, Figure 1], where the model is "simpler" than the true function generating the data).

In the case where X' contains only a single observation, i.e. $X' = \{x'\}$, the above procedure leads to estimates which try to find a subset of observations in X and a weighting scheme such that the error at x' is approximated well. In practice, this leads to a local sample weighting scheme, and consequently an algorithm for local learning [39]. Our key advantage, however, is that we do not need to define the shape of the neighborhood in which we approximate the error at x'. Instead, this is automatically taken care of via the choice of the Hilbert space $\mathcal{H}$ and the location of x' relative to X.

2.3   Independence Measures

A third application of our mean mapping arises in measures of whether two random variables x and y are independent. Assume that pairs of random variables (x_i, y_i) are jointly drawn from some distribution $\mathbf{P}_{x,y}$. We wish to determine whether this distribution factorizes.

Having a measure of (in)dependence between random variables is a very useful tool in data analysis. One application is in independent component analysis [40], where the goal is to find a linear mapping of the observations x_i to obtain mutually independent outputs. One of the first algorithms to gain popularity was InfoMax, which relies on information theoretic quantities [41]. Recent developments using cross-covariance or correlation operators between Hilbert space representations have since improved on these results significantly [42, 43, 44]; in particular, a faster and more accurate quasi-Newton optimization procedure for kernel ICA is given in [45]. In the following we re-derive one of the above kernel independence measures using mean operators instead.

We begin by defining

$$\mu[\mathbf{P}_{xy}] := \mathbf{E}_{x,y} [v((x, y), \cdot)]$$
$$\text{and } \mu[\mathbf{P}_x \times \mathbf{P}_y] := \mathbf{E}_x \mathbf{E}_y [v((x, y), \cdot)].$$

Here we assumed that $\mathcal{V}$ is an RKHS over the space $\mathcal{X} \times \mathcal{Y}$ with kernel $v((x,y),(x',y'))$. If x and y *are* dependent, the equality $\mu[\mathbf{P}_{xy}] = \mu[\mathbf{P}_x \times \mathbf{P}_y]$ will not hold. Hence we may use $\Delta := \|\mu[\mathbf{P}_{xy}] - \mu[\mathbf{P}_x \times \mathbf{P}_y]\|$ as a measure of dependence.

Now assume that $v((x,y),(x',y')) = k(x,x')l(y,y')$, i.e. that the RKHS $\mathcal{V}$ is a direct product $\mathcal{H} \otimes \mathcal{G}$ of the RKHSs on $\mathcal{X}$ and $\mathcal{Y}$. In this case it is easy to see that

$$
\begin{aligned}
\Delta^2 &= \left\| \mathbf{E}_{xy}\left[k(x,\cdot)l(y,\cdot)\right] - \mathbf{E}_x\left[k(x,\cdot)\right]\mathbf{E}_y\left[l(y,\cdot)\right] \right\|^2 \\
&= \mathbf{E}_{xy}\mathbf{E}_{x'y'}\left[k(x,x')l(y,y')\right] - 2\mathbf{E}_x\mathbf{E}_y\mathbf{E}_{x'y'}\left[k(x,x')l(y,y')\right] \\
&\quad + \mathbf{E}_x\mathbf{E}_y\mathbf{E}_{x'}\mathbf{E}_{y'}\left[k(x,x')l(y,y')\right]
\end{aligned}
$$

The latter, however, is exactly what [43] show to be the Hilbert-Schmidt norm of the covariance operator between RKHSs: this is zero if and only if x and y are independent, for universal kernels. We have the following theorem:

Theorem 6. *Denote by C_{xy} the covariance operator between random variables x and y, drawn jointly from $\mathbf{P}_{xy}$, where the functions on $\mathcal{X}$ and $\mathcal{Y}$ are the reproducing kernel Hilbert spaces $\mathcal{F}$ and $\mathcal{G}$ respectively. Then the Hilbert-Schmidt norm $\|C_{xy}\|_{\mathrm{HS}}$ equals Δ.*

Empirical estimates of this quantity are as follows:

Theorem 7. *Denote by K and L the kernel matrices on X and Y respectively. Moreover, denote by $H = I - 1/m$ the projection matrix onto the subspace orthogonal to the vector with all entries set to 1. Then $m^{-2}\operatorname{tr}HKHL$ is an estimate of Δ^2 with bias $O(m^{-1})$. With high probability the deviation from Δ^2 is $O(m^{-\frac{1}{2}})$.*

See [43] for explicit constants. In certain circumstances, including in the case of RKHSs with Gaussian kernels, the empirical Δ^2 may also be interpreted in terms of a smoothed difference between the joint empirical characteristic function (ECF) and the product of the marginal ECFs [46, 47]. This interpretation does not hold in all cases, however, e.g. for kernels on strings, graphs, and other structured spaces. An illustration of the witness function of the equivalent optimization problem in Eq. 7 is provided in Figure 3. We observe that this is a smooth function which has large magnitude where the joint density is most different from the product of the marginals.

Note that if $v((x,y),\cdot)$ does *not* factorize we obtain a more general measure of dependence. In particular, we might not care about all types of interaction between x and y to an equal extent, and use an ANOVA kernel. Computationally efficient recursions are due to [48], as reported in [49]. More importantly, this representation will allow us to deal with *structured* random variables which are *not* drawn independently and identically distributed, such as time series.

For instance, in the case of EEG (electroencephalogram) data, we have both spatial and temporal structure in the signal. That said, few algorithms take full advantage of this when performing independent component analysis [50]. The pyramidal kernel of [51] is one possible choice for dependent random variables.

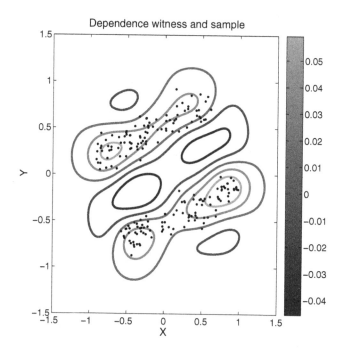

Fig. 3. Illustration of the function maximizing the mean discrepancy when MMD is used as a measure of independence. A sample from dependent random variables x and y is shown in black, and the associated function f that witnesses the MMD is plotted as a contour. The latter was computed empirically on the basis of 200 samples, using a Gaussian kernel with $\sigma = 0.2$.

2.4   Feature Extraction

Kernel measures of statistical dependence need not be applied *only* to the analysis of independent components. To the contrary, we may also use them to extract highly dependent random variables, i.e. features. This procedure leads to variable selection algorithms with very robust properties [52].

The idea works as follows: given a set of patterns X and a set of labels Y, find a subset of features from X which maximizes $m^{-2} \operatorname{tr} HKHL$. Here L is the kernel matrix on the labels. In the most general case, the matrix K will arise from an arbitrary kernel k, for which no efficient decompositions exist. In this situation [52] suggests the use of a greedy feature removal procedure, i.e. to remove subsets of features iteratively such that $m^{-2} \operatorname{tr} HKHL$ is maximized for the remaining features.

In general, for particular choices of k and l, it is possible to recover well known feature selection methods, such as Pearson's correlation, shrunken centroid, or signal-to-noise ratio selection. Below we give some examples, mainly when a linear kernel $k(x, x') = \langle x, x' \rangle$. For more details see [53].

Pearson's Correlation is commonly used in microarray analysis [54, 55]. It is defined as

$$R_j := \frac{1}{m} \sum_{i=1}^{m} \left(\frac{x_{ij} - x_j}{s_{x_j}} \right) \left(\frac{y_i - y}{s_y} \right) \text{ where} \tag{10}$$

$$x_j = \frac{1}{m} \sum_{i=1}^{m} x_{ij} \text{ and } y = \frac{1}{m} \sum_{i=1}^{m} y_i$$

$$s_{x_j}^2 = \frac{1}{m} \sum_{i=1}^{m} (x_{ij} - x_j)^2 \text{ and } s_y^2 = \frac{1}{m} \sum_{i=1}^{m} (y_i - y)^2. \tag{11}$$

This means that all features are individually centered by x_j and scaled by their coordinate-wise variance s_{x_j} as a preprocessing step. Performing those operations before applying a linear kernel yields the formulation:

$$\text{tr}\, KHLH = \text{tr} \left(XX^\top Hyy^\top H \right) = \left\| HX^\top Hy \right\|^2 \tag{12}$$

$$= \sum_{j=1}^{d} \left(\sum_{i=1}^{m} \left(\frac{x_{ij} - x_j}{s_{x_j}} \right) \left(\frac{y_i - y}{s_y} \right) \right)^2 = \sum_{j=1}^{d} R_j^2. \tag{13}$$

Hence $\text{tr}\, KHLH$ computes the sum of the squares of the Pearson Correlation (pc) coefficients. Since the terms are additive, feature selection is straightforward by picking the list of best performing features.

Centroid. The difference between the means of the positive and negative classes at the jth feature, $(x_{j+} - x_{j-})$, is useful for scoring individual features. With different normalization of the data and the labels, many variants can be derived.

To obtain the centroid criterion [56] use $v_j := \lambda x_{j+} - (1 - \lambda)x_{j-}$ for $\lambda \in (0, 1)$ as the score[1] for feature j. Features are subsequently selected according to the absolute value $|v_j|$. In experiments the authors typically choose $\lambda = \frac{1}{2}$.

For $\lambda = \frac{1}{2}$ we can achieve the same goal by choosing $L_{ii'} = \frac{y_i y_{i'}}{m_{y_i} m_{y_{i'}}}$ ($y_i, y_{i'} \in \{\pm 1\}$), in which case $HLH = L$, since the label kernel matrix is already centered. Hence we have

$$\text{tr}\, KHLH = \sum_{i,i'=1}^{m} \frac{y_i y_{i'}}{m_{y_i} m_{y_{i'}}} x_i^\top x_{i'} \tag{14}$$

$$= \sum_{j=1}^{d} \left(\sum_{i,i'=1}^{m} \frac{y_i y_{i'} x_{ij} x_{i'j}}{m_{y_i} m_{y_{i'}}} \right) = \sum_{j=1}^{d} (x_{j+} - x_{j-})^2. \tag{15}$$

This proves that the centroid feature selector can be viewed as a special case of BAHSIC in the case of $\lambda = \frac{1}{2}$. From our analysis we see that other values of λ amount to effectively rescaling the patterns x_i *differently* for different classes, which may lead to undesirable features being selected.

---

[1] The parameterization in [56] is different but it can be shown to be equivalent.

t-**Statistic.** The normalization for the jth feature is computed as

$$\bar{s}_j = \left[\frac{s_{j+}^2}{m_+} + \frac{s_{j-}^2}{m_-} \right]^{\frac{1}{2}} \tag{16}$$

In this case we define the t-statistic for the jth feature via $t_j = (x_{j+} - x_{j-})/\bar{s}_j$. Compared to the Pearson correlation, the key difference is that now we normalize each feature not by the overall sample standard deviation but rather by a value which takes each of the two classes separately into account.

Signal to noise ratio is yet another criterion to use in feature selection. The key idea is to normalize each feature by $\bar{s}_j = s_{j+} + s_{j-}$ instead. Subsequently the $(x_{j+} - x_{j-})/\bar{s}_j$ are used to score features.

Moderated t-score is similar to t-statistic and is used for microarray analysis [57]. Its normalization for the jth feature is derived via a Bayes approach as

$$\tilde{s}_j = \frac{m\bar{s}_j^2 + m_0\bar{s}_0^2}{m + m_0} \tag{17}$$

where $\bar{s}_j$ is from (16), and $\bar{s}_0$ and m_0 are hyperparameters for the prior distribution on $\bar{s}_j$ (all $\bar{s}_j$ are assumed to be iid). $\bar{s}_0$ and m_0 are estimated using information from all feature dimensions. This effectively borrows information from the ensemble of features to aid with the scoring of an individual feature. More specifically, $\bar{s}_0$ and m_0 can be computed as [57]

$$m_0 = 2\Gamma'^{-1}\left(\frac{1}{d}\sum_{j=1}^{d}(z_j - \bar{z})^2 - \Gamma'\left(\frac{m}{2}\right) \right), \tag{18}$$

$$\bar{s}_0^2 = \exp\left(\bar{z} - \Gamma\left(\frac{m}{2}\right) + \Gamma\left(\frac{m_0}{2}\right) - \ln\left(\frac{m_0}{m}\right) \right), \tag{19}$$

where $\Gamma(\cdot)$ is the gamma function, $'$ denotes derivative, $z_j = \ln(\bar{s}_j^2)$ and $\bar{z} = \frac{1}{d}\sum_{j=1}^{d} z_j$.

B-statistic is the logarithm of the posterior odds (lods) that a feature is differentially expressed. [58, 57] show that, for large number of features, the B-statistic is given by

$$B_j = a + b\tilde{t}_j^2, \tag{20}$$

where both a and b are constant ($b > 0$), and $\tilde{t}_j$ is the moderated-t statistic for the jth feature. Here we see that B_j is monotonic increasing in $\tilde{t}_j$, and thus results in the same gene ranking as the moderated-t statistic.

2.5   Density Estimation

General setting. Obviously, we may also use the connection between mean operators and empirical means for the purpose of estimating densities. In fact, [59, 17, 60] show that this may be achieved in the following fashion:

$$\text{maximize}_{\mathbf{P}_x} H(\mathbf{P}_x) \text{ subject to } \|\mu[X] - \mu[\mathbf{P}_x]\| \leq \epsilon. \tag{21}$$

Here H is an entropy-like quantity (e.g. Kullback Leibler divergence, Csiszar divergence, Bregmann divergence, Entropy, Amari divergence) that is to be maximized subject to the constraint that the expected mean should not stray too far from its empirical counterpart. In particular, one may show that this approximate maximum entropy formulation is the dual of a maximum-a-posteriori estimation problem.

In the case of conditional probability distributions, it is possible to recover a raft of popular estimation algorithms, such as Gaussian Process classification, regression, and conditional random fields. The key idea in this context is to identify the sufficient statistics in generalized exponential families with the map $x \to k(x, \cdot)$ into a reproducing kernel Hilbert space.

Mixture Model. In problem (21) we try to find the optimal $\mathbf{P}_x$ over the entire space of probability distributions on $\mathcal{X}$. This can be an exceedingly costly optimization problem, in particular in the nonparametric setting. For instance, computing the normalization of the density itself may be intractable, in particular for high-dimensional data. In this case we may content ourselves with finding a suitable mixture distribution such that $\|\mu[X] - \mu[\mathbf{P}_x]\|$ is minimized with respect to the mixture coefficients. The diagram below summarizes our approach:

$$\text{density } p \longrightarrow \text{ sample } X \longrightarrow \text{ emp. mean } \mu[X] \longrightarrow \text{ estimate via } \mu[\widehat{\mathbf{P}}_x] \tag{22}$$

The connection between $\mu[\mathbf{P}_x]$ and $\mu[X]$ follows from Theorem 2. To obtain a density estimate from $\mu[X]$ assume that we have a set of candidate densities $\mathbf{P}_x^i$ on $\mathcal{X}$. We want to use these as basis functions to obtain $\widehat{\mathbf{P}}_x$ via

$$\widehat{\mathbf{P}}_x = \sum_{i=1}^{M} \beta_i \mathbf{P}_x^i \text{ where } \sum_{i=1}^{M} \beta_i = 1 \text{ and } \beta_i \geq 0. \tag{23}$$

In other words we wish to estimate $\mathbf{P}_x$ by means of a mixture model with mixture densities $\mathbf{P}_x^i$. The goal is to obtain good estimates for the coefficients β_i and to obtain performance guarantees which specify how well $\widehat{\mathbf{P}}_x$ is capable of estimating $\mathbf{P}_x$ in the first place. This is possible using a very simple optimization problem:

$$\text{minimize}_{\beta} \left\| \mu[X] - \mu[\widehat{\mathbf{P}}_x] \right\|_{\mathcal{H}}^2 \text{ subject to } \beta^\top \mathbf{1} = 1 \text{ and } \beta \geq 0. \tag{24}$$

To ensure good generalization performance we add a regularizer $\Omega[\beta]$ to the optimization problem, such as $\frac{1}{2} \|\beta\|^2$. It follows using the expansion of $\widehat{\mathbf{P}}_x$ in (23) that the resulting optimization problem can be reformulated as a quadratic program via

$$\text{minimize}_{\beta} \frac{1}{2} \beta^\top [Q + \lambda \mathbf{1}] \beta - l^\top \beta \text{ subject to } \beta^\top \mathbf{1} = 1 \text{ and } \beta \geq 0. \tag{25}$$

Here $\lambda > 0$ is a regularization constant, and the quadratic matrix $Q \in \mathbb{R}^{M \times M}$ and the vector $l \in \mathbb{R}^M$ are given by

$$Q_{ij} = \langle \mu[\mathbf{P}_x^i], \mu[\mathbf{P}_x^j] \rangle = \mathop{\mathbf{E}}_{x^i, x^j} \left[k(x^i, x^j) \right] \tag{26}$$

$$\text{and } l_j = \langle \mu[X], \mu[\mathbf{P}_x^j] \rangle = \frac{1}{m} \sum_{i=1}^m \mathop{\mathbf{E}}_{x^j} \left[k(x_i, x^j) \right]. \tag{27}$$

By construction $Q \succeq 0$ is positive semidefinite, hence the quadratic program (25) is convex. For a number of kernels and mixture terms $\mathbf{P}_x^i$ we are able to compute Q, l in closed form.

Since $\widehat{\mathbf{P}}_x$ is an empirical estimate it is quite unlikely that $\widehat{\mathbf{P}}_x = \mathbf{P}_x$. This raises the question of how well expectations with respect to $\mathbf{P}_x$ are approximated by those with respect to $\widehat{\mathbf{P}}_x$. This can be answered by an extension of the Koksma-Hlawka inequality [61].

Lemma 1. *Let $\epsilon > 0$ and let $\epsilon' := \left\| \mu[X] - \mu[\widehat{\mathbf{P}}_x] \right\|$. Under the assumptions of Theorem 2 we have that with probability at least $1 - \exp(-\epsilon^2 m R^{-2})$,*

$$\sup_{\|f\|_{\mathcal{H}} \leq 1} \left| \mathbf{E}_{x \sim \mathbf{P}_x}[f(x)] - \mathbf{E}_{x \sim \widehat{\mathbf{P}}_x}[f(x)] \right| \leq 2 R_m(\mathcal{H}, \mathbf{P}_x) + \epsilon + \epsilon'. \tag{28}$$

Proof We use that in Hilbert spaces, $\mathbf{E}_{x \sim \mathbf{P}_x}[f(x)] = \langle f, \mu[\mathbf{P}_x] \rangle$ and $\mathbf{E}_{x \sim \widehat{\mathbf{P}}_x}[f(x)] = \left\langle f, \mu[\widehat{\mathbf{P}}_x] \right\rangle$ both hold. Hence the LHS of (28) equates to $\sup_{\|f\|_{\mathcal{H}} \leq 1} \left| \left\langle \mu[\mathbf{P}_x] - \mu[\widehat{\mathbf{P}}_x], f \right\rangle \right|$, which is given by the norm of $\left\| \mu[\mathbf{P}_x] - \mu[\widehat{\mathbf{P}}_x] \right\|$. The triangle inequality, our assumption on $\mu[\widehat{\mathbf{P}}_x]$, and Theorem 2 complete the proof. ■

This means that we have good control over the behavior of expectations of random variables, as long as they belong to "smooth" functions on $\mathcal{X}$ — the uncertainty increases with their RKHS norm.

The above technique is useful when it comes to representing distributions in message passing and data compression. Rather than minimizing an information theoretic quantity, we can choose a Hilbert space which accurately reflects the degree of smoothness required for any subsequent operations carried out by the estimate. For instance, if we are only interested in linear functions, an accurate match of the first order moments will suffice, without requiring a good match in higher order terms.

2.6   Kernels on Sets

Up to now we used the mapping $X \to \mu[X]$ to compute the distance between two distributions (or their samples). However, since $\mu[X]$ itself is an element of an RKHS we can define a kernel on sets (and distributions) directly via

$$k(X, X') := \langle \mu[X], \mu[X'] \rangle = \frac{1}{mm'} \sum_{i,j}^{m, m'} k(x_i, x_j'). \tag{29}$$

In other words, $k(X, X')$, and by analogy $k(\mathbf{P}_x, \mathbf{P}_{x'}) := \langle \mu[\mathbf{P}_x], \mu[\mathbf{P}_{x'}] \rangle$, define kernels on sets and distributions, and obviously also between sets and distributions. If we have multisets and sample weights for instances we may easily include this in the computation of $\mu[X]$. It turns out that (29) is exactly the set kernel proposed by [62], when dealing with multiple instance learning. This notion was subsequently extended to deal with intermediate density estimates by [63]. We have therefore that in situations where estimation problems are well described by distributions we inherit the consistency properties of the underlying RKHS simply by using a universal set kernel for which $\mu[X]$ converges to $\mu[\mathbf{P}_x]$. We have the following corollary:

Corollary 2. *If k is universal the kernel matrix defined by the set/distribution kernel (29) has full rank as long as the sets/distributions are not identical.*

Note, however, that the set kernel may not be ideal for all multi instance problems: in the latter one assumes that at least a *single instance* has a given property, whereas for the use of (29) one needs to assume that at least a certain *fraction of instances* have this property.

3   Summary

We have seen that Hilbert space embeddings of distributions are a powerful tool to deal with a broad range of estimation problems, including two-sample tests, feature extractors, independence tests, covariate shift, local learning, density estimation, and the measurement of similarity between sets. Given these successes, we are very optimistic that these embedding techniques can be used to address further problems, ranging from issues in high dimensional numerical integration (the connections to lattice and Sobol sequences are apparent) to more advanced nonparametric property testing.

Acknowledgments. We thank Karsten Borgwardt, Kenji Fukumizu, Jiayuan Huang, Quoc Le, Malte Rasch, and Vladimir Vapnik for helpful discussions. NICTA is funded through the Australian Government's *Baking Australia's Ability* initiative, in part through the Australian Research Council. This work was supported in part by the IST Programme of the European Community, under the PASCAL Network of Excellence, IST-2002-506778.

References

[1] Vapnik, V.: The Nature of Statistical Learning Theory. Springer, New York (1995)
[2] Schölkopf, B., Smola, A.: Learning with Kernels. MIT Press, Cambridge (2002)
[3] Joachims, T.: Learning to Classify Text Using Support Vector Machines: Methods, Theory, and Algorithms. Kluwer Academic Publishers, Boston (2002)
[4] Rasmussen, C.E., Williams, C.K.I.: Gaussian Processes for Machine Learning. MIT Press, Cambridge (2006)

[5] Cover, T.M., Thomas, J.A.: Elements of Information Theory. John Wiley and Sons, New York (1991)

[6] Amari, S., Nagaoka, H.: Methods of Information Geometry. Oxford University Press (1993)

[7] Krause, A., Guestrin, C.: Near-optimal nonmyopic value of information in graphical models. Uncertainty in Artificial Intelligence UAI'05 (2005)

[8] Slonim, N., Tishby, N.: Agglomerative information bottleneck. In: Solla, S.A., Leen, T.K., Müller, K.R. (eds.) Advances in Neural Information Processing Systems, vol. 12, pp. 617–623. MIT Press, Cambridge (2000)

[9] Stögbauer, H., Kraskov, A., Astakhov, S., Grassberger, P.: Least dependent component analysis based on mutual information. Phys. Rev. E 70(6), 66123 (2004)

[10] Nemenman, I., Shafee, F., Bialek, W.: Entropy and inference, revisited. In: Neural Information Processing Systems, vol. 14, MIT Press, Cambridge (2002)

[11] Shawe-Taylor, J., Cristianini, N.: Kernel Methods for Pattern Analysis. Cambridge University Press, Cambridge (2004)

[12] Schölkopf, B., Tsuda, K., Vert, J.P.: Kernel Methods in Computational Biology. MIT Press, Cambridge (2004)

[13] Hofmann, T., Schölkopf, B., Smola, A.J.: A review of kernel methods in machine learning. Technical Report 156, Max-Planck-Institut für biologische Kybernetik (2006)

[14] Steinwart, I.: The influence of the kernel on the consistency of support vector machines. Journal of Machine Learning Research 2 (2002)

[15] Fukumizu, K., Bach, F.R., Jordan, M.I.: Dimensionality reduction for supervised learning with reproducing kernel hilbert spaces. J. Mach. Learn. Res. 5, 73–99 (2004)

[16] Gretton, A., Borgwardt, K., Rasch, M., Schölkopf, B., Smola, A.J.: A kernel method for the two-sample-problem. In: Schölkopf, B., Platt, J., Hofmann, T. (eds.) Advances in Neural Information Processing Systems, vol. 19, MIT Press, Cambridge (2007)

[17] Altun, Y., Smola, A.: Unifying divergence minimization and statistical inference via convex duality. In: Simon, H., Lugosi, G. (eds.) Proc. Annual Conf. Computational Learning Theory, pp. 139–153. Springer, Heidelberg (2006)

[18] Bartlett, P.L., Mendelson, S.: Rademacher and gaussian complexities: Risk bounds and structural results. J. Mach. Learn. Res. 3, 463–482 (2002)

[19] Koltchinskii, V.: Rademacher penalties and structural risk minimization. IEEE Trans. Inform. Theory 47, 1902–1914 (2001)

[20] Vapnik, V., Chervonenkis, A.: On the uniform convergence of relative frequencies of events to their probabilities. Theory Probab. Appl. 16(2), 264–281 (1971)

[21] Vapnik, V., Chervonenkis, A.: The necessary and sufficient conditions for the uniform convergence of averages to their expected values. Teoriya Veroyatnostei i Ee Primeneniya 26(3), 543–564 (1981)

[22] Wainwright, M.J., Jordan, M.I.: Graphical models, exponential families, and variational inference. Technical Report 649, UC Berkeley, Department of Statistics (September 2003)

[23] Ravikumar, P., Lafferty, J.: Variational chernoff bounds for graphical models. In: Uncertainty in Artificial Intelligence UAI04 (2004)

[24] Altun, Y., Smola, A.J., Hofmann, T.: Exponential families for conditional random fields. In: Uncertainty in Artificial Intelligence (UAI), Arlington, Virginia, pp. 2–9. AUAI Press (2004)

[25] Hammersley, J.M., Clifford, P.E.: Markov fields on finite graphs and lattices (unpublished manuscript, 1971)

[26] Besag, J.: Spatial interaction and the statistical analysis of lattice systems (with discussion). J. Roy. Stat. Soc. Ser. B Stat. Methodol. 36(B), 192–326 (1974)

[27] Hein, M., Bousquet, O.: Hilbertian metrics and positive definite kernels on probability measures. In: Ghahramani, Z., Cowell, R. (eds.) Proc. of AI & Statistics, vol. 10 (2005)

[28] Serfling, R.: Approximation Theorems of Mathematical Statistics. Wiley, New York (1980)

[29] Hoeffding, W.: Probability inequalities for sums of bounded random variables. Journal of the American Statistical Association 58, 13–30 (1963)

[30] Gretton, A., Borgwardt, K., Rasch, M., Schölkopf, B., Smola, A.: A kernel method for the two-sample-problem. In: Advances in Neural Information Processing Systems 19, MIT Press, Cambridge (2007)

[31] McDiarmid, C.: On the method of bounded differences. In: Surveys in Combinatorics, pp. 148–188. Cambridge University Press, Cambridge (1969)

[32] Anderson, N., Hall, P., Titterington, D.: Two-sample test statistics for measuring discrepancies between two multivariate probability density functions using kernel-based density estimates. Journal of Multivariate Analysis 50, 41–54 (1994)

[33] Grimmet, G.R., Stirzaker, D.R.: Probability and Random Processes, 3rd edn. Oxford University Press, Oxford (2001)

[34] Arcones, M., Giné, E.: On the bootstrap of u and v statistics. The Annals of Statistics 20(2), 655–674 (1992)

[35] Johnson, N.L., Kotz, S., Balakrishnan, N.: Continuous Univariate Distributions, 2nd edn., vol. 1. John Wiley and Sons, Chichester (1994)

[36] Borgwardt, K.M., Gretton, A., Rasch, M.J., Kriegel, H.P., Schölkopf, B., Smola, A.J.: Integrating structured biological data by kernel maximum mean discrepancy. Bioinformatics 22(14), e49–e57 (2006)

[37] Huang, J., Smola, A., Gretton, A., Borgwardt, K., Schölkopf, B.: Correcting sample selection bias by unlabeled data. In: Schölkopf, B., Platt, J., Hofmann, T. (eds.) Advances in Neural Information Processing Systems, vol. 19, MIT Press, Cambridge (2007)

[38] Shimodaira, H.: Improving predictive inference under convariance shift by weighting the log-likelihood function. Journal of Statistical Planning and Inference 90 (2000)

[39] Bottou, L., Vapnik, V.N.: Local learning algorithms. Neural Computation 4(6), 888–900 (1992)

[40] Comon, P.: Independent component analysis, a new concept? Signal Processing 36, 287–314 (1994)

[41] Lee, T.W., Girolami, M., Bell, A., Sejnowski, T.: A unifying framework for independent component analysis. Comput. Math. Appl. 39, 1–21 (2000)

[42] Bach, F.R., Jordan, M.I.: Kernel independent component analysis. J. Mach. Learn. Res. 3, 1–48 (2002)

[43] Gretton, A., Bousquet, O., Smola, A., Schölkopf, B.: Measuring statistical dependence with Hilbert-Schmidt norms. In: Jain, S., Simon, H.U., Tomita, E. (eds.) Proceedings Algorithmic Learning Theory, pp. 63–77. Springer, Heidelberg (2005)

[44] Gretton, A., Herbrich, R., Smola, A., Bousquet, O., Schölkopf, B.: Kernel methods for measuring independence. J. Mach. Learn. Res. 6, 2075–2129 (2005)

[45] Shen, H., Jegelka, S., Gretton, A.: Fast kernel ICA using an approximate newton method. In: AISTATS 11 (2007)

[46] Feuerverger, A.: A consistent test for bivariate dependence. International Statistical Review 61(3), 419–433 (1993)

[47] Kankainen, A.: Consistent Testing of Total Independence Based on the Empirical Characteristic Function. PhD thesis, University of Jyväskylä (1995)

[48] Burges, C.J.C., Vapnik, V.: A new method for constructing artificial neural networks. Interim technical report, ONR contract N00014-94-c-0186, AT&T Bell Laboratories (1995)

[49] Vapnik, V.: Statistical Learning Theory. John Wiley and Sons, New York (1998)

[50] Anemuller, J., Duann, J.R., Sejnowski, T.J., Makeig, S.: Spatio-temporal dynamics in fmri recordings revealed with complex independent component analysis. Neurocomputing 69, 1502–1512 (2006)

[51] Schölkopf, B.: Support Vector Learning. R. Oldenbourg Verlag, Munich (1997), http://www.kernel-machines.org

[52] Song, L., Smola, A., Gretton, A., Borgwardt, K., Bedo, J.: Supervised feature selection via dependence estimation. In: Proc. Intl. Conf. Machine Learning (2007)

[53] Song, L., Bedo, J., Borgwardt, K., Gretton, A., Smola, A.: Gene selection via the BAHSIC family of algorithms. In: Bioinformatics (ISMB) (to appear, 2007)

[54] van't Veer, L.J., Dai, H., van de Vijver, M.J., He, Y.D., Hart, A.A.M., et al.: Gene expression profiling predicts clinical outcome of breast cancer. Nature 415, 530–536 (2002)

[55] Ein-Dor, L., Zuk, O., Domany, E.: Thousands of samples are needed to generate a robust gene list for predicting outcome in cancer. Proc. Natl. Acad. Sci. USA 103(15), 5923–5928 (2006)

[56] Bedo, J., Sanderson, C., Kowalczyk, A.: An efficient alternative to svm based recursive feature elimination with applications in natural language processing and bioinformatics. Artificial Intelligence (2006)

[57] Smyth, G.: Linear models and empirical bayes methods for assessing differential expressionin microarray experiments. Statistical Applications in Genetics and Molecular Biology 3 (2004)

[58] Lönnstedt, I., Speed, T.: Replicated microarray data. Statistica Sinica 12, 31–46 (2002)

[59] Dudík, M., Schapire, R., Phillips, S.: Correcting sample selection bias in maximum entropy density estimation. Advances in Neural Information Processing Systems 17 (2005)

[60] Dudík, M., Schapire, R.E.: Maximum entropy distribution estimation with generalized regularization. In: Lugosi, G., Simon, H.U. (eds.) Proc. Annual Conf. Computational Learning Theory, Springer, Heidelberg (2006)

[61] Hlawka, E.: Funktionen von beschränkter variation in der theorie der gleichverteilung. Annali di Mathematica Pura ed Applicata 54 (1961)

[62] Gärtner, T., Flach, P.A., Kowalczyk, A., Smola, A.J.: Multi-instance kernels. Proc. Intl. Conf. Machine Learning (2002)

[63] Jebara, T., Kondor, I.: Bhattacharyya and expected likelihood kernels. In: Schölkopf, B., Warmuth, M.K. (eds.) COLT/Kernel 2003. LNCS (LNAI), vol. 2777, pp. 57–71. Springer, Heidelberg (2003)

Simple Algorithmic Principles of Discovery, Subjective Beauty, Selective Attention, Curiosity and Creativity*

Jürgen Schmidhuber

TU Munich, Boltzmannstr. 3, 85748 Garching bei München, Germany &
IDSIA, Galleria 2, 6928 Manno (Lugano), Switzerland
juergen@idsia.ch
http://www.idsia.ch/~juergen

I postulate that human or other intelligent agents function or should function as follows. They store all sensory observations as they come—the data is 'holy.' At any time, given some agent's current coding capabilities, part of the data is compressible by a short and hopefully fast program / description / explanation / world model. In the agent's subjective eyes, such data is more regular and more *beautiful* than other data [2,3]. It is well-known that knowledge of regularity and repeatability may improve the agent's ability to plan actions leading to external rewards. In absence of such rewards, however, *known* beauty is boring. Then *interestingness* becomes the *first derivative* of subjective beauty: as the learning agent improves its compression algorithm, formerly apparently random data parts become subjectively more regular and beautiful. Such progress in data compression is measured and maximized by the *curiosity* drive [1,4,5]: create action sequences that extend the observation history and yield previously unknown / unpredictable but quickly learnable algorithmic regularity. We discuss how all of the above can be naturally implemented on computers, through an extension of passive unsupervised learning to the case of active data selection: we reward a general reinforcement learner (with access to the adaptive compressor) for actions that improve the subjective compressibility of the growing data. An unusually large data compression breakthrough deserves the name *discovery*. The *creativity* of artists, dancers, musicians, pure mathematicians can be viewed as a by-product of this principle. Good observer-dependent art deepens the observer's insights about this world or possible worlds, unveiling previously unknown regularities in compressible data, connecting previously disconnected patterns in an initially surprising way that makes the combination of these patterns subjectively more compressible, and eventually becomes known and less interesting. Several qualitative examples support this hypothesis.

---

* The full version of this paper is published in the Proceedings of the 10th International Conference on Discovery Science, Lecture Notes in Artificial Intelligence Vol. 4755.

M. Hutter, R.A. Servedio, and E. Takimoto (Eds.): ALT 2007, LNAI 4754, pp. 32–33, 2007.

References

1. Schmidhuber, J.: Curious model-building control systems. In: Proc. Intl. Joint Conference on Neural Networks, Singapore, vol. 2, pp. 1458–1463. IEEE press (1991)
2. Schmidhuber, J.: Low-complexity art. Leonardo, Journal of the International Society for the Arts, Sciences, and Technology 30(2), 97–103 (1997)
3. Schmidhuber, J.: Facial beauty and fractal geometry. Technical Report TR IDSIA-28-98, IDSIA, Published in the Cogprint Archive (1998),
   http://cogprints.soton.ac.uk
4. Schmidhuber, J.: Exploring the predictable. In: Ghosh, A., Tsuitsui, S. (eds.) Advances in Evolutionary Computing, pp. 579–612. Springer, Heidelberg (2002)
5. Schmidhuber, J.: Developmental robotics, optimal artificial curiosity, creativity, music, and the fine arts. Connection Science 18(2), 173–187 (2006)

Feasible Iteration of Feasible Learning Functionals[*]

John Case[1], Timo Kötzing[1], and Todd Paddock[2]

[1] Department of Computer and Information Sciences, University of Delaware,
Newark, DE 19716-2586,USA
case@cis.udel.edu, koetzing@cis.udel.edu
[2] Majestic Research,
1270 Avenue of the Americas, Suite 1900, New York, NY 10020
todd@majesticresearch.com

Abstract. For learning functions in the limit, an algorithmic learner obtains successively more data about a function and calculates trials each resulting in the output of a corresponding program, where, hopefully, these programs eventually converge to a correct program for the function. The authors desired to provide a feasible version of this learning in the limit — a version where each trial was conducted feasibly *and* there was some feasible limit on the number of trials allowed. Employed were *basic feasible functionals* which query an input function as to its values and which provide each trial. An additional tally argument 0^i was provided to the functionals for their execution of the i-th trial. In this way more time resource was available for each successive trial. The mechanism employed to feasibly limit the number of trials was to feasibly count them down from some feasible notation for a constructive ordinal. Since all processes were feasible, their termination was feasibly detectable, and, so, it was possible to wait for the trials to terminate and suppress all the output programs but the last. Hence, although there is still an iteration of trials, the learning was a special case of what has long been known as total **Fin**-learning, i.e., learning in the limit, where, on each function, the learner always outputs exactly one conjectured program. Our general main results provide for strict learning hierarchies where the trial count down involves all and only notations for infinite limit ordinals. For our hierarchies featuring finitely many limit ordinal jumps, we have upper and lower total run time bounds of our feasible **Fin**-learners in terms of finite stacks of exponentials. We provide, though, an example of how to regain feasibility by a suitable parameterized complexity analysis.

[*] Case and Paddock were supported in part by NSF grant number NSF CCR-0208616. We are also grateful to anonymous referees for many helpful suggestions. One such referee provided hints about the truth and truth and proof, respectively, of what became, then, Lemmas 6 and 7; hence, these results are joint work with that referee. This same referee suggested, for the future, team learning as an approach to studying some probabilistic variants of our learning criteria.

M. Hutter, R.A. Servedio, and E. Takimoto (Eds.): ALT 2007, LNAI 4754, pp. 34–48, 2007.

1   Introduction and Motivation

One-shot (algorithmic) learners, on input data about a function, output at most a single (hopefully correct) conjectured program [JORS99]. *Feasible* (deterministic) one-shot function learning can be modeled by the polytime multi-tape Oracle Turing machines (OTMs) as used in [IKR01] (see also [KC96, Meh76]). We call the corresponding functionals *basic feasible functionals*.

In the context of learning in the limit, i.e., learning with a succession of one-shots, where only the final shots are hoped to be correct, we are interested, then, in how one *might* define *feasible* for *limiting*-computable functionals. We next discuss the concepts we require for such a definition.

Intuitively *ordinals* [Sie65] are representations of well-orderings. 0 represents the empty ordering, 1 represents the ordering of 0 by itself, 2 the ordering $0 < 1$, 3 the ordering $0 < 1 < 2$, ... The ordinal ω represents the standard ordering of all of $\mathbb{N}$. $\omega + 1$ represents the ordering of $\mathbb{N}$ consisting of the positive integers in standard order *followed by* 0. The *successor ordinals* are those of the form $\alpha + 1$ which have a single element laid out after a copy of another ordinal α. $\omega + \omega$ can be thought of as two copies of ω laid end to end — much bigger than ω. $\omega \cdot 3$ represents three copies of ω laid end to end. By contrast, $3 \cdot \omega$ represents ω copies of 3 — which is just ω. We see, for ordinals, $+, \cdot$ are not commutative. $\omega \cdot \omega$ is ω copies of ω laid out end to end. We can iterate this and define exponentiation for ordinals. *Limit ordinals* are those, like ω, $\omega + \omega$, $\omega \cdot \omega$, and ω^ω, which are not 0 and are not successor ordinals. All of them are infinite. Importantly, the *constructive ordinals* are just those that have a program (called a *notation*) in some system which specifies how to build them (lay them out end to end, so to speak). Everyone knows how to use the natural numbers for counting, including for counting *down*. Freivalds and Smith [FS93], as well as [ACJS04], employed in learning theory *notations for constructive ordinals* as devices for algorithmic counting down. Herein we need to count down iterations of applications of feasible learning functionals. For example, for us, as we will see more formally in Section 4 below, algorithmic counting down iterations from any notation u for $\omega + 1$ is roughly equivalent to counting down one iteration and, then, deciding dynamically how many further but finite number of iterations will be allowed. Herein, though, we want the counting down process itself to be feasible. Hence, in Section 3, we introduce *feasibly related feasible systems of ordinal notations*, where, basically, the definition of a system of ordinal notations (as in [Rog67]) is restricted to those systems where all necessary operations and decision processes are *feasibly* computable. In Section 3, by Theorem 8, for each constructive ordinal α, we have such a system containing a notation for α and all its predecessors.

In Section 4, we present our proposed definition (Definition 11) for feasible iteration of feasible learning functionals. Then we present our general main results providing for strict learning hierarchies at all and only notations for (infinite) limit ordinals. First, Theorem 14 provides the learning hierarchy collapse between feasible notations for α and for $\alpha + 1$. Importantly, Theorem 17, provides a *strict* learning hierarchy between feasible notations for successive feasible *limit* ordinals.

In Section 5, our main results involve upper and lower runtime bounds for learning hierarchies featuring feasibly counting down from feasible notations for the successive initial limit ordinals $\omega \cdot n$, $n = 1, 2, 3, \ldots$. These runtime bounds are expressed in terms of exponential polynomials $\mathbf{q}$. In Theorem 20, for learning featuring feasible counting down from feasible notations for $\omega \cdot n$, the stacking of exponentials in the *upper* bound $\mathbf{q}$ is no more than n. Theorem 21 says there are classes learnable featuring feasible counting down from feasible notations for $\omega \cdot n$, where the stacking of exponentials in the *lower* bound $\mathbf{q}$ is at least n. In Section 5, we provide, though, an *example* of how to regain feasibility by a suitable parameterized complexity analysis [DF98].

Due to space constraints some portions of proofs are omitted. Complete proofs are in [CPK07].

2 Mathematical Preliminaries

$\mathbb{N}$ denotes the set of natural numbers, $\{0,1,2,\ldots\}$. We do not distinguish between natural numbers and their *dyadic* representation.[1] $\text{card}(D)$ denotes the cardinality of a set D.

The symbols $\subseteq, \subset, \supseteq, \supset$ respectively denote the subset, proper subset, superset and proper superset relation between sets.

We sometimes denote a function f of $n > 0$ arguments $x_1, \ldots, x_n$ in lambda notation (as in Lisp) as $\lambda x_1, \ldots, x_n . f(x)$. For example, with $c \in \mathbb{N}$, $\lambda x.c$ is the constantly c function of one argument. From now on, by convention, f and g with or without decoration range over functions $\mathbb{N} \to \mathbb{N}$, x, y with or without decorations range over $\mathbb{N}$, 0^i, 0^j range over $\{0\}^*$.

We use 'string' and 'finite sequence' synonymously, and, for each sequence s, we will denote the first element of that sequence by $s(0)$, (or, equivalently, with s_0,) the second with $s(1)$ (or s_1) and so on.

Similarly we will consider infinite sequences s as functions with domain $\mathbb{N}$ (or $\mathbb{N} \cup \{-1\}$, as the case may be), and denote them at position a in the domain by $s(a)$ or s_a.

For each string w, define $\text{len}(w)$ to be the length of the string. As we identify each natural number x with its dyadic representation, $\text{len}(x)$ denotes the length of the dyadic representation of x. For all strings w, we define $|w|$ to be $\max\{1, \text{len}(w)\}$.[2]

Following [LV97], we define for all $x \in \mathbb{N}$: $\bar{x} = 1^{\text{len}(x)}0x$. Using this notation we can define a function $\langle \cdot \rangle$ coding tuples of natural numbers of arbitrary size ($k \geq 0$) into $\mathbb{N}$ such that $\langle v_1, \ldots, v_k \rangle := \overline{v_1} \ldots \overline{v_k}$.

For example the tuple $(4, 7, 10)_{decimal} = (01, 000, 011)_{dyadic}$ would be coded as $11001\ 1110000\ 1110011$ (but without the spaces added for ease of parsing).

---

[1] The *dyadic* representation of a natural number $x :=$ the x-th finite string over $\{0, 1\}$ in *lexicographical order*, where the counting of strings starts with zero [RC94]. Hence, unlike with binary representation, lead zeros matter.

[2] ε denotes the empty string. This convention about $|\varepsilon| = 1$ helps with runtime considerations.

Obviously $\langle \cdot \rangle$ is 1-1. The time to encode tuples, that is, to compute $\lambda v_1, \ldots, v_k. \langle v_1, \ldots, v_k \rangle$ is $\in \mathcal{O}(\lambda v_1, \ldots, v_k. \sum_{i=1}^{k} |v_i|)$. Therefore the size of the codeword is also linear in the size of the components: $\lambda v_1, \ldots, v_k. |\langle v_1, \ldots, v_k \rangle| \in \mathcal{O}(\lambda v_1, \ldots, v_k. \sum_{i=1}^{k} |v_i|)$. Decoding is linear in the length of the codeword: For all $k, i \leq k$, we have that $\lambda \langle v_1, \ldots, v_k \rangle. v_i$ is computable in linear time, so is $\lambda \langle v_1, \ldots, v_k \rangle. k$.

A function ψ is *partial computable* iff there is a Turing machine computing ψ. φ^{TM} is the fixed programming system from [RC94, Chapter 3] for the partial computable functions. This system is based on deterministic, multi-tape Turing machines (TMs). In this system the TM-programs are *efficiently* given numerical names or codes.[3] Φ^{TM} denotes the TM step counting complexity measure also from [RC94, Chapter 3] and associated with φ^{TM}. In the present paper, we employ a number of complexity bound results from [RC94, Chapters 3 & 4] regarding $(\varphi^{\mathrm{TM}}, \Phi^{\mathrm{TM}})$. These results will be clearly referenced as we use them. For simplicity of notation, hereafter we write (φ, Φ) for $(\varphi^{\mathrm{TM}}, \Phi^{\mathrm{TM}})$. φ_p denotes the partial computable function computed by the TM-program with code number p in the φ-system, and Φ_p denotes the partial computable *runtime* function of the TM-program with code number p in the φ-system.

Whenever we consider tuples of natural numbers as input to TMs, it is understood that the general coding function $\langle \cdot \rangle$ is used to code the tuples into appropriate TM-input. We say that a function from k-tuples of natural numbers into $\mathbb{N}$ is *feasibly computable* iff, for some p, it is computed by TM p in polytime in the lengths of its inputs.[4]

The next definitions provide the formal details re the polytime constraint on basic feasible functionals.

The *length* of $f : \mathbb{N} \to \mathbb{N}$ is the function $|f| : \mathbb{N} \to \mathbb{N}$ such that $|f| = \lambda n. \max(\{|f(x)| \mid |x| \leq n\})$.

A *second-order polynomial* over type-1 variables $g_0, \ldots, g_m$ and type-0 variables $y_0, \ldots, y_n$ (in this paper simply referred to as a polynomial) is an expression of one of the following five forms.

$$a; \quad y_i; \quad \mathbf{q}_1 + \mathbf{q}_2; \quad \mathbf{q}_1 \cdot \mathbf{q}_2; \quad g_j(\mathbf{q}_1)$$

where $a \in \mathbb{N}$, $i \leq n$, $j \leq m$, and $\mathbf{q}_1$ and $\mathbf{q}_2$ are second-order polynomials over $\overrightarrow{g}$ and $\overrightarrow{y}$.

A *subpolynomial* of $\mathbf{q}$ is, recursively, any polynomial which is used in the construction of $\mathbf{q}$, or which is a subpolynomial of a polynomial that is used in the construction of $\mathbf{q}$.

We understand each such polynomial $\mathbf{q}$ as a symbolic object and for functions $f_0, \ldots, f_m : \mathbb{N} \to \mathbb{N}$, $x_0, \ldots, x_n \in \mathbb{N}$ we write $\mathbf{q}(f_0, \ldots, f_m, x_0, \ldots, x_n)$ as the obvious evaluation of $\mathbf{q}$ to an element in $\mathbb{N}$.

---

[3] This numerical coding guarantees that many simple operations involving the coding run in linear time. This is by contrast with historically more typical codings featuring prime powers and corresponding at least exponential costs to do simple things.

[4] We are mostly not considering herein interesting polytime probablistic or quantum computing variants of the deterministic feasibility case.

For two functions $h_1, h_2 : \mathbb{N} \to \mathbb{N}$ we write $h_1 \leq h_2 :\Leftrightarrow \forall n \in \mathbb{N} : h_1(x) \leq h_2(x)$ and we say that h_2 *majorizes* h_1. It is easy to see from the definition of a second-order polynomial $\mathbf{q}$ that $\lambda f_0, \ldots, f_m, x_0, \ldots, x_n.\mathbf{q}(f_0, \ldots, f_m, x_0, \ldots, x_n)$ is non-decreasing in each argument, given that all function-arguments are non-decreasing and order on functions is as defined just above.

An *Oracle Turing Machine* (OTM) is a multi-tape Turing Machine that also has a query tape and a reply tape. To query an oracle f, an OTM writes the dyadic representation of an $x \in \mathbb{N}$ on the query tape and enters its query state. The query tape is then erased, and the dyadic representation of $f(x)$ appears on the reply tape. This model is extended to the case of multiple oracles in the obvious way. The (time) cost model is the same as for non-oracle Turing machines, *except* for the additional cost of a query to the oracle. This is handled with the length-cost model, where the cost of a query is $|f(x)|$, where $|f(x)|$ is the length of the string on the reply tape.

Suppose $k \geq 1$ and $l \geq 0$. Then $F : (\mathbb{N} \to \mathbb{N})^k \times \mathbb{N}^l \to \mathbb{N}$ is a *basic feasible functional* if and only if there is an OTM $\mathbf{M}$ and a second-order polynomial $\mathbf{q}$, such that, for each input $(f_1, \ldots, f_k, x_1, \ldots, x_l)$,

(a) $\mathbf{M}$ outputs $F(f_1, \ldots, f_k, x_1, \ldots, x_l)$, and
(b) $\mathbf{M}$ runs within $\mathbf{q}(|f_1|, \ldots, |f_k|, |x_1|, \ldots, |x_l|)$ time steps (we will then say that $\mathbf{q}$ *majorizes the runtime of* F).

Any unexplained computability-theoretic notions are from [Rog67].

3   Feasible Systems of Ordinal Notations

In this section we begin with some definitions regarding systems of ordinal notations. The first definition is quite technically useful in our proofs in Section 4 below.

Definition 1. For a system of ordinal notations S as, for example, in [Rog67], a pair (l_S, n_S) is a *decompose pair for* S iff l_S and n_S are functions $\mathbb{N} \to \mathbb{N}$ and for all notations $u \in S$ for an ordinals α, $l_S(u)$ denotes a notation for the biggest (limit ordinal or 0) $\lambda \leq \alpha$, and $n_S(u)$ is such that $\alpha = \lambda + n_S(u)$.

Definition 2. (Feasible System of Ordinal Notations) For an ordinal α, a *feasible system of ordinal notations* for all and only the ordinals $< \alpha$ is a tuple $(S, \nu_S, \lim_S, +_S, \cdot_S, l_S, n_S)$ where $S \subseteq \mathbb{N}$, ν_S maps $\mathbb{N}$ onto the set of all ordinals $< \alpha$, $\lim_S : \mathbb{N} \times \{0\}^* \to \mathbb{N}$, $+_S$ and $\cdot_S$ are ordinal sum and multiplication on notations respectively[5] and (l_S, n_S) is a decompose pair for S.[6] Additionally we require:
The following predicates over $u \in S$ are feasibly decidable.

(a) "u is a notation for 0",

---

[5] Therefore, each feasible system of ordinal notations will give notation to an additively and multiplicatively closed set of ordinals.

[6] We will sometime ambiguously refer to $(S, \nu_S, \lim_S, +_S, \cdot_S, l_S, n_S)$ as S.

(b) "u is a notation for a successor ordinal" and
(c) "u is a notation for a limit ordinal".

And:

(d) There is a feasibly computable function pred_S such that for all u notations for a successor ordinal $\alpha + 1$, $\mathrm{pred}_S(u)$ is a notation for α.
(e) $\lim_S$ is a feasible function and for all limit-ordinals $\lambda < \alpha$ and notations l for λ we have that $(\nu_S(\lim_S(l, 0^i)))_{i<\omega}$ is a strictly increasing sequence of ordinals with limit λ.

Up to this point in this definition, we have a modification of Rogers' concept of *system of ordinal notations* [Rog67], *where*, when we require feasible computability, Rogers requires only partial computability. Additionally we require

(f) $+_S$ is feasibly computable,
(g) $\cdot_S$ is feasibly computable,
(h) from any natural number n, a notation $\underline{n}_S$ for n is feasibly computable and
(i) l_S, n_S are feasibly computable.

Definition 3. Following Rogers [Rog67], we say that a system of ordinal notations S is *univalent* iff ν_S is 1-1; we define the relation $\leq_S$ on natural numbers such that: $u \leq_S v \Leftrightarrow [u, v \in S \land \nu_S(u) \leq \nu_S(v)]$. Also following Rogers, we say a system of ordinal notations S is *computably related* iff $\leq_S$ is computably decidable, and *computably decidable* iff the set of notations S is computably decidable. Analogously, we define a system S to be *feasibly related* iff $\leq_S$ is feasibly decidable, and *feasibly decidable* iff the set S is feasibly decidable.

Remark 4. In Definition 2 above we have that feasible relatedness, together with (f), (h) and (i) implies (a)-(d). Every feasibly related feasible system of ordinal notations S is feasibly decidable, as we have: $u \in S \Leftrightarrow u \leq_S u$. Every feasibly related feasible system of ordinal notations is a computably related system of ordinal notations.[7] For a feasibly related or univalent feasible system of ordinal notations S, it is feasibly decidable whether two notations are notations for the same ordinal.[8]

Lemma 5. Suppose S is a system of ordinal notations in which a notation in S for the successor ordinal is feasibly computable from a given notation in S. Let $\lim_S : \mathbb{N} \times \{0\}^* \to \mathbb{N}$ be a computable function satisfying the analog of (e) where "feasible" is replaced by "partial computable". Then there is a *feasibly* computable function $\lim'_S : \mathbb{N} \times \{0\}^* \to \mathbb{N}$ satisfying (e).

---

[7] Therefore, all theorems for computably related systems of ordinal notations hold. For example, there cannot be a feasibly related feasible system of ordinal notations for all constructive ordinals (see [Rog67]).

[8] For univalent systems there are of course no two different notations for the same ordinal. For a feasibly related systems of ordinal notations, $u, v \in \mathbb{N}$ are notations in S for the same ordinal iff $[u \leq_S v$ and $v \leq_S u]$.

Proofsketch. Define $\lim'_S$ thus. On input $(u, 0^i)$, run $\lim_S$ on inputs $(u, 0^j)$ for all $j \leq i$, each for up to i steps. If none converges, output i — a notation in S for i. Otherwise, for some $j \leq i$, $\lim_S(u, 0^j)$ converges. In this case, for the maximal such j, compute the i-times successor of $\lim_S(u, 0^j)$ and output the result — a notation for $\nu_S(\lim_S(u, 0^j)) + i$. Importantly, thanks to [RC94, Corollary 3.7], the algorithm just provided for $\lim'_S$ *is* feasible. We omit the remaining details of the proof.   □

Lemma 6. Suppose S is a computably related system of ordinal notation for all and only the ordinals $< \alpha$ for some ordinal α. Then there is a *feasibly* related system S' of ordinal notations for all and only the ordinals $< \alpha$.

Proof. Define S' thus. Let e be the numerical name for a program deciding $\leq_S$. Define $t : \mathbb{N} \to \mathbb{N}, u \mapsto \max(\{\Phi_e(i, j) \mid i, j \leq u\})$. Let S' be the system of notations where for all β given a notation u in S, we have that $\langle 0^{t(u)}, 0^u \rangle$ is a notation for β. Obviously, $\forall m, n, u, v \in \mathbb{N} : \langle 0^m, 0^u \rangle \leq_{S'} \langle 0^n, 0^v \rangle \Leftrightarrow [\varphi_e(u, v) = 1$ in $\leq \max\{n, m\}$ steps and $m = t(u)$ and $n = t(v)]$. It follows from [RC94, Lemma 3.2(f) and Corollary 3.7] that $\lambda 0^m, 0^u . t(u) = m$ is feasibly decidable. Therefore, on the resulting notations we have that order is feasibly decidable.   □

Lemma 7. Suppose S is a feasibly related system of ordinal notations giving a notation to all and only the ordinals $< \alpha$ for some ordinal α. Then there is a feasibly related system of ordinal notations S' fulfilling (a)-(f) and (h)-(i) as in Definition 2, giving a notation at least to all ordinals $< \alpha$. In fact, S' gives a notation to all and only the ordinals $< \omega^\alpha$. If S is univalent, so is S'.

Proofsketch. Assume without loss of generality that 0 is the only notation for 0 in S. Let $\langle \rangle$ be a notation in S' for 0. By the Cantor Normal Form theorem, each ordinal γ, $0 < \gamma < \omega^\alpha$ has exactly one representation such that $\gamma = \sum_{i=k}^0 \omega^{\delta_i} \times n_i$, where $\alpha > \delta_k > \ldots > \delta_0 \geq 0$ and $n_k, \ldots, n_0 \in \mathbb{N} \setminus \{0\}$ (see [Sie65], Theorem 2, Chapter XIV.19, page 323]). Define a system S' by the following assignment of notations. For each γ with $0 < \gamma < \omega^\alpha$, the representation as above and $d_k, \ldots, d_0$ notations in S for $\delta_k, \ldots, \delta_0$, respectively, let

$$\langle d_k, n_k, \ldots, d_0, n_0 \rangle \text{ be a notation in } S' \text{ for } \gamma.$$

From here we omit most remaining details. To show (e) for S': We apply Lemma 5.   □

Theorem 8. Suppose S is a feasibly related system of ordinal notations giving a notation to all and only the ordinals $< \alpha$. Then there is a feasibly related *feasible* system of ordinal notations S' giving a notation at least to all ordinals $< \alpha$. In fact, S' gives a notation to all and only the ordinals $< \omega^{\omega^\alpha}$. If S is univalent, so is S'.

Proofsketch. Apply the construction of the proof of Lemma 7 *twice* to S. The resulting system will also allow for feasible multiplication.   □

Corollary 9. *Let α be a constructive ordinal. Then there is a univalent, feasibly related feasible system of ordinal notations giving a notation to α.*

Proof. By [Rog67, Theorem 11.XIX], there is a univalent, computably related system of ordinal notations giving a notation to α. The result follows now from first applying Lemma 6 and then Theorem 8.   □

Assumption 10. For the rest of this paper, fix an arbitrary univalent feasibly related feasible system of ordinal notations S. We furthermore make the following assumption.

$$\forall u \in S, n \in \mathbb{N} : |n| \leq |\underline{n}| \leq |u +_S \underline{n}|. \tag{1}$$

This reasonable assumption holds for all systems constructed in the proof of Corollary 9. (1) above also shows that for all $u \in S$ we have $|n_S(u)| \leq |l_S(u) + n_S(u)| = |u|$; therefore, we get

$$\forall u \in S : n_S(u) \leq u. \tag{2}$$

4   Hierarchies at Limit Ordinal Jumps

Next is our proposed definition of feasible iteration of feasible learning functionals.

Definition 11. Suppose $u \in S$. A set of functions $\mathcal{S}$ is $\text{Itr}_u\text{BffFin}$-*identifiable* (we write $\mathcal{S} \in \text{Itr}_u\text{BffFin}$) iff there exist basic feasible functionals $H : (\mathbb{N} \to \mathbb{N}) \times \{0\}^* \to \mathbb{N}$ and $F : (\mathbb{N} \to \mathbb{N}) \times \{0\}^* \to \mathbb{N}$ such that for all $f \in \mathcal{S}$ there exists $k \in \mathbb{N}$ such that

(a) $F(f, 0^t) <_S u$ for all $t < k$,[9]
(b) $F(f, 0^{t+1}) <_S F(f, 0^t)$ for all $t < k$,
(c) $F(f, 0^k) = \underline{0}$ and
(d) $\varphi_{H(f,0^k)} = f$.

Lemma 12. Without loss of generality the count down function F in Definition 11 can be chosen such that for all computable functions f there is a $k \in \mathbb{N}$ such that (a) and (b) in Definition 11 hold, as well as $\forall t \geq k : F(f, 0^t) = \underline{0}$.

Proof. Let q be a polynomial upper-bounding the runtime of F. Then there is F' such that F' on input $(f, 0^t)$ computes for all $w \leq t$ $F(f, 0^w)$ (taking time in $\mathcal{O}(\sum_{w=0}^{t} q(|f|, w)) \subseteq \mathcal{O}(t \cdot q(|f|, t)))$. If we have $F(f, 0^0) <_S u$ and for all $w < t$ $F(f, 0^{w+1}) <_S F(f, 0^w)$ (t comparisons decidable in polytime), then output $F(f, 0^t)$, otherwise output $\underline{0}$.   □

---

[9] Earlier papers using count down functions, such as for example [ACJS04], usually use instead, at this point in the definition, $\leq_S u$. This earlier way of starting count downs can be recovered in the version presented herein by using $<_S u +_S \underline{1}$. Our present version has additional expressibility for u being a notation for a limit ordinal, which is not available in a version starting with $\leq_S u$. However, it is a theorem in this paper (Theorem 14 below) that, for our way herein of starting count downs, no resultant extra learning power class exists.

Assumption 13. From now on, all witnesses for a set to be in $\mathrm{Itr}_u\mathrm{BffFin}$ to have these additional properties as stated in Lemma 12. Witnesses explicitly constructed might not have this property.

Note that for all $u \in S$, as it is also the case for many other identification criteria [JORS99], $\mathrm{Itr}_u\mathrm{BffFin}$ is closed under taking subsets.

The first theorem shows that there is no difference in learning power for an ordinal and its successor:

Theorem 14. Suppose $u \in S$. Then $\mathrm{Itr}_u\mathrm{BffFin} = \mathrm{Itr}_{(u+_s\underline{1})}\mathrm{BffFin}$.

Proof. Trivial for u a notation for 0. Otherwise, let $\mathcal{S} \in \mathrm{Itr}_{(u+_s\underline{1})}\mathrm{BffFin}$ as witnessed by (H, F). We have for all $f \in \mathcal{S}$: $F(f, 0^0) <_S u \ \vee \ F(f, 0^1) <_S u$. Let $P := \lambda f . \mu i < 2.(F(f, 0^i) <_S u)$; $H' := \lambda f, 0^i . H(f, 0^{i+P(f)})$; $F' := \lambda f, 0^i . F(f, 0^{i+P(f)})$. So (H', F') witnesses $\mathcal{S} \in \mathrm{Itr}_u\mathrm{BffFin}$. $\qquad\square$

Recall that in the mathematical preliminaries it has been mentioned that all polynomials as defined in this paper fulfill several monotonicity constraints. Furthermore, we can find a single polynomial upper bounding the runtime all functions of a given, finite set of BFFs (for example by adding all polynomials for each single BFF up). The next definition gives a desirable property of polynomials. The following remark will imply that we can – for all uses of polynomials in this paper – suppose without loss of generality that our polynomials have this property.

Definition 15. A polynomial $\mathbf{q}$ is called *request-bounding* iff for all polynomials $\mathbf{q}'$ such that $g(\mathbf{q}')$ is a subpolynomial of $\mathbf{q}$ we have that $\mathbf{q}$ majorizes $\mathbf{q}'$.

Remark 16. For all polynomials $\mathbf{q}$ there is a request-bounding polynomial majorizing $\mathbf{q}$.

Proof. Let $\mathbf{q}$ be a polynomial, $\mathbf{q}'$ a polynomial such that $g(\mathbf{q}')$ is a subpolynomial of $\mathbf{q}$ and $\mathbf{q}$ does not majorize $\mathbf{q}'$. We have that $\overline{\mathbf{q}} := \mathbf{q} + \mathbf{q}'$ majorizes $\mathbf{q}$ and $\mathbf{q}'$ such that $\{\mathbf{r} \ | \ g(\mathbf{r}) \text{ is a subpolynomial of } \overline{\mathbf{q}}\} = \{\mathbf{r} \ | \ g(\mathbf{r}) \text{ is a subpolynomial of } \overline{\mathbf{q}}\}$. Iterating this construction of a bigger polynomial will finally yield a polynomial with the desired properties. $\qquad\square$

Next is the main theorem of this section. It provides a strict learning power hierarchy.

Theorem 17. Suppose $u <_S v \in S$ represent non-successor ordinals. Then $\mathrm{Itr}_u\mathrm{BffFin} \subset \mathrm{Itr}_v\mathrm{BffFin}$.

Proofsketch. The inclusion is trivial, so that the separation remains to be shown. This is done by constructing a suitable set of total computable functions which belongs to the right set, but not to the left. Let $\mathcal{S}^*$ be the set of all functions f such that: There is a sequence r of notations for non-successor ordinals, strictly decreasing (with respect to $<_S$), where $r(0) < v$. There is a strictly increasing sequence s of natural numbers of length $\mathrm{len}(r) + 1$ such that:

(a) $s(0) = 1$,

(b) for all $i < \text{len}(s) - 1 : s(i+1) \in \{2^{f(s(i))}, \ldots, 2^{f(s(i))} + |f(0)| - 1\}$,[10]

(c) for all $i < \text{len}(s) - 1 : f(s(i)) \in S$,

(d) for all $i < \text{len}(s) - 1 : l_S(f(s(i))) = r(i)$,

(e) $r(\text{len}(r) - 1) = \underline{0}$ and

(f) for all $i \in \mathbb{N}: i \in (\text{range}(s) \cup \{0\}) \Leftrightarrow f(i) \neq 0$.

That our $\mathcal{S}^*$ witnesses the separation of Theorem 17 just above provides the reason its use in Proposition 22 (in Section 5 below) is interesting.

Define $\mathcal{S} := \{f \in \mathcal{S}^* \mid |f(0)| = 1\}$. To save space in this proof, we will actually show instead that $\mathcal{S}$, a proper subset of $\mathcal{S}^*$, witnesses the separation. Of course, the negative part of the separation trivially applies to supersets.

Claim: $\mathcal{S} \in \text{Itr}_v\text{BffFin}$

Proof (of claim). Let e be such that for all finite functions σ (treated as strings of size $\text{len}(\sigma)$, coded onto the tape by $\langle \rangle$), and for all $x \in \mathbb{N}$,

$$\varphi_e(\sigma, x) = \begin{cases} \sigma(i) \, , & \text{if } i < \text{len}(\sigma) \text{ such that } x = 2^i; \\ 0 \, , & \text{otherwise.} \end{cases}$$

Runtime in $\mathcal{O}$

F on $(f, 0^t)$:

```
query f for σ := λi ≤ t.f(2ⁱ)
determine biggest index i ≤ t such that f(2ⁱ) > 0
if such an index does not exist, output 0
if two such indices exist, let l < i be the biggest two
if l_S(f(2ⁱ)) = 0 redefine i := l
output l_S(f(2ⁱ)) +_S (f(2ⁱ) − t)
```

$|f|(t) \cdot t$
$|f|(t) \cdot t$

The next functional makes use of a linear time instance of an s-m-n-function.[11]

H on $(f, 0^t)$:

```
query f for σ := λi ≤ t.f(2ⁱ)
output s-m-n(e, σ)
```

$|f|(t) \cdot t$
linear time s-m-n

(H, F) shows the claim.

$\square$ (OF CLAIM)

Claim: $\mathcal{S} \notin \text{Itr}_v\text{BffFin}$.

Proof (of claim). Suppose by way of contradiction otherwise, as witnessed by (H, F). Let $\mathbf{q}(g, x)$ be a polynomial with the following properties. $\mathbf{q}$ strictly majorizes the runtime of F and H; $\mathbf{q}$ is request-bounding; and, for technical reasons, for all $c, n \in \mathbb{N}$, $\mathbf{q}(\lambda x.c, n) \geq c$.[12]

Define now two functions $a, b : \mathbb{N} \times \mathbb{N} \to \mathbb{N}$ such that $\forall x, y : a(x, y) = \mathbf{q}(\lambda z.|x|, y), b(x, y) = \mathbf{q}(\lambda z.|x|, a(x, y))$. For all $x, y \in \mathbb{N}$, $a(x, y)$ is, then, an

---

[10] This entails, as s is required to be strictly increasing, that $f(s(i)) > 0$ for all $i < \text{len}(s) - 1$.

[11] Linear time s-m-n is a function s running in linear time such that $\forall e, x, y : \varphi_{s(e,x)}(y) = \varphi_e(x, y)$; see [RC94, Theorem 4.7(b)].

[12] For each polynomial $\mathbf{q}'$, $\mathbf{q}' + g(0)$ is an example polynomial fulfilling this property.

upper runtime-bound for the computation of F on second argument 0^y, if the first argument is never requested at anything yielding something bigger then x. For all $x, y \in \mathbb{N}$, $b(x, y)$ is an upper runtime-bound for the computation of F on second argument $0^{a(x,y)}$ with the same restriction on the first argument.

Let w be a notation for ω. We define, by multiply recursive calls, for $i \geq 0$, infinite sequences s, v and a computable function f as follows.

$$v_{-1} = 0$$
$$s(0) = 1$$
$$f(s(0)) = v_0 = u +_S \underline{x},$$
$$\text{where } x \text{ minimal so that } v_0 > b(v_0, v_{-1}) \text{ and}$$
$$|v_0| > 2^{b(v_{-1}, 0)} + 1$$
$$s(i+1) = \begin{cases} 2^{v_i}, & \text{if } v_i \neq 0 \\ \max\{s(j) \mid j \leq i\} + 1, & \text{otherwise} \end{cases}$$
$$f(s(i+1)) = v_{i+1} = \begin{cases} 0, & \text{if there is } j < i-1 \text{ such that } l_S(v_j) = w \\ 1, & \text{if } l_S(v_{i-1}) = w \\ ((v_i + 1) + x), & \text{if } l_S(v_i) = w \\ (F(f, 0^{a(v_i, v_{i-1})}) +_S w) +_S \underline{((v_i + 1) + x)}, & \text{otherwise} \end{cases}$$
$$\text{where } x \text{ minimal so that } v_{i+1} > b(v_{i+1}, v_i) \text{ and}$$
$$|v_{i+1}| > 2^{b(v_i, v_{i-1})} + a(v_i, v_{i-1}) + 1$$

Define f on all so far undefined values as 0.
Notes on the construction:

- To show that the condition on v_0 is possible: On the one hand, by Assumption 10 in Section 3, $\lambda x. u +_S \underline{x}$ grows at least as fast as $\lambda x.x$, and $\lambda x.x$ grows *exponentially* in the *length* of its argument, and, on the other hand, $\lambda x. b(u +_S \underline{x}, 0)$ grows only *polynomially* in the *length* of its argument. Similar reasoning holds for the condition on v_{i+1}.
- As part (f) of the proof of the next subclaim we will show that, in the computation of $F(f, 0^{a(v_i, v_{i-1})})$, the value of f at $s(i+1)$ will never be requested. This avoids $f(s(i+1))$ calling itself.

Let m be maximal such that $v_m \neq 0$. Clearly, from above, for all i such that $0 \leq i \leq m$, $v_i \neq 0$. Abbreviate for all $i \leq m$: $a_i := a(v_i, v_{i-1})$ and $b_i := b(v_i, v_{i-1})$.

Subclaim: f, s, v are well defined.
Proof (of subclaim). We proof this by induction on $i < m$. We have the following induction invariants.

(a) if $0 \leq i$, then $b_i < v_i$,
(b) if $0 \leq i$, then there are $> v_i$ steps required to query f at $s(i+1)$,
(c) if $0 \leq i$, then f is circle-free defined up through $s(i+1) - 1$,
(d) if $0 \leq i$, then $\forall t \leq b_i : |f|(t) < v_i$,

(e) if $0 \leq i$, then b_i is an upper runtime bound for F on $(f, 0^{a_i})$,

(f) if $0 \leq i$, then $f(s(i+1))$ is not requestable in the computation of F on $(f, 0^{a_i})$.

(g) v_{i+1} is well defined

These invariants hold obviously for $i = -1$. Let now $i < m$ be such that $0 \leq i$ and these induction invariants hold for for all $k < i$ such that $-1 \geq k$. We show now that the invariants hold for i.

To show (a): By construction we have $v_i > b(v_i, v_{i-1}) = b_i$.

To show (b): We have $|s(i+1)| = |2^{v_i}| = v_i$. Therefore, v_i steps are required to write $s(i+1)$ on the query tape; at least an additional step is used to complete this query process.

To show (c): Follows from for all $k < i$, v_{k+1} well defined (that is, $\forall k \leq i$, v_k is well defined).

To show (d): Let $t \leq b_i$. We have:

$$|f|(t) \leq |f|(b_i) = \max_{|x| \leq b_i} |f(x)| \leq \max_{|x| \leq v_i} |f(x)| \leq \max_{x < 2^{v_i}} |f(x)| = |v_i| < v_i \ .$$

To show (e): By induction on the polynomials $\mathbf{q}'$ such that $g(\mathbf{q}')$ is a subpolynomial of $\mathbf{q}$. Using invariant (d) and $\mathbf{q}$ being request-bounding it is easy to show that all such polynomials $\mathbf{q}'$ evaluate on $(|f|, a_i)$ to something $\leq b_i$. Then we can conclude that $\mathbf{q}(|f|, a_i) \leq b_i$.

To show (f): (b), (e) and (a) show (f).

To show (g): From (f).   □ (OF SUBCLAIM)

It is now easy to verify that $f \in \mathcal{S}$, and, then, a simple adversary argument (as, for example, in [DZ01]) shows that either f or f modified at $s(m)$ (so that the modified version is still in $\mathcal{S}$) is not properly identified by (H, F).

□ (OF CLAIM)   □ (OF THEOREM)

5   Upper and Lower Bounds on Runtime

For this section assume S gives a notation to at least all ordinals $< \omega^2$ (the existence of such an S is guaranteed by Corollary 9). In this section we will characterize the hierarchy of finitely many limit ordinal jumps in terms of explicit total runtime bounds. The next definition introduces polynomials with exponential terms and the exponential nesting depth of such.

Definition 18 (Polynomials with Exponentials). We define recursively in parallel the set $Q[g]$ of symbolic polynomials with exponentials, as well as the exponential nesting depth of $\mathbf{q} \in Q[g]$, $rk(\mathbf{q})$, (read: rank of $\mathbf{q}$):

for all $a \in \mathbb{N}$: $a \in Q[g]$, $rk(a) = 0$,

for all $\mathbf{q}_1, \mathbf{q}_2 \in Q[g]$: $(\mathbf{q}_1 + \mathbf{q}_2) \in Q[g]$, $rk((\mathbf{q}_1 + \mathbf{q}_2)) = \max(rk(\mathbf{q}_1), rk(\mathbf{q}_2))$,

$\mathbf{q}_1, \mathbf{q}_2 \in Q[g]$: $\mathbf{q}_1 \cdot \mathbf{q}_2 \in Q[g]$, $rk(\mathbf{q}_1 \cdot \mathbf{q}_2) = \max(rk(\mathbf{q}_1), rk(\mathbf{q}_2))$,

for all $\mathbf{q} \in Q[g]$: $g(\mathbf{q}) \in Q[g]$, $rk(g(\mathbf{q})) = rk(\mathbf{q})$,

for all $\mathbf{q} \in Q[g]$: $2\hat{\ }(\mathbf{q}) \in Q[g]$, $rk(2\hat{\ }(\mathbf{q})) = rk(\mathbf{q}) + 1$.

The following definition will enable us to study the runtime of functionals beyond the feasible.

Definition 19. Suppose $k \geq 1$ and $l \geq 0$. Then $F : (\mathbb{N} \to \mathbb{N})^k \times \mathbb{N}^l \to \mathbb{N}$ is a *computable functional* if and only if there is an OTM $\mathbf{M}$ such that, for each input $(f_1, \ldots, f_k, x_1, \ldots, x_l)$, $\mathbf{M}$ outputs $F(f_1, \ldots, f_k, x_1, \ldots, x_l)$.

The two theorems below are our main results of the present section.

Theorem 20 (Learning Time – Upper Bound). Let $n > 0$. Let $\mathcal{S} \in$ Itr$_{w \cdot \underline{n}}$BffFin. Then there is a computable functional h such that

- $\forall f \in \mathcal{S} : \varphi_{h(f)} = f$
- the runtime of h is bounded above by some $\mathbf{q} \in Q[g]$ such that $rk(\mathbf{q}) \leq n$.

Proof. Let $\mathcal{S} \in$ Itr$_{w \cdot \underline{n}}$BffFin as witnessed by (H, F). Define t, h such that for all f computable functions: $t(f) = \mu x. F(f, 0^x) = \underline{0}$, $h(f) = H(f, 0^{t(f)})$. Obviously h fulfills the first requirement.

Let $\mathbf{p}$ be a polynomial upper-bounding the runtime of H and F. Let $\mathbf{q}_0 = 0$. Define recursively for all $i < n$: $\mathbf{q}_{i+1} = 2^{\mathbf{P}(g, \mathbf{q}_i)} + \mathbf{q}_i$. It is clear that $rk(\mathbf{q}_i) = i$.

We have now, for all $f \in \mathcal{S}$ and $i < n$, that the calculation of F on $(f, 0^{\mathbf{q}_i(|f|)})$ is bounded above by $\mathbf{p}(|f|, \mathbf{q}_i(|f|))$; therefore, $F(f, 0^{\mathbf{q}_i(|f|)}) < 2^{\mathbf{P}(|f|, \mathbf{q}_i(|f|))}$, and, hence, by (2) from the very end of Section 3:

$$n_S(F(f, 0^{\mathbf{q}_i(|f|)})) <_S 2^{\mathbf{P}(|f|, \mathbf{q}_i(|f|))} = \mathbf{q}_{i+1}(|f|) - \mathbf{q}_i(|f|) . \tag{3}$$

(3) is to be read as follows. After $\mathbf{q}_i(|f|)$ iterations of F, F will output a notation for an ordinal with natural-number part less then $\mathbf{q}_{i+1}(|f|) - \mathbf{q}_i(|f|)$ – therefore, after no more then $\mathbf{q}_{i+1}(|f|) - \mathbf{q}_i(|f|)$ additional iterations (after a total of $\mathbf{q}_{i+1}(|f|)$ iterations), there has to be a limit ordinal jump in the output of F. A simple induction shows now that we have, for all $f \in \mathcal{S}$ and $i < n$, $F(f, 0^{\mathbf{q}_i(|f|)}) <_S w \cdot \underline{n} - \underline{i}$, and, therefore, $F(f, 0^{\mathbf{q}_n(|f|)}) = \underline{0}$ (this makes use of assumption 13).

Let $f \in \mathcal{S}$. The above shows that we have $t(f) \leq \mathbf{q}_n(|f|)$; therefore, an algorithm for computing t could run F on all arguments $(f, 0^i)$ in increasing order for all $i \leq \mathbf{q}_n(|f|)$, checking each output for equaling $\underline{0}$. This takes a total time $\leq \sum_{i=0}^{\mathbf{q}_n(|f|)} \mathbf{p}(|f|, i) \leq (\mathbf{q}_n(|f|) + 1) \cdot \mathbf{p}(|f|, \mathbf{q}_n(|f|))$. Therefore, h can be computed in $\leq (\mathbf{q}_n(|f|) + 1) \cdot \mathbf{p}(|f|, \mathbf{q}_n(|f|)) + \mathbf{p}(|f|, \mathbf{q}_n(|f|)) = (\mathbf{q}_n(|f|) + 2) \cdot \mathbf{p}(|f|, \mathbf{q}_n(|f|))$ steps, where $rk((\mathbf{q}_n + 2) \cdot \mathbf{p}(g, \mathbf{q}_n)) = rk(\mathbf{q}_n) = n$. □

Theorem 21 (Learning Time – Lower Bound). Let $n > 0$. Let $\mathcal{S}$ be as in the proof of the limit ordinal jump hierarchy (Theorem 17) for the special case of Itr$_{w \cdot (n-1)}$BffFin $\subset$ Itr$_{w \cdot \underline{n}}$BffFin. Define $\mathcal{S}' := \{f \in \mathcal{S} \mid \text{card}(\{x > 0 \mid f(x) > 0\}) = n + 1\}$. Let h be a computable functional such that $\forall f \in \mathcal{S}' : \varphi_{h(f)} = f$ and fix an OTM M computing h. Define $\mathbf{q}_0 := g(1)$, define for all $i < n$, $\mathbf{q}_{i+1} := g(2\hat{\ }(\mathbf{q}_i))$. Then $rk(\mathbf{q}_n) = n$ and we have, for all $f \in \mathcal{S}'$, $\mathbf{q}_n(|f|)$ is a *lower* bound on the runtime of M on argument f.

Proof. For $f \in \mathcal{S}$ we have by induction that $|\max(\{x > 0 \mid f(x) > 0\})| \geq \mathbf{q}_n(|f|)$, which shows by way of a simple adversary argument (as, for example, in [DZ01]) the claim.   □

The following proposition illustrates *one* possibility to analyze a set in $\mathrm{Itr}_v\mathrm{BffFin}$ (for any $v \in \mathcal{S}$ representing a limit ordinal) in terms of parametrized complexity as in [DF98].

Proposition 22. Let $v \in \mathcal{S}$ be a notation for a limit ordinal, let $\mathcal{S}^* \in \mathrm{Itr}_v\mathrm{BffFin}$ be as in the proof of Theorem 17. Define for all $k \in \mathbb{N}$, $\mathcal{S}_k^* := \{f \in \mathcal{S}^* \mid \forall x > 0 : f(x) < k\}$. Then we have

(a) $\bigcup_{k \in \mathbb{N}} \mathcal{S}_k^* = \mathcal{S}^*$,
(b) for all $k \in \mathbb{N}$, $\mathcal{S}_k^* \in \mathrm{Itr}_0\mathrm{BffFin}$ (that is, $\mathcal{S}_k^*$ is one-shot learnable by a basic feasible functional) and
(c) there is a k_0 such that for all $k > k_0$, $\mathcal{S}_k^*$ is infinite.

Proofsketch. (a) is trivially true, as all $f \in \mathcal{S}^*$ are finite variants of the constant 0 function.

(b) Let $k \in \mathbb{N}$. Among the numbers $< k$ there are of course at most k notations for ordinals. Let $f \in \mathcal{S}_k^*$. Let $C := \{x > 0 \mid f(x) \neq 0\} = \{s(0) < s(1) < \ldots < s(m)\}$. We have now, for distinct $x, y \in C$, that $f(x) \neq f(y)$, as they have to be notations for ordinals with different limit parts. Furthermore, as $f \in \mathcal{S}^*$, for all $i < m$, we have $s(i+1) < 2^{f(s(i))} + |f(0)| < 2^k + |f(0)|$. Now we have for all $x \geq 2^k + |f(0)|$, $f(x) = 0$. Checking all other positions $< 2^k + |f(0)|$ and creating an appropriate output with linear time s-m-n (as it was also done in the positive part of the proof of Theorem 17) takes therefore $\mathcal{O}(|f|(0))$ time.

(c) Let $k_0 := \underline{3}$. We omit the remaining detailed verification.   □

6   Conclusions and Future Work

In this paper we showed *one* possible approach to putting feasibilty restrictions on learning in the limit learning. However, our strict learning hierarchies are at the price of some infeasibility. Furthermore, our particular scheme of feasibly iterating basic feasible learning functionals requires the count down function to bottom out at $\underline{0}$, so one can tell when the iterations are done (and can and do suppress all the programs output but the last). We were initially surprised that, for a scheme like ours, we get a learning hierarchy result as in our Theorem 17 (in Section 4 above). We are interested in the future investigation of more ways for feasibly iterating feasible learning functionals. *We'd like variant definitions and results where one cannot suppress all the output programs but the last.* It seems this may be difficult if we retain strict determinism. In this interest, then, we would also like to see studied *probabilistic* variants of feasibly iterated feasible learners – this toward producing practical generalizations of Valiant's *PAC* learning [KV94] and Reischuk and Zeugmann's [RZ00] *stochastically finite learning*. These latter involve, probabilistic, one-shot learners. [RZ00], intriguingly for our purposes, compiles the multiple trials of a special case of *deterministic limit learning* into a feasible *probabilistic one-shot* variant.

References

[ACJS04]　Ambainis, A., Case, J., Jain, S., Suraj, M.: Parsimony hierarchies for inductive inference. Journal of Symbolic Logic 69, 287–328 (2004)

[CPK07]　Case, J., Paddock, T., Kötzing, T.: Feasible iteration of feasible learning functionals (expanded version). Technical report, University of Delaware (2007), At http://www.cis.udel.edu/~case/papers/FeasibleLearningTR.pdf and contains complete proofs

[DF98]　Downey, R., Fellows, M.: Parameterized Complexity. In: Downey, R., Fellows, M. (eds.) Monographs in Computer Science. Springer, Heidelberg (1998)

[DZ01]　Dor, D., Zwick, U.: Median selection requires $(2 + \epsilon)n$ comparisons. SIAM Journal on Discrete Mathematics 14(3), 312–325 (2001)

[FS93]　Freivalds, R., Smith, C.: On the role of procrastination in machine learning. Information and Computation 107(2), 237–271 (1993)

[IKR01]　Irwin, R., Kapron, B., Royer, J.: On characterizations of the basic feasible functional, Part I. Journal of Functional Programming 11, 117–153 (2001)

[JORS99]　Jain, S., Osherson, D., Royer, J., Sharma, A.: Systems that Learn: An Introduction to Learning Theory. 2nd edn. MIT Press, Cambridge (1999)

[KC96]　Kapron, B., Cook, S.: A new characterization of type 2 feasibility. SIAM Journal on Computing 25, 117–132 (1996)

[KV94]　Kearns, M., Vazirani, U.: An Introduction to Computational Learning Theory. MIT Press, Cambridge (1994)

[LV97]　Li, M., Vitanyi, P.: An Introduction to Kolmogorov Complexity and Its Applications, 2nd edn. Springer, Heidelberg (1997)

[Meh76]　Mehlhorn, K.: Polynomial and abstract subrecursive classes. Journal of Computer and System Sciences 12, 147–178 (1976)

[RC94]　Royer, J., Case, J.: Subrecursive Programming Systems: Complexity and Succinctness. Research monograph in Progress in Theoretical Computer Science. Birkhäuser Boston (1994)

[Rog67]　Rogers, H.: Theory of Recursive Functions and Effective Computability. McGraw Hill, New York, 1967. MIT Press (Reprinted, 1987)

[RZ00]　Reischuk, R., Zeugmann, T.: An average-case optimal one-variable pattern language learner. Journal of Computer and System Sciences 60(2), 302–335 (2000), Special Issue for COLT'98

[Sie65]　Sierpinski, W.: Cardinal and ordinal numbers. Second revised edn. PWN –Polish Scientific Publishers (1965)

Parallelism Increases Iterative Learning Power

John Case and Samuel E. Moelius III

Department of Computer & Information Sciences
University of Delaware
103 Smith Hall
Newark, DE 19716
{case,moelius}@cis.udel.edu

Abstract. *Iterative learning* (**It**-learning) is a Gold-style learning model in which each of a learner's output conjectures may depend *only* upon the learner's *current* conjecture and the *current* input element. Two extensions of the **It**-learning model are considered, each of which involves parallelism. The first is to run, in parallel, distinct instantiations of a single learner on each input element. The second is to run, in parallel, n individual learners *incorporating the first extension*, and to allow the n learners to communicate their results. In most contexts, parallelism is only a means of improving efficiency. However, as shown herein, learners incorporating the first extension are more powerful than **It**-learners, and, *collective* learners resulting from the second extension increase in learning power as n increases. Attention is paid to how one would actually implement a learner incorporating each extension. Parallelism is the underlying mechanism employed.

1   Introduction

Iterative learning (**It**-learning) [Wie76, LZ96, CJLZ99, CCJS06, CM07a] is a mathematical model of language learning in the style of Gold [Gol67].[1] In this model, the learner (commonly denoted by **M**, for *machine*) is an algorithmic device that is repeatedly fed elements from an infinite sequence. The elements of the sequence consist of numbers and, possibly, pauses (#). The set of all such numbers represents a *language*. After being fed each element, the learner either: outputs a *conjecture*, or diverges.[2] A conjecture may be either: a *grammar*, possibly for the language represented by the sequence, or '?'.[3] Most importantly, the learner may *only* consider its *current* conjecture and the *current* input element when forming a new conjecture.

For the remainder of this section, let **M** be a fixed learner. For now, **M** may be thought of as an **It**-learner. Later in this section, we will treat **M** as a instance of

---

[1] In this paper, we focus exclusively on language learning, as opposed to, say, function learning [JORS99].

[2] Intuitively, if a learner **M** diverges, then **M** *goes into an infinite loop*.

[3] N.B. Outputting '?' is *not* the same as diverging. Outputting '?' requires only *finitely many* steps; whereas, diverging requires *infinitely many* steps.

M. Hutter, R.A. Servedio, and E. Takimoto (Eds.): ALT 2007, LNAI 4754, pp. 49–63, 2007.
© Springer-Verlag Berlin Heidelberg 2007

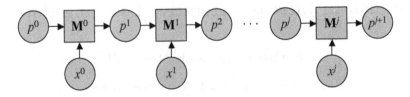

Fig. 1. The iterative learning process. The jth instantiation of learner $\mathbf{M}$, $\mathbf{M}^j$, is fed the current conjecture p^j and current input element x^j. From these, $\mathbf{M}^j$ produces a new conjecture p^{j+1}.

a more general type of learner. Let $x^0, x^1, \ldots$ be an arbitrary input sequence. Let p^0 be $\mathbf{M}$'s initial conjecture (i.e., $\mathbf{M}$'s conjecture having been fed no data), and, for all j, let p^{j+1} be the result of $\mathbf{M}^j$, where $\mathbf{M}^j$ is the computation performed by running $\mathbf{M}$ on inputs p^j and x^j. In the event that $\mathbf{M}^j$ diverges, we let $p^{j+1} = \bot$. (By convention, p^0 can*not* be $\bot$.) We shall refer to $\mathbf{M}^j$ as the jth *instantiation* of $\mathbf{M}$. See Figure 1.

An **It**-learner $\mathbf{M}$ is *successful* at learning the language represented by x^0, x^1, ... $\overset{\text{def}}{\Longleftrightarrow}$

- *none* of $\mathbf{M}^0, \mathbf{M}^1, \ldots$ diverge (i.e., *none* of $p^1, p^2, \ldots$ is $\bot$);
- for some index j_0, each of $\mathbf{M}^{j_0}, \mathbf{M}^{j_0+1}, \ldots$ results in p^{j_0+1}; *and*,
- p^{j_0+1} correctly describes the language represented by $x^0, x^1, \ldots$.

We say that $\mathbf{M}$ *identifies* a language L, or, L is *identifiable* by $\mathbf{M}$ $\overset{\text{def}}{\Longleftrightarrow}$ $\mathbf{M}$ is successful at learning L from any input sequence representing L.

The *pattern languages* are an example of a class of languages that are **It**-learnable, i.e., there exists an **It**-learner capable of identifying every language in the class. A pattern language is (by definition) the language generated by all positive length substitution instances in a *pattern* (e.g., `abXYcbbZXa`, where the variables/*non*terminals are depicted in uppercase, and the constants/terminals are depicted in lowercase). The pattern languages and their learnability were first considered by Angluin [Ang80]. Since then, much work has been done on the learnability of pattern languages [Sal94a, Sal94b, CJK+01] and finite unions thereof [Shi83, Wri89, KMU95, BUV96]. The class of pattern languages, itself, was shown to be **It**-learnable by Lange and Wiehagen [LW91]. Subsequently, this result was extended by Case, *et al.* [CJLZ99] who showed that, for each k, the class formed by taking the union of all choices of k pattern languages is **It**-learnable. Nix [Nix83], as well as Shinohara and Arikawa [SA95], outline interesting applications of pattern inference algorithms.

It-learning is a *memory limited* special case of the more general *explanatory learning* (**Ex**-learning) [Gol67, JORS99][4] and *behaviorally correct learning* (**Bc**-learning) [CL82, JORS99].[5] **Ex** and **Bc**-learners are *not*, in general, limited

---

[4] **Ex**-learning is the model that was actually studied by Gold [Gol67].
[5] Other memory limited learning models are considered in [OSW86, FJO94, CJLZ99, CCJS06].

to just the current conjecture and current input element when forming a new conjecture. Rather, such learners can refer to conjectures and/or input elements arbitrarily far into the past.[6]

Many **It**-learnable classes of languages are of practical interest. For example, the pattern languages, mentioned above, are a class whose learnability has applications to problems in molecular biology [AMS+93, SSS+94, SA95]. There is benefit in knowing that a class of languages is **It**-learnable, in that **It**-learners satisfy the following informal property.

Property 1. Each element of an input sequence may be discarded (and any associated resources freed) immediately after the element is fed to the learner.

Clearly, **Ex** and **Bc**-learners do *not* satisfy Property 1. In general, an implementation of an **Ex** or **Bc**-learner would have to store each element of an input sequence indefinitely. Thus, from a practical perspective, showing a class of languages to be **It**-learnable is far more desirable than showing it to be merely **Ex** or **Bc**-learnable.

Herein, we consider two extensions of the **It**-learning model, each of which involves parallelism. The first is to run, in parallel, distinct instantiations of a single learner on each input element (see Section 1.1 below). We call a learner incorporating this extension a 1-**ParIt**-*learner*. Our second extension is to run, in parallel, n distinct learners *incorporating the first extension*, and to allow the n learners to communicate their results (see Section 1.2 below).[7] We call a *collective* learner resulting from this latter extension, an n-**ParIt**-*learner*.

Each extension is described in further detail below.

1.1   First Extension

As mentioned previously, for an **It**-learner **M** to be *successful* at learning a language, *none* of its instantiations $\mathbf{M}^0, \mathbf{M}^1, \ldots$ may diverge. Thus, a most obvious implementation of **M** would run $\mathbf{M}^j$ only after $\mathbf{M}^{j-1}$ has converged. We can put each such $\mathbf{M}^j$ squarely into one of two categories: those that *need* p^j to compute p^{j+1}, and those that do *not*. For those that do *not*, there is *no* reason to wait until $\mathbf{M}^{j-1}$ has converged, *nor* is there reason to require that $\mathbf{M}^{j-1}$ converge at all.

Thus, our first extension is to allow $\mathbf{M}^0, \mathbf{M}^1, \ldots$ to run in parallel. We do *not* require that each of $\mathbf{M}^0, \mathbf{M}^1, \ldots$ converge, as is required by **It**-learning. However, we do require that if $\mathbf{M}^j$ *needs* p^j to compute p^{j+1}, *and*, $\mathbf{M}^{j-1}$ diverges, then $\mathbf{M}^j$ also diverges. This is an informal way of saying that **M** must be *monotonic* [Win93]. This issue is discussed further in Section 1.2.

We call a learner incorporating our first extension a 1-**ParIt**-*learner*. We say that such a learner is *successful* at learning the language represented by $x^0, x^1, \ldots$ $\overset{\text{def}}{\Longleftrightarrow}$ for some index j_0,

---

[6] **Bc**-learners differ from **Ex**-learners in that, beyond some point, all of the conjectures output by a **Bc**-learner must correctly (semantically) describe the input language, but those conjectures need *not* be (syntactically) identical.

[7] The reader should *not* confuse this idea with *team learning* [JORS99].

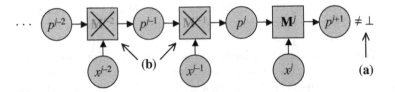

Fig. 2. How a 1-**ParIt**-learner **M** may be implemented. Once $\mathbf{M}^j$ has converged (i.e., has resulted in something other than $\bot$) (a), any previous instantiations of **M** that are still running may be forcibly terminated (b).

– each of $\mathbf{M}^{j_0}, \mathbf{M}^{j_0+1}, \ldots$ converges;
– each of $\mathbf{M}^{j_0}, \mathbf{M}^{j_0+1}, \ldots$ results in p^{j_0+1}; *and,*
– p^{j_0+1} correctly describes the language represented by $x^0, x^1, \ldots$.

A 1-**ParIt**-learner may be implemented in the following manner. Successively, for each j, start running $\mathbf{M}^j$. Simultaneously, watch for each $\mathbf{M}^j$ that is currently running to converge. Whenever j is such that $\mathbf{M}^j$ converges, forcibly terminate any currently running instantiations of the form $\mathbf{M}^0, \ldots, \mathbf{M}^{j-1}$. (The idea is that once $\mathbf{M}^j$ has converged, the results of any previous instantiations of **M** are no longer needed. See Figure 2.)

Clearly, a learner implemented in this way will *not* satisfy Property 1. However, if $x^0, x^1, \ldots$ represents a language *identifiable* by **M**, then, for some index j_0, each of $\mathbf{M}^{j_0}, \mathbf{M}^{j_0+1}, \ldots$ will converge. Thus, on such an input sequence, each instantiation $\mathbf{M}^j$ will eventually either: converge or be forced to terminate. Once either has occurred, the inputs of $\mathbf{M}^j$ may be discarded. As such, every 1-**ParIt**-learner satisfies the following weakened version of Property 1.

Property 2. If an input sequence represents a language identifiable by the learner, then each element of the sequence may be discarded *eventually*.

Clearly, **Ex** and **Bc**-learners do *not* satisfy even the weaker Property 2. Thus, from a practical perspective, 1-**ParIt**-learners are more attractive than **Ex** or **Bc**-learners.

Our first main result, Theorem 1 in Section 3, is that 1-**ParIt**-learners are strictly more powerful than **It**-learners.

1.2   Second Extension

An obvious parallel generalization of the preceding ideas is to run, in parallel, distinct, individual learners incorporating the first extension. Clearly, nothing is gained if each such learner runs in isolation. But, if the learners are allowed to communicate their results, then the resulting *collective* learner can actually be more powerful than each of its individual learners.

For the remainder of this section, let $n \geq 1$ be fixed, and let $\mathbf{M}_0, \ldots, \mathbf{M}_{n-1}$ be n learners incorporating the first extension. Let p_i^0 be $\mathbf{M}_i$'s initial conjecture, and, for each $i < n$, and each j, let p_i^{j+1} be the result of $\mathbf{M}_i^j$.

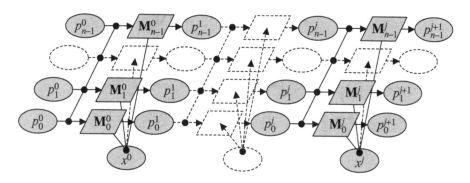

Fig. 3. A *collective* learner resulting from our second extension. For each $i < n$, and each j, individual learner $\mathbf{M}_i$ may consider conjectures $p_0^j, ..., p_{n-1}^j$ and input element x^j when forming conjecture p_i^{j+1}.

Our second extension is to allow $\mathbf{M}_0, ..., \mathbf{M}_{n-1}$ to run in parallel. For each $i < n$, and each j, we allow $\mathbf{M}_i^j$ to consider $p_0^j, ..., p_{n-1}^j$ and x^j when forming conjecture p_i^{j+1} (see Figure 3). However, as in the 1-ary case, we require that each $\mathbf{M}_i$ be monotonic.[8] So, if $\mathbf{M}_i^j$ *needs* $p_{i'}^j$ to compute p_i^{j+1}, *and*, $\mathbf{M}_{i'}^{j-1}$ diverges, then $\mathbf{M}_i^j$ also diverges. The following examples give some intuition as to which strategies $\mathbf{M}_i^j$ may employ, and which strategies $\mathbf{M}_i^j$ may *not* employ, in considering $p_0^j, ..., p_{n-1}^j$. *Exactly* which such strategies $\mathbf{M}_i^j$ may employ is made formal by Definition 2 in Section 3.

Example 1. $\mathbf{M}_i^j$ *may* employ any of the following strategies in considering $p_0^j, ..., p_{n-1}^j$.

(a) $\mathbf{M}_i^j$ does *not* wait for any of $\mathbf{M}_0^{j-1}, ..., \mathbf{M}_{n-1}^{j-1}$ to converge; $\mathbf{M}_i^j$ uses just x^j to compute p_i^{j+1}.
(b) $\mathbf{M}_i^j$ waits for $\mathbf{M}_{i'}^{j-1}$ to converge. Then, $\mathbf{M}_i^j$ uses $p_{i'}^j$ to compute p_i^{j+1}.
(c) $\mathbf{M}_i^j$ waits for $\mathbf{M}_{i'}^{j-1}$ to converge. Then, $\mathbf{M}_i^j$ performs some computable test on $p_{i'}^j$, and, based on the outcome, either: uses just $p_{i'}^j$ to compute p_i^{j+1}; or, waits for $\mathbf{M}_{i''}^{j-1}$ to converge, and uses both $p_{i'}^j$ and $p_{i''}^j$ to compute p_i^{j+1}.
(d) $\mathbf{M}_i^j$ waits for each of $\mathbf{M}_0^{j-1}, ..., \mathbf{M}_{n-1}^{j-1}$ to converge, in some predetermined order. Then, $\mathbf{M}_i^j$ uses each of $p_0^j, ..., p_{n-1}^j$ to compute p_i^{j+1}.

Example 2. In general, $\mathbf{M}_i^j$ may *not* employ the following strategy in considering $p_0^j, ..., p_{n-1}^j$ when $n \geq 2$.

(∗) $\mathbf{M}_i^j$ waits for *any* of $\mathbf{M}_0^{j-1}, ..., \mathbf{M}_{n-1}^{j-1}$ to converge. Then, for that $i' < n$ such that $\mathbf{M}_{i'}^{j-1}$ converges *first*, $\mathbf{M}_i^j$ uses $p_{i'}^{j+1}$ to compute p_i^{j+1}.

Example 2 is revisited following Definition 2 in Section 3.

---

[8] In this context, monotonicity is equivalent to *continuity* [Win93], since each $\mathbf{M}_i^j$ operates on only *finitely much* data.

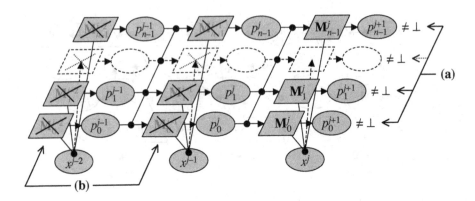

Fig. 4. How an n-**ParIt**-learner $\mathbf{M} = (\mathbf{M}_0, ..., \mathbf{M}_{n-1})$ may be implemented. Once *each* of $\mathbf{M}_0^j, ..., \mathbf{M}_{n-1}^j$ has converged (a), any previous instantiations of $\mathbf{M}_0, ..., \mathbf{M}_{n-1}$ that are still running may be forcibly terminated (b).

Let $\mathbf{M} = (\mathbf{M}_0, ..., \mathbf{M}_{n-1})$. We call such a *collective* learner $\mathbf{M}$ an n-**ParIt**-*learner*. We say that such a learner is *successful* at learning the language represented by $x^0, x^1, ... \overset{\text{def}}{\Longleftrightarrow}$ for some index j_0, and each $i < n$,

- each of $\mathbf{M}_i^{j_0}, \mathbf{M}_i^{j_0+1}, ...$ converges;
- each of $\mathbf{M}_i^{j_0}, \mathbf{M}_i^{j_0+1}, ...$ results in $p_i^{j_0+1}$; *and,*
- $p_i^{j_0+1}$ correctly describes the language represented by $x^0, x^1, ...$.

A strategy for running instantiations of an n-**ParIt**-learner can easily be generalized from the 1-ary case. Instantiations may be terminated using the following strategy. Whenever j is such that *each* of $\mathbf{M}_0^j, ..., \mathbf{M}_{n-1}^j$ converges, forcibly terminate any currently running instantiations of the form $\mathbf{M}_i^0, ..., \mathbf{M}_i^{j-1}$, where $i < n$. (The idea is that once *each* of $\mathbf{M}_0^j, ..., \mathbf{M}_{n-1}^j$ has converged, the results of any previous instantiations of $\mathbf{M}_0, ..., \mathbf{M}_{n-1}$ are no longer needed. See Figure 4.)

Clearly, if $x^0, x^1, ...$ represents a language *identifiable* by $\mathbf{M}$, then, for all but finitely many j, each of $\mathbf{M}_0^j, ..., \mathbf{M}_{n-1}^j$ will converge. It follows that an n-**ParIt**-learner implemented as described in the just previous paragraph satisfies Property 2. Thus, from a practical perspective, n-**ParIt**-learners are more attractive than **Ex** or **Bc**-learners.

Our second main result, Theorem 2 in Section 3, is that, for all $n \geq 1$, $(n + 1)$-**ParIt**-learners are strictly more powerful than n-**ParIt**-learners.

1.3   Summary of Results

Our results are summarized by the following diagram, where the arrows represent proper inclusions.

$$\textbf{It} \longrightarrow 1\text{-}\textbf{ParIt} \longrightarrow 2\text{-}\textbf{ParIt} \longrightarrow \cdots$$

That is, 1-**ParIt**-learners are strictly more powerful than **It**-learners (Theorem 1). Furthermore, for all $n \geq 1$, $(n+1)$-**ParIt**-learners are strictly more powerful than n-**ParIt**-learners (Theorem 2). Thus, we think it fair to say: parallelism increases iterative learning power.

The remainder of this paper is organized as follows. Section 2 covers notation and preliminaries. Section 3 gives the formal definition of n-**ParIt**-learning and presents our results.

2   Notation and Preliminaries

Computability-theoretic concepts not explained below are treated in [Rog67].

$\mathbb{N}$ denotes the set of natural numbers, $\{0, 1, 2, \ldots\}$. $\mathbb{N}_? \stackrel{\text{def}}{=} \mathbb{N} \cup \{?\}$. $\mathbb{N}_{?,\perp} \stackrel{\text{def}}{=} \mathbb{N}_? \cup \{\perp\}$. $\mathbb{N}_\# \stackrel{\text{def}}{=} \mathbb{N} \cup \{\#\}$. Lowercase Roman letters *other than* f, g, p, *and* q, with or without decorations, range over elements of $\mathbb{N}$. f and g will be used to denote (possibly partial) functions of various types. The exact type of f and g will be made clear whenever they are introduced. p and q, with or without decorations, range over $\mathbb{N}_{?,\perp}$. $\boldsymbol{p}$ and $\boldsymbol{q}$ will be used to denote tuples whose elements are drawn from $\mathbb{N}_{?,\perp}$. The *size* of $\boldsymbol{p}$ and $\boldsymbol{q}$ will be made clear whenever they are introduced. For all n, all $\boldsymbol{p} \in \mathbb{N}_{?,\perp}^n$, and all $i < n$, $(\boldsymbol{p})_i$ denotes the ith element of $\boldsymbol{p}$, where the first element is considered the 0th. $D_0, D_1, \ldots$ denotes a fixed, canonical enumeration of all finite subsets of $\mathbb{N}$ such that $D_0 = \emptyset$ [Rog67]. Uppercase Roman letters, with or without decorations, range over *all* (finite and infinite) subsets of $\mathbb{N}$. $\mathcal{L}$ ranges over collections of subsets of $\mathbb{N}$.

$\langle \cdot, \cdot \rangle : \mathbb{N} \times \mathbb{N} \to \mathbb{N}$ denotes any fixed, 1-1, onto, computable function. In some cases, we will write $A \times B$ for $\{\langle a, b \rangle : a \in A \wedge b \in B\}$.

For all p and q, $p \sqsubseteq q \stackrel{\text{def}}{\Leftrightarrow} [p = \perp \vee p = q]$. For all n, and all $\boldsymbol{p}, \boldsymbol{q} \in \mathbb{N}_{?,\perp}^n$, $\boldsymbol{p} \sqsubseteq \boldsymbol{q} \stackrel{\text{def}}{\Leftrightarrow} (\forall i < n)[(\boldsymbol{p})_i \sqsubseteq (\boldsymbol{q})_i]$. For all n, and all $\boldsymbol{p} \in \mathbb{N}_{?,\perp}^n$, $|\boldsymbol{p}|_{\neq\perp} \stackrel{\text{def}}{=} |\{i < n : (\boldsymbol{p})_i \neq \perp\}|$. So, for example, $|(0, 1, \perp, \perp, ?)|_{\neq\perp} = 3$.

$\varphi_0, \varphi_1, \ldots$ denotes any fixed, acceptable numbering of all *partial* computable functions of type $\mathbb{N} \to \mathbb{N}$ [Rog67]. For each i, we will treat φ_i as a *total* function of type $\mathbb{N} \to \mathbb{N}_\perp$, where $\perp$ denotes the value of a divergent computation.[9] For all i, $W_i \stackrel{\text{def}}{=} \{x \in \mathbb{N} : \varphi_i(x) \neq \perp\}$. Thus, for all i, W_i is the ith recursively enumerable set [Rog67].

$\mathbb{N}_\#^*$ denotes the set of all finite initial segments of total functions of type $\mathbb{N} \to \mathbb{N}_\#$. $\mathbb{N}_\#^{\leq\omega}$ denotes the set of *all* (finite and infinite) initial segments of total functions of type $\mathbb{N} \to \mathbb{N}_\#$. λ denotes the empty initial segment. ρ, σ, and τ, with or without decorations, range over elements of $\mathbb{N}_\#^*$.

For all $f \in \mathbb{N}_\#^{\leq\omega}$, content$(f) \stackrel{\text{def}}{=} \{y \in \mathbb{N} : (\exists x)[f(x) = y]\}$. For all $f \in \mathbb{N}_\#^{\leq\omega}$ and L, f *represents* $L \stackrel{\text{def}}{\Leftrightarrow} f$ is total and content$(f) = L$.[10] For all σ, $|\sigma|$ denotes the length of σ, i.e., the number of elements in σ. For all $f \in \mathbb{N}_\#^{\leq\omega}$, and all n, $f[n]$ denotes the initial segment of f of length n, if it exists; f, otherwise. For all σ, all $f \in \mathbb{N}_\#^{\leq\omega}$, and all i,

---

[9] N.B. It can*not*, in general, be determined whether $\varphi_i(x) = \perp$, for arbitrary i and x.

[10] Such an f is often called a *text* (for L) [JORS99].

$$(\sigma \diamond f)(i) \stackrel{\text{def}}{=} \begin{cases} \sigma(i), & \text{if } i < |\sigma|; \\ f(i - |\sigma|), & \text{otherwise.} \end{cases} \tag{1}$$

$\mathbf{M}$ will be used to denote *partial* computable functions of type $\mathbb{N}_{\#}^* \rightharpoonup \mathbb{N}_?^n$, for various n. However, as with $\varphi_0, \varphi_1, \ldots$, we will treat each $\mathbf{M}$ as a *total* function of type $\mathbb{N}_{\#}^* \to \mathbb{N}_{?,\perp}^n$. The exact type of $\mathbf{M}$ will be made clear whenever it is introduced. For all n, all $\mathbf{M} : \mathbb{N}_{\#}^* \to \mathbb{N}_{?,\perp}^n$, all $i < n$, and all ρ, $\mathbf{M}_i(\rho) \stackrel{\text{def}}{=} \left(\mathbf{M}(\rho) \right)_i$.

The following is the formal definition of **It**-learning.[11]

Definition 1

(a) For all $\mathbf{M} : \mathbb{N}_{\#}^* \to \mathbb{N}_{?,\perp}$ and L, $\mathbf{M}$ **It**-*identifies* L $\Leftrightarrow$ (i) and (ii) below.
    (i) For all f representing L, there exist j and $p \in \mathbb{N}$ such that $(\forall j' \geq j)$
        $\left[\mathbf{M}(f[j']) = p \right]$ and $W_p = L$.[12]
    (ii) For all ρ, σ, and τ such that $\text{content}(\rho) \cup \text{content}(\sigma) \cup \text{content}(\tau) \subseteq L$,
        (α) and (β) below.
        (α) $\mathbf{M}(\rho) \neq \perp$.
        (β) $\mathbf{M}(\rho) = \mathbf{M}(\sigma) \Rightarrow \mathbf{M}(\rho \diamond \tau) = \mathbf{M}(\sigma \diamond \tau)$.
(b) For all $\mathbf{M} : \mathbb{N}_{\#}^* \to \mathbb{N}_{?,\perp}$, $\mathbf{It}(\mathbf{M}) = \{ L : \mathbf{M} \text{ } \mathbf{It}\text{-identifies } L \}$.
(c) $\mathbf{It} = \{ \mathcal{L} : (\exists \mathbf{M} : \mathbb{N}_{\#}^* \to \mathbb{N}_{?,\perp})[\mathcal{L} \subseteq \mathbf{It}(\mathbf{M})] \}$.

Some of our proofs make use of the **Operator Recursion Theorem (ORT)** [Cas74]. **ORT** represents a form of infinitary self-reference, similar to the way in which Kleene's Recursion Theorem [Rog67, page 214, problem 11-4] represents a form of individual self-reference. That is, **ORT** provides a means of forming an infinite computable sequence of programs $e_0, e_1, \ldots$ such that each program e_i *knows all* programs in the sequence *and* its own index i. The sequence can also be assumed monotone increasing. The first author gives a thorough explanation of **ORT** in [Cas94].

3 Results

This section gives the formal definition of n-**ParIt**-learning and presents our results. Namely, this section shows that 1-**ParIt**-learners are strictly more powerful than **It**-learners (Theorem 1). It also shows that, for all $n \geq 1$, $(n + 1)$-**ParIt**-learners are strictly more powerful than n-**ParIt**-learners (Theorem 2).

Definition 2. For all $n \geq 1$, (a)-(c) below.

(a) For all $\mathbf{M} : \mathbb{N}_{\#}^* \to \mathbb{N}_{?,\perp}^n$ and L, $\mathbf{M}$ n-**ParIt**-*identifies* L $\Leftrightarrow$ (i) and (ii) below.
    (i) For all f representing L, there exist j and $\boldsymbol{p} \in \mathbb{N}^n$ such that $(\forall j' \geq j)$
        $\left[\mathbf{M}(f[j']) = \boldsymbol{p} \right]$ and $(\forall i < n)[W_{(\boldsymbol{p})_i} = L]$.

---

[11] **It**-learners are often given a formal definition more in line with their description in Section 1. The definition given herein was inspired, in part, by the Myhill-Nerode Theorem [DSW94]. A proof that this definition is equivalent to the more common definition can be found in [CM07b].

[12] Condition (a)(i) in Definition 1 is equivalent to: $\mathbf{M}$ **Ex**-identifies L [Gol67, JORS99].

(ii) $(\forall \rho, \sigma, \tau)[\mathbf{M}(\rho) \sqsubseteq \mathbf{M}(\sigma) \Rightarrow \mathbf{M}(\rho \diamond \tau) \sqsubseteq \mathbf{M}(\sigma \diamond \tau)]$.

(b) For all $\mathbf{M} : \mathbb{N}_\#^* \to \mathbb{N}_{?,\perp}^n$, n-**ParIt**$(\mathbf{M}) = \{L : \mathbf{M}\ n$-**ParIt**-identifies $L\}$.

(c) n-**ParIt** $= \{\mathcal{L} : (\exists \mathbf{M} : \mathbb{N}_\#^* \to \mathbb{N}_{?,\perp}^n)[\mathcal{L} \subseteq n$-**ParIt**$(\mathbf{M})]\}$.

Example 3 (Example 2 revisited). Suppose that $\mathbf{M} : \mathbb{N}_\#^* \to \mathbb{N}_{?,\perp}^2$, $\rho, \sigma, \boldsymbol{p} \in \mathbb{N}_{?,\perp}^2$, and x are such that (a)-(e) below.

(a) $\mathbf{M}(\rho) = \mathbf{M}(\sigma) = \boldsymbol{p}$.

(b) $|\boldsymbol{p}|_{\neq \perp} = 2$.

(c) $(\boldsymbol{p})_0 \neq (\boldsymbol{p})_1$.

(d) In the computation of $\mathbf{M}_0(\rho \diamond x)$, $\mathbf{M}_0$ waits for *either* of $\mathbf{M}_0(\rho)$ or $\mathbf{M}_1(\rho)$ to converge. Then, for the $i \leq 1$ such that $\mathbf{M}_i(\rho)$ converges *first*, $\mathbf{M}_0(\rho \diamond x) = \mathbf{M}_i(\rho)$. Similarly, in the computation of $\mathbf{M}_0(\sigma \diamond x)$, $\mathbf{M}_0$ waits for either of $\mathbf{M}_0(\sigma)$ or $\mathbf{M}_1(\sigma)$ to converge. Then, for the $i \leq 1$ such that $\mathbf{M}_i(\sigma)$ converges *first*, $\mathbf{M}_0(\sigma \diamond x) = \mathbf{M}_i(\sigma)$.

(e) In the computation of $\mathbf{M}(\rho)$, $\mathbf{M}_0(\rho)$ converges before $\mathbf{M}_1(\rho)$; in computation of $\mathbf{M}(\sigma)$, $\mathbf{M}_1(\sigma)$ converges before $\mathbf{M}_0(\sigma)$.

Then, for all L, $\mathbf{M}$ does *not* 2-**ParIt**-identify L, i.e., $\mathbf{M}$ is *not* a 2-**ParIt**-learner.

Proof. By (a) above, $\mathbf{M}(\rho) \sqsubseteq \mathbf{M}(\sigma)$. By (c)-(e) above, $\mathbf{M}_0(\rho \diamond x) = (\boldsymbol{p})_0 \neq (\boldsymbol{p})_1 = \mathbf{M}_0(\sigma \diamond x)$. Thus, by (b) above, $\mathbf{M}(\rho \diamond x) \not\sqsubseteq \mathbf{M}(\sigma \diamond x)$. But this contradicts condition (a)(ii) in Definition 2. $\square$ *(Example 3)*

Intuitively, the $\mathbf{M}$ described in Example 3 violates Definition 2 because: (1) $\mathbf{M}$ makes use of, not just the *value* of a conjecture, but also the *time* used to compute it; and, (2) the elements of $\mathbb{N}_{?,\perp}$ do *not* capture this information. To overcome this difficulty would require that a learner be defined as object with a more complex range than $\mathbb{N}_{?,\perp}^n$. It would be interesting to explore generalizations of Definition 2 that do this.

The following straightforward variant of **It**-learning is used in the proof of Theorem 1.

Definition 3

(a) For all $\mathbf{M} : \mathbb{N}_\#^* \to \mathbb{N}_{?,\perp}$ and L, $\mathbf{M}$ **TotIt**-*identifies* $L \Leftrightarrow \mathbf{M}$ **It**-identifies L, and, for all ρ, $\mathbf{M}(\rho) \neq \perp$.

(b) For all $\mathbf{M} : \mathbb{N}_\#^* \to \mathbb{N}_{?,\perp}$, **TotIt**$(\mathbf{M}) = \{L : \mathbf{M}$ **TotIt**-identifies $L\}$.

(c) **TotIt** $= \{\mathcal{L} : (\exists \mathbf{M} : \mathbb{N}_\#^* \to \mathbb{N}_{?,\perp})[\mathcal{L} \subseteq$ **TotIt**$(\mathbf{M})]\}$.

Recall that if a learner $\mathbf{M}$ **It**-identifies language L, then it is only required that $\mathbf{M}(\rho) \neq \perp$ for those ρ such that content$(\rho) \subseteq L$. However, if $\mathbf{M}$ **TotIt**-identifies L, then, for *all* ρ, $\mathbf{M}(\rho) \neq \perp$.

The following is a basic fact relating **It** and **TotIt**.

Proposition 1. For all $\mathcal{L} \in$ **It**, if $\mathbb{N} \in \mathcal{L}$, then $\mathcal{L} \in$ **TotIt**.

Proof. Straightforward. $\square$ *(Proposition 1)*

The following lemma is used in the proof of Theorem 1.

Lemma 1. Let $\mathcal{L}$ be the class of languages consisting of each L satisfying (a)-(c) below.

(a) $(\forall e \in L)[\varphi_e(0) \neq \perp]$.
(b) $\{\varphi_e(0) : e \in L\}$ is finite.
(c) $L = \bigcup_{e \in L} W_{\varphi_e(0)}$.

Then, $\mathcal{L} \in \mathbf{It} - \mathbf{TotIt}$.

Proof that $\mathcal{L} \in \mathbf{It}$. Let $f : \mathbb{N} \to \mathbb{N}$ be a 1-1, computable function such that, for all a,

$$W_{f(a)} = \bigcup_{e \in D_a} W_e. \tag{2}$$

Let $\mathbf{M} : \mathbb{N}_{\#}^* \to \mathbb{N}_{?,\perp}$ be such that $\mathbf{M}(\lambda) = f(0)$, and, for all ρ, a, and e, if $\mathbf{M}(\rho) = f(a)$, then

$$\mathbf{M}(\rho \diamond e) = \begin{cases} f(a), & \text{if } \varphi_e(0) \in D_a; \\ f(b), & \text{if } \varphi_e(0) \in (\mathbb{N} - D_a), \\ & \quad \text{where } b \text{ is such that } D_b = D_a \cup \{\varphi_e(0)\}; \\ \perp, & \text{if } \varphi_e(0) = \perp. \end{cases} \tag{3}$$

Clearly, $\mathcal{L} \subseteq \mathbf{It}(\mathbf{M})$.

Proof that $\mathcal{L} \notin \mathbf{TotIt}$. By way of contradiction, suppose that $\mathbf{M} : \mathbb{N}_{\#}^* \to \mathbb{N}_{?,\perp}$ is such that $\mathcal{L} \subseteq \mathbf{TotIt}(\mathbf{M})$. By **ORT**, there exist distinct φ-programs $e_0, e_1, \ldots$ such that, for all i and x,

$$W_{e_0} = \{e_{j+2} : \varphi_{e_{j+2}}(0) = e_0\}; \tag{4}$$

$$W_{e_1} = \{e_{j+2} : \varphi_{e_{j+2}}(0) = e_1\}; \tag{5}$$

$$\varphi_{e_{i+2}}(x) = \begin{cases} e_1, & \text{if } i \text{ is } \textit{least} \text{ such that} \\ & \quad \mathbf{M}(e_2 \diamond \cdots \diamond e_{i+2}) = \mathbf{M}(e_2 \diamond \cdots \diamond e_{i+1}); \\ e_0, & \text{otherwise.} \end{cases} \tag{6}$$

Consider the following cases.

CASE $(\forall i)[\varphi_{e_{i+2}}(0) = e_0]$. Then, clearly, $W_{e_0} = \{e_{j+2} : j \in \mathbb{N}\}$ and $W_{e_0} \in \mathcal{L}$. By the case, for all i, $\mathbf{M}(e_2 \diamond \cdots \diamond e_{i+2}) \neq \mathbf{M}(e_2 \diamond \cdots \diamond e_{i+1})$. But then, clearly, $W_{e_0} \notin \mathbf{It}(\mathbf{M})$.

CASE $(\exists i)[\varphi_{e_{i+2}}(0) = e_1]$. Then, clearly, $W_{e_0} = \{e_{j+2} : j \neq i\}$ and $(\forall j \neq i)$ $[\varphi_{e_{j+2}}(0) = e_0]$. Furthermore, $W_{e_1} = \{e_{i+2}\}$ and $\varphi_{e_{i+2}}(0) = e_1$. Thus, $W_{e_0} \cup W_{e_1}$ and W_{e_0} are *distinct* languages in $\mathcal{L}$. Let f and f^- be as follows.

$$f = e_2 \diamond e_3 \diamond \cdots . \tag{7}$$

$$f^- = e_2 \diamond e_3 \diamond \cdots \diamond e_{i+1} \diamond e_{i+3} \diamond e_{i+4} \diamond \cdots . \tag{8}$$

Clearly, f represents $W_{e_0} \cup W_{e_1}$, and f^- represents W_{e_0}. Let k be such that $\mathbf{M}(f[k]) \in \mathbb{N}$, $\mathbf{M}(f^-[k]) \in \mathbb{N}$, and

$$(\forall k' \geq k)[\mathbf{M}(f[k']) = \mathbf{M}(f[k]) \wedge \mathbf{M}(f^-[k']) = \mathbf{M}(f^-[k])]. \tag{9}$$

From the case, it follows that $\mathbf{M}(f[k]) = \mathbf{M}(f^-[k])$. But, clearly, this is a contradiction. □ *(Lemma 1)*

Theorem 1. Let $\mathcal{L}$ be as in Lemma 1. Let $\mathcal{L}'$ be such that $\mathcal{L}' = \mathcal{L} \cup \{\mathbb{N}\}$. Then, $\mathcal{L}' \in 1\text{-}\mathbf{ParIt} - \mathbf{It}$.

Proof (Sketch) that $\mathcal{L}' \in 1\text{-}\mathbf{ParIt}$. By Lemma 1, there exists $\mathbf{M} : \mathbb{N}_{\#}^* \to \mathbb{N}_{?,\perp}$ such that $\mathcal{L} \subseteq \mathbf{It}(\mathbf{M})$. Let z_0 be such that $\varphi_{z_0}(0) = \perp$. Clearly, for all $L \in \mathcal{L}$, $z_0 \notin L$. Consider an $\mathbf{M}' : \mathbb{N}_{\#}^* \to \mathbb{N}_{?,\perp}$ described informally as follows. On any given input sequence, $\mathbf{M}'$ simulates $\mathbf{M}$ until, if ever, $\mathbf{M}'$ is fed z_0. Upon being fed z_0, $\mathbf{M}'$ stops simulating $\mathbf{M}$, and starts outputting a conjecture for $\mathbb{N}$. Clearly, for such an $\mathbf{M}'$, $\mathcal{L}' \subseteq 1\text{-}\mathbf{ParIt}(\mathbf{M}')$.

Proof that $\mathcal{L}' \notin \mathbf{It}$. By way of contradiction, suppose that $\mathcal{L}' \in \mathbf{It}$. Then, by Proposition 1, $\mathcal{L}' \in \mathbf{TotIt}$. Let $\mathbf{M} : \mathbb{N}_{\#}^* \to \mathbb{N}_{?,\perp}$ be such that $\mathcal{L}' \subseteq \mathbf{TotIt}(\mathbf{M})$. Then, $\mathcal{L} \subset \mathcal{L}' \subseteq \mathbf{TotIt}(\mathbf{M})$. But this contradicts Lemma 1. $\approx \square$ (*Theorem 1*)

Theorem 2. Let $n \geq 1$ be fixed. For each $i < n$, let z_i be any fixed φ-program such that $W_{\varphi_{z_i}(0)} = \{\langle i, z_i \rangle\}$. Let $\mathcal{L}_n$ be the class of languages consisting of each $L \subseteq \{0, ..., n-1\} \times \mathbb{N}$ satisfying *either* (a) *or* (b) below.

(a) (i) and (ii) below.
    (i) $L \cap (\{0, ..., n-1\} \times \{z_0, ..., z_{n-1}\}) = \emptyset$.
    (ii) For each $i < n$, if E is such that $E = \{e \in \mathbb{N} : \langle i, e \rangle \in L\}$, then (α)-(γ) below.
        (α) $(\forall e \in E)[\varphi_e(0) \in \mathbb{N}]$.
        (β) $\{\varphi_e(0) : e \in E\}$ is finite.
        (γ) $L = \bigcup_{e \in E} W_{\varphi_e(0)}$.
(b) There exists $i < n$ such that (i) and (ii) below.
    (i) $L \cap (\{0, ..., n-1\} \times \{z_0, ..., z_{n-1}\}) = \{\langle i, z_i \rangle\}$.
    (ii) If E is such that $E = \{e \in \mathbb{N} : \langle i, e \rangle \in L\}$, then (α)-(γ) as in (a)(ii) above *for this E*.

Then, for all $n \geq 1$, $\mathcal{L}_{n+1} \in (n+1)\text{-}\mathbf{ParIt} - n\text{-}\mathbf{ParIt}$.

Proof (Sketch) that $\mathcal{L}_{n+1} \in (n+1)\text{-}\mathbf{ParIt}$. Let $n \geq 1$ be fixed. Let $f : \mathbb{N}^2 \to \mathbb{N}$ be a 1-1, computable function such that, for all j and a,

$$W_{f(j,a)} = \bigcup_{e \in D_a} W_e. \tag{10}$$

Let $\mathbf{M} : \mathbb{N}_{\#}^* \to \mathbb{N}_{?,\perp}^{n+1}$ be such that, for each $i \leq n$, $\mathbf{M}_i(\lambda) = f(i, 0)$, and, for all ρ, k, and e, $\mathbf{M}_i(\rho \diamond \langle k, e \rangle)$ is

$$
\begin{cases}
\mathbf{M}_k(\rho), & \text{if } e = z_k; \\
\mathbf{M}_j(\rho), & \text{if } e \neq z_k \wedge [j \neq i \vee k \neq i \vee \varphi_e(0) \in D_a], \\
& \text{where } j \text{ and } a \text{ are such that } \mathbf{M}_i(\rho) = f(j, a); \\
f(i, b), & \text{if } e \neq z_k \wedge j = i \wedge k = i \wedge \varphi_e(0) \in (\mathbb{N} - D_a), \\
& \text{where } j, a, \text{ and } b \text{ are such that } \mathbf{M}_i(\rho) = f(j, a) \\
& \text{and } D_b = D_a \cup \{\varphi_e(0)\}; \\
\perp, & \text{if } e \neq z_k \wedge [[j = i \wedge k = i \wedge \varphi_e(0) = \perp] \vee \mathbf{M}_i(\rho) = \perp], \\
& \text{where } j \text{ is such that } \mathbf{M}_i(\rho) = f(j, a), \text{ for some } a.
\end{cases}
\tag{11}
$$

STAGE $s = 0$.

   1. For each $i \leq n$, set $\varphi_{e_{i+1}}(0) = e_0$.
   2. Set $W_{e_0}^1 = \{\langle 0, e_1 \rangle, ..., \langle n, e_{n+1} \rangle\}$.
   3. Set $\rho^1 = \langle 0, e_1 \rangle \diamond \cdots \diamond \langle n, e_{n+1} \rangle$.

STAGE $s \geq 1$.

   1. Find ρ', *if any*, such that $\rho^s \subseteq \rho' \subset \rho^s \diamond \#^\omega$ and $|\mathbf{M}(\rho')|_{\neq\perp} = n$.
   2. For k from n down through -1, do:
      Wait until, *if ever*, it is discovered that one of the following two conditions applies.
      COND. (α): $(\exists q : |q|_{\neq\perp} = n - k)(\forall \sigma \in \{\langle 0, e_{f(s)} \rangle \diamond \cdots \diamond \langle k, e_{f(s)+k} \rangle, \lambda\})$
$$[q \sqsubseteq \mathbf{M}(\rho' \diamond \sigma \diamond \langle k+1, e_{f(s)+k+1} \rangle \diamond \cdots \diamond \langle n, e_{f(s)+n} \rangle)].$$
      a. Set $\varphi_{e_{f(s)+k}}(0) =$ any φ-program p such that $W_p = \{\langle k, e_{f(s)+k} \rangle\}$.
      b. Proceed to the next value of k.
      COND. (β): $\mathbf{M}(\rho' \diamond \langle 0, e_{f(s)} \rangle \diamond \cdots \diamond \langle k, e_{f(s)+k} \rangle) \not\sqsubseteq \mathbf{M}(\rho')$.
      a. For each $i \leq k$, set $\varphi_{e_{f(s)+i}}(0) = e_0$.
      b. Set $W_{e_0}^{s+1} = W_{e_0}^s \cup \{\langle 0, e_{f(s)} \rangle, ..., \langle k, e_{f(s)+k} \rangle\}$.
      c. Set $\rho^{s+1} = \rho' \diamond \langle 0, e_{f(s)} \rangle \diamond \cdots \diamond \langle k, e_{f(s)+k} \rangle$.
      d. Go to stage $s + 1$.
   (Note that the iteration of the loop in which $k = n$ is always exited. Also, note that if k reaches the value -1, then the construction *intentionally* goes into an infinite loop.)

Fig. 5. The behavior of φ-programs $e_0, e_1, ...$ in the proof of Theorem 2

It can be shown that $\mathcal{L}_{n+1} \subseteq (n+1)$-**ParIt**$(\mathbf{M})$ (details omitted).

*Proof that $\mathcal{L}_{n+1} \notin n$-**ParIt**.* By way of contradiction, suppose that $n \geq 1$ and $\mathbf{M} : \mathbb{N}_{\#}^* \to \mathbb{N}_{?,\perp}^n$ are such that $\mathcal{L}_{n+1} \subseteq n$-**ParIt**$(\mathbf{M})$. Let $f : \mathbb{N} \to \mathbb{N}$ be such that, for all s,

$$f(s) = s \cdot (n+1) + 1. \tag{12}$$

By **ORT**, there exist distinct φ-programs $e_0, e_1, ...$, *none* of which are $z_0, ..., z_n$, and whose behavior is as in Figure 5.

Claim 1. For all $s \geq 1$, if stage s is entered, then (a)-(c) below.
(a) $(\forall \langle i, e \rangle \in W_{e_0}^s)[i \leq n \wedge e \notin \{z_0, ..., z_n\} \wedge \varphi_e(0) = e_0]$.
(b) $\text{content}(\rho^s) = W_{e_0}^s$.
(c) $\rho^s \diamond \#^\omega$ represents $W_{e_0}^s$.
Proof of Claim. (a) is clear by construction. (b) is proven by a straightforward induction. (c) follows immediately from (b).   □ (*Claim 1*)

Claim 2. For all $s \geq 1$, if stage s is exited, then there exist ρ' and ρ'' such that $\rho^s \subseteq \rho' \subset \rho'' \subseteq \rho^{s+1}$ and $\mathbf{M}(\rho'') \not\sqsubseteq \mathbf{M}(\rho')$.
Proof of Claim. Clear by construction.   □ (*Claim 2*)

If every stage s is exited, then, by Claim 1(a) and Claim 2, $W_{e_0} \in \mathcal{L}_{n+1} - n\text{-}\mathbf{ParIt}(\mathbf{M})$ (a contradiction). So, for the remainder of the proof, suppose that stage s is entered but *never* exited.

If stage s is never exited because there is *no* ρ' such that $\rho^s \subseteq \rho' \subset \rho^s \diamond \#^\omega$ and $|\mathbf{M}(\rho')|_{\neq\perp} = n$, then, by (a) and (c) of Claim 1, $W_{e_0}^s \in \mathcal{L}_{n+1} - n\text{-}\mathbf{ParIt}(\mathbf{M})$ (a contradiction). So, suppose that stage s is never exited because there exists k such that $-1 \leq k < n$ and $(\neg\alpha)$ and $(\neg\beta)$ below.

$(\neg\alpha)$ $(\forall \boldsymbol{q} : |\boldsymbol{q}|_{\neq\perp} = n - k)(\exists \sigma \in \{\langle 0, e_{f(s)}\rangle \diamond \cdots \diamond \langle k, e_{f(s)+k}\rangle, \lambda\})$
$\qquad [\boldsymbol{q} \not\sqsubseteq \mathbf{M}(\rho' \diamond \sigma \diamond \langle k+1, e_{f(s)+k+1}\rangle \diamond \cdots \diamond \langle n, e_{f(s)+n}\rangle)].$

$(\neg\beta)$ $\mathbf{M}(\rho' \diamond \langle 0, e_{f(s)}\rangle \diamond \cdots \diamond \langle k, e_{f(s)+k}\rangle) \sqsubseteq \mathbf{M}(\rho').$

Claim 3.
$\mathbf{M}(\rho' \diamond \langle 0, e_{f(s)}\rangle \diamond \cdots \diamond \langle n, e_{f(s)+n}\rangle) \sqsubseteq \mathbf{M}(\rho' \diamond \langle k+1, e_{f(s)+k+1}\rangle \diamond \cdots \diamond \langle n, e_{f(s)+n}\rangle).$
Proof of Claim. Follows from $(\neg\beta)$. $\qquad\qquad\qquad\qquad\square$ *(Claim 3)*

By the choice of k, there exists $\boldsymbol{p}$ such that $|\boldsymbol{p}|_{\neq\perp} = n - k - 1$ and

$$(\forall \sigma \in \{\langle 0, e_{f(s)}\rangle \diamond \cdots \diamond \langle k+1, e_{f(s)+k+1}\rangle, \lambda\}) \tag{13}$$
$$[\boldsymbol{p} \sqsubseteq \mathbf{M}(\rho' \diamond \sigma \diamond \langle k+2, e_{f(s)+k+2}\rangle \diamond \cdots \diamond \langle n, e_{f(s)+n}\rangle)].$$

Claim 4. $\boldsymbol{p} = \mathbf{M}(\rho' \diamond \langle 0, e_{f(s)}\rangle \diamond \cdots \diamond \langle n, e_{f(s)+n}\rangle).$
Proof of Claim. By way of contradiction, suppose otherwise. By (13), it must be the case that

$$\boldsymbol{p} \sqsubset \mathbf{M}(\rho' \diamond \langle 0, e_{f(s)}\rangle \diamond \cdots \diamond \langle n, e_{f(s)+n}\rangle). \tag{14}$$

But (14) together with Claim 3 contradicts $(\neg\alpha)$. $\qquad\qquad\square$ *(Claim 4)*

Claim 5.
$\mathbf{M}(\rho' \diamond \langle 0, e_{f(s)}\rangle \diamond \cdots \diamond \langle n, e_{f(s)+n}\rangle) \sqsubseteq \mathbf{M}(\rho' \diamond \langle k+2, e_{f(s)+k+2}\rangle \diamond \cdots \diamond \langle n, e_{f(s)+n}\rangle).$
Proof of Claim. Immediate by Claim 4 and (13). $\qquad\qquad\square$ *(Claim 5)*

Let $p = \varphi_{e_{f(s)+k+1}}(0)$. Thus, by construction, $W_p = \{\langle k+1, e_{f(s)+k+1}\rangle\}$. Let $e' \notin \{z_0, ..., z_n, e_0, e_1, ...\}$ and p' be as follows.

$$\varphi_{e'}(0) = p'. \tag{15}$$
$$W_{p'} = \begin{Bmatrix} \langle k+2, e_{f(s)+k+2}\rangle, ..., \langle n, e_{f(s)+n}\rangle, \\ \langle 0, e_{f(s)}\rangle, ..., \langle k, e_{f(s)+k}\rangle, \langle k+1, e'\rangle \end{Bmatrix}. \tag{16}$$

Let L and L^- be as follows.

$$L = W_{e_0}^s \cup W_p \cup W_{p'} \cup \{\langle k+1, z_{k+1}\rangle\}. \tag{17}$$
$$L^- = W_{e_0}^s \cup W_{p'} \cup \{\langle k+1, z_{k+1}\rangle\}. \tag{18}$$

Clearly, L and L^- are *distinct* languages in $\mathcal{L}_{n+1}$. Let g and g^- be as follows.

$$g = \rho' \diamond \langle 0, e_{f(s)}\rangle \diamond \cdots \diamond \langle n, e_{f(s)+n}\rangle \tag{19}$$
$$\diamond \langle 0, e_{f(s)}\rangle \diamond \cdots \diamond \langle k, e_{f(s)+k}\rangle \diamond \langle k+1, e'\rangle \diamond \langle k+1, z_{k+1}\rangle \diamond \#^\omega.$$

$$g^- = \rho' \diamond \langle k+2, e_{f(s)+k+2}\rangle \diamond \cdots \diamond \langle n, e_{f(s)+n}\rangle \tag{20}$$
$$\diamond \langle 0, e_{f(s)}\rangle \diamond \cdots \diamond \langle k, e_{f(s)+k}\rangle \diamond \langle k+1, e'\rangle \diamond \langle k+1, z_{k+1}\rangle \diamond \#^\omega.$$

Clearly, g represents L, and g^- represents L^-. Let ℓ be such that $\mathbf{M}(g[\ell]) \in \mathbb{N}^n$, $\mathbf{M}(g^-[\ell]) \in \mathbb{N}^n$, and

$$(\forall \ell' \geq \ell)\big[\mathbf{M}(g[\ell']) = \mathbf{M}(g[\ell]) \;\wedge\; \mathbf{M}(g^-[\ell']) = \mathbf{M}(g^-[\ell])\big]. \tag{21}$$

From Claim 5, and the fact that $\mathbf{M}(g[\ell]) \in \mathbb{N}^n$ and $\mathbf{M}(g^-[\ell]) \in \mathbb{N}^n$, it follows that $\mathbf{M}(g[\ell]) = \mathbf{M}(g^-[\ell])$. But, clearly, this is a contradiction. $\approx \square$ (*Theorem 2*)

Acknowledgments. We are grateful to several anonymous referees for their *meticulous* reading of an earlier draft of this paper.

References

[AMS+93] Arikawa, S., Miyano, S., Shinohara, A., Kuhara, S., Mukouchi, Y., Shinohara, T.: A machine discovery from amino-acid-sequences by decision trees over regular patterns. New Generation Computing 11, 361–375 (1993)

[Ang80]   Angluin, D.: Finding patterns common to a set of strings. Journal of Computer and System Sciences 21, 46–62 (1980)

[BUV96]   Brazma, A., Ukkonen, E., Vilo, J.: Discovering unbounded unions of regular pattern languages from positive examples. In: Nagamochi, H., Suri, S., Igarashi, Y., Miyano, S., Asano, T. (eds.) ISAAC 1996. LNCS, vol. 1178, Springer, Heidelberg (1996)

[Cas74]   Case, J.: Periodicity in generations of automata. Mathematical Systems Theory 8, 15–32 (1974)

[Cas94]   Case, J.: Infinitary self-reference in learning theory. Journal of Experimental and Theoretical Artificial Intelligence 6, 3–16 (1994)

[CCJS06] Carlucci, L., Case, J., Jain, S., Stephan, F.: Memory-limited U-shaped learning. In: Lugosi, G., Simon, H.U. (eds.) COLT 2006. LNCS (LNAI), vol. 4005, pp. 244–258. Springer, Heidelberg (2006)

[CJK+01] Case, J., Jain, S., Kaufmann, S., Sharma, A., Stephan, F.: Predictive learning models for concept drift. Theoretical Computer Science, Special Issue for ALT'98, 268, 323–349 (2001)

[CJLZ99] Case, J., Jain, S., Lange, S., Zeugmann, T.: Incremental concept learning for bounded data mining. Information and Computation 152, 74–110 (1999)

[CL82]   Case, J., Lynes, C.: Machine inductive inference and language identification. In: Nielsen, M., Schmidt, E.M. (eds.) Automata, Languages, and Programming. LNCS, vol. 140, pp. 107–115. Springer, Heidelberg (1982)

[CM07a]   Case, J., Moelius, S.E.: U-shaped, iterative, and iterative-with-counter learning. In: COLT 2007. LNCS(LNAI), vol. 4539, pp. 172–186. Springer, Berlin (2007)

[CM07b]   Case, J., Moelius, S.E.: U-shaped, iterative, and iterative-with-counter learning (expanded version). Technical report, University of Delaware (2007), Available at http://www.cis.udel.edu/~moelius/publications

[DSW94]   Davis, M., Sigal, R., Weyuker, E.: Computability, Complexity, and Languages, 2nd edn. Academic Press, London (1994)

[FJO94]   Fulk, M., Jain, S., Osherson, D.: Open problems in Systems That Learn. Journal of Computer and System Sciences 49(3), 589–604 (1994)

[Gol67]   Gold, E.: Language identification in the limit. Information and Control 10, 447–474 (1967)

[JORS99]   Jain, S., Osherson, D., Royer, J., Sharma, A.: Systems that Learn: An Introduction to Learning Theory, 2nd edn. MIT Press, Cambridge (1999)

[KMU95]   Kilpeläinen, P., Mannila, H., Ukkonen, E.: MDL learning of unions of simple pattern languages from positive examples. In: Vitányi, P.M.B. (ed.) EuroCOLT 1995. LNCS, vol. 904, pp. 252–260. Springer, Heidelberg (1995)

[LW91]   Lange, S., Wiehagen, R.: Polynomial time inference of arbitrary pattern languages. New Generation Computing 8, 361–370 (1991)

[LZ96]   Lange, S., Zeugmann, T.: Incremental learning from positive data. Journal of Computer and System Sciences 53, 88–103 (1996)

[Nix83]   Nix, R.: Editing by examples. Technical Report 280, Department of Computer Science, Yale University, New Haven, CT, USA (1983)

[OSW86]   Osherson, D., Stob, M., Weinstein, S.: Systems that Learn: An Introduction to Learning Theory for Cognitive and Computer Scientists. MIT Press, Cambridge (1986)

[Rog67]   Rogers, H.: Theory of Recursive Functions and Effective Computability. MIT Press, Cambridge (1967) (Reprinted, MIT Press, 1987)

[SA95]   Shinohara, T., Arikawa, A.: Pattern inference. In: Lange, S., Jantke, K.P. (eds.) Algorithmic Learning for Knowledge-Based Systems. LNCS, vol. 961, pp. 259–291. Springer, Heidelberg (1995)

[Sal94a]   Salomaa, A.: Patterns (The Formal Language Theory Column). EATCS Bulletin 54, 46–62 (1994)

[Sal94b]   Salomaa, A.: Return to patterns (The Formal Language Theory Column). EATCS Bulletin 55, 144–157 (1994)

[Shi83]   Shinohara, T.: Inferring unions of two pattern languages. Bulletin of Informatics and Cybernetics 20, 83–88 (1983)

[SSS+94]   Shimozono, S., Shinohara, A., Shinohara, T., Miyano, S., Kuhara, S., Arikawa, S.: Knowledge acquisition from amino acid sequences by machine learning system BONSAI. Trans. Information Processing Society of Japan 35, 2009–2018 (1994)

[Wie76]   Wiehagen, R.: Limes-erkennung rekursiver funktionen durch spezielle strategien. Electronische Informationverarbeitung und Kybernetik 12, 93–99 (1976)

[Win93]   Winskel, G.: The Formal Semantics of Programming Languages: An Introduction. In: Foundations of Computing Series, MIT Press, Cambridge (1993)

[Wri89]   Wright, K.: Identification of unions of languages drawn from an identifiable class. In: Rivest, R., Haussler, D., Warmuth, M. (eds.) Proceedings of the Second Annual Workshop on Computational Learning Theory, Santa Cruz, California, pp. 328–333. Morgan Kaufmann, San Francisco (1989)

Prescribed Learning of R.E. Classes

Sanjay Jain[1,*], Frank Stephan[2,*], and Nan Ye[2]

[1] Department of Computer Science,
National University of Singapore, Singapore 117590, Republic of Singapore
sanjay@comp.nus.edu.sg
[2] Department of Computer Science and Department of Mathematics,
National University of Singapore, Singapore 117543, Republic of Singapore
fstephan@comp.nus.edu.sg, u0407028@nus.edu.sg

Abstract. This work extends studies of Angluin, Lange and Zeugmann on the dependence of learning on the hypotheses space chosen for the class. In subsequent investigations, uniformly recursively enumerable hypotheses spaces have been considered. In the present work, the following four types of learning are distinguished: class-comprising (where the learner can choose a uniformly recursively enumerable superclass as hypotheses space), class-preserving (where the learner has to choose a uniformly recursively enumerable hypotheses space of the same class), prescribed (where there must be a learner for every uniformly recursively enumerable hypotheses space of the same class) and uniform (like prescribed, but the learner has to be synthesized effectively from an index of the hypothesis space). While for explanatory learning, these four types of learnability coincide, some or all are different for other learning criteria. For example, for conservative learning, all four types are different. Several results are obtained for vacillatory and behaviourally correct learning; three of the four types can be separated, however the relation between prescribed and uniform learning remains open. It is also shown that every (not necessarily uniformly recursively enumerable) behaviourally correct learnable class has a prudent learner, that is, a learner using a hypotheses space such that it learns every set in the hypotheses space. Moreover the prudent learner can be effectively built from any learner for the class.

1 Introduction

The intuition behind learning in inductive inference [10] is that a learner sees more and more data and while reading the data produces conjectures about the concept to be learned which eventually stabilize on a correct description. The learning task is not arbitrary, but stems from a given class of concepts. Angluin [1] considered the important case that such a class is given by an indexed family, that is, the class is uniformly recursive. She has given a characterization when such a class is explanatorily learnable and introduced also important variants such as conservative learning. In the present work, the more general case of

---

* Supported in part by NUS grant number R252-000-212-112 and 251RES070107.

M. Hutter, R.A. Servedio, and E. Takimoto (Eds.): ALT 2007, LNAI 4754, pp. 64–78, 2007.

uniformly r.e. classes is addressed. Previously learnability of uniformly r.e. classes had been considered by de Jongh, Kanazawa [7] and Zilles [26,27].

Remark 1. First some basic notation. Let $W_0, W_1, W_2, \ldots$ be an acceptable enumeration of all r.e. subsets of the set of natural numbers $\mathbb{N}$. A language is a r.e. subset of natural numbers. Let φ_e denote the e-th partial recursive function, again from an acceptable numbering. For more information on recursion theory, the reader is referred to standard text books like the ones of Odifreddi [18] and Soare [21]. The function $\langle e, x \rangle = \frac{1}{2} \cdot (e + x)(e + x + 1) + x$ is Cantor's pairing function. A family $L_0, L_1, L_2, \ldots$ is uniformly recursively enumerable iff $\{\langle e, x \rangle : x \in L_e\}$ is a recursively enumerable set. For ease of notation, uniformly r.e. classes are just called *r.e. classes*. Note that in this paper, notations like $\{L_0, L_1, L_2, \ldots\}$ are used as a short-hand for both, the family as well as for the class of the sets; so set-theoretic comparisons like $\{L_0, L_1, L_2, \ldots\} \subseteq \{H_0, H_1, H_2, \ldots\}$ and $\{L_0, L_1, L_2, \ldots\} = \{H_0, H_1, H_2, \ldots\}$ ignore the ordering of the sets inside the class. Furthermore, let $W_{e,s}, L_{e,s}, H_{e,s}$ be the elements enumerated within time s into W_e, L_e, H_e, respectively. Without loss of generality, $W_{e,s}, L_{e,s}, H_{e,s}$ are subsets of $\{0, 1, \ldots, s\}$.

Let σ, τ range over $(\mathbb{N} \cup \{\#\})^*$. Furthermore, let $\sigma \subseteq \tau$ denote that τ is an extension of σ as a string. content(σ) denotes the set of natural numbers in the range of σ. T is a text if T maps $\mathbb{N}$ to $\mathbb{N} \cup \{\#\}$ and T is a text for L_a iff the numbers occurring in T are exactly those in L_a. content(T) denotes the set of natural numbers in the range of T. $T[n]$ denotes the string consisting of the first n elements of the text T, so $T[0]$ is the empty string and $T[2] = T(0)T(1)$.

Remark 2. A learner is a recursive function from $(\mathbb{N} \cup \{\#\})^*$ to $\mathbb{N} \cup \{?\}$. In the following, let M be a learner and let $\{L_0, L_1, L_2, \ldots\}$, $\{H_0, H_1, H_2, \ldots\}$ be r.e. classes. Here $\{L_0, L_1, L_2, \ldots\}$ is the class M should learn and $\{H_0, H_1, H_2, \ldots\}$ is the hypotheses space used by M.

The learner M *converges* on T to b if there is an n with $M(T[m]) = b$ for all $m \geq n$.

The learner M is *finite* [10] if for every text T there is one index e such that for all n, either $M(T[n]) = ?$ or $M(T[n]) = e$.

The learner M is *confident* [19] if M converges on every text T to a hypothesis.

The learner M is *conservative* [1] if for all σ, τ with $M(\sigma\tau) \neq M(\sigma)$ there is an x occurring in $\sigma\tau$ such that $x \notin H_{M(\sigma)}$.

The learner M *semantically identifies* L_a if, given any text T for L, $H_{M(T[n])} = L_a$ for almost all n. The learner M *syntactically identifies* L_a if, given any text T for L, there is a b with $H_b = L_a$ and $M(T[n]) = b$ for almost all n.

The learner M is a *behaviourally correct learner* for $\{L_0, L_1, L_2, \ldots\}$ iff M semantically identifies every L_a [3,6]; M is an *explanatory learner* for $\{L_0, L_1, L_2, \ldots\}$ if M syntactically identifies every L_a [4,10]. M is a *vacillatory learner* for $\{L_0, L_1, L_2, \ldots\}$ iff M is a behaviourally correct learner for $\{L_0, L_1, L_2, \ldots\}$ which on every text for a language L_a outputs only finitely many syntactically different hypotheses [5].

The learner M is *prudent* [9,19] if it learns all languages in its hypotheses space $\{H_0, H_1, H_2, \ldots\}$.

In the first three sections, all classes considered are recursively enumerable, only in Section 4 learnability of general classes is investigated.

Remark 3. Let M be a learner for $\{L_0, L_1, L_2, \ldots\}$ using hypotheses space $\{H_0, H_1, H_2, \ldots\}$. A sequence σ is called *syntactic stabilizing sequence* for M on a set L iff $\sigma \in (L \cup \{\#\})^*$ and for all $\tau \in (L \cup \{\#\})^*$, $M(\sigma\tau) = M(\sigma)$. A sequence σ is called *semantic stabilizing sequence* for M on a set L iff $\sigma \in (L \cup \{\#\})^*$ and for all $\tau \in (L \cup \{\#\})^*$, $H_{M(\sigma\tau)} = H_{M(\sigma)}$. Stabilizing sequences are called *locking sequences* for M on L, if in addition to the above conditions it holds that $H_{M(\sigma)} = L$. Note that, if M learns L then stabilizing sequences for M on L are also locking sequences for M on L.

Let K denote the halting problem. There is a partial K-recursive function Γ which assigns to each e the length-lexicographically least syntactic stabilizing sequence for M on L_e; $\Gamma(e)$ is defined iff such a sequence exists. Γ has a two-place approximation $\gamma(e, t)$ which converges to $\Gamma(e)$ if $\Gamma(e)$ is defined and diverges otherwise. Note that Γ and γ can be obtained effectively from an index for M and an index e' with $W_{e'} = \{\langle e, x \rangle : x \in L_e\}$. Blum and Blum [4] introduced the notion of locking sequences and Fulk [9] introduced the notion of stabilizing sequences.

Angluin [1], Lange, Kapur and Zeugmann [15,16,23,24,25] studied the dependence between the family $\{L_0, L_1, L_2, \ldots\}$ to be learned and the hypotheses space $\{H_0, H_1, H_2, \ldots\}$ used by the learner. To formalize this, they introduced the notions of exact, class-preserving and class-comprising learning. In addition to this, new notions like uniform and prescribed are introduced. Here I ranges over properties of the learner as defined in Remark 2, so I stands for "finite", "explanatory", "conservatively explanatory", "confidently explanatory", "vacillatory" and "behaviourally correct".

Definition 4. $\{L_0, L_1, L_2, \ldots\}$ is *class-comprisingly I learnable* iff it is I learnable with respect to some hypotheses space $\{H_0, H_1, H_2, \ldots\}$; note that learnability automatically implies $\{L_0, L_1, L_2, \ldots\} \subseteq \{H_0, H_1, H_2, \ldots\}$.

$\{L_0, L_1, L_2, \ldots\}$ is *class-preservingly I learnable* iff it is I learnable with respect to some hypotheses space $\{H_0, H_1, H_2, \ldots\}$ satisfying $\{H_0, H_1, H_2, \ldots\} = \{L_0, L_1, L_2, \ldots\}$.

$\{L_0, L_1, L_2, \ldots\}$ is *prescribed I learnable* iff it is I learnable with respect to every hypotheses space $\{H_0, H_1, H_2, \ldots\}$ such that $\{L_0, L_1, L_2, \ldots\} = \{H_0, H_1, H_2, \ldots\}$.

$\{L_0, L_1, L_2, \ldots\}$ is *uniformly I learnable* iff there is a recursive enumeration of partial-recursive functions $M_0, M_1, M_2, \ldots$ such that the following holds: Whenever $\{H_0, H_1, H_2, \ldots\} = \{L_0, L_1, L_2, \ldots\}$ and $W_e = \{\langle d, x \rangle : x \in H_d\}$ then M_e is total and an I learner for $\{L_0, L_1, L_2, \ldots\}$ with respect to this hypotheses space $\{H_0, H_1, H_2, \ldots\}$.

Remark 5. Lange and Zeugmann [15,23] considered besides class-preserving and class-comprising also the following notion: $\{L_0, L_1, L_2, \ldots\}$ is *exactly I learnable* iff it is I learnable with $\{L_0, L_1, L_2, \ldots\}$ itself taken as hypotheses space. Note that this notion needs that the ordering of the languages in $\{L_0, L_1, L_2, \ldots\}$ is taken into account, while all other definitions hold without paying attention to the specific ordering of the sets inside $\{L_0, L_1, L_2, \ldots\}$. The relation to prescribed learning is that a class $\{L_0, L_1, L_2, \ldots\}$ is prescribed I learnable iff every family $\{H_0, H_1, H_2, \ldots\}$ with $\{H_0, H_1, H_2, \ldots\} = \{L_0, L_1, L_2, \ldots\}$ is exactly I learnable.

The question whether a class can be learned using any given representation is quite natural. It reflects the situation where a company building learners cannot enforce its representation of the data/hypothesis on the clients but has to make for each client a learning algorithm using the client's representation. The difference between prescribed and uniform learning would then be that in the first case the programmers have to adjust for each client the learning program by hand, while in the second case there is some synthesizer which reads the clients requirements from some file and then adapts the learner automatically.

Remark 6. Note that in the case of learning with respect to r.e. families, uniform learning and prescribed learning are defined in a class-preserving way. Jain and Stephan [13] showed that there is a one-one numbering of all r.e. sets (that is, a Friedberg Numbering [8]) such that only classes with finitely many infinite sets can be behaviourally correct learned with respect to this numbering as hypotheses space.

Furthermore, above result can be strengthened to uniform learning by showing that only classes consisting of finite sets are class-comprising-uniformly behaviourally correct learnable. To see this, let $\{H_0, H_1, H_2, \ldots\}$ be a Friedberg numbering [8]. For a given parameter e, a family $\{G_0, G_1, G_2, \ldots\}$ is constructed from $\{H_0, H_1, H_2, \ldots\}$ such that the following holds for all a:

- For all b, $G_{\langle a,b \rangle} \subseteq H_a$;
- $G_{\langle a,b \rangle} = H_a$ if either $b = 0 \wedge |W_e| = \infty$ or $b = |W_e| + 1$;
- $G_{\langle a,b \rangle}$ is finite if either $b > 0 \wedge |W_e| = \infty$ or $b \neq |W_e| + 1 \wedge |W_e| < \infty$.

Suppose by way of contradiction that there is an r.e. infinite set H_a such that some class containing H_a can be class-comprising-uniformly behaviourally correctly learned. Note that for any fixed e and the class $\{G_0, G_1, G_2, \ldots\}$ with parameter e built as above, there exists exactly one index $\langle f(e), g(e) \rangle$ with $G_{\langle f(e), g(e) \rangle} = H_a$. By construction, $f(e) = a$. By the assumption on uniform learnability, there is a recursive enumeration of learners $N_0, N_1, N_2, \ldots$ such that each N_e learns the given class with respect to the hypotheses space $\{G_0, G_1, G_2, \ldots\}$ built with parameter e. As there is a fixed recursive text T for H_a and one can simulate N_e on T, the function g is limit-recursive (that is, there exists a recursive function h such that $g(x) = \lim_{t \to \infty} h(x, t)$). Note that W_e is infinite iff $g(e) = 0$. As $\{e : |W_e| = \infty\} \not\leq_T K$, this gives a contradiction. So class-comprising uniform behaviourally correct learning only permits to learn classes of finite sets.

Thus it is reasonable to restrict oneself to the class-preserving versions of prescribed and uniform learning; this convention has already been adapted in Definition 4.

The next result is obvious from the definitions.

Proposition 7. *For any notion I of learning and any class $\mathcal{L}$, the following implications hold: $\mathcal{L}$ is uniformly I-learnable $\Rightarrow$ $\{L_0, L_1, L_2, \ldots\}$ is prescribed I-learnable $\Rightarrow$ $\mathcal{L}$ is class-preservingly I-learnable $\Rightarrow$ $\mathcal{L}$ is class-comprisingly I-learnable.*

It depends on the chosen learning criterion I, which of the implications can be reversed. For finite and explanatory learning, all four notions are the same, as shown in Theorems 8 and 9. A lot of research [11] deals with requiring additional constraints on how hypotheses are chosen during explanatory learning. Such requirements change also the relations between the four types of learning. For confident learning, Theorem 10 shows that the uniform, prescribed and class-preserving type coincide while class-comprising confident learning is more general. For conservative learning, Example 11 gives classes which separate all four types of conservative learning. Theorems 12, 13, 15 and 16 deal with vacillatory and behaviourally correct learning. They give classes which, for these criteria, are class-comprisingly but not class-preservingly learnable as well as classes which are class-preservingly but not prescribed learnable. The separation of prescribed from uniform is open for these two criteria.

The importance of prudence is that the hypotheses space and the class of learned sets coincide; so the learner never conjectures some set it cannot learn. Fulk [9] showed that prudence is not restrictive for explanatory learning. Jain and Sharma [12] showed that prudence is not restrictive for vacillatory learning. In Theorem 17 it is shown that prudence is not restrictive for behaviourally correct learning. The prudent behaviourally correct learner can be constructed effectively from the original learner; it is still open whether prudence for explanatory and vacillatory learning can be effectivized. Note that Kurtz and Royer [14] had claimed to have this result but their proof had a bug and the problem had remained open since then.

2   Finite and Explanatory Learning

Finite learnable classes can be learnt uniformly, because finite learning is determined by a finite subset of the target language.

Theorem 8. *Every class-comprisingly finitely learnable class is also uniformly finitely learnable.*

Proof. Let M be a finite learner for $\{L_0, L_1, L_2, \ldots\}$ using a class-comprising hypotheses space. Let e be an index for a hypothesis space $\{H_0, H_1, H_2, \ldots\}$. That is, $W_e = \{\langle b, x \rangle : x \in H_b\}$. Further suppose $\{H_0, H_1, H_2, \ldots\} = \{L_0, L_1, L_2, \ldots\}$. Then a learner M_e is defined as follows. $M_e(T[n])$ is defined by the first case below which applies:

-   If there is an $m < n$ with $M_e(T[m]) \neq ?$ then $M_e(T[n]) = M_e(T[m])$ for the least such m;

- If there are $m \leq n$ and $b \leq n$ with $M(T[m]) \neq ?$ and content$(T[m]) \subseteq H_{b,n}$ then $M_e(T[n]) = b$;
- Otherwise $M_e(T[n]) = ?$.

The first condition guarantees that M_e outputs on T at most one hypothesis besides the symbol ?. Hence every M_e is a finite learner. It follows from the definition of finite learning that $H_b = H_c$ whenever $M(T[m]) \neq ?$, content$(T[m]) \subseteq H_b$ and content$(T[m]) \subseteq H_c$. Hence the b chosen in the second case is a correct hypothesis whenever this case applies. Furthermore, this case eventually applies on texts for languages in $\{L_0, L_1, L_2, \ldots\}$. This completes the proof that $\{L_0, L_1, L_2, \ldots\}$ is uniformly finitely learnable. □

The same result holds for explanatory learning.

Theorem 9. *Every class-comprisingly explanatorily learnable class is also uniformly explanatorily learnable.*

Proof. Let $\mathcal{L}$ be given and let M be a learner using a hypotheses space $\{L_0, L_1, L_2, \ldots\}$ containing $\mathcal{L}$ and perhaps other languages. Choose i such that $W_i = \{\langle a, x \rangle : x \in L_a\}$.

Fix any j and assume that j is an index of a hypotheses space $\{H_0, H_1, H_2, \ldots\}$ for $\mathcal{L}$, that is, assume $\{H_0, H_1, H_2, \ldots\} = \mathcal{L}$ and $W_j = \{\langle b, x \rangle : x \in H_b\}$. Let Γ_j be the function from Remark 3 which assigns to the members of $\{H_0, H_1, H_2, \ldots\}$ the length-lexicographically least syntactic stabilizing sequences with respect to the learner M. $\gamma_j(b, t)$ is then the t-th approximation of $\Gamma_j(b)$ as defined in Remark 3.

The learner M_j is constructed as follows: $M_j(\sigma)$ is the least b such that either $\gamma_i(M(\sigma), |\sigma|) = \gamma_j(b, |\sigma|)$ or $b = |\sigma|$. The latter condition is just to make M_j total and to terminate the search.

Assume that M converges on some text T to an index a of a language $L_a \in \mathcal{L}$. As $L_a \in \mathcal{L}$, there is a b with $H_b = L_a$; assume that b is the least such index. As $\{H_0, H_1, H_2, \ldots\} = \mathcal{L}$ and M is a learner for $\{H_0, H_1, H_2, \ldots\}$, an index c satisfies $\Gamma_j(c) = \Gamma_i(a)$ iff $H_c = L_a$. Hence M_j converges on T to b as, for all $c < b$ and almost all s, $\gamma_j(b, s) = \gamma_i(a, s)$ and $\gamma_j(c, s) \neq \gamma_i(a, s)$. It follows that M_j learns $\mathcal{L}$ using the hypotheses space $\{H_0, H_1, H_2, \ldots\}$. □

The next result shows that class-preserving confident learning coincides with uniform confident learning. The proof of the second part shows that class-preserving confident learning is not closed under taking subclasses.

Theorem 10. (a) *Every class-preservingly confidently learnable class $\mathcal{L}$ is also uniformly confidently learnable.*

(b) *The class $\{D : |D| = 2 \vee (|D| = 1 \wedge D \subseteq K')\}$ is class-comprisingly but not class-preservingly confidently learnable.*

Proof. (a) Reviewing the proof of Theorem 9, the additional constraints to those given there on M and $\{L_0, L_1, L_2, \ldots\}$ are that $\{L_0, L_1, L_2, \ldots\} = \mathcal{L}$ and M converges on every text to some index. Assume again that j and $\{H_0, H_1,$

$H_2, \ldots\}$ satisfy $\{L_0, L_1, L_2, \ldots\} = \{H_0, H_1, H_2, \ldots\}$ and $W_j = \{\langle b, x \rangle : x \in H_b\}$. Assume that T is any text. Then M converges on T to some index a as M is confident. By construction, M_j converges then to the least index b with $L_a = H_b$. Hence M_j also converges on all texts and hence M_j is confident. Furthermore, M_j learns $\mathcal{L}$ explanatorily with respect to the hypotheses space $\{H_0, H_1, H_2, \ldots\}$.

(b) The class $\{D : |D| = 2 \vee (|D| = 1 \wedge D \subseteq K')\}$ is class-comprisingly confidently learnable as follows. On a text for a set with up to two elements, the learner converges to an index for this set using $\{W_0, W_1, W_2, \ldots\}$ as hypotheses space. The learner does not revise its hypothesis after seeing three elements in the input, in order to obtain confidence.

Note that $\{D : |D| = 2 \vee (|D| = 1 \wedge D \subseteq K')\}$ is an r.e. class. To see this, note that there is a two-place recursive function g with $x \in K'$ iff $g(x, y) = 1$ for almost all y and $x \notin K'$ iff $g(x, y) = 0$ for infinitely many y. Now let

$$L_{2\langle x, y \rangle} = \{x, x + y + 1\} \text{ and}$$

$$L_{2\langle x, y \rangle + 1} = \begin{cases} \{x, x + z + 1\} & \text{if } z \text{ is the least number with} \\ & z > y \text{ and } g(x, z) \neq 1; \\ \{x\} & \text{if } g(x, z) = 1 \text{ for all } z > y. \end{cases}$$

It is easy to verify that $\{L_0, L_1, \ldots\} = \{D : |D| = 2 \vee (|D| = 1 \wedge D \subseteq K')\}$. Now assume that some confident learner M for $\{L_0, L_1, L_2, \ldots\}$ uses some hypotheses space $\{H_0, H_1, H_2, \ldots\}$ with $\{H_0, H_1, H_2, \ldots\} = \{L_0, L_1, L_2, \ldots\}$. Then one can define the K-recursive function f with $f(x)$ being the hypothesis to which M converges on the text x^∞. If $x \in K'$ then $H_{f(x)} = \{x\}$ as M learns this set. If $x \notin K'$ then $H_{f(x)} \neq \{x\}$ as no member of $\{H_0, H_1, H_2, \ldots\}$ equals $\{x\}$. The test whether $H_{f(x)} = \{x\}$ is also K-recursive. This would give a contradiction to $K' \not\leq_T K$. Thus there is no class-preserving confident learner for $\{L_0, L_1, L_2, \ldots\}$. $\qquad\square$

For conservative learning, a full hierarchy can be established. Note that the following example can be transferred to many related notions like monotonic [22] and non U-shaped learning [2] without giving more insight. Therefore, these learning criteria are not considered in the present work.

Example 11. (a) *The class* $\{D : |D| \leq 1\}$ *is prescribed conservatively but not uniformly conservatively learnable.*

(b) *The class* $\{D : |D| < \infty\}$ *is class-preservingly conservatively but not prescribed conservatively learnable.*

(c) *The class* $\{D : |D| = 2 \vee (|D| = 1 \wedge D \subseteq K')\}$ *is class-comprisingly conservatively but not class-preservingly conservatively learnable.*

Proof. (a) The prescribed learner knows the index a of $\emptyset$ in the given numbering $\{H_0, H_1, H_2, \ldots\}$. So it conjectures H_a until a number x occurs in the input and an index b is found with $x \in H_b$. Then the learner makes one mind change to b and keeps this index forever. This learner is conservative and correct as $\{x\}$ is the only set in $\{H_0, H_1, H_2, \ldots\}$ containing x. For the second part, let S be a simple set [20], $S^e = S \cup \{0, 1, \ldots, e\}$, define class-preserving hypotheses spaces $\mathcal{H}^0, \mathcal{H}^1, ...$, where $\mathcal{H}^e = \{H_0^e, H_1^e, ...\}$ with $H_x^e(y) = 1$ if $x \in S_y^e - S_{y-1}^e$ and $H_x^e(y) = 0$ if $x \notin S_y^e - S_{y-1}^e$. If $\{D : |D| \leq 1\}$ is uniformly conservatively learnable, then there exists a recursive family of learners $N_0, N_1, N_2, ...$ such that for all $e \in \mathbb{N}$, N_e conservatively learns the class $\{D : |D| \leq 1\}$ with respect to $\mathcal{H}^e$. The r.e. set $A = \{x : \text{for some } e, N_e \text{ outputs } x \text{ on } \#^\infty\}$ is infinite (as for all e, N_e outputs an index larger than e) and disjoint to S. This contradicts the fact that S is simple.

(b) The class of all finite sets is clearly conservatively learnable in the canonical numbering of the finite sets. Now let $I_0, I_1, I_2, ...$ be a recursive partition of the natural numbers into intervals such that there is a simple set A with $I_n \not\subseteq A$ for all n. Let $\{L_0, L_1, L_2, ...\}$ be the canonical numbering of the finite sets and let $H_m = L_n$ for $m \in I_n - A$ and $H_m = L_n \cup \{m+n+t, m+n+t+1\}$ for $m \in I_n \cap A$, with $m \in A_t - A_{t-1}$. It is easy to see that $\{H_0, H_1, H_2, ...\}$ is also a numbering of all finite sets. Assume now that M is a learner using the hypotheses space $\{H_0, H_1, H_2, ...\}$. Then one defines a recursive function f as follows: $f(x) = b$ for the first b found such that $x \in H_b$ and $M(x^k) = b$ for some k. As all H_b are finite, the set $\{f(0), f(1), f(2), ...\}$ contains infinitely many indices and is recursively enumerable. Hence there is an x with $f(x) \in A$. It follows that $\{x\} \subset H_{f(x)}$ as $H_{f(x)}$ contains at least two elements. So the learner M overgeneralizes on x^k and is not conservative.

(c) In Theorem 10, it has been shown that the class $\{D : |D| = 2 \vee (|D| = 1 \wedge D \subseteq K')\}$ is an r.e. class. The class-comprising confident learner given there is also conservative. Now assume that some conservative learner M for this class uses some class-preserving hypotheses space $\{H_0, H_1, H_2, ...\}$. Then one can again define $f(x)$, this time only partial-recursive, to be the b found such that M outputs b on the text x^∞ and $x \in H_b$. Now $x \in K'$ iff $f(x)$ is defined and $H_{f(x)} = \{x\}$. This condition can be checked with oracle K although $K' \not\leq_T K$. From this contradiction follows that there is no class-preserving conservative learner for $\{D : |D| = 2 \vee (|D| = 1 \wedge D \subseteq K')\}$. □

3   Vacillatory and Behaviourally Correct Learning

For vacillatory and behaviourally correct learning, a strict hierarchy from prescribed to class-preserving to class-comprising learning can be established. It remains open whether uniform learning is more restrictive than prescribed learning.

Theorem 12. *Let $L_{2a} = \{\langle a, b \rangle : b \in \mathbb{N}\}$ and $L_{2a+1} = \{\langle a, b \rangle : b \leq |W_a|\}$. Then $\{L_0, L_1, L_2, ...\}$ is uniformly behaviourally correct learnable and class-preservingly vacillatorily learnable but neither prescribed vacillatorily learnable nor class-comprisingly explanatorily learnable.*

Proof. Assume that $\{H_0, H_1, H_2, ...\} = \{L_0, L_1, L_2, ...\}$ and $W_e = \{\langle b, x \rangle : x \in H_b\}$. Let s be the length and D be the content of the input. Now a learner M_e is constructed. M_e first computes the sets

- $A = \{c \leq s : D = H_{c,s}\}$ and
- $B = \{c \leq s : D \cap H_{c,s} \neq \emptyset\}$;

then M_e follows the first of the following cases which applies:

– If $D = \emptyset$ then M_e outputs ?;
– If $A \neq \emptyset$ then M_e outputs $\min(A)$;
– If $B \neq \emptyset$ then M_e outputs some $c \in B$ for which $H_{c,s}$ has largest number of elements;
– Otherwise M_e repeats the previous conjecture.

The first case, together with the last, make sure that M_e is total, starts with ? and never returns to ? once it has taken another hypothesis. Assume now that M_e sees a text for a language $H_b \in \{L_{2a}, L_{2a+1}\}$ and that b is the least index of H_b in $\{H_0, H_1, H_2, \ldots\}$. Furthermore, assume that so much data has been observed such that the following four conditions hold:

– $s \geq b$;
– The datum $\langle a, 0 \rangle$ is in both, D and $H_{b,s}$;
– If $H_b \neq L_{2a+1}$ then $|H_{b,s}| > |L_{2a+1}|$ and $|D| > |L_{2a+1}|$;
– If H_b is finite then $H_b = H_{b,s} = D$ and for all $d < b$ and $t \geq s$, $H_{d,t} \neq D$.

Note that $D \neq \emptyset$ and $B \neq \emptyset$ and therefore M_e outputs a hypothesis c different from ?. Now it is shown that $H_c = H_b$: First note that $\langle a, 0 \rangle \in D$ and $b \in B$, hence the algorithm chooses c either by the second or the third condition in the algorithm. It follows that $H_c = L_{2a}$ or $H_c = L_{2a+1}$. If H_b is finite, it follows directly from the learning algorithm that $b = \min(A)$ for the set A considered there and hence $c = b$. If H_b is infinite and L_{2a+1} is finite, then $|H_c| \geq |H_{b,s}| > |L_{2a+1}|$ and $H_c = L_{2a} = H_b$. If H_b and L_{2a+1} are both infinite then $H_b = L_{2a} = L_{2a+1}$ and $H_c = H_b$. So M_e is a behaviourally correct learner for $\{L_0, L_1, L_2, \ldots\}$ using the hypotheses space $\{H_0, H_1, H_2, \ldots\}$.

To see that $\{L_0, L_1, L_2, \ldots\}$ is class-preservingly vacillatorily learnable, take $H_b = L_b$ for all b. For each language there are at most 2 indices in $\{H_0, H_1, H_2, \ldots\}$ and therefore the above described behaviorally correct learner is also a vacillatory one.

To see that $\{L_0, L_1, L_2, \ldots\}$ is not prescribed vacillatory learnable, one constructs a suitable hypotheses space as follows:

$$H_{\langle a,b \rangle} = \begin{cases} L_{2a+1} & \text{if } b = \min(\{s : |W_{a,s}| = |W_a|\}); \\ L_{2a} & \text{otherwise.} \end{cases}$$

For each a there is a b with $H_{\langle a,b \rangle} = L_{2a+1}$; if W_a is finite then one can take b as the minimum of the nonempty set $\{s : |W_{a,s}| = |W_a|\}$; if W_a is infinite then one can take $b = 0$. The reason for the latter case is that then $L_{2a} = L_{2a+1}$. Furthermore, all but at most one of the b satisfy $L_{2a} = H_{\langle a,b \rangle}$. Hence $\{H_0, H_1, H_2, \ldots\}$ is a hypotheses space for $\{L_0, L_1, L_2, \ldots\}$. If there were a prescribed vacillatory learner using $\{H_0, H_1, H_2, \ldots\}$ as the hypothesis space then there would also be a K-recursive function f such that $f(a)$ is the maximal element output by this learner on the canonical text for L_{2a+1}. It would follow that W_a is finite iff $W_{a,f(a)} = W_a$; note that $f(a) \geq \langle a, b \rangle \geq b$ for the least b such that $L_{2a+1} = H_{\langle a,b \rangle}$. But then a K-recursive procedure could check, given a, whether

W_a is finite. As such a procedure does not exist [21], $\{L_0, L_1, L_2, \ldots\}$ is not vacillatorily learnable with respect to the hypotheses space $\{H_0, H_1, H_2, \ldots\}$.

As just seen, $\{L_0, L_1, L_2, \ldots\}$ is not prescribed vacillatorily learnable and hence also not prescribed explanatorily learnable. It follows using Theorem 9 that $\{L_0, L_1, L_2, \ldots\}$ is also not class-comprisingly explanatorily learnable. $\square$

Theorem 13. *For all a, b let*

$$L_{\langle a,b \rangle} = \begin{cases} \{\langle a,c \rangle : c \in \mathbb{N}\} & \text{if } b = 0; \\ \{\langle a,c \rangle : c \leq |W_a|\} & \text{if } b = 1; \\ \{\langle a,c \rangle : c \leq |W_{a,d}|\} \cup \{\langle a+1, |W_{a,d}| + e + 1 \rangle\} & \text{if } b = 2 + \langle d,e \rangle. \end{cases}$$

The class $\{L_0, L_1, L_2, \ldots\}$ is class-preservingly behaviourally correct learnable but not prescribed behaviourally correct learnable.

Proof. Recall that $|W_{a,d}| \leq d + 1$ for all d. It is easy to see that $\{L_0, L_1, L_2, \ldots\}$ is a uniformly r.e. class. Assume that an input of length s and content D is given. A behaviourally correct learner takes now the first case which applies.

- If there is a pair $\langle a,b \rangle$ such that $\langle a+1, a+b+2 \rangle < s$ and $L_{\langle a,b \rangle, s} = D$ then output $\langle a,b \rangle$ for the least pair where these conditions are true.
- If there is an a such that $\{\langle a,0 \rangle\} \subseteq D \subseteq L_{\langle a,0 \rangle}$ then output $\langle a,0 \rangle$.
- Otherwise output ?

In this context it is assumed that for $b > 1$ and $s > \langle a+1, a+b+2 \rangle$, $L_{\langle a,b \rangle, s} = L_{\langle a,b \rangle}$ as one can compute all members directly from the parameters a, b. It is easy to see that this learner succeeds on all finite sets from $\{L_0, L_1, L_2, \ldots\}$. So assume that an infinite set $L_{\langle a,0 \rangle}$ is given. If $L_{\langle a,1 \rangle} = L_{\langle a,0 \rangle}$ then the learner will eventually vacillate between these two indices. If $L_{\langle a,1 \rangle} \subset L_{\langle a,0 \rangle}$ then $L_{\langle a,1 \rangle}$ is finite and as the learner eventually sees an element of $L_{\langle a,0 \rangle} - L_{\langle a,1 \rangle}$, it will converge to $\langle a,0 \rangle$. So $\{L_0, L_1, L_2, \ldots\}$ is class-preservingly behaviourally correct learnable.

Now a hypotheses space is constructed using which $\{L_0, L_1, L_2, \ldots\}$ cannot be behaviourally correct learned. For all a, b let

$$H_{\langle a,0 \rangle} = L_{\langle a,0 \rangle};$$
$$H_{\langle a,2b+1 \rangle} = L_{\langle a,b+2 \rangle};$$
$$H_{\langle a,2b+2 \rangle} = \begin{cases} \{\langle a,c \rangle : c \leq |W_{a,b}|\} & \text{if } W_{a,b} = W_a; \\ \{\langle a,c \rangle : c \leq |W_{a,b}|\} \cup \{\langle a+1, |W_{a,b}| + s + 1 \rangle\} & \text{if } s \text{ is the least} \\ & \text{number with} \\ & W_{a,b} \subset W_{a,s}. \end{cases}$$

It is easy to check that this class is an indexed family, that is, $\{H_0, H_1, H_2, \ldots\}$ is uniformly recursive. Thus, if one could behaviourally correct learn $\{L_0, L_1, L_2, \ldots\}$ using $\{H_0, H_1, H_2, \ldots\}$ as the hypotheses space, one could also explanatorily learn $\{L_0, L_1, L_2, \ldots\}$ using $\{H_0, H_1, H_2, \ldots\}$ (this folklore result is based on the observation that, for hypotheses space being an indexed family, the mind changes can be delayed until it can be verified that the later hypothesis differs from the earlier one). Using Theorem 9, this would imply that the class from Theorem 12 (which is contained in $\{L_0, L_1, L_2, \ldots\}$) is prescribed explanatorily learnable and hence prescribed vacillatory learnable. This contradicts Theorem 12. So $\{L_0, L_1, L_2, \ldots\}$ is not prescribed behaviourally correct learnable. $\square$

Corollary 14. *Let* $\{L_0, L_1, L_2, \ldots\}$ *be as in Theorem 13. Then* $\{L_0, L_1, L_2, \ldots\}$ $\cup \{\mathbb{N}\}$ *is class-preserving behaviourally correct learnable. Furthermore, no* $\{F_0, F_1, F_2, \ldots\} \supseteq \{L_0, L_1, L_2, \ldots\} \cup \{\mathbb{N}\}$ *is prescribed behaviourally correct learnable.*

For the next result, let $I_n = \{2^n - 1, 2^n, 2^n + 1, \ldots, 2^{n+1} - 3, 2^{n+1} - 2\}$ form a partition of the natural numbers into intervals of length 2^n and let C denote the plain Kolmogorov complexity [17]. Furthermore, let

$$A = \{m : \exists n \, [m \in I_n \wedge C(m) < 0.4n]\} \text{ and}$$
$$B = \{m : \exists n \, [m \in I_n \wedge C(m) > 0.8n]\}$$

be the sets of numbers of small and large Kolmogorov complexity, respectively.

Theorem 15. *Let A and B be the sets of numbers of small and large Kolmogorov complexity as above. Then the class consisting of $\mathbb{N}$, A and all sets $A \cup \{b\}$ with $b \in B$ is uniformly r.e. and is class-comprisingly but not class-preservingly behaviourally correct learnable.*

Proof. Note that A is recursively enumerable and B is co-r.e.; an indexing of the class is now given by fixing one index $a \in A$ and then letting $L_a = A$, $L_b = A \cup \{b\}$ for all $b \in B$ and $L_b = \mathbb{N}$ for all $b \in \mathbb{N} - B - \{a\}$.

Note that $0 \notin A \cup B$. Hence $\mathbb{N}$ is the only member of $\{L_0, L_1, L_2, \ldots\}$ containing 0. Furthermore, let $D_0, D_1, \ldots$ be a canonical enumeration of all finite sets. Now let

$$H_b = \begin{cases} \mathbb{N} & \text{if } 0 \in D_b; \\ D_b \cup A & \text{if } 0 \notin D_b. \end{cases}$$

Furthermore, one can build a behaviourally correct learner using the hypotheses space $\{H_0, H_1, H_2, \ldots\}$ by conjecturing H_b for the unique b with $D_b = \text{content}(\sigma)$ on input σ. It is easy to verify that this learner succeeds on all languages in $\{H_0, H_1, H_2, \ldots\}$. Therefore $\{L_0, L_1, L_2, \ldots\}$ is class-comprisingly behaviourally correct learnable.

Now assume that M is a class-preserving behaviourally correct learner for $\{L_0, L_1, L_2, \ldots\}$. There is a family $T_0, T_1, \ldots$ of texts and an n such that

- $T_x[n]$ is a fixed semantic locking sequence for M on A;
- $T_x(n) = x$;
- for all x, the subsequence $T_x(n+1), T_x(n+2), T_x(n+3), \ldots$ of T_x is the same recursive enumeration of A.

Now one defines two sets X and Y according to the behaviour of M on T_x.

- X is the set of all x such that, for some $m > n$, $M(T_x[m])$ conjectures a set containing x;

- Y is the set of all x such that, for some $m > n$, $M(T_x[m])$ conjectures a set containing 0.

Both sets are recursively enumerable. The set Y is disjoint to A as, for all $x \in A$ and all $m > n$, $M(T_x[m])$ is an index of A. As A is a simple set [17], Y is finite. As $A \cup B \subseteq X \subseteq A \cup B \cup Y$, the set $A \cup B$ is recursively enumerable. For each sufficiently large n, at least half of elements of I_n are in $A \cup B$. Now let J_n be the first $2^{0.6n}$ elements of I_n to be enumerated into $A \cup B$. The J_n are uniformly r.e. and due to Kolmogorov-complexity considerations, for all sufficiently large n, $J_n \cap B = \emptyset$. Hence $J_n \subseteq A \cap I_n$ in contradiction to the fact that $|A \cap I_n| \le 2^{0.4n}$. This shows that the learner M cannot exist and $\{L_0, L_1, L_2, \ldots\}$ is not class-preservingly behaviourally correct learnable. $\qquad \square$

Theorem 16. *There exists an r.e. class $\mathcal{L}$ which is class-comprisingly but not class-preservingly vacillatorily learnable.*

4 Prudence for Behaviourally Correct Learning

Osherson, Stob and Weinstein [19] were interested in the question whether every learnable class is prudently learnable. Fulk [9] showed that every explanatory learnable class is prudently explanatory learnable. Jain and Sharma [12] showed the corresponding result for vacillatory learning. The next theorem shows this result for behaviourally correct learning. In 1988, Kurtz and Royer [14] had claimed to have this result, but their proof had a bug and the problem had remained open since then. Furthermore, the construction of the prudent learner in the next theorem is effective in the original learner. It is still open whether prudence for explanatory and vacillatory learning can be effectivized.

Theorem 17. *If $\mathcal{L}$ is a (not necessarily uniformly r.e.) behaviourally correct learnable class then $\mathcal{L}$ is a subclass of an r.e. class which is class-preservingly behaviourally correct learnable.*

Proof. For any set A, let T_A be the ascending text which is given by $T_A(x) = x$ for all $x \in A$ and $T_A(x) = \#$ for all $x \notin A$. Furthermore, let $\delta_\emptyset$ be the empty string and $\delta_A = T_A[\max(A) + 1]$ for all finite non-empty sets A. For example, $\delta_{\{0,2,3\}} = 0 \, \# \, 2 \, 3$.

There is a behaviourally correct learner for the class $\mathcal{L}$ using the acceptable numbering $\{W_0, W_1, W_2, \ldots\}$ as hypotheses space and satisfying the following constraints:

- M is consistent, that is, content$(\sigma) \subseteq W_{M(\sigma)}$ for all σ;
- M is rearrangement-independent, that is, $W_{M(\sigma)} = W_{M(\tau)}$ whenever σ, τ have the same content and length;
- $W_{M(\sigma)}$ is finite whenever σ is not a semantical locking sequence for M on $W_{M(\sigma)}$.

Kurtz and Royer [14] showed that the first two conditions can be satisfied and such a learner can be found effectively from any given learner. The third condition can also be effectively added since the complement of the set of semantical locking sequences is K-r.e.; that is, σ is not a semantical locking sequence iff there is a τ in $(W_{M(\sigma)} \cup \{\#\})^*$ and an $x \in \mathbb{N}$ with $x \in W_{M(\sigma\tau)} \Leftrightarrow x \notin W_{M(\sigma)}$. For that reason, M is a behaviourally correct learner for all infinite sets for which some index is output by M. So, to prove the theorem, one has mainly to take care of finite sets.

Now the following new learner N is constructed. N is defined by mapping σ to a hypothesis H_σ; thus the hypotheses space is given directly instead of N. H_σ takes the first case which applies.

Case (1): $H_{\#^s} = \emptyset$ for all s.

Case (2): H_{δ_D} first enumerates all elements of D.

   Let $D' = \{0, 1, \ldots, \max(D)\} - D$. Let $S = \{s : W_{M(\delta_D \#^{\max(D)}),s} \cap D' = \emptyset\}$. For all $s \in S$, enumerate all elements of $W_{M(\delta_D \#^{\max(D)}),s}$ into H_{δ_D}.
   If $W_{M(\delta_D \#^{\max(D)})} \cap D' \neq \emptyset$ then let $s = \max(S)$, let $E = D \cup W_{M(\delta_D \#^{\max(D)}),s}$, let $x = \min(W_{M(\delta_D \#^{\max(D)}),s+1} \cap D')$ and let $F = D \cap \{0, 1, \ldots, x\}$.
   Now, if $H_{\delta_F} \supseteq E$ then $H_{\delta_D} = H_{\delta_F}$ else $H_{\delta_D} = E$.

Case (3): $H_{\delta_D \#^s}$ with $s > 0$ is defined as follows. If there is an x such that $H_{\delta_{E_x},s} = H_{\delta_{E_x}} = D$ for the set $E_x = D \cap \{0, 1, \ldots, x\}$ or if $W_{M(\delta_D \#^t)} = D$ for all $t \geq s$ then $H_{\delta_D \#^s} = D$ else $H_{\delta_D \#^s} = H_{\delta_D}$.

Case (4): $H_\sigma = H_{\delta_D \#^s}$ if H_σ is not defined by Cases (1), (2), (3), $s = \max(\{|\sigma| - \max(D) - 1, 0\})$ and $D = \text{content}(\sigma)$.

Note that the only infinite sets in the hypothesis space are the ones which are conjectured by M. So M learns all the infinite sets in the hypotheses space. Furthermore, for any A in the hypotheses space, if $E_x = \{0, 1, \ldots, x\} \cap A$ and $\delta_{E_x} \#^{\max(E_x)}$ is a semantic locking sequence for M on A, then for all finite D such that $E_x \subseteq D \subseteq A$, $H_{\delta_D} = A$. This can be easily seen by induction on cardinality of $D - E_x$, as in Case (2), either H_{δ_D} is made equal to A or H_{δ_D} would simulate H_{δ_F} for some F such that $E_x \subseteq F \subset D$.

It will be shown first that the hypotheses space covers all sets learned by M and then it will be shown that all sets in the hypotheses space are learned by N.

Clearly if M learns a finite set D then $H_{\delta_D \#^s} = D$ for almost all s. Now consider an infinite set A learned by M. Let $E_x = A \cap \{0, 1, 2, \ldots, x\}$ for all x. As M learns A there is a semantic locking sequence τ for M on A. Now let $x \in A$ be such that $x > |\tau| + \max(\text{content}(\tau))$. Then, for the sequence $\delta_{E_x} \#^{\max(E_x)}$, there is an $\eta \in (E_x \cup \{\#\})^*$ such that $|\tau\eta| = |\delta_{E_x} \#^{\max(E_x)}|$ and $\text{content}(\tau\eta) = \text{content}(\delta_{E_x} \#^{\max(E_x)}) = E_x$. As M is rearrangement-independent, one has that $W_{M(\delta_{E_x} \#^{\max(E_x)})} = A$. Hence $H_{\delta_{E_x}} = A$ as well. This completes the first part of the verification.

For the second part of the verification consider any set A occurring in the hypotheses space of N. There are three cases, those where A is empty, where A is finite but not empty and where A is infinite.

Case (a): $A = \emptyset$. N learns A as $H_{\#^s} = \emptyset$ for all s by Case (1) in the algorithm to enumerate the hypotheses space.

Case (b): A is finite but not empty. Let D be smallest set such that $H_{\delta_D \#^s} = A$ for some s. By Case (1) in the algorithm for H_σ, D is not empty.

Assume the subcase $A = H_{\delta_D \#^s} \subset H_{\delta_D}$. By Case (3) and D being smallest set such that $H_{\delta_D \#^t} = A$ for some t, this can happen only if $A = D$ and $W_{M(\delta_D \#^t)} = D$ for all $t \geq s$. So $H_{\delta_D \#^t} = D$ for all $t \geq s$ and hence N learns A in this subcase as well.

Assume the subcase $A = H_{\delta_D \#^s} = H_{\delta_D}$. Hence, by Case (2) it follows that there is no element in $A - D$ below $\max(D)$ since otherwise $H_{\delta_F} = A$ for some $F \subset D$. Thus, $D = A \cap \{0, 1, \ldots, \max(D)\}$. Therefore, $H_{\delta_A \#^t} = A$ for almost all t and N learns A.

Case (c): A is infinite. Again, let $E_z = A \cap \{0, 1, \ldots, z\}$ for all z. As M is rearrangement-independent, there is a semantic locking sequence for M on A of the form $\delta_{E_x} \#^{\max(E_x)}$. Hence only finitely many sets $H_{\delta_{E_z}}$ are finite. So there is an $y \in A$ such that $y > x$ and y is an upper bound on all elements of these finite sets $H_{\delta_{E_z}}$. Let F be any finite set with $E_y \subseteq F \subseteq A$. Let $G_z = F \cap \{0, 1, \ldots, z\}$. If $z \geq y$ then $H_{\delta_{G_z}} = A$ (as $E_x \subseteq G_z \subseteq A$) and $H_{\delta_{G_z}} \neq F$. If $z < y$ then $G_z = E_z$ and $H_{\delta_{G_z}} \neq F$ again. Furthermore, M does not learn F. Hence $H_{\delta_F \#^s} = H_{\delta_F} = A$ for all s. So δ_{E_y} is a semantic locking sequence for N on A. It follows that N learns A. This completes the verification that N is a behaviourally correct learner for all the languages in its hypotheses space.   □

Acknowledgment. We thank J. Royer and S. Zilles for correspondence.

References

1. Angluin, D.: Inductive inference of formal languages from positive data. Information and Control 45, 117–135 (1980)
2. Baliga, G., Case, J., Merkle, W., Stephan, F., Wiehagen, R.: When unlearning helps. Technical Report TRA5/06, School of Computing, National University of Singapore (2005), Preliminary version appeared as "Unlearning helps" by Baliga, G., Case, J., Merkle, W., Stephan, F., Welzl, E., Montanari, U., Rolim, J.D.P. (eds.) ICALP 2000. LNCS, vol. 1853, pp. 844–855. Springer, Heidelberg (2000)
3. Bārzdiņš, J.: Two theorems on the limiting synthesis of functions. In: Theory of Algorithms and Programs– Volume 1, vol. 210, pp. 82–88. Latvian State University, Riga (1974)
4. Blum, L., Blum, M.: Toward a mathematical theory of inductive inference. Information and Control 28, 125–155 (1975)
5. Case, J.: The power of vacillation in language learning. SIAM Journal on Computing 28, 1941–1969 (1999)
6. Case, J., Lynes, C.: Inductive inference and language identification. In: Nielsen, M., Schmidt, E.M. (eds.) Ninth International Colloquium on Automata, Languages and Programming (ICALP). LNCS, vol. 140, pp. 107–115. Springer, Heidelberg (1982)
7. de Jongh, D., Kanazawa, M.: Angluin's theorem for indexed families of r.e. sets and applications. In: Proceedings of the Ninth Annual Conference on Computational Learning Theory, pp. 193–204. ACM Press, New York (1996)
8. Friedberg, R.: Three theorems on recursive enumeration. Journal of Symbolic Logic 23, 309–316 (1958)

9. Fulk, M.: Prudence and other conditions on formal language learning. Information and Computation 85, 1–11 (1990)
10. Gold, E.M.: Language identification in the limit. Information and Control 10, 447–474 (1967)
11. Jain, S., Osherson, D., Royer, J.S., Sharma, A.: Systems That Learn: An Introduction to Learning Theory. MIT-Press, Boston (1999)
12. Jain, S., Sharma, A.: Prudence in vacillatory language identification. Mathematical Systems Theory 28, 267–279 (1995)
13. Jain, S., Stephan, F.: Learning in Friedberg Numberings. In: Hutter, M., Servedio, R., Takimoto, E. (eds.) Algorithmic Learning Theory: 18th International Conference, ALT 2007, Sendai, Japan (to appear 2007)
14. Kurtz, S., Royer, J.S.: Prudence in language learning. In: Proceedings of the First Annual Workshop on Computational Learning Theory, pp. 143–156. MIT, Cambridge (1988)
15. Lange, S., Zeugmann, T.: Language learning in dependence on the space of hypotheses. In: Proceedings of the Sixth Annual Conference on Computational Learning Theory, Santa Cruz, California, United States, pp. 127–136 (1993)
16. Lange, S., Zeugmann, T., Kapur, S.: Monotonic and dual monotonic language learning. Theoretical Computer Science 155, 365–410 (1996)
17. Li, M., Vitányi, P.: An Introduction to Kolmogorov Complexity and Its Applications. Springer, Heidelberg (1993)
18. Odifreddi, P.: Classical Recursion Theory, North-Holland, Amsterdam (1989)
19. Osherson, D., Stob, M., Weinstein, S.: Systems That Learn, An Introduction to Learning Theory for Cognitive and Computer Scientists. Bradford — The MIT Press, Cambridge, Massachusetts (1986)
20. Post, E.: Recursively enumerable sets of positive integers and their decision problems. Bulletin of the American Mathematical Society 50, 284–316 (1944)
21. Soare, R.: Recursively Enumerable Sets and Degrees. Springer, Heidelberg (1987)
22. Wiehagen, R.: A thesis in inductive inference. In: Dix, J., Schmitt, P.H., Jantke, K.P. (eds.) Proceedings First International Workshop on Nonmonotonic and Inductive Logic. LNCS, vol. 543, pp. 184–207. Springer, Heidelberg (1990)
23. Zeugmann, T.: Algorithmisches Lernen von Funktionen und Sprachen. Habilitationsschrift, Technische Hochschule Darmstadt (1993)
24. Zeugmann, T., Lange, S.: A guided tour across the boundaries of learning recursive languages. In: Lange, S., Jantke, K.P. (eds.) Algorithmic Learning for Knowledge-Based Systems. LNCS, vol. 961, pp. 193–262. Springer, Heidelberg (1995)
25. Zeugmann, T., Lange, S., Kapur, S.: Characterizations of monotonic and dual monotonic language learning. Information and Computation 120, 155–173 (1995)
26. Zilles, S.: Separation of uniform learning classes. Theoretical Computer Science 313, 229–265 (2004)
27. Zilles, S.: Increasing the power of uniform inductive learners. Journal of Computer and System Sciences 70, 510–538 (2005)

Learning in Friedberg Numberings

Sanjay Jain[1],[*] and Frank Stephan[2],[*]

[1] Department of Computer Science,
National University of Singapore, Singapore 117590, Republic of Singapore
`sanjay@comp.nus.edu.sg`
[2] Department of Computer Science and Department of Mathematics,
National University of Singapore, Singapore 117543, Republic of Singapore
`fstephan@comp.nus.edu.sg`

Abstract. In this paper we consider learnability in some special numberings, such as Friedberg numberings, which contain all the recursively enumerable languages, but have simpler grammar equivalence problem compared to acceptable numberings. We show that every explanatorily learnable class can be learnt in some Friedberg numbering. However, such a result does not hold for behaviourally correct learning or finite learning. One can also show that some Friedberg numberings are so restrictive that all classes which can be explanatorily learnt in such Friedberg numberings have only finitely many infinite languages. We also study similar questions for several properties of learners such as consistency, conservativeness, prudence, iterativeness and non U-shaped learning. Besides Friedberg numberings, we also consider the above problems for programming systems with K-recursive grammar equivalence problem.

1 Introduction

Consider the following model of learning languages, first studied by Gold [14]. A learner is receiving, one element at a time, all and only the sentences of a language (such a presentation of data is called text of the language). As the learner is receiving the elements of the language, it conjectures hypotheses about what the input language might be. The conjecture about the input language may change over time, as more and more data becomes available. In inductive inference, we use indices from some underlying numbering or programming system as hypotheses. Following conventions from formal languages, we refer to these indices as grammars. One can say that the learner is successful if the sequence of grammars output as above converges to a grammar for the input language. This is essentially the model of **TxtEx**-learning (= explanatory learning) as proposed by Gold [14] and subsequently studied by several researchers [1,5,10,16,26,30].

One of the important issues in learning has been the hypotheses space which a learner uses for making its conjectures. A natural hypotheses space, as considered by Gold [14], is an acceptable programming system. However, there have also been several studies which consider special programming systems [30]. For

---

[*] Supported in part by NUS grant number R252-000-212-112 and 251RES070107.

example, in the context of learning indexed families of languages (an indexed family is a uniformly recursive family of languages), the hypotheses space often considered are themselves indexed families (where the hypotheses space might be class-preserving or class-comprising; a class-preserving hypotheses space contains exactly the languages in the class being learnt while a class-comprising hypotheses space may contain some other languages in addition to the languages of the class being learnt). Furthermore, considering special hypotheses spaces have also been useful in obtaining various characterizations of learnability — see, for example, [17,28,30].

Testing grammar equivalence in acceptable numberings is a difficult problem [24]. In this paper we consider learnability in some special numberings, which contain all the recursively enumerable languages, but with simpler grammar equivalence problem. Friedberg numberings [11] are numberings which contain exactly one grammar for each recursively enumerable language. Besides their historical importance, Friedberg numberings may be considered as a natural hypotheses space, as they do not contain any redundancy. Another natural class of numberings is the Ke-numberings in which grammar equivalence problem is recursive in the halting problem. Freivalds, Kinber and Wiehagen [12] considered learnability of recursive functions in Friedberg and other 1–1 numberings (for the criteria of explanatory and finite learning). We extend their study by considering how the learnability in various common criteria are effected when one uses hypotheses spaces as above.

We show (Theorem 9) that for **TxtEx**-model of learning, as described above, one can learn every **TxtEx**-learnable class in some Friedberg numbering. However, no Friedberg numbering is omnipotent. More precisely, for every Friedberg numbering η, there exists a **TxtEx**-learnable class which cannot be learnt using hypotheses space η. Furthermore, there are Friedberg numberings η which are trivial in the sense that any class **TxtEx**-learnable in η contains only finitely many infinite languages (Theorem 26).

In finite learning [14], denoted **TxtFin**, one requires that the learner outputs just one hypothesis, which must be correct. In contrast to the result for **TxtEx**-learning, there are **TxtFin**-learnable classes which cannot be learnt in any Friedberg numbering (Theorem 10). However, Ke-numberings are not so restrictive, as every **TxtFin**-learnable class can be learnt in some Ke-numbering (Theorem 14). Theorem 12 gives a characterization of the recursively enumerable classes which can be learnt in Friedberg numberings.

Several properties of learners have been considered in the literature. For example a consistent learner [1,4] is a learner whose hypotheses always generate the data seen up to the point an hypothesis is made. A conservative learner does not change a hypothesis which is consistent with the input [2,30]. A prudent learner [22] only outputs hypotheses for the languages which it is able to learn. A confident learner [22] always converges on any input text, even on texts for languages outside the class being learnt. A non U-shaped learner is a learner which does not have a sequence of hypotheses of form "..., correct hypothesis, ..., wrong hypothesis, ..., correct hypothesis, ..." [3,7,8]. We show that, though confident and consistent learning are not restrictive for learning in Friedberg numberings (Theorems 15 and 24), non U-shaped, conservative and prudent learning are restrictive (Theorems 18 and 19). On the other hand, none of the above properties are restrictive for learning in Ke-numberings (Theorems 20, 21 and 22 along with Theorems 15 and 24).

Behaviourally correct learning [10,23] is similar to **TxtEx**-learning except that one does not require syntactic convergence, but only semantic convergence: the hypotheses conjectured by the learner are correct beyond some time. For Friedberg numberings, notion of **TxtBc** collapses to **TxtEx** due to trivial grammar equivalence problem. It is open at present whether every **TxtBc**-learnable class can be learnt in some Ke-numberings — though we can show that every class which can be **TxtFEx**-learnt can be **TxtBc**-learnt in some Ke-numbering (**TxtFEx**-learning [9] is **TxtBc**-learning where the learner only outputs finitely many distinct hypotheses). We can though show that there exists a non U-shaped behaviourally learnable class, which cannot be learnt in non U-shaped behaviourally correct manner in any Ke-numbering (Theorem 30).

2   Notation and Preliminaries

Any unexplained recursion-theoretic notions are from [21,24].

$\mathbb{N}$ denotes the set of natural numbers, $\{0,1,2,\ldots\}$. $\emptyset$ denotes empty set. $card(S)$ denotes the cardinality of set S. $\max(S)$ and $\min(S)$, respectively, denote the maximum and minimum of a set S, where $\max(\emptyset)$ is 0 and $\min(\emptyset)$ is ∞. The symbols $\subseteq, \supseteq, \subset, \supset$ respectively denote the subset, superset, proper subset and proper superset relation between sets. $A \triangle B$ denotes the symmetric difference of A and B: $(A \cup B) - (A \cap B)$. The quantifiers $\forall^\infty$ and $\exists^\infty$ mean "for all but finitely many" and "there exist infinitely many", respectively.

A pair $\langle i, j \rangle$ stands for an arbitrary, computable one-to-one encoding of all pairs of natural numbers onto $\mathbb{N}$ [24]. Similarly we can define $\langle \cdot, \ldots, \cdot \rangle$ for encoding n-tuples of natural numbers, for $n > 1$, onto $\mathbb{N}$.

Any partial recursive function of two arguments is called a numbering. For a numbering ψ, $\psi_i(x)$ denotes $\psi(i, x)$. We let Ψ denote a Blum complexity measure [6] associated with the numbering ψ. We let $\psi_{i,s}(x) = \psi_i(x)$, if $x < s$ and $\Psi_i(x) < s$; $\psi_{i,s}(x)$ is undefined if $x \geq s$ or $\Psi_i(x) \geq s$. We let $W_i^\psi = domain(\psi_i)$ and $W_{i,s}^\psi = domain(\psi_{i,s})$. We call i a ψ-grammar for W_i^ψ.

For numberings ψ and η, $\psi \leq \eta$ denotes that there exists a recursive function g such that $W_i^\psi = W_{g(i)}^\eta$ for all i. $\psi \leq^A \eta$ denotes that there exists an A-recursive function g such that $W_i^\psi = W_{g(i)}^\eta$ for all i.

$\mathcal{E}$ denotes the class of all recursively enumerable (r.e.) subsets of the natural numbers [24]; an r.e. set is also called a *language*. $\mathcal{F}$ is the class of all finite sets and $\mathcal{I}$ is the class $\{\emptyset, \{0\}, \{0,1\}, \{0,1,2\}, \ldots, \{0,1,\ldots,n\}, \ldots\}$. A *universal numbering* [24] ψ is a numbering such that, for all $L \in \mathcal{E}$, there exists a ψ-grammar for L. An *acceptable numbering* [24] ψ is a numbering such that, for all numberings η, $\eta \leq \psi$. Acceptable numberings are also called Gödel numberings.

φ denotes a fixed acceptable programming system for the partial computable functions [24]. We let $W_e = W_e^{\varphi} = domain(\varphi_e)$. $K = \{e : e \in W_e\}$, the diagonal halting problem, is a standard example for a nonrecursive r.e. set.

Friedberg [11] showed that there exist numberings in which every r.e. language has exactly one index (grammar). Hence the equivalence problem for grammars is obviously recursive in such numberings; furthermore, one can easily translate every numbering with a recursive equivalence problem into a Friedberg numbering. It might be important to relax this condition and to consider numberings where the equivalence problem is only K-recursive. K-recursive equivalence and translations have already received some attention; for example Goncharov [15] showed that if two Friedberg numberings of a given family of r.e. sets are not equivalent but can be K-recursively translated into each other, then this family has infinitely many non-equivalent numberings.

We are not aware of any common name for numberings with a K-recursive equivalence problem; thus we refer to them as Ke-numberings, "Ke" standing for "K-recursive equivalence".

Definition 1. A *Friedberg-numbering* is a universal numbering in which every recursively enumerable set has exactly one grammar. A *Ke-numbering* is a universal numbering for which the grammar equivalence problem is K-recursive.

A class $\mathcal{L}$ is said to be recursively enumerable if there exists an r.e. set S such that $\mathcal{L} = \{W_i : i \in S\}$. Note that for a non-empty recursively enumerable class $\mathcal{L}$, there exists a recursive function h such that $\mathcal{L} = \{W_{h(i)} : i \in \mathbb{N}\}$. A class $\mathcal{L}$ is said to be 1–1 recursively enumerable iff $\mathcal{L}$ is finite or there exists a recursive function h such that $\mathcal{L} = \{W_{h(i)} : i \in \mathbb{N}\}$ and, for all different i, j, $W_{h(i)} \neq W_{h(j)}$.

We now introduce the basic definitions of inductive inference, that is, of Gold-style computational learning theory.

Definition 2. A *sequence* σ is a mapping from an initial segment of $\mathbb{N}$ into $\mathbb{N} \cup \{\#\}$. The content of a finite sequence σ is the set of natural numbers occurring in σ and is denoted by content(σ). The length of a sequence σ is the number of elements in the domain of σ and is denoted by $|\sigma|$. For a subset L of $\mathbb{N}$, $Seg(L)$ denotes the set of sequences σ with content(σ) $\subseteq L$. An infinite sequence T is a mapping from $\mathbb{N}$ to $\mathbb{N} \cup \{\#\}$. Furthermore, content(T) denotes the set of natural numbers in the range of T. T is a *text for L* iff $L = $ content(T).

Concatenation of two sequences σ and τ is denoted by $\sigma\tau$. If $x \in (\mathbb{N} \cup \{\#\})$, then σx means $\sigma\tau$ where τ is the sequence consisting of exactly one element which is x. $\sigma \subseteq \tau$ means that σ is an initial segment of τ and $\sigma \subset \tau$ means that σ is a proper initial segment of τ.

Intuitively, a *text* for a language L is an infinite stream or sequential presentation of *all* the elements of the language L in any order and with the #'s representing pauses in the presentation of the data. For example, the only text for the empty language is an infinite sequence of #'s. We let T, with possible subscripts and superscripts, range over texts. $T[n]$ denotes the finite initial segment of T with length n, that is $T[n]$ is $T(0)T(1)\ldots T(n-1)$. $\sigma \subset T$ denotes the fact that σ is an initial segment of T. Observe that in this case we have $\sigma = T[|\sigma|]$. Note that one can effectively produce a text for a language L, from its grammar in a given numbering. Canonical text for W_j (W_j^ψ) denotes such an effective text.

A learner is an algorithmic mapping from finite sequences to $\mathbb{N} \cup \{?\}$. Output of ? denotes the fact that the learner does not wish to issue a conjecture on the input. The elements of $\mathbb{N}$ in the output of a learner are interpreted as a grammar in some prefixed numbering (also called hypotheses space). M, with possible superscripts and subscripts, is intended to range over language learning machines. We say that $M(T)\downarrow$ iff there exists an i such that, for all but finitely many n, $M(T[n]) = i$. In this case we say that $M(T)\downarrow = i$; in the case that there is no such i we say that $M(T)\uparrow$.

We now give the formal definitions of explanatory (**TxtEx**) learning, finite (**TxtFin**) learning and behaviourally correct (**TxtBc**) learning.

Definition 3. [10,14,23] Suppose ψ is a numbering and let **I** be a variable ranging over the criteria **TxtEx**, **TxtFin** and **TxtBc** which are defined now.

(a) M **TxtEx**$_\psi$-*identifies a text* T just in case $(\exists i : W_i^\psi = \text{content}(T))$ $(\forall^\infty n)[M(T[n]) = i]$.

(b) M **TxtFin**$_\psi$-*identifies a text* T just in case $(\exists i : W_i^\psi = \text{content}(T))$ $(\exists n)[(\forall m \geq n)[M(T[m]) = i] \text{ and } (\forall m < n)[M(T[m]) = ?]]$.

(c) M **TxtBc**$_\psi$-*identifies a text* T just in case $(\forall^\infty n)[W_{M(T[n])}^\psi = \text{content}(T)]$.

(d) M **I**$_\psi$-*identifies an r.e. language* L (written: $L \in \mathbf{I}_\psi(M)$) just in case M **I**$_\psi$-identifies each text for L.

(e) M **I**$_\psi$-*identifies a class* $\mathcal{L}$ *of r.e. languages* (written: $\mathcal{L} \subseteq \mathbf{I}_\psi(M)$) just in case M **I**$_\psi$-identifies each language from $\mathcal{L}$.

(f) $\mathbf{I}_\psi = \{\mathcal{L} \subseteq \mathcal{E} : (\exists M)[\mathcal{L} \subseteq \mathbf{I}_\psi(M)]\}$ and $\mathbf{I} = \bigcup_\psi \mathbf{I}_\psi$.

Note that parts (d), (e) and (f) are not specific to $\mathbf{I} \in \{\mathbf{TxtEx}, \mathbf{TxtFin}, \mathbf{TxtBc}\}$ but also done for other learning criteria introduced later. Furthermore, as φ is acceptable numbering, it holds for all numberings ψ that $\mathbf{TxtEx}_\psi \subseteq \mathbf{TxtEx}_\varphi$, $\mathbf{TxtFin}_\psi \subseteq \mathbf{TxtFin}_\varphi$ and $\mathbf{TxtBc}_\psi \subseteq \mathbf{TxtBc}_\varphi$. Thus, $\mathbf{I} = \mathbf{I}_\varphi$ for $\mathbf{I} \in \{\mathbf{TxtEx}, \mathbf{TxtBc}, \mathbf{TxtFin}\}$. For this reason, we often use the notation **I**-identification for $\mathbf{I}_\varphi$-identification.

Blum and Blum [5] introduced the notion of locking sequences and Fulk [13] generalized this notion to stabilizing sequences. We use these notions often in our proofs.

Definition 4. (a) [13] We say that σ is a **TxtEx**-*stabilizing sequence* for a learner M on a set L iff $\sigma \in \text{Seg}(L)$ and $M(\sigma\tau) = M(\sigma)$ for all $\tau \in \text{Seg}(L)$.

(b) [5] σ is called a **TxtEx**$_\psi$-*locking sequence* for M on L iff σ is a stabilizing sequence for M on L and $W_{M(\sigma)}^\psi = L$.

Lemma 5. [5] *Suppose* M **TxtEx**$_\psi$-*identifies* L. *Then,*

(a) *there exists a* **TxtEx**$_\psi$-*locking sequence for* M *on* L;

(b) *for every* $\sigma \in \text{Seg}(L)$, *there exists a* $\tau \in \text{Seg}(L)$ *such that* $\sigma\tau$ *is a* **TxtEx**$_\psi$-*locking sequence for* M *on* L;

(c) *every* **TxtEx**-*stabilizing sequence* σ *for* M *on* L *is also a* **TxtEx**$_\psi$-*locking sequence for* M *on* L.

Note that the definitions for stabilizing and locking sequence, as well as Lemma 5, can be generalized to other learning criteria such as **TxtBc**. We often omit the term like "**TxtEx**$_\psi$" from **TxtEx**$_\psi$-locking (stabilizing) sequence, when it is clear from context.

We assume some fixed 1–1 ordering of all the finite sequences, $\sigma_0, \sigma_1, \ldots$; thus, one can talk about the least stabilizing sequence and so on.

Definition 6. (a) [5] M is *order independent* iff for all texts T, if $M(T)\downarrow = i$, then for all T' such that content(T') = content(T), $M(T')\downarrow = i$.

(b) [13,25] M is *rearrangement independent* iff for all σ and τ such that content(σ) = content(τ) and $|\sigma| = |\tau|$, $M(\sigma) = M(\tau)$.

Given any learner M, one can construct a learner M' such that **TxtEx**$(M) \subseteq$ **TxtEx**(M') and M' is rearrangement and order independent [5,13].

In this paper we are mainly interested in learnability in Friedberg numberings and Ke-numberings. To this end, for any learning criterion $\mathbf{I}$, we let **FrI** denote the union of $\mathbf{I}_\psi$, where ψ is a Friedberg numbering and let **KeI** denote the union of $\mathbf{I}_\psi$, where ψ is a Ke-numbering.

3 Ke-Numberings and Friedberg Numberings

In this section, some basic learnability properties are established for Ke-numberings and Friedberg numberings. The next result shows that there are quite natural examples of Ke-numberings:

Proposition 7. *If* ψ *is a universal numbering such that every infinite r.e. language has only one* ψ-*grammar, then* ψ *is a Ke-numbering.*

Proof. Given two different indices i, j, search with help of the oracle K until an x is found such that one of the following conditions hold:

– $x \in W_i^\psi \triangle W_j^\psi$;
– $(\forall y \in W_i^\psi \cup W_j^\psi)[y \leq x]$.

The search terminates as either the two sets are different or both are finite and equal. Having determined x,

$$W_i = W_j \Leftrightarrow W_i \cap \{0, 1, \ldots, x\} = W_j \cap \{0, 1, \ldots, x\}.$$

The above can be checked using the oracle K. $\square$

Theorem 8. *Suppose* ψ *is a Ke-numbering. Then, there exists a Friedberg numbering* η *such that* $\psi \leq^K \eta$ *and* $\eta \leq^K \psi$.

Proof. We use a construction similar to that of Kummer [20, pages 29–30]. Let ψ be a Ke-numbering. There is a recursive $\{0, 1\}$-valued function F such that

- $F(i,0) = 0$ for all i;
- $(\forall^\infty t)\,[F(i,t) = 1]$ iff $(\forall j < i)\,[W_j^\psi \neq W_i^\psi]$ and $(\exists x)\,[x + 1 \in W_i^\psi \wedge x \notin W_i^\psi]$;

Now let

$$W_0^\eta = \mathbb{N};$$

$$W_{\langle i,t\rangle+1}^\eta = \begin{cases} W_i^\psi & \text{if } F(i,t) = 0 \text{ and } F(i,s) = 1 \text{ for all } s > t; \\ \{x : x < \langle i, t-1\rangle\} & \text{if } F(i,t) = 1; \\ \{x : x < \langle i, s-1\rangle\} & \text{if } s \text{ is the least number with} \\ & s > t \text{ and } F(i,t) = F(i,s) = 0. \end{cases}$$

Intuitively, for i being the minimal ψ-grammar for an r.e. language not in $\{\mathbb{N}\}\cup\mathcal{I}$, $\langle i,t\rangle + 1$ is the (only) η-grammar for W_i^ψ, where t is the unique number such that $F(i,t) = 0$ and $F(i,s) = 1$ for all $s > t$. All the other η-grammars are for languages in $\{\mathbb{N}\}\cup\mathcal{I}$, where one makes sure that there is exactly one η-grammar for each of these languages.

It is easy to verify that η is a Friedberg numbering. Moreover, $W_j^\psi = W_r^\eta$ can be checked using oracle K as follows. As ψ is a Ke-numbering, one can find using the oracle K the minimal i with $W_j^\psi = W_i^\psi$. Then $W_i^\psi = W_r^\eta$ iff one of the following four conditions holds:

- $W_i^\psi = \mathbb{N}$ and $r = 0$;
- $r = \langle k,t\rangle + 1$, $F(k,t) = 0$, $k = i$ and for all $s > t$, $F(i,t) = 1$;
- $r = \langle k,t\rangle + 1$, $F(k,t) = 1$ and $W_i^\psi = \{x : x < \langle i, t-1\rangle\}$;
- $r = \langle k,t\rangle + 1$, $F(k,t) = 0$, $s = \min(\{u > t : F(k,u) = 0\})$ exists and $W_i^\psi = \{x : x < \langle k, s-1\rangle\}$.

The k and t in the last three conditions are computed from r, thus not quantified. Hence each of the above conditions can be determined K-recursively. It also follows that one can find, using oracle K, for any given j the corresponding r with $W_r^\eta = W_j^\psi$ and for any given r the minimal i with $W_i^\psi = W_r^\eta$. Thus, the theorem follows. □

Note that for Friedberg numberings, the grammar equivalence problem is recursive. Thus, **FrTxtBc** = **FrTxtEx**. Theorem 8 implies that **KeTxtEx** = **FrTxtEx** as indices can be translated in the limit from a given Ke-numbering to a chosen Friedberg numbering. Theorem 20 below shows that **TxtEx** = **KeTxtEx**; note that the proof is delayed to that place as the theorem actually shows a bit more than just **TxtEx** = **KeTxtEx**. These two results together give the following as our first result. Here note that, for function learning, Freivalds, Kinber and Wiehagen [12] showed that every explanatorily learnable class of recursive functions is learnable in some Friedberg numbering.

Theorem 9. TxtEx ⊆ FrTxtEx.

4   Finite Learning

Freivalds, Kinber and Wiehagen [12] showed that in the context of learning recursive functions, every finitely learnable class of recursive functions can be learnt in some Friedberg numbering. In contrast, our next result shows that for **TxtFin**, requiring learning in some Friedberg numbering is restrictive. Note that the following result holds, even if one considers learnability of only infinite languages (which can be proved by easy cylinderification of the languages in the class considered in the following proof).

Theorem 10. TxtFin $\not\subseteq$ FrTxtFin.

Proof. Let $\mathcal{L} = \{L : (\forall x \in L)[W_x = L]\}$. Clearly, $\mathcal{L} \in$ **TxtFin**. Suppose by way of contradiction that M **TxtFin**-identifies $\mathcal{L}$ in Friedberg numbering ψ. Without loss of generality assume that M does not output more than one conjecture on any text. Then, by Smullyan's double recursion theorem [24], there exist distinct e_1, e_2 such that W_{e_1}, W_{e_2} may be defined as follows.

$W_{e_1} = W_{e_2} = \{e_1, e_2\}$, if there exist τ_1, τ_2 such that content$(\tau_i) \subseteq \{e_i\}$, $M(\tau_1)\downarrow \neq?, M(\tau_2)\downarrow \neq?$ and $M(\tau_1)\downarrow \neq M(\tau_2)\downarrow$; otherwise, $W_{e_i} = \{e_i\}$. It is easy to verify that $W_{e_i} \in \mathcal{L}$. Furthermore, if for some p, M outputs either ? or p, on all sequences in $Seg(\{e_1\}) \cup Seg(\{e_2\})$, then clearly $W_{e_1} \neq W_{e_2}$ and thus M does not **TxtFin**$_\psi$-identify $\mathcal{L}$. On the other hand, if there exist τ_1, τ_2 such that $\tau_i \in Seg(\{e_i\})$, $M(\tau_1)\downarrow \neq?, M(\tau_2)\downarrow \neq?$, and $M(\tau_1)\downarrow \neq M(\tau_2)\downarrow$, then $W_{e_1} = W_{e_2}$ and M does not **TxtFin**$_\psi$-identify $\mathcal{L}$ (as ψ is a Friedberg numbering). In either case, M does not **TxtFin**$_\psi$-identify $\mathcal{L}$. $\square$

A learner is prudent [22] if it only outputs grammars (in a given numbering used as hypotheses space) for the languages it learns (according to a given criterion). We denote prudent learning by attaching "**Prudent**" to the name of the criteria. One can strengthen the above proof to show that **PrudentTxtFin** $\not\subseteq$ **FrTxtFin**. This can be done by using $\mathcal{L} = \{W_{e_1(M)}, W_{e_2(M)} : M$ is a learning machine$\}$, where $e_1(M)$ and $e_2(M)$ denote the values of e_1 and e_2 as in the proof above, obtained effectively from the learner M.

Remark 11. In contrast to Theorem 10, one can show that several natural classes are finitely learnable in Friedberg numberings. The main idea is to use the even indices to provide a one-one numbering of a natural class of sets and to use the odd indices to make a Friedberg numbering of all remaining r.e. sets. Hence, for every $n \in \mathbb{N}$, $\{S : card(S) = n\} \in$ **FrTxtFin**. Furthermore, $\{\{\langle i, j \rangle : j \in \mathbb{N}\} : i \in \mathbb{N}\} \in$ **FrTxtFin**. Another natural class in **FrTxtFin** is $\{S : (\exists i) [S \subseteq \{\langle i, j \rangle : j \in \mathbb{N}\}$ and $card(S) = f(i)]\}$ for some recursive function f where only non-empty sets S are considered.

Our next result gives a characterization of **FrTxtFin**-learning for uniformly recursively enumerable classes.

Theorem 12. *A recursively enumerable class is in* **FrTxtFin** *iff it is 1–1 recursively enumerable and in* **TxtFin**.

Proof. Suppose $\mathcal{L}$ is r.e. and $\mathcal{L} \in$ **FrTxtFin**. Let M and Friedberg numbering ψ be such that $\mathcal{L} \subseteq$ **TxtFin**$_\psi(M)$. If $\mathcal{L}$ is finite, then the theorem immediately follows. So assume $\mathcal{L}$ is infinite. Let red be a recursive function such that $W_i^\psi =$

$W_{red(i)}$, for all i. Let $S = \{red(i) : (\exists L \in \mathcal{L})(\exists \sigma \in Seg(L))[M(\sigma) = i]\}$. Let $h(j)$ denote the $(j + 1)$-th element in some 1–1 enumeration of S. It is easy to verify that h witnesses that $\mathcal{L}$ is 1–1 recursively enumerable.

Now suppose $\mathcal{L}$ is 1–1 recursively enumerable and $\mathcal{L} \in \mathbf{TxtFin}$ as witnessed by M. Without loss of generality assume $\mathcal{L}$ is infinite. Let h be such that $\mathcal{L} = \{W_{h(i)} : i \in \mathbb{N}\}$ and, for all different i, j, $W_{h(i)} \neq W_{h(j)}$. Without loss of generality assume that M only outputs conjectures of form $h(j)$ on any input (whether from or outside the class $\mathcal{L}$).

Before defining the numbering ψ, we need to introduce an auxiliary function F which converges to 1 on minimal indices of non-members of $\mathcal{L} \cup \mathcal{I} \cup \{\mathbb{N}\}$ and outputs infinitely many zeroes on other inputs. More precisely, there is a $\{0, 1\}$-valued recursive function F satisfying the following requirements:

- $F(i, 0) = 0$ for all i;
- $(\forall^\infty t)\,[F(i, t) = 1]$ iff $(\forall j < i)\,[W_j \neq W_i]$ and $(\exists x)\,[x + 1 \in W_i \wedge x \notin W_i]$ and either $(\forall \sigma \in Seg(W_i))\,[M(\sigma) = ?]$ or $(\exists \sigma \in Seg(W_i))\,[M(\sigma) \neq ? \wedge W_{M(\sigma)} \neq W_i]$.

It is easy to verify that the second condition is a Σ_2 condition. Hence such a function F exists. Now the numbering ψ is defined as follows.

- $W_{3e}^{\psi} = W_{h(e)}$.
- $W_{3\langle i,t\rangle+1}^{\psi} = W_i$, if $F(i, t) = 0$ and for all $s > t$, $F(i, s) = 1$. Otherwise, $W_{3\langle i,t\rangle+1}^{\psi}$ will be spoiled and becomes some set from $\mathcal{I}$ not assigned to any other value.
- W_{3e+2}^{ψ} is either $\mathbb{N}$ or a member of $\mathcal{I}$.

We assume that the W_{3e+1}^{ψ} which are spoiled and W_{3e+2}^{ψ} together enumerate $\mathcal{I} \cup \{\mathbb{N}\}$ in 1–1 fashion (except for the unique element of $\mathcal{I} \cup \{\mathbb{N}\}$, if any, which belongs to $\mathcal{L}$).

It is now easy to verify that ψ is a Friedberg numbering and one can $\mathbf{TxtFin}_\psi$-identify $\mathcal{L}$ by outputting $3h(e)$, whenever M outputs $h(e)$. $\qquad\square$

The above does not give a characterization of $\mathbf{FrTxtFin}$, as the following theorem shows that there does exist a class in $\mathbf{FrTxtFin}$ which is not contained in any $\mathbf{TxtFin}$-learnable recursively enumerable class.

Theorem 13. *There exists a class $\mathcal{L} \in \mathbf{FrTxtFin}$ which is not contained in any r.e. class in $\mathbf{TxtFin}$.*

In contrast to this, finite learning is preserved when all Ke-numberings are permitted as hypotheses spaces.

Theorem 14. $\mathbf{TxtFin} \subseteq \mathbf{KeTxtFin}$.

5   Explanatory Learning with Additional Constraints

A learner is said to be *confident* [22] if it converges on all input texts, irrespective of whether the text is for a language in the class to be learnt or not. We denote confident learning by attaching "**Conf**" to the name of the criteria. The following theorem shows that confident learning in some Friedberg numbering can be achieved for every confident learnable class.

Theorem 15. ConfTxtEx = ConfFrTxtEx.

Even though every class which is Confidently learnable can be learnt in Friedberg numberings, there is still a subtle difference between learning in Friedberg numberings and acceptable numberings.

Remark 16. Let $\mathcal{L}_1 = \{L : L \neq \emptyset$ and $W_{\min(L)} = L\}$. Let $\mathcal{L}_2 = \{L : card(L) \geq 2$ and $W_{\min_2(L)} = L\}$, where $\min_2(L)$ denotes the second least element of L, if any. It is easy to see that both $\mathcal{L}_1$ and $\mathcal{L}_2$ are in **ConfTxtEx**. However, $\mathcal{L}_1 \cup \mathcal{L}_2 \notin \textbf{TxtEx}$ as can be shown by using the idea of the proof of Case [9] that **TxtFEx$_2$** $\not\subseteq$ **TxtEx** (here **TxtFEx$_2$** learning allows a learner to eventually vacillate among up to 2 grammars for the language being learnt — we refer the reader to [9] for details). So **ConfTxtEx** is not closed under union for acceptable numberings.

However, confident learning is closed under union, if a Friedberg numbering or Ke-numbering is used.

Proposition 17. *Let ψ be a Ke-numbering and $\mathcal{L}_1, \mathcal{L}_2 \in$ **ConfTxtEx$_\psi$**. Then $\mathcal{L}_1 \cup \mathcal{L}_2 \in$ **ConfTxtEx$_\psi$**.*

In contrast to confidence, several other properties do not preserve their full learning power when using Friedberg numberings instead of Gödel numberings as hypotheses spaces.

A learner is said to be *U-shaped on L* (see [3,7,8]), if on some text T for L, for some n, m, k with $n < m < k$, $M(T[n])$ and $M(T[k])$ are grammars for L (in the numbering being used as hypotheses space), but $M(T[m])$ is not a grammar for L. A learner is said to be *non U-shaped on L* if it is not U-shaped on L. A learner **NUShI**-identifies a class $\mathcal{L}$ if it **I**-identifies $\mathcal{L}$ and is non U-shaped on each $L \in \mathcal{L}$.

The following theorem shows that even simple classes such as $\mathcal{F}$ fail to be **NUShTxtEx**-identified in Friedberg numberings.

Theorem 18. $\mathcal{F} \notin$ **NUShFrTxtEx**.

Conservative learning [2,30] requires that a learner does not abandon a hypothesis which is consistent with the input seen so far. Strong monotonicity [18] is a requirement that learners always output larger and larger hypothesis: for all texts T and m, n with $m < n$, $W^\psi_{M(T[m])} \subseteq W^\psi_{M(T[n])}$ (where ψ is the numbering used as hypotheses space).

Theorem 19. *The class $\mathcal{F}$ of all finite sets is not conservatively, prudently or strong monotonically learnable in Friedberg numberings.*

However, prudence is not restrictive for Ke-numberings.

Theorem 20. TxtEx $\subseteq$ PrudentKeTxtEx.

Proof. Suppose a **TxtEx**-learner M is given. Without loss of generality assume that either M **TxtEx**-identifies $\mathbb{N}$ or M **TxtEx**-identifies each member of $\mathcal{I}$ (see [13]).

Let $F(\cdot, \cdot)$ be a recursive function such that $\lim_{t \to \infty} F(i, t)$ converges to σ, if σ is the least stabilizing sequence for M on W_i; $\lim_{t \to \infty} F(i, t)$ does not converge, if there exists no such σ.

Let $G(\cdot, \cdot)$ be a recursive function such that $\lim_{t \to \infty} G(i, t)$ converges to 1 iff i is the least φ-grammar for W_i; $\lim_{t \to \infty} G(i, t)$ does not converge if i is not the least φ-grammar for W_i.

Note that there exist such F and G. Let $Y = \mathbb{N}$ if M **TxtEx**-identifies $\mathbb{N}$. Otherwise, $Y = \emptyset$. Thus, M **TxtEx**-identifies $Y \cup S$, for each $S \in \mathcal{I}$.

$W^{\psi}_{2\langle j,m,t\rangle} = W_j$, if the following properties hold for all $s \in \mathbb{N}$:

- $M(\sigma_m) = j$;
- if $s = t - 1$, then $F(j, s) \neq F(j, t)$;
- if $s \geq t$, then $F(j, s) = \sigma_m$.

Otherwise, $W^{\psi}_{2\langle j,m,t\rangle} = Y \cup \{x : x < s\}$ for the least s where one of the above properties fails.

Intuitively, the above properties checked if $M(\sigma_m) = W_j$, σ_m is the least stabilizing sequence for M on W_j and t is the convergence point for $F(j, \cdot)$.

Let $W^{\psi}_{2\langle j,m,t\rangle+1} = W_j$, if the following properties hold for all $s \in \mathbb{N}$:

- if $s = t - 1$, then $G(i, s) = 0$;
- if $s \geq t$, then $G(i, s) = 1$;
- if $m = 0$, then there exists an $s' > s$ such that $F(i, s') \neq F(i, s)$;
- if $m = \langle v, w\rangle + 1 \wedge s = v - 1$, then $F(i, s) \neq F(i, v)$;
- if $m = \langle v, w\rangle + 1 \wedge s > v$, then $F(i, s) = F(i, v)$;
- if $m = \langle v, w\rangle + 1$, then there is an $s' \geq s$ such that $[w = \min(W_{M(F(i,v)),s'} \triangle W_{j,s'})]$.

Otherwise, $W^{\psi}_{2\langle j,m,t\rangle+1} = W_{j,s}$, for the least s for which one of the above properties fails.

Intuitively, first two properties above check if $G(i, \cdot)$ converges to 1, with t being the convergence point for $G(i, \cdot)$. Third property checks, for $m = 0$, whether $F(i, \cdot)$ diverges. Fourth and sixth properties check, for $m = \langle v, w\rangle + 1$, whether v is the convergence point for $F(i, \cdot)$ and $w = \min(W_{M(F(i,v))} \triangle W_j)$.

Claim. (a) If M has a least stabilizing sequence on L which is also a locking sequence for M on L, then $2\langle j, m, t\rangle$ is a ψ-grammar for L, where $M(\sigma_m) = j$, and σ_m is the least stabilizing sequence for M on L and t is the convergence point for $F(j, \cdot)$.

(b) ψ is a universal numbering (though not acceptable).

(c) every infinite recursively enumerable language L, except possibly for $\mathbb{N}$, has exactly one ψ-grammar.

(d) $\mathbb{N}$ has exactly one ψ-grammar, except possibly for grammars of the form $2\langle j, m, t\rangle$ which eventually follow the otherwise clause in the definition of W^ψ above.

(e) M has a least stabilizing sequence for each W^ψ_{2i} which is also a locking sequence for M on W^ψ_{2i}.

We now prove the claim and then continue with the main proof.

Part (a) follows from the definition of $W^\psi_{2\langle j,m,t\rangle}$.

For (b), suppose L is r.e., If M has a least stabilizing sequence on L, which is also a locking sequence for M on L, then part (a) gives a ψ-grammar for L.

Otherwise, let i be the least φ-grammar for L. Let t be the convergence point for $G(i, \cdot)$. If M does not have a least stabilizing sequence on L, then $2\langle i, 0, t\rangle + 1$ is the ψ-grammar for L. Otherwise, let v be the convergence point of $F(i, \cdot)$. Let $w = \min(W_{M(\sigma)} \triangle W_j)$, where $\sigma = F(i, v)$. Then, $2\langle i, \langle v, w\rangle + 1, t\rangle + 1$ is a ψ-grammar for L.

For (c) note that if M has a least stabilizing sequence on L, which is also a locking sequence for M on L, then the proof of part (a) gives the only ψ-grammar for L. Otherwise the proof of part (b) gives the only ψ-grammar for L.

Part (d) can be proved similarly to part (c).

Part (e) follows directly from the definition of $W^\psi_{2\langle j,m,t\rangle}$: either σ_m is the least stabilizing sequence for M on W_j with t being convergence point for $F(i, \cdot)$ and $M(\sigma_m) = j$ (thus, $W^\psi_{2\langle j,m,t\rangle} = W_j$) or $W^\psi_{2\langle j,m,t\rangle} = Y \cup S$ for some $S \in \mathcal{I}$. Hence, (e) holds.

This completes the proof of the claim. Note that either all ψ-grammars $2\langle j, m, t\rangle$ which follow the otherwise clause in the definition are grammars for $\mathbb{N}$, or all of these ψ-grammars are for finite sets. Thus, essentially Proposition 7 can be used to show that ψ is Ke-numbering. Using part (a) and (e) of the claim, prudent learning of $\mathbf{TxtEx}(M)$ follows easily as, on input σ, a learner can search for the least t and m such that the following two conditions hold:

- $M(\sigma_m) = M(\sigma_m \tau)$ for all τ such that $|\tau| \le |\sigma|$ and $\tau \in Seg(\text{content}(\sigma))$,
- for all t' such that $t \le t' \le |\sigma|$, $F(M(\sigma_m), t') = \sigma_m$.

If t and m are found, then the learner outputs $2\langle M(\sigma_m), m, t\rangle$, else the learner outputs 0. Note that learner only uses grammars of form $2i$. It is easy to verify that M learns all languages of form $W^\psi_{2\langle j,m,t\rangle}$ (which, by part (a) of the above claim, includes all languages $\mathbf{TxtEx}$-identified by M). Thus, M is a prudent learner. $\qquad\square$

Similar proofs can be used to show that non U-shaped learning and conservativeness are not restrictive for Ke-numberings.

Theorem 21. TxtEx $\subseteq$ NUShKeTxtEx.

Theorem 22. *Every class which can be conservatively* **TxtEx** *learnt can be conservatively learnt in some Ke-numbering.*

Remark 23. An iterative learner [26,27] does not remember its history, but bases its conjecture on just the latest input and its previous conjecture. It can be shown that $\mathcal{F}$ cannot be iteratively learnt in any Friedberg numbering. It is open at present whether every iteratively **TxtEx**-learnable class can be learnt iteratively in some Ke-numbering.

A learner is said to be *consistent* [1,4,29] if for all σ, content(σ) $\subseteq W^{\psi}_{M(\sigma)}$, where ψ is the numbering used for hypotheses space. There have been three different versions of consistency studied in the literature. The notion considered here is often referred to as **TCons** (see [29]) where the "T" indicates that the learner has to be consistent on all total functions. **RCons** (see [19]) refers to consistent learning when the learners are total, but may not be consistent on inputs outside the class. In **Cons** learning (see [4]) the requirement is further relaxed to allow the learners to be partial: the learner may be defined and consistent only on inputs from the class being learnt. Theorem 24 can be extended to **Cons** too. We do not yet know if the result extends to **RCons**.

Theorem 24. *Every consistently learnable class can be learnt consistently in some Friedberg numbering.*

6   Learning with Respect to a Fixed Friedberg Numbering

We now investigate how powerful it is to learn with respect to one fixed Friedberg numbering. While **TxtEx** = **TxtEx**$_\varphi$ for every acceptable numbering φ, there is no optimal Friedberg numbering in this sense. This result can also be shown using the result of [12] that for every Friedberg numbering η (for partial functions), one can find an explanatory learnable class of functions, which is not explanatory learnable using η as hypothesis space. Theorem 26 and Remark 27 below show that there is an adversary Friedberg numbering ψ such that **TxtEx**$_\psi \subseteq$ **TxtEx**$_\eta$ for every universal numbering η. This is language learning counterpart of the result from [12] that, for function learning, there exists a Friedberg numbering in which only finite classes of recursive functions can be learnt.

Proposition 25. *Let η be a Ke-numbering and $\mathcal{L}_1, \mathcal{L}_2$ be as in Remark 16. Then either $\mathcal{L}_1 \notin$ **TxtEx**$_\eta$ or $\mathcal{L}_2 \notin$ **TxtEx**$_\eta$. In particular, **TxtEx** $\neq$ **TxtEx**$_\eta$.*

Theorem 26. *There exists a Friedberg numbering ψ such that every class in **TxtEx**$_\psi$ contains only finitely many infinite languages.*

Remark 27. If $\mathcal{L}$ is a **TxtEx**-learnable class containing only finitely many infinite languages, then $\mathcal{L}$ is in **TxtEx**$_\eta$ for every universal numbering η.

7   Behaviourally Correct Learning and Its Variants

TxtFEx-learning [9] denotes **TxtBc**-learning with the additional constraint that the learner outputs only finitely many distinct conjectures on a text for an input language from the class to be learnt.

Theorem 28. TxtFEx ⊆ KeTxtBc.

Note that **FrTxtBc = FrTxtFEx = FrTxtEx** and **KeTxtFEx = KeTxtEx**. As a corollary to Theorem 28, we obtain **FrTxtBc ⊂ KeTxtBc**.

Theorem 29. NUShKeTxtBc ⊂ KeTxtBc.

Theorem 30. NUShKeTxtBc ⊂ NUShTxtBc.

Acknowledgements. We thank S. Goncharov, C. Jockusch, B. Khoussainov, M. Kummer, S. Lempp, R. Wiehagen and S. Zilles for correspondence.

References

1. Angluin, D.: Finding patterns common to a set of strings. Journal of Computer and System Sciences 21, 46–62 (1980)
2. Angluin, D.: Inductive inference of formal languages from positive data. Information and Control 45, 117–135 (1980)
3. Baliga, G., Case, J., Merkle, W., Stephan, F., Wiehagen, R.: When unlearning helps. Technical Report Technical Report TRA5/06, School of Computing, National University of Singapore (2005)
4. Bārzdin, J.: Inductive inference of automata, functions and programs. In: International Congress of Mathematicians, Vancouver, pp. 771–776 (1974)
5. Blum, L., Blum, M.: Toward a mathematical theory of inductive inference. Information and Control 28, 125–155 (1975)
6. Blum, M.: A machine-independent theory of the complexity of recursive functions. Journal of the ACM 14, 322–336 (1967)
7. Carlucci, L., Case, J., Jain, S., Stephan, F.: U-shaped learning may be necessary. Journal of Computer and System Sciences (to appear)
8. Carlucci, L., Jain, S., Kinber, E., Stephan, F.: Variations on U-shaped learning. Information and Computation 204(8), 1264–1294 (2006)
9. Case, J.: The power of vacillation in language learning. SIAM Journal on Computing 28(6), 1941–1969 (1999)
10. Case, J., Lynes, C.: Machine inductive inference and language identification. In: Nielsen, M., Schmidt, E.M. (eds.) Proceedings of the 9th International Colloquium on Automata, Languages and Programming. LNCS, vol. 140, pp. 107–115. Springer, Heidelberg (1982)
11. Friedberg, R.: Three theorems on recursive enumeration. Journal of Symbolic Logic 23(3), 309–316 (1958)
12. Wiehagen, R., Kinber, E., Freivalds, R.: Inductive Inference and Computable One-one Numberings. Zeitschrift für mathematische Logik und Grundlagen der Mathematik 28, 463–479 (1982)
13. Fulk, M.: Prudence and other conditions on formal language learning. Information and Computation 85, 1–11 (1990)
14. Gold, E.M.: Language identification in the limit. Information and Control 10, 447–474 (1967)
15. Goncharov, S.: Nonequivalent constructivizations. In: Proceedings of the Mathematical Institute, Siberian Branch of Russian Academy of Sciences, Nauka, Novosibirsk (1982)

16. Jain, S., Osherson, D., Royer, J., Sharma, A.: Systems that Learn: An Introduction to Learning Theory, 2nd edn. MIT Press, Cambridge (1999)
17. Jain, S., Sharma, A.: Characterizing language learning in terms of computable numberings. Annals of Pure and Applied Logic 84(1), 51–72 (1997), Special issue on Asian Logic Conference (1993)
18. Jantke, K.-P.: Monotonic and non-monotonic inductive inference. New Generation Computing 8, 349–360 (1991)
19. Jantke, K.-P., Beick, H.-R.: Combining postulates of naturalness in inductive inference. Journal of Information Processing and Cybernetics (EIK) 17, 465–484 (1981)
20. Kummer, M.: Beiträge zur Theorie der Numerierungen: Eindeutige Numerierungen. PhD Thesis, Karlsruhe (1989)
21. Odifreddi, P.: Classical Recursion Theory, North-Holland, Amsterdam (1989)
22. Osherson, D., Stob, M., Weinstein, S.: Systems that Learn: An Introduction to Learning Theory for Cognitive and Computer Scientists. MIT Press, Cambridge (1986)
23. Osherson, D., Weinstein, S.: Criteria of language learning. Information and Control 52, 123–138 (1982)
24. Rogers, H.: Theory of Recursive Functions and Effective Computability. McGraw-Hill, New York (1967), Reprinted by MIT Press in (1987)
25. Schäfer-Richter, G.: Some results in the theory of effective program synthesis - learning by defective information. In: Bibel, W., Jantke, K.P. (eds.) Mathematical Methods of Specification and Synthesis of Software Systems 1985. LNCS, vol. 215, pp. 219–225. Springer, Heidelberg (1986)
26. Wexler, K., Culicover, P.W.: Formal Principles of Language Acquisition. MIT Press, Cambridge (1980)
27. Wiehagen, R.: Limes-Erkennung rekursiver Funktionen durch spezielle Strategien. Journal of Information Processing and Cybernetics (EIK) 12, 93–99 (1976)
28. Wiehagen, R.: A thesis in inductive inference. In: Dix, J., Schmitt, P.H., Jantke, K.P. (eds.) Nonmonotonic and Inductive Logic, 1st International Workshop. LNCS, vol. 543, pp. 184–207. Springer, Heidelberg (1991)
29. Wiehagen, R., Liepe, W.: Charakteristische Eigenschaften von erkennbaren Klassen rekursiver Funktionen. Journal of Information Processing and Cybernetics (EIK) 12, 421–438 (1976)
30. Zeugmann, T., Lange, S.: A guided tour across the boundaries of learning recursive languages. In: Lange, S., Jantke, K.P. (eds.) Algorithmic Learning for Knowledge-Based Systems. LNCS, vol. 961, pp. 190–258. Springer, Heidelberg (1995)

Separating Models of Learning with Faulty Teachers*

Vitaly Feldman, Shrenik Shah, and Neal Wadhwa

Harvard University,
Cambridge MA 02138, USA
vitaly@eecs.harvard.edu, {sshah, nwadhwa}@fas.harvard.edu

Abstract. We study the power of two models of faulty teachers in Angluin's exact learning model. The first model we consider is learning from equivalence and incomplete membership query oracles introduced by Angluin and Slonim [1]. In this model, the answers to a random subset of the learner's membership queries may be missing. The second model we consider is random persistent classification noise in membership queries introduced by Goldman et al. [2]. In this model, the answers to a random subset of the learner's membership queries are flipped.

We show that the incomplete membership query oracle is strictly stronger than the membership query oracle with persistent noise under the assumption that the problem of PAC learning parities with noise is intractable.

We also show that under the standard cryptographic assumptions the incomplete membership query oracle is strictly weaker than the perfect membership query oracle. This strengthens the result of Simon [3] and resolves an open question of Bshouty and Eiron [4].

Our techniques are based on ideas from coding theory and cryptography.

1 Introduction

Modeling and handling of faulty information is one of the most important and well-studied topics in learning theory. In this paper we study two natural models of a faulty teacher in Angluin's exact model of learning. In the first model the teacher answers "I don't know" with some probability p to every membership query of the learner. Furthermore, if the learner asks the same membership query again the answer will be the same (in other words, it *persists*). This model was introduced by Angluin and Slonim [1] and the faulty membership query oracle is referred to as *incomplete*. Angluin and Slonim showed that monotone DNF formulas are learnable even with incomplete membership queries for constant p. This result was improved by Bshouty and Eiron who gave an algorithm that can learn monotone DNF even when only an inverse polynomial fraction of membership queries is answered [4]. Bshouty and Owshanko showed learnability

---

* Supported by grants from the National Science Foundation NSF 0432037 and NSF 0427129.

M. Hutter, R.A. Servedio, and E. Takimoto (Eds.): ALT 2007, LNAI 4754, pp. 94–106, 2007.

of regular sets in this model [5], Goldman and Mathias showed learnability of k-term DNF [6], and Chen showed learnability of some restricted classes of DNF in this model [7]. Given a number of strong positive results for this model a natural question to ask is whether this model is equivalent to learning with perfect membership queries [4]. This question was addressed by Simon who answered it in the negative for exact learning with proper equivalence queries (that is the hypothesis in the equivalence query has to belong to the concept class that is learned) [3]. In this work (Theorem 3) we give a more general version of this result that also applies to unrestricted equivalence queries. Our result shows that if there exists a concept not learnable in the exact model, then learning with MQs is stronger then learning with incomplete MQs[1]. In particular, if one-way functions exist, then incomplete MQs are strictly weaker than perfect ones.

The other model of a faulty teacher we study is random persistent noise in membership queries defined by Goldman, Kearns and Schapire [2]. In this model, the teacher flips the label of the answer to every membership query with some probability p. As in the incomplete MQ model, the answers persist. It is easy to see that learning is this model is at least as hard as learning in the incomplete MQ model. Among the few techniques that manage to exploit noisy MQs is the result of Goldman et al. who prove that certain classes of read-once formulas are exactly learnable in this model [2]. It is also not hard to see that concept classes that are exactly learnable using Kushilevitz-Mansour algorithm [8] can be learned from noisy MQs by using noise tolerant versions of the Kushilevitz-Mansour algorithm given by Jackson et al. [9] and Feldman [10]. These classes include juntas and $\log n$-depth decision trees [8]. Learnability of monotone DNF in this model is an open problem [1].

In the main result of this work, we demonstrate that under the assumption that parities are not learnable with random noise, the incomplete membership query oracle is strictly stronger than the noisy one. Formally, we prove the following result.

Theorem 1. *If the problem of PAC learning parities from random and uniform examples with random classification noise of rate η is intractable then there exists a concept class C that for any polynomial $p(n)$ is learnable with equivalence and incomplete membership queries with error rate $1 - \frac{1}{p(n)}$, but not learnable from equivalence and noisy membership queries with error rate η.*

Learning of parities from noisy random and uniform examples (which we refer to as the *noisy parity problem*) is a notoriously hard open problem [11]. Feldman et al. show that this problem is central to PAC learning with respect to the uniform distribution by reducing a number of other well-known open problems to it [12]. Furthermore, it is known to be equivalent to decoding of binary linear codes generated randomly – a long-standing open problem in coding theory (cf. [10]). For example the McEliece cryptosystem is based, among other assumptions, on the hardness of this problem [13]. While the average-case hardness of decod-

---

[1] The main idea of this simple result is similar to that of Simon and we include it primarily for completeness.

ing binary linear codes is unknown, a number of related worst-case problems are known to be NP-hard (*cf.* [14,15]). Other evidence of its hardness includes non-learnability in the statistical query model of Kearns [16] and hardness of a generalized version of this problem that was shown by Regev [17]. The only known non-trivial algorithm for learning parities with noise is a $2^{O(n/\log n)}$-time algorithm by Blum *et al.* [18]. Our separation is optimal in the sense that it separates learning with a rate of "I don't know"s bounded by an inverse of a polynomial from 1, from learning with the constant rate of noise (or even less if under the corresponding strengthening of the assumption on the noisy parity problem). This separation is based on a way to convert learning with membership queries to learning from random and uniform examples via a suitable cryptographic primitive. We hope that this tool will find other applications.

1.1   Organization

We define the relevant models in Section 2. Separation of the incomplete membership query oracle from the usual one is presented in Section 3. Separation of learning with incomplete membership query oracle from the noisy one is presented in Section 4.

2   Preliminaries

The exact learning model was proposed by Angluin [19]. In her model, a learning algorithm is trying to identify a target concept $c : X \to \{0,1\}$ in concept class $\mathcal{C}$. X is called the instance space and in this work we will assume $X = \{0,1\}^n$. The learning algorithm has access to a membership oracle MQ and an equivalence oracle EQ. On a query to the MQ oracle, the learning algorithm submits a point $x \in X$ and is given the value of $c(x)$. On a query to EQ oracle, the algorithm submits an efficiently evaluable hypothesis h. If $h \equiv c$, then the response YES is returned. Otherwise, $x \in X$ such that $h(x) \neq c(x)$ is returned. Note that x like that can be chosen in an adversarial way.

Definition 1. *We say that a concept class $\mathcal{C}$ is efficiently exactly learnable from membership and equivalence queries if there exists a polynomial $p(\cdot,\cdot)$ and an algorithm A with access to a membership oracle and an equivalence oracle, such that for any target concept c, A outputs a hypothesis h in time $p(size(c),n)$ with $h(x) = c(x)$ for all $x \in X$.*

A variant of this model introduced by Angluin and Slonim [1] is exact learning with an incomplete membership oracle of probability p IMQ^p and an equivalence oracle EQ. This model addresses the fact that the teacher (modeled by MQ) might not be omniscient. Whenever IMQ^p is queried, it will give one of 3 responses $0, 1, \perp$. Before it is used, IMQ^p flips a biased coin that comes up heads with probability p for each point in the instance space X. If for a point x, the coin comes up heads, it replies with $\perp$ whenever x is queried in the future.

Otherwise, it will reply with the correct value of $c(x)$. The response $\perp$ corresponds to "I don't know". Note that it is possible that the oracle may not know the answer to any question asked of it. Therefore we only require a learning algorithm to succeed with high probability $1 - \Delta$ over the coin flips of IMQ^p for some negligible Δ. Learnability in this model is defined as follows.

Definition 2. *We say that a concept class C is exactly learnable ¿from equivalence queries and incomplete membership queries of rate p and if there exists an algorithm A such that for any target concept $c \in C$, any $0 < \delta < 1$ and any $p < 1$, with probability $1 - \Delta - \delta$ over the coin flips of IMQ^p and A, A outputs a hypothesis h equivalent to c. A is efficient if it runs in time polynomial in $size(c)$, n, $\frac{1}{1-p}$ and $\frac{1}{\delta}$.*

For example, the algorithm for learning monotone DNF of Bshouty and Eiron satisfies this definition [4]. We remark that this definition is different from the definition of Angluin and Slonim as they define learnability in a more restricted proper exact model and do not require polynomial dependence on $\frac{1}{1-p}$ [1].

Another model we explore in this paper is exact learning with a persistently noisy membership oracle and equivalence oracle. Prior to use, the oracle NMQ$^\eta$ goes through the entire instance space and flips a biased coin that comes up heads with probability $\eta < \frac{1}{2}$ (the flips are independent). If the coin comes up heads on a point x, then the oracle will always return $\neg c(x)$ when queried for x. This oracle was introduced by Goldman *et al.* in the context of exact identification [2]. We refer to this model as the *persistent noise* model of exact learning.

Definition 3. *We say that a concept class C is exactly learnable from equivalence queries and persistently noisy membership if there exists an algorithm A that for every $c \in C$, $\delta < 1$ and $\eta < \frac{1}{2}$, with probability $1 - \Delta - \delta$ over the coin flips of NMQ$^\eta$ and A, A outputs a hypothesis $h \equiv c$. A is efficient if it runs in time polynomial in $size(c), n, \frac{1}{1-2\eta}$ and $\frac{1}{\delta}$.*

In addition to these models, we also consider malicious noise models introduced by Angluin *et al.* [20]. In these models, the oracle has control over which queries it corrupts but the total number of corruptions is limited. The learning algorithm is allowed to run in time polynomial time in the number of corrupted queries and the standard learning parameters.

3   Separation of Incomplete from Perfect MQ Models

In this section, we show that the incomplete model is strictly weaker than exact learning, using a cryptographic assumption. An analogue of this result for the proper exact learning was given Simon [3] and we include our version primarily for completeness.

First, we present a result regarding the malicious incomplete MQ model.

Theorem 2. *Suppose a concept class $C = \cup_n C_n$ with concepts of size bounded by $s(n)$ is not learnable with a perfect membership oracle and equivalence oracle. Then for error rate of $\geq s(n)$, the malicious incomplete model is strictly weaker than the equivalence and query model.*

Proof. We construct a concept class $\mathcal{D}$ as follows. The class $\mathcal{D}_{n+1}$ on $\{0,1\}^{n+1}$ contains elements d_c parameterized by $c \in \mathcal{C}$. Define $d_c(x, 0) = c(x)$, and $d_c(x, 1) = \phi_c(x)$, where ϕ_c is defined as follows. Split the input set $\{0,1\}^n$ into $\{0,1\}^{\log s(n)} \times \{0,1\}^{n-\log s(n)}$. Then define $\phi_c(i, j) = c_i$ if $j = 0^{n-\log s(n)}$ and $\phi_c(i, j) = 0$ if $j \neq 0^{n-\log s(n)}$, where c_i is the ith bit in the representation of c. Thus $\mathcal{D}$ is clearly learnable in the equivalence query and membership query model using only membership queries on $d_c(i, 0^{n-\log s(n)}, 1)$. However, in the malicious incomplete noise model, with $s(n)$ errors, one can simply hide the $s(n)$ outputs of the form $d_c(i, 0^{n-\log s(n)}, 1)$, and the algorithm is reduced to learning the original language in the usual equivalence and membership query model, which we assumed was not learnable. Note that the algorithm is given no extra time, since the concept size changed from $s(n)$ to $2s(n) + O(1)$, and the time allowed originally was polynomial in $s(n)$.

A simple modification of this idea allows us to separate the random incomplete membership query model from the usual exact learning model.

Theorem 3. *Suppose a concept class $\mathcal{C} = \cup_n \mathcal{C}_n$ is hard to exactly learn. For any polynomial $p(n)$, we can construct a class $\mathcal{F} = \cup_n \mathcal{F}_n$ that is exactly learnable, but is not exactly learnable with probability $1 - \delta$ from incomplete membership oracle $\text{IMQ}^{\frac{1}{p(n)}}$ and equivalence queries where $\delta = \Omega(2^{-n})$.*

Proof. Given a polynomial $p(n)$, we design a concept class $\mathcal{F} = \cup_n \mathcal{F}_n$ so that it is easy to learn with membership queries alone but not with equivalence queries and incomplete membership queries with probability $\frac{1}{p(n)}$ of failure, or, in the above notation, the oracles EQ and $\text{IMQ}^{\frac{1}{p(n)}}$. Let $\mathcal{C} = \cup_n \mathcal{C}_n$ be a concept class that is not learnable in the exact model, with some polynomial bound $s(n)$ on the size of the descriptions of the hypotheses. Let $c : \{0,1\}^n \to \{0,1\}$ be a concept in c. Also let c_i denote the ith bit of the description of c. We define $\mathcal{F}$ in the following manner. A concept $d_{c,u} \in \mathcal{F}_{n+1}$ is defined by a concept $c \in \mathcal{C}_n$ and data $u = \{u_{i,j}\}_{i \in [s(n)], j \in [t(n)]}$, $u_{i,j} \in \{0,1\}$, satisfying the condition $c_i = \bigoplus_{j=1}^{t(n)} u_{i,j}$, for some polynomial $t(n)$ we'll specify later. The concept $d_{c,u}$ maps $(x, 0) \mapsto c(x)$, $(x, 1) \mapsto \phi_u(x)$, where $\phi_u : \{0,1\}^{n-1} \to \{0,1\}$ is defined as follows. Partition $\{0,1\}^{n-1} = \{0,1\}^{\log s(n)} \times \{0,1\}^{\log t(n)} \times \{0,1\}^{n-1-\log s(n)t(n)}$, and write $\phi_u(i, j, k)$ instead of $\phi_u(x)$, where i, j, k are elements of each set in this product. Let $\phi_u(i, j, k) = 0$ if $k \neq 0^{n-\log s(n)t(n)}$, and $\phi_u(i, j, 0^{n-1-\log s(n)t(n)}) = u_{i,j}$. For every possible u meeting these conditions, there is a corresponding concept $d_{c,u} \in \mathcal{F}$.

This defines the concept class $\mathcal{F}$. We will first prove that it can be learned with membership queries alone, and then that it cannot be learned with equivalence queries and incomplete membership queries with success probability $\leq 1 - \frac{1}{p(n)}$. To see the first statement, the algorithm makes membership queries on $(i, j, 0^{n-\log s(n)t(n)}, 1)$ to find the values $u_{i,j}$ for all $1 \leq i \leq s(n)$ and $1 \leq j \leq t(n)$. Then, by computing $c_i = \bigoplus_{j=1}^{t(n)} u_{i,j}$, the algorithm can easily learn the concept c. This gives the algorithm an encoding of all the values of the target concept d

on inputs of the form $(x, 0)$, just by evaluating c. For inputs of the form $(x, 1)$, the algorithm just uses the value of u to compute this as well, as it is clearly efficiently evaluatable, and u only has polynomial length.

To see why this cannot be efficiently learned with access to EQ and $\text{IMQ}^{\frac{1}{p(n)}}$, we'll prove that learning in this model is equivalent to learning of $\mathcal{C}$, which we assumed to be hard. Let $\text{EQ}_{\mathcal{C}}$ be an equivalence query oracle for $\mathcal{C}$, such that with the responses of this oracle plus a membership oracle, the class $\mathcal{C}$ is hard to learn. Suppose that an algorithm A is trying to learn the concept $d_{c,u} \in \mathcal{F}$. Every time that A asks an equivalence query on some hypothesis h, we reason as follows. If $h(x, 0) \neq d_{c,u}(x, 0)$ for some value of x, we use $\text{EQ}_{\mathcal{C}}$ to tell us which counter example y to use, giving it the function $h(x, 0)$ as a hypothesis for guessing $c(x) := d_{c,u}(x, 0)$, and returning $(y, 0)$ to A. If A were to never query membership $(x, 1)$ for any x, we would be essentially done. This is because without such queries, A's behavior is exactly as if it were trying to learn $\mathcal{C}$. Also note the possibility that $h(x, 0) = d_{c,u}(x, 0) = c(x)$ for all x. In this case, then A has learned c, so it is irrelevant what response we give at this point. However, since $\mathcal{C}$ was hard to learn, this happens with low probability.

Unfortunately, however, A can indeed perform membership queries on $(x, 1)$. We will show, however, that with probability $1 - \delta$ this gives, information-theoretically, no information, where δ may be a constant or some function of n (it may, in fact, be exponentially small, as will be evident from the definition of $t(n)$ below). First note that querying $(i, j, k, 1)$ with $k \neq 0^{n - \log s(n) t(n)}$ provides no information, since it always returns 0 regardless of the concept. For some fixed value of i, the probability that there does not exist j such that a query of $(i, j, 0^{n - \log s(n) t(n)}, 1)$ is unavailable can be (lower) bounded by a Chernoff bound. Let X_j be the event that $(i, j, 0^{n - \log s(n) t(n)}, 1)$ is unavailable, which happens with probability $E[X_j] = \frac{1}{p(n)}$. Let $X = \sum_{j=1}^{t(n)} X_j$, and note that we just need $X \geq 1$ for such a j to exist. Define $t(n) = p(n)\left(1 + 3\log\frac{s(n)}{\delta}\right)$. By a multiplicative Chernoff bound, we have

$$
\begin{aligned}
\Pr[X < 1] &= \Pr\left[X < \frac{1}{p(n)} \cdot t(n)(1 - (1 - \frac{p(n)}{t(n)}))\right] \\
&\leq \exp\left(-\frac{t(n)}{2p(n)}(1 - \frac{p(n)}{t(n)})^2\right) = \exp\left(-\frac{1}{2p(n)t(n)}(t(n) - p(n))^2\right) \\
&= \exp\left(-\frac{9\log\frac{s(n)}{\delta}}{2(1 + 3\log\frac{s(n)}{\delta})}\right) \leq \exp\left(-\log\frac{s(n)}{\delta}\right) = \frac{\delta}{s(n)}
\end{aligned}
$$

Thus, by a union bound over the $s(n)$ values of i, there is a probability of at most δ that there exists some i where all of the values $u_{i,j}$ are available to the algorithm. If there is an unavailable $u_{i,j}$ for each i, since any proper subset of the $u_{i,j}$ for fixed i can attain any possible values without constraint, there is information-theoretically no information available from any membership query of the form $(x, 1)$, as desired. In other words, there are values for u that allow any value for the encoding of c that look identical to the algorithm A, so it can gain no information about c from the values it reads.

Thus the information that A can gain when trying to learn $\mathcal{F}$ is entirely from, except with probability δ, membership queries on $(x, 0)$ (which give no more information, and often less, than in the case of $\mathcal{C}$) and equivalence queries that are answered by the same oracle as for $\mathcal{C}$. Thus hardness of learning $\mathcal{C}$ implies hardness of learning $\mathcal{F}$, with a loss of δ (which can be taken to be exponentially small) in hardness.

This separation result is optimal in the following sense. For every concept class that is efficiently learnable in the exact model with perfect MQ is also efficiently exactly learnable with EQ and $\text{IMQ}^{\frac{1}{p(n)}}$ for sufficiently large polynomial p. This is true since for a low enough rate of "I don't know"s, with high probability, the learner will not encounter any of them in the answers to a polynomial number of membership queries.

Corollary 1. *If one-way functions exist, then learning from incomplete membership queries and equivalence queries is strictly harder than learning from membership queries and equivalence queries.*

Proof. Valiant observed that if one-way functions exist then polynomial size circuits are hard to learn in the PAC model with membership queries [21]. This implies that circuits are hard to learn with equivalence queries and membership queries, which gives us the desired result by Theorem 3.

4   Separation of Incomplete from Noisy MQ Models

We will now show that learning with noisy membership queries is strictly weaker than learning with incomplete membership queries. First note that if a concept class is exactly learnable with noisy membership queries and equivalence queries, then it can be learned with incomplete membership queries and equivalence queries. This follows from the fact that NMQ^{η} can be simulated using $\text{IMQ}^{2\eta}$ by returning the outcome of a fair coin whenever $\text{IMQ}^{2\eta}$ return "I don't know" and $c(x)$ otherwise (and giving the same label if the same query is made).

We will now show that there exists a concept class that is learnable in the incomplete model with an error rate of $1 - \frac{1}{poly(n)}$, but not in the noisy model with any constant rate of noise, using additional cryptographic assumptions. Specifically, we will assume that parities are not PAC learnable with respect to the uniform distribution in the presence of random classification noise. We start by providing several relevant definitions and key facts.

Definition 4. *A* noisy example oracle *for a function f with respect to a distribution $\mathcal{D}$ and noise rate η is the oracle that on each call, draws x according to $\mathcal{D}$, and returns $\langle x, f(x) \rangle$ with probability η and $\langle x, \neg f(x) \rangle$ with probability $1 - \eta$. We denote it by $EX_{\mathcal{D}}^{\eta}(f)$.*

A parity function $\chi_a(x)$ for a vector $a \in \{0,1\}^n$ is defined as $\chi_a(x) = a \cdot x = \sum_i x_i y_i \pmod 2$. We refer to the vector associated with a parity function as its *index*. We denote the concept class of parity functions $\{\chi_a \mid a \in \{0,1\}^n\}$ by PAR.

Definition 5. *The noisy parity problem for noise rate η is the problem of finding a given access to $EX_{\mathcal{U}}^{\eta}(\chi_a)$, where $\mathcal{U}$ is the uniform distribution.*

It is well-known that learning a parity with respect to $\mathcal{U}$ in the PAC sense (that is up to accuracy ϵ) is equivalent to finding its index (*cf.* [10]). Another simple observation made by Blum *et al.* [11] is that the noisy parity problem is randomly self-reducible. That is,

Lemma 1 ([11]). *Assume that there exists an efficient algorithm that can solve the noisy parity problem for noise rate η when the target parity belongs to subset S where, $|S|/2^n \geq 1/p(n)$ for some polynomial p. Then there exists an efficient (randomized) algorithm that can solve the noisy parity problem for noise rate η.*

Blum *et al.* also prove that if parities are not learnable efficiently then there exist pseudo-random generators [11]. Namely they prove the following result.

Lemma 2 ([11]). *Assume that there exists η such that noisy parity problem is intractable for noise rate η and $\frac{1}{1-H(\eta)} \leq p(n)$ for some polynomial p and binary entropy function H. Then there exist pseudo-random generators.*

In particular, by the result of Goldreich *et al.* intractability of the noisy parity problem implies existence of *pseudo-random function (PRF) families* [22] that will be a key part of our construction.

Definition 6. *A function family $F_{k,n} = \{\sigma_z\}_{z \in \{0,1\}^k}$ (where the key length is taken to be the security parameter and each σ_z is an efficiently evaluable function from $\{0,1\}^n$ to $\{0,1\}^n$) is a* pseudorandom function family *if any adversary M (whose resources are bounded by a polynomial in n and k) can distinguish between a function σ_z (where $z \in \{0,1\}^k$ is chosen randomly and kept secret) and a totally random function only with negligible probability. That is, for every probabilistic polynomial time M with an oracle access to a function from $\{0,1\}^n$ to $\{0,1\}^n$ and a negligible function $\nu(k)$,*

$$|\mathbf{Pr}[M^{\mathcal{F}_{k,n}}(1^n) = 1] - \mathbf{Pr}[M^{\mathcal{H}_n}(1^n) = 1]| \leq \nu(k),$$

where $\mathcal{F}_{k,n}$ is the random variable produced by choosing $\sigma_z \in F_{k,n}$ for a random and uniform $z \in \{0,1\}^k$ and $\mathcal{H}_n$ is the random variable produced by choosing randomly and uniformly a function from $\{0,1\}^n$ to $\{0,1\}^n$. The probability is taken over the random choice of $\mathcal{F}_{k,n}$ (or $\mathcal{H}_{k,n}$) and the coin flips of M.

The idea behind our separation is the following. It is easy to see that parities are learnable from "incomplete random examples", that is random examples where the learner does not get the label with some probability p. This is true since the learner can just ignore incomplete examples and only use the random examples with labels (which will still be random and uniform). Our goal is, in a sense, to transform membership queries into random examples. This is done by creating a function that maps x to $\sigma_z(x), \chi_a(\sigma_z(x))$ where σ_z is a function in a pseudorandom function family. Note that this function is not Boolean but can can be converted to a Boolean one via a simple trick. The problem with this construction is that in order to learn the given function, the learner would also need to learn σ_z (which is not possible since σ_z is a pseudorandom function). A way to avoid this problem is to have a encode an address in another part of the domain at which one can find the parameter z (one cannot just have $a = z$ since then the adversary could potentially use information about χ_z to "break" the pseudorandom function). One can use redundant encoding (or any other encoding that tolerates erasures) to make sure that the incomplete MQ will suffice to read $\sigma_z(x)$ and z (at location a). We are now ready to prove Theorem 1 which we restate more precisely here.

Theorem 4 ($= 1$). *If the noisy parity problem for noise rate η is intractable and $\frac{1}{1-H(\eta)}$ is upper-bounded by some polynomial, then there exists a concept class C that for any polynomial $p(n)$ is learnable with equivalence and incomplete membership queries with error rate $1 - \frac{1}{p(n)}$, but not learnable from equivalence and noisy membership queries with error rate η.*

Proof. We define the concept class $C = \cup_n C_n$ as follows. Let $F_{\frac{n}{2},\frac{n}{2}}$ be a pseudorandom family of functions whose existence is implied by Lemma 2. Let $a \in \{0,1\}^{\frac{n}{2}}$ and χ_a be the corresponding parity on $\frac{n}{2}$ variables. For each a and $z \in \{0,1\}^{n/2}$, define a function $c_{z,a} : \{0,1\}^n \to \{0,1\}$ as follows. We split the input x into 4 parts $b,y,j,$ and k where $b \in \{0,1\}$, $y \in \{0,1\}^{n/2}$, $k \in \{0,1\}^\ell$ for $\ell = \lceil \log(n/2 + 1) \rceil$ and $j \in \{0,1\}^{n/2-1-\ell}$. For $b = 0$ we encode a parity on pseudorandomly permuted points and for $b = 1$ we encode z, the secret key to a pseudorandom function family in a "hidden" location uncovering which requires knowing a. Formally,

$$c_{z,a}(0,y,j,k) = \begin{cases} \chi_a(\sigma_z(y)) & \text{if } j = k = 0 \\ k\text{-th bit of } \sigma_z(y) & \text{if } 1 \le k \le n/2 \\ 0 & \text{otherwise} \end{cases} \qquad (2)$$

$$c_{z,a}(1,y,j,k) = \begin{cases} k\text{-th bit of } z & \text{if } y = a \text{ and } 1 \le k \le n/2 \\ 0 & \text{otherwise} \end{cases}$$

Lemma 3. *The concept class C is learnable from $IMQ^{1-\frac{1}{p(n)}}$.*

Proof. The learning algorithm chooses $y \in \{0,1\}^{\frac{n}{2}}$ randomly and attempts to get $\sigma_z(y)$ by querying $c_{z,a}(0,y,0,0)$. Then for every $1 \le k \le n/2$, it finds the k-th bit of $\chi_a(\sigma_z(y))$ by querying $c_{z,a}(0,y,j,k)$ for $j = 0,1,\ldots,t$ where t is chosen to be large enough to make sure that at least one of the queries is answered with probability at least $1/n$. Each query is answered with probability $1/p(n)$ and therefore by Chernoff bound a polynomial t will suffice. This procedure allows us to find $\sigma_z(y)$ and $\chi_a(\sigma_z(y))$ with probability at least $\frac{1}{2p(n)}$. Repeating it polynomial in n and $p(n)$ times, with high probability, yields values of $\chi(a)$ on enough random and uniform points to recover a (via the usual Gaussian elimination). Given a the algorithm can query $c_{z,a}(1, a, j, k)$ for every $1 \leq k \leq n/2$ and a polynomial number j's to recover each bit of z with high probability. Given a and z the algorithm outputs $c_{z,a}$ (which is clearly an efficiently evaluable concept).

Lemma 4. *Under the assumption of Theorem 4, the concept class C is not learnable from EQ and NMQ^{η}.*

Proof. We claim that if C can be efficiently learned from EQ and NMQ^{η} by an algorithm L, then we can learn parities with noise η or distinguish a function in our PRF family from a randomly chosen function. The latter would also imply learning of parities with noise by Lemma 2 of Blum *et al.* [11]. Our distinguishing test M with oracle access to a function $\sigma : \{0,1\}^{n/2} \to \{0,1\}^{n/2}$ works as follows. Choose a random $a \in \{0,1\}^{n/2}$ and run the algorithm L for $\delta = 1/5$. When L asks for a membership query $x = (b, y, j, k)$ for $b = 0$ we compute $c_{z,a}(0, y, j, k)$ (as given by Equation 2) using σ in place of σ_z, and use randomness to simulate the noise. If L asks for a membership query for $b = 1$ then, if $y = a$ we stop and output 1. Otherwise we reply with 0. If L asks for an equivalence query, then we do one of two things depending on the hypothesis submitted. If L submits a hypothesis h for which $h(0, y, j, 1)$ equals to the first bit of $\sigma(y)$ for at least $3/4$ of randomly chosen y's and j's on a sample of polynomial size, then M stops and outputs 1. Otherwise, we return the counterexample $\langle 0, y', 0, 1, \sigma(y')_1 \rangle$ for a value of y' on which the submitted hypothesis is wrong. If L runs for more time than its promised polynomial upper bound, we return 0. We now claim that M returns 1 with probability $1 - \delta - \Delta$ when it has oracle access to σ_z randomly chosen from $\mathcal{F}_{\frac{n}{2},\frac{n}{2}}$. Note that the answers provided by the simulation are valid answers for $c_{z,a}$ until a membership query with $b = 1$ and $y = a$ is made at which point the distinguisher returns 1. If such a query is never asked then, with probability at least $1 - \Delta - \delta$, L has to correctly learn the concept $c_{z,a}$. Therefore M will return 1 when L asks its final equivalence query, since $\sigma_z(y)_1$ will be consistent with the corresponding values of the hypothesis for all y's. By the definition, Δ is negligible and we have that $\mathbf{Pr}[M^{\mathcal{F}_{\frac{n}{2},\frac{n}{2}}}(1^{n/2}) = 1] > 4/5 - \Delta > 3/4$.

Now let $\mathcal{H}_{\frac{n}{2}}$ be the uniform distribution over functions ¿from $\{0,1\}^{\frac{n}{2}}$ to $\{0,1\}^{\frac{n}{2}}$ and let σ' be a function randomly chosen according to this distribution (that is a truly random function). For a randomly chosen σ', L cannot hope to discover the first bit of $\sigma'(y)$ for $3/4$ of all the values of y in polynomial time. Thus to an equivalence query our simulator will return the value of $\sigma'(y)_1$ on a randomly chosen y where L does not predict $\sigma'(y)_1$ correctly (their fraction is at least $1/4$). Therefore, in fact, each equivalence query can be replaced by membership queries on $(0, y, j, 1)$ on enough j's to discover it with arbitrarily high probability (by Chernoff bound taking a majority over a polynomial in $\frac{1}{1-2\eta}$ number of point will suffice). We can therefore ignore equivalence queries altogether. Noisy membership queries for the first part of the domain ($b = 0$) can at best give us randomly chosen point $\sigma'(y)$ and a value of $\chi_a(\sigma'(y))$

corrupted by noise of rate η. For a truly random σ' this is equivalent to the noisy parity problem. All membership queries to the second part of the domain (that is $b = 1$) return value 0 unless $y = a$. If this happens with non-negligible probability then L has effectively managed to find a and solved the instance of the noisy parity problem we have generated with non-negligible probability over the choice of a and the coin flips of M. By Lemma 1, this implies that there is an efficient algorithm that solves the noisy parity problem for noise rate η, contradicting our assumption. This implies that, with probability close to 1, L cannot find a and hence M will not output 1 when used with σ' oracle. In particular, $\mathbf{Pr}[M^{\mathcal{H}_{\frac{n}{2}}}(1^{n/2}) = 1] \leq 1/4$. It is easy to see that if L is efficient then M is efficient and thus, we have obtained an efficient distinguisher for the PRF family of functions $F_{\frac{n}{2},\frac{n}{2}}$.

Theorem 4 can be also easily extended to PAC learning with respect to the uniform distribution. That is, we can show that C cannot be learned even approximately from correct random and uniform examples and persistently noisy membership queries.

Corollary 2. *Under the assumptions of Theorem 4 there exists a concept class learnable in the PAC model with incomplete membership queries model but not in the PAC model with persistently noisy membership queries of rate η, both with the uniform distribution.*

Proof. The concept class C is learnable in the PAC model with incomplete membership queries model, since this model is stronger than the equivalence and incomplete membership query model. To see that it cannot be learned in the PAC model with persistently noisy membership queries we observe that in the proof of Theorem 4 we required that L produce a hypothesis h for which $h(0, y, j, 1)$ equals to the first bit of $\sigma(y)$ for at least $3/4$ of randomly chosen y's and j's. There are $2^{n-\ell-1} \geq \frac{2^n}{2(n+2)}$ points of this form and therefore by setting $\epsilon = \frac{1}{8(n+1)}$ we force the learning algorithm to have the desired consistency condition. Simulating random examples that return correct labels is also easy since with very high probability they do not reveal any information that cannot be obtained using noisy membership queries. This is true since for a uniformly chosen $x = (b, y, j, k)$, the probability that $j = 0$ equals $2^{-n/2+\ell+1}$ and therefore random examples will not reveal any labels. Similarly probability that $y = a$ is $2^{-n/2}$ and therefore simulating random examples can be done without knowing the bits of z.

Setting $\epsilon = \frac{1}{8(n+1)}$ implies that the running time of the PAC learning algorithm (and hence our distinguisher) will still be polynomial in n and therefore we will reach the same contradiction. We also note that, by replacing the consistency with only the first bit of $\sigma(y)$ with consistency with a randomly chosen bit, and slightly modifying the encoding of the information on the domain, it is easy to produce a concept class C that is not even *weakly* PAC learnable ¿from noisy membership queries (while still learnable from the incomplete MQs).

5   Concluding Remarks

In this paper, we gave two separation results for exact learning with faulty membership queries. Perhaps the most interesting aspect of the second separation result is a surprising connection to learning of parities in the PAC model with noise. It appears to be the first result that is based on the intractability of the noisy parity problem. An interesting related question is whether this assumption can be replaced by a more general complexity theoretic assumption.

Acknowledgments

We would like to thank Nader Bshouty and Leslie Valiant for useful discussions on this research. We are grateful to Salil Vadhan for contributing an important insight to the proof of Theorem 1.

References

1. Angluin, D., Slonim, D.K.: Randomly fallible teachers: Learning monotone DNF with an incomplete membership oracle. Machine Learning 14, 7–26 (1994)
2. Goldman, S., Kearns, M., Schapire, R.: Exact identification of read-once formulas using fixed points of amplification functions. SIAM Journal on Computing 22, 705–726 (1993)
3. Simon, H.U.: How many missing answers can be tolerated by query learners? Theory of Computing Systems 37, 77–94 (2004)
4. Bshouty, N.H., Eiron, N.: Learning monotone DNF from a teacher that almost does not answer membership queries. The Journal of Machine Learning Research 3, 49–57 (2003)
5. Bshouty, N.H., Owshanko, A.: Learning regular sets with an incomplete membership oracle. In: Helmbold, D., Williamson, B. (eds.) COLT 2001 and EuroCOLT 2001. LNCS (LNAI), vol. 2111, pp. 574–588. Springer, Heidelberg (2001)
6. Goldman, S.A., Mathias, H.D.: Learning -term dnf formulas with an incomplete membership oracle. In: Proceedings of COLT, pp. 77–84 (1992)
7. Chen, Z.: A note on learning dnf formulas using equivalence and incomplete membership queries. In: Arikawa, S., Jantke, K.P. (eds.) AII 1994 and ALT 1994. LNCS, vol. 872, pp. 272–281. Springer, Heidelberg (1994)
8. Kushilevitz, E., Mansour, Y.: Learning decision trees using the Fourier spectrum. In: Proceedings of STOC, pp. 455–464 (1991)
9. Jackson, J., Shamir, E., Shwartzman, C.: Learning with queries corrupted by classification noise. In: Proceedings of the Fifth Israel Symposium on the Theory of Computing Systems, pp. 45–53 (1997)
10. Feldman, V.: Attribute Efficient and Non-adaptive Learning of Parities and DNF Expressions. Journal of Machine Learning Research, 1431–1460 (2007)
11. Blum, A., Furst, M., Kearns, M., Lipton, R.J.: Cryptographic primitives based on hard learning problems. In: Stinson, D.R. (ed.) CRYPTO 1993. LNCS, vol. 773, pp. 278–291. Springer, Heidelberg (1993)
12. Feldman, V., Gopalan, P., Khot, S., Ponuswami, A.: New Results for Learning Noisy Parities and Halfspaces. In: Proceedings of FOCS, pp. 563–574 (2006)

13. McEliece, R.J.: A public-key cryptosystem based on algebraic coding theory. DSN progress report, 42–44 (1978)
14. Barg, A.: Complexity issues in coding theory. Electronic Colloquium on Computational Complexity (ECCC) 4 (1997)
15. Vardy, A.: Algorithmic complexity in coding theory and the minimum distance problem. In: Proceedings of STOC, pp. 92–109 (1997)
16. Kearns, M.: Efficient noise-tolerant learning from statistical queries. Journal of the ACM 45, 983–1006 (1998)
17. Regev, O.: On lattices, learning with errors, random linear codes, and cryptography. In: Proccedings of STOC, pp. 84–93 (2005)
18. Blum, A., Kalai, A., Wasserman, H.: Noise-tolerant learning, the parity problem, and the statistical query model. In: Proceedings of STOC, pp. 435–440. ACM Press, New York (2000)
19. Angluin, D.: Queries and concept learning. Machine Learning 2, 319–342 (1988)
20. Angluin, D., Krikis, M., Sloan, R., Turán, G.: Malicious Omissions and Errors in Answers to Membership Queries. Machine Learning 28, 211–255 (1997)
21. Valiant, L.: A theory of the learnable. Communications of the ACM 27, 1134–1142 (1984)
22. Goldreich, O., Goldwasser, S., Micali, S.: How to construct random functions. Journal of the ACM 33, 792–807 (1986)

Vapnik-Chervonenkis Dimension of Parallel Arithmetic Computations*

César L. Alonso[1] and José Luis Montaña[2]

[1] Centro de Inteligencia Artificial, Universidad de Oviedo
Campus de Viesques, 33271 Gijón, Spain
calonso@aic.uniovi.es
[2] Departamento de Matemáticas, Estadística y Computación,
Universidad de Cantabria, 39005 Santander, Spain
montanjl@unican.es

Abstract. We provide upper bounds for the Vapnik-Chervonenkis dimension of concept classes parameterized by real numbers whose membership tests are programs described by bounded-depth arithmetic networks. Our upper bounds are of the kind $O(k^2 d^2)$, where d is the depth of the network (representing the parallel running time) and k is the number of parameters needed to codify the concept. This bound becomes $O(k^2 d)$ when membership tests are described by Boolean-arithmetic circuits. As a consequence we conclude that families of concepts classes having parallel polynomial time algorithms expressing their membership tests have polynomial VC dimension.

Keywords: Concept learning, Vapnik-Chervonenkis dimension, Milnor-Thom bounds, parallel computation, formula size.

1 Introduction

We deal with general concept classes whose concepts and instances are represented by tuples of real numbers. For such a concept class $\mathcal{C}$, let $\mathcal{C}_{k,n}$ be $\mathcal{C}$ restricted to concepts represented by k real values and instances represented by n real values. Following [9], the *membership test* of a concept class $\mathcal{C}$ over domain X takes as input a concept $C \in \mathcal{C}$ and an instance $x \in X$, and returns the Boolean value "$x \in C$". The membership test of a concept class can be thought of in two common ways: either as a formula, or as an algorithm taking as input representations of a concept and an instance, and evaluating to the Boolean value indicating membership.

Throughout this paper, the membership test for a concept class $\mathcal{C}_{k,n}$ is assumed to be expressed as a parallel algorithm $\mathcal{N}_{k,n}$ taking $k + n$ real inputs, representing a concept $C \in \mathbb{R}^k$ and an instance $x \in X = \mathbb{R}^n$, which uses exact real arithmetic and returns the truth value $x \in \mathcal{C}$.

We seek general conditions on parallel algorithms $\mathcal{N}_{k,n}$ that guarantee that VC dimension of $C_{k,n}$ be polynomial in k and n. This approach follows the

* Partially supported by spanish grant TIN2007-67466-C02-02.

M. Hutter, R.A. Servedio, and E. Takimoto (Eds.): ALT 2007, LNAI 4754, pp. 107–119, 2007.
© Springer-Verlag Berlin Heidelberg 2007

same pattern as the work by Goldberg and Jerrum ([9]), who exhibit sufficient conditions on the size of formulas in the first order theory of the real numbers $\Phi_{k,n}$, and on the sequential running time of algorithms $\mathcal{N}_{k,n}$ to define a class of polynomial VC dimension in k and n.

In this same spirit the seminal paper by Karpinski and Macintyre ([11]) deserves special mention: here polynomial bounds on the VC dimension of sigmoidal networks, and networks with general Pfaffian activation functions, are derived.

For classes defined by algorithms that are allowed to perform conditional statements (conditioned on equality and inequality of real values) and the standard arithmetic operations $+, -, *, /$, we prove the following results.

- For a hierarchy of concept classes $\mathcal{C}_{k,n}$, defined by algorithms $\mathcal{N}_{k,n}$ which run in "parallel time" $d = d(k, n)$ the VC dimension of $\mathcal{C}_{k,n}$ is at most $O(k(k + n)d^2)$. In particular, if $k \geq n$, then the VC dimension of $\mathcal{C}_{k,n}$ is at most $O(k^2 d^2)$.
- For a hierarchy of concept classes $\mathcal{C}_{k,n}$, defined by algorithms $\mathcal{N}_{k,n}$ which run in "parallel time" polynomial in k and n, the VC dimension of $\mathcal{C}_{k,n}$ is also polynomial in k and n.

Note that there are many examples of boolean formulas of the first order theory of real numbers that can be described with short parallel complexity but large sequential representation (see for instance [7] for examples coming from elimination theory). On the contrary there are also many cases of formulas represented by algebraic computation trees ([3],[16]), straight-line programs ([1],[2], linear search programs ([12]) or Turing machines over the field of real numbers ([5] and [4]) that cannot be efficiently parallelized - in poly-logarithmic depth- by arithmetic networks, as it is shown in [15]. We must remark that in these last cases our results remains meaningless.

The paper is organized as follows. In Section 2 we present some known results on the VC dimension of formulas and sequential programs. Section 3 describes in detail the parallel model of computation given by arithmetic networks. In Section 4 we study the formula size of arithmetic networks. Section 5 contains the proof of our main results. Finally in Sections 6 and 7 we analyze the optimality of our upper bounds.

2   Known Results on the VC Dimension of Formulas and Sequential Algorithms

The following definition of VC dimension is standard. See for instance [21].

Definition 1. *Le $\mathcal{F}$ be a class of subsets of a set X. We say that $\mathcal{F}$ shatters a set $A \subset X$ if for every subset $E \subset A$ there exists $S \in \mathcal{F}$ such that $E = S \cap A$. The VC dimension of $\mathcal{F}$ is the cardinality of the largest set that is shattered by $\mathcal{F}$.*

Following familiar approaches, we deal with concept classes $\mathcal{C}_{k,n}$ such that concepts are represented by k real numbers, $w = (w_1, \ldots, w_k)$, instances are represented by n real numbers, $x = (x_1, \ldots, x_n)$, and the membership test to the family $\mathcal{C}_{k,n}$ is expressed either by a formula $\Phi_{k,n}(w, x)$ or by a program $\mathcal{N}_{k,n}$ taking as inputs the pair concept/instance (w, x) and returning the value 1 if "x belongs to the concept represented by w" and 0 otherwise.

In both situations, we can think of $\Phi_{k,n}$ or $\mathcal{N}_{k,n}$ as a function from $\mathbb{R}^{k+n}$ to $\{0, 1\}$. So for each concept w, define:

$$C_w := \{x \in \mathbb{R}^n : \Phi_{k,n}(w, x) = 1\}, \tag{1}$$

in case the membership test is expressed by formula $\Phi_{k,n}$ or

$$C_w := \{x \in \mathbb{R}^n : \mathcal{N}_{k,n}(w, x) = 1\}, \tag{2}$$

in the case membership test is expressed by a program $\mathcal{N}_{k,n}$.

The objective is to obtain an upper bound on the VC dimension of the collection of sets

$$\mathcal{C}_{k,n} = \{C_w : w \in \mathbb{R}^k\}. \tag{3}$$

Now assume that formula $\Phi_{k,n}$ is a Boolean combination of s atomic formulas, each of them being of one of the following forms:

$$\tau_i(w, x) > 0 \tag{4}$$

or

$$\tau_i(w, x) = 0 \tag{5}$$

where $\{\tau_i(w, x)\}_{1 \le i \le s}$ are infinitely differentiable functions from $\mathbb{R}^{k+n}$ to $\mathbb{R}$. Next, make the following assumptions about the functions τ_i. Let $\alpha_1, \ldots, \alpha_v \in \mathbb{R}^n$. Form the sv functions $\tau_i(w, \alpha_j)$ from $\mathbb{R}^k$ to $\mathbb{R}$. Choose $\Theta_1, \ldots, \Theta_r$ among these, and let

$$\Theta : \mathbb{R}^k \to \mathbb{R}^r \tag{6}$$

be defined by

$$\Theta(w) := (\Theta_1(w), \ldots, \Theta_r(w)) \tag{7}$$

Assume there is a bound B independent of the α_i, r and the $\epsilon_1, \ldots, \epsilon_r$ such that if $\Theta^{-1}(\epsilon_1, \ldots, \epsilon_r)$ is an $(k-r)$-dimensional $\mathcal{C}^\infty$- sub-manifold of $\mathbb{R}^k$ then $\Theta^{-1}(\epsilon_1, \ldots, \epsilon_r)$ has at most B connected components.

With the above set-up, the following result is proved in [11].

Theorem 2. *The VC dimension V of a family of concepts $\mathcal{C}_{k,n}$ whose membership test can be expressed by a formula $\Phi_{k,n}$ satisfying the above conditions satisfies:*

$$V \le 2log_2 B + 2klog_2(2es) \tag{8}$$

Using the classical result by Milnor ([13]), Thom([20]), Oleinik and Petrovsky ([18], [17]) and also Warren ([22]) we have:

Lemma 3. *Assume* $\Theta_1, \ldots, \Theta_r$ *are polynomials in* k *variables with degree at most* d *and* $\Theta^{-1}(\epsilon_1, \ldots, \epsilon_r)$ *is an* $(k-r)$*-dimensional* C^∞*- submanifold of* $\mathbb{R}^k$*. Then, the preimage* $\Theta^{-1}(\epsilon_1, \ldots, \epsilon_r)$*, has at most* B *connected components, where*

$$B \leq (2d)^k \tag{9}$$

Now we can derive the following results proved in [9].

Theorem 4 ([9], Theorem 2.2). *Suppose* $\mathcal{C}_{k,n}$ *is a class of concepts whose membership test can be expressed by a formula* $\Phi_{k,n}$ *involving a total of* s *polynomial equalities and inequalities, where each polynomial has degree no larger than* d*. Then the VC dimension* V *of* $\mathcal{C}_{k,n}$ *satisfies*

$$V \leq 2k log_2(4eds) \tag{10}$$

Theorem 5 ([9], Theorem 2.3). *Suppose* $\mathcal{C}_{k,n}$ *is a class of concepts whose membership test can be expressed by an algebraic computation tree* $\mathcal{T}_{k,n}$ *of height bounded by* $t = t(k, n)$ *(representing sequential time). Then the VC dimension* V *of* $\mathcal{C}_{k,n}$ *is* $V = O(kt)$*.*

Remark 6. Theorem 4 is a direct consequence of Theorem 2 and Lemma 3. Theorem 5 follows from Theorem 4 and the fact, proven in [3], that the set accepted by an algebraic computation tree of height bounded by t can be expressed by a formula of the first order theory of real numbers having at most $t2^t$ atomic predicates, where each predicate is a polynomial (in)equality of degree bounded by 2^t.

3   Arithmetic Networks

Our model of parallel computation is that of arithmetic networks. An arithmetic network $\mathcal{N}$ is an arithmetic circuit (or straight line program) augmented with a special kind of gates, called *sign gates*. A sign gate outputs 1 if their input is greater or equal than 0 and 0 otherwise.

Definition 7. *An arithmetic network* $\mathcal{N}$ *over* $\mathbb{R}$ *is a directed acyclic graph where each node has indegree 0, 1 or 2. Nodes with indegree 0 are labelled as inputs or with elements of* $\mathbb{R}$*. Nodes with indegree 2 are labelled with a binary operation of* $\mathbb{R}$*, that is* $+, -, *, /$*. Nodes of indegree 1 are sign gates.*

To each gate v we inductively associate a function as follows.

- if v is an input or constant gate then f_v is the label of v.
- if v has indegree 2 and v_1 and v_2 are the ancestors of v then $f_v = f_{v_1} op_v f_{v_2}$ where $op_v \in \{+, -, *, /\}$ is the label of v.
- if v is a sign gate then $f_v = sign(f'_v)$ where v' is the ancestor of v in the graph.

In particular the function associated to the output gate is the function computed by the arithmetic network. We say that a subset W is accepted by an arithmetic network $\mathcal{N}$ if the function computed by $\mathcal{N}$ is the characteristic function of W. In this case it is assumed that the output gate is a sign gate.

Given an arithmetic network $\mathcal{N}$, the size $s(\mathcal{N})$ is the number of gates in $\mathcal{N}$. The depth $d(\mathcal{N})$ is the length of the longest path from some input gate to some output gate. As usual, we shall refer to $d(\mathcal{N})$ as parallel time.

Remark 8. There is an alternative definition of arithmetic networks that combines arithmetic gates with Boolean gates using an interface between them (see [8]). This interface is given by two special gates, *Boolean sign gates* and *selection gates*. For a fixed sign condition, $\epsilon \in \{>, =, <\}$, the Boolean sign gate $sign(f, \epsilon)$ outputs the Boolean value $TRUE$ if $f\epsilon 0$ is satisfied and $FALSE$ otherwise.

The selection gates choose a particular input according to a Boolean instance. They have associated a function

$$sel(f, g, b)$$

where $f, g \in \mathbb{R}$ and $b \in \{TRUE, FALSE\}$, defined by

$$sel(f, g, b) = f$$

if the Boolean value $b = TRUE$ and

$$sel(f, g, b) = g$$

otherwise.

Arithmetic networks with arithmetic operations and sign gates as described in Definition 7 are able to simulate arithmetic networks with Boolean sign gates and selection gates, defining equivalent computation models.

Remark 9. If L is any language and ϕ is a quantifier free L formula, the size of ϕ is the number of atomic predicates it contains. Observe that the combination of arithmetic gates with sign gates may increase the number of polynomials involved in the computation (the formula size) up to a number which is doubly exponential in the depth (parallel time). This is not strange: if d is parallel time, sequential time could be, in the worst case, $t = 2^d$. On the other hand, if t is sequential time the number of polynomials appearing in the formula, could be at worst 2^t. This last consideration is implicit in the proof of [9], Theorem 2.3. Accordingly, we see that, using this straightforward argument, the best we can expect either from [9], Theorem 2.2 and Theorem 2.3, is an $O(k2^d)$ upper bound for the VC dimension of concept classes $\mathcal{C}_{k,n}$ whose membership test is represented by an algorithm $\mathcal{N}_{k,n}$ working within parallel time $d = d(n, k)$.

Example 10. An example of the feature pointed out in Remark 9 can be constructed as follows. Consider an arithmetic network $\mathcal{N}(l)$ expressing the membership to a concept class $\mathcal{C}(l)$ in which concepts can be represented by three real numbers $\{w_i\}_{1 \leq i \leq 3}$ and instances are represented by two real numbers (x, y) as indicated below.

- (1) The input gates of $\mathcal{N}(l)$ are the variables x, y, w_1, w_2, w_3.

- (2) Compute the set of 32^l powers $\{p_i\}_i$ given by

$$p_i = w_1^i, \tag{11}$$

$$p_{2^l+i} = w_2^i, \tag{12}$$

and

$$p_{2^{l+1}+i} = w_3^i, \tag{13}$$

for $1 \leq i \leq 2^l$.
By elementary results of algebraic complexity theory this step can be completed by an arithmetic network of size $O(2^l)$ and depth $O(l)$.

- (3) Compute the polynomials

$$g_i(x, y, w_1, w_2, w_3) = x + y - p_i. \tag{14}$$

This can be done by an arithmetic network having constant depth and size $O(2^l)$. Note that the degree of the polynomial g_i is at most 2^l.

- (4) In constant depth and size $O(2^l)$, build 2^l gates v_i^0, $1 \leq i \leq 2^l$, as follows: the output $f_{v_i^0}$ is the polynomial g_{3i-2}, when $g_{3i} = 0$, or g_{3i-1}, when $g_{3i} \neq 0$.

- (5) Within depth $l+1$ and size $2^{l+1} - 1$, add product gates $v_1^i, \cdots, v_{2^{l-i+1}}^i$ where

$$f_{v_k^i} = f_{v_{2k-1}^{i-1}} * f_{v_{2k}^{i-1}}. \tag{15}$$

In this latter definition, the superscript index i indicates the depth level and ranges in $1...l+1$, and the subscript index k indicates the gate number at level i; moreover k ranges in $1...2^{l-i+1}$.

- (6) Finally, add an output gate v whose output is given by

$$f_v = sign(f_{v_1^{l+1}}). \tag{16}$$

Now, note that the membership test to the class $\mathcal{C}(l)$ can be expressed by a formula $\Phi(l) = \Phi(l)(w_1, w_2, w_3, x, y)$ defined by

$$\Phi(l) = (g_3 = 0, g_6 = 0, \ldots, g_{32^l} = 0, g_1 g_4 \cdots g_{32^l-2} \geq 0) \vee \ldots \tag{17}$$

$$\ldots \vee (g_3 \neq 0, g_6 \neq 0, \ldots, g_{3.2^l} \neq 0, g_2 g_5 \cdots g_{32^l-1} \geq 0) \tag{18}$$

Indeed, we are involving all possible products

$$\prod_{i=1}^{2^l} f_{k(i)}, \tag{19}$$

where $k(i)$ is either $3i - 1$ or $3i - 2$. This means that formula $\Phi(l)$ has size $\Theta(2^{2^l})$ in 5 free variables with polynomials of degree at most 2^l. Note also that network $\mathcal{N}(l)$ has depth $O(l)$.

If we apply directly Theorem 4 we obtain an upper bound for the VC dimension of $\mathcal{C}(l)$ of the kind $O(2^l)$.

However, a finer argument provides a better relation between depth and formula size.

Proposition 11. *The class $\mathcal{C}(l)$ has formula size $2^{O(l)}$ and VC-dimension $O(l)$.*

Proof. Let us observe that formula $\Phi(l)$ can be given as a finite disjunction

$$\Phi(l) = \bigvee_i (\Phi_{i,1}(l) \wedge \Phi_{i,2}(l)), \tag{20}$$

where $\Phi_{i,1}(l)$ is a sign assignment to the polynomials $g_3, ..., g_{32^l}$, and $\Phi_{i,2}(l)$ is given by the condition $\prod_{i=1}^{2^l} g_{k(i)} \geq 0$, where the selection of $g_{k(i)}$ depends on g_{3i}. Now using Lemma 12 below we see that the number of non-empty sign assignments to the polynomials $g_3, ..., g_{3.2^l}$ is of the order $(1 + D)^5$, where

$$D = \sum deg(g_{3i}) = 2^{O(l)}.$$

This implies that $\Phi(l)$ has size $2^{O(l)}$. Since the degree of the polynomials is at most 2^l we obtain an upper bound for the VC-dimension of $\mathcal{C}(l)$ of the kind $O(l)$.

4   Formula Size of Arithmetic Networks

Throughout this section we are interested in upper bounds on the minimum size of formulas $\Phi_{k,n}$ expressing membership test to concept classes $\mathcal{C}_{k,n}$. We shall also refer to this minimum as the formula size of the class $\mathcal{C}_{k,n}$. Bounds on the formula size, combined with Theorem 2 and Lemma 3, are the key point in the proof of our results.

A result that we will use when analyzing the formula size of concept classes is an upper bound on the number of consistent sign assignments to a set of multivariate polynomials. A sign assignment to polynomial f is one of the (in)equalities

$$f > 0 \text{ or } f = 0 \text{ or } f < 0.$$

A sign assignment to a set of s polynomials is consistent if all s (in)equalities can be satisfied simultaneously by some assignment of real numbers to the variables. We substitute the commonly used bound due to Warren ([22]) by another one used in algebraic complexity theory. This bound can be found, for instance, in [10] and is stated as follows.

Lemma 12 ([10]). *Let $\mathcal{F}$ be a finite family of n-variate polynomials having real coefficients. Let D be the sum of the degrees of the polynomials in the family. Then the number of consistent sign assignments to polynomials of the family $\mathcal{F}$ is at most $D^{O(n)}$.*

Lemma 13. *Let $C_{k,n}$ be a family of concept classes whose membership test can be expressed by a family of arithmetic networks $\mathcal{N}_{k,n}$ having $k + n$ real variables representing the concept and the instance and depth $d = d(k,n)$. Then, the membership test to $C_{k,n}$ can be expressed by a family of formulas $\Phi_{k,n}$ in $k + n$ free variables having the following properties.*

(1) Formula $\Phi_{k,n}$ has size at most $2^{O((k+n)d^2))}$.

(2) The polynomials in $\Phi_{k,n}$ have degree at most 2^d.

Proof. We transform $\mathcal{N}_{k,n}$ into a formula $\Phi_{k,n}$ having the desired properties. Let $d = d(k,n)$ be the parallel time of $\mathcal{N}_{k,n}$. Let

$$\{sign(i,1), \ldots, sign(i, l_i)\}$$

be the collection of sign gates of the network $\mathcal{N}_{k,n}$ whose depth is $i \leq d = d(k,n)$ (here we are not considering the parts of the network that do not affect the output, i. e. having no path to the output unit). Now, for each pair (i,j), $1 \leq j \leq l_i$, let $y_{i,j}$ be the function of $(x_1, \ldots, x_n, w_1, \ldots, w_k)$ that the sign gate $sign(i,j)$ receives as input.

Since the indegree of the gates is bounded by 2 it easily follows by induction that $y_{i,j}$ is a piecewise rational function of $(x_1, \ldots, x_n, w_1, \ldots, w_k)$ of formal degree bounded by 2^i and, at level i the number l_i is bounded above by 2^{d-i}.

Now, for each sign assignment $\epsilon = (\epsilon_{i,j}) \in \{>, =, <, \}^{\sum_{1 \leq i \leq d} l_i}$ let Φ_ϵ be the formula:

$$\Phi_\epsilon = \bigwedge_{1 \leq i \leq d, 1 \leq j \leq l_i} (y_{i,j} \epsilon_{i,j} 0), \tag{21}$$

and observe that

$$\Phi_\epsilon = \bigwedge_{1 \leq i \leq d} \Phi_{\epsilon_i} \tag{22}$$

where

$$\Phi_{\epsilon_i} = \bigwedge_{1 \leq j \leq l_i} (y_{i,j} \epsilon_{i,j} 0) \tag{23}$$

Claim 1. For every $\epsilon \in \{>, =, <, \}^{\sum_{1 \leq i \leq d} l_i}$ there are rational functions $r_{i,j}$ of $(x_1, \ldots, x, w_1, \ldots, w_n)$ of degree bounded by 2^i such that formula Φ_ϵ is equivalent to the formula

$$\bigwedge_{1 \leq i \leq d, 1 \leq j \leq l_i} (r_{i,j} \epsilon_{i,j} 0) \tag{24}$$

Proof. The proof is by finite induction on the number of conjunctions k in Equation 22: if $k = 1$, the ancestors of gate $sgn(1,j)$ are arithmetic gates computed by an arithmetic circuit (i.e. an arithmetic network without sign gates) of depth ≤ 1 and the result trivially follows. Assume now that $\bigwedge_{1 \leq i \leq k-1} \Phi_{\epsilon_i}$ satisfies the required condition. In this case the result follows by noting that the role played by sign gates which are ancestors of some sign gate (k,j), $1 \leq j \leq l_k$, on inputs satisfying formula $\bigwedge_{1 \leq i \leq k-1} \Phi_{\epsilon_i}$ is superfluous.

In what follows formula in Equation 24 will be also denoted by Φ_ϵ. Notice that the set of inputs $(x_1, ..., x_n, w_1, ..., w_k)$ accepted by the arithmetic network $\mathcal{N}_{k,n}$ can be described by a disjunction of some of the formulae Φ_ϵ. Hence, the proof of Lemma 13 follows from the degree bounds of Claim 1 and Lemma 12 if we show the following.

Lemma 14. *The number of tuples ϵ such that formula Φ_ϵ represents a consistent sign assignment is bounded by $2^{O((k+n)d^2)}$.*

Proof. To show this assertion we proceed again by finite induction on the number of conjunctions k in Equation 22. Since $l_1 \leq 2^{d-1}$ and the degree of rational functions $r_{1,j}$ is bounded by 2, we conclude from Claim 1 and Lemma 12 that there are at most $2^{O((n+k)d)}$ values $\epsilon_1 \in \{>, =, <\}^{l_1}$ such that formula Φ_{ϵ_1} represents a consistent sign assignment.

From Claim 1, each consistent sign assignment ϵ_1 for the l_1 rational functions $r_{1,j}$ determines specific rational functions $\{r_{2,j}\}_{1 \leq j \leq l_2}$ in the variables $(x_1, ..., x_n, w_1, ..., w_k)$ as inputs for the sign gates at depth level 2. Since the number of these gates $l_2 \leq 2^{d-2}$ and the degree is bounded by 2^2, we conclude, applying Lemma 12, that for each l_1-tuple $\epsilon_1 \in \{>, =, <\}^{l_1}$ such that Φ_{ϵ_1} represents a consistent assignment, there are at most $2^{O(n+k)d}$ sign assignments $\epsilon_2 \in \{>, =, <\}^{l_2}\}$ such that Φ_{ϵ_2} is consistent. Hence we see that the number of pairs (ϵ_1, ϵ_2) such that $\Phi_{\epsilon_1} \bigwedge \Phi_{\epsilon_2}$ represents a consistent sign assignment is bounded by

$$2^{O((n+k)d)}2^{O(n+k)d}.$$

Iterating h times this argument, each $(l_1 + \cdots + l_{h-1}) - tuple$

$$(\epsilon_1, ..., \epsilon_{h-1}) \in \{>, =, <\}^{\sum_{1 \leq i \leq h-1} l_i}$$

such that the formula

$$\Phi_{\epsilon_1} \bigwedge \cdots \bigwedge \Phi_{\epsilon_{h-1}}$$

represents a consistent sign assignment determines a specific set of rational functions $\{r_{h,j}\}_{1 \leq j \leq l_h}$ in the variables $(x_1, ..., x_n, w_1, ..., w_k)$ as input for the sign gates $sign(h, j)$ at depth level h. Since the number of these gates $l_h \leq 2^{d-h}$ and degree is bounded by 2^h, we conclude, applying Lemma 12, that for each satisfiable $(l_1 + \cdots + l_{h-1})$-tuple as above there are at most $2^{O(n+k)d}$ tuples $\epsilon_h \in \{>, =, <\}^{l_h}\}$ such that Φ_{ϵ_h} represents a consistent sign assignment.

Finally, one gets that the number of tuples

$$(\epsilon_1, ..., \epsilon_h) \in \{>, =, <\}^{\sum_{1 \leq i \leq h} l_i}$$

such that $\Phi_{\epsilon_1} \bigwedge \cdots \bigwedge \Phi_{\epsilon_h}$ represents a consistent sign assignment is bounded by

$$\prod_{1 \leq i \leq h} 2^{O((n+k)d)} = 2^{O((n+k)dh)}$$

Setting $h = d$ finishes the proof.

5   Upper Bounds on the VC Dimension of Arithmetic Networks

Throughout this section we give precise statements and proofs of our results.

Theorem 15. *Let $\mathcal{C}_{k,n}$ a family of concept classes whose membership test can be expressed by arithmetic networks $\mathcal{N}_{k,n}$ having depth $d = d(k,n)$. Then the VC dimension V of $\mathcal{C}_{k,n}$ is at most*

$$V = O(k(k+n)d^2) \tag{25}$$

In particular, if $k \geq n$, then the VC dimension of $\mathcal{C}_{k,n}$ is at most $O(k^2d^2)$.

Proof. Let $\mathcal{C}_{k,n}$ be a family concept class whose membership test can be defined by arithmetic networks $\mathcal{N}_{k,n}$ with depth $d = d(k,n)$. According to Lemma 13 the formula size s of the concept class $\mathcal{C}_{k,n}$ is

$$s = 2^{O((n+k)d^2)} \tag{26}$$

Again from Lemma 13 the degree of the polynomials involved in the description of $\mathcal{N}_{k,n}$ is bounded by 2^d hence, applying Lemma 3 we see that the log_2B term in Theorem 2 satisfies

$$log_2B \leq k(d+1). \tag{27}$$

From equation 26 the second operand in Theorem 2 is

$$2klog_2(2es) = O(k(n+k)d^2), \tag{28}$$

and the theorem follows.

Corollary 16. *For a family of concept classes $\mathcal{C}_{k,n}$, whose membership test can be defined by arithmetic networks $\mathcal{N}_{k,n}$ with depth polynomial in k and n, the VC dimension of $\mathcal{C}_{k,n}$ is also polynomial in k and n.*

6   Analysis of the d^2 Factor

Throughout this section we analyze the factor d^2 in Theorem 15. We give some conditions that allow an upper bound on VC dimension of the kind $O(k^2d)$. To this end we introduce the following notion which is motivated by Example 10.

Definition 17. *An arithmetic network $\mathcal{N}$ is called a Boolean-arithmetic circuit if it is built from*

- *arithmetic gates $+, -, *, /$,*
- *Boolean gates $(\wedge, \vee, \neg)$,*
- *and Boolean sign gates $sign(f, \epsilon)$ as described in Remark 8.*

Remark 18. Note that if we add selection gates, according to Remark 8, we recover the devices described in Definition 7.

We state the following result that bounds the formula size of Boolean-arithmetic circuits.

Lemma 19. *Let $C_{k,n}$ be a family of concept classes whose membership test can be expressed by a family of Boolean-arithmetic circuits $N_{k,n}$ having $k + n$ real variables representing the concept and the instance and depth $d = d(k,n)$. Then, the membership test to $C_{k,n}$ can be expressed by a family of formulas $\Phi_{k,n}$ in $k + n$ free variables having the following properties.*

(1) Formula $\Phi_{k,n}$ has size at most $2^{O((k+n)d)}$.

(2) The polynomials in $\Phi_{k,n}$ have degree at most 2^d.

Proof. Since all gates have indegree at most 2, the size of $N_{k,n}$ is bounded by 2^{d+1}. The total number of Boolean sign gates $h \leq 2^{d+1}$. Enumerate the Boolean sign gates $sign(f_i, \epsilon_i)$, where $\epsilon_i \in \{>, =, <\}$. Note that for $1 \leq i \leq h$, f_i must be the output of an arithmetic gate.

Since all ancestors of a Boolean sign gate are arithmetic gates, f_i are computed by an arithmetic circuit of depth $\leq d$. This means that f_i is a rational function of degree $deg(f_i) \leq 2^d$.

Observe that a Boolean sign gate is ancestor of Boolean gates, hence the subset of $\mathbb{R}^{k+n}$ accepted by the Boolean-arithmetic circuit $N_{k,n}$ can be described by a formula:

$$\Phi_{k,n} = \bigvee_{j \in S} \Phi_j \tag{29}$$

S finite , where Φ_j represents a sign assignment on rational functions in the family $\{f_i\}_{1 \leq i \leq h}$.

Now, note that

$$\sum_{1 \leq i \leq h} deg(f_i) = 2^{O(d)} \tag{30}$$

Finally, using Lemma 12 we conclude that formula $\Phi_{k,n}$ has size at most

$$2^{O(dn)}$$

as wanted.

Combining Lemma 19 and Theorem 2 we conclude the following bound on the VC dimension of Boolean-arithmetic circuits.

Theorem 20. *Let $C_{k,n}$ a family of concept classes whose membership test can be expressed by Boolean-arithmetic circuits $N_{k,n}$ having depth $d = d(k,n)$. Then the VC dimension V of $C_{k,n}$ is at most*

$$V = O(k(k + n)d) \tag{31}$$

In particular, if $k \geq n$, then the VC dimension of $C_{k,n}$ is at most $O(k^2d)$.

7   Analysis of the $k(k + n)$ Factor

We have shown in previous Section that factor d^2 is due to the presence of selection gates or, in other words, due to the combination of arithmetic gates with sign gates. To finish we investigate the component $k(n + k)$ of our bound.

About this question, as a consequence of a construction in [9], one can show the following.

Proposition 21. *For every $k, n \in \mathbb{N}^+$ there is an arithmetic network $\mathcal{N}_{k,n}$ with depth $n + k$ which defines a concept class $\mathcal{C}_{k,n}$ of VC dimension $\Omega(k(n + k))$.*

According to the previous proposition we can conclude that our upper bounds are optimal "modulo a square root". We remark that this situation is the same as when studying bounds on the VC dimension of sigmoidal and Pfaffian neural networks where there is still a gap between the $O(l^4)$ upper bounds and the $O(l^2)$ quadratic lower bounds in the number, l, of programable parameters.

Acknowledgments

We thank Jesús Araujo for his careful reading of this manuscript. We also thank the anonymous referees for their useful corrections and suggestions that have improved the quality of our manuscript.

References

1. Aldaz, M., Heintz, J., Matera, G., Montaña, J.L., Pardo, L.M.: Time-space trade-offs in algebraic complexity theory. Real computation and complexity (Schloss Dagstuhl, 1998). J. Complexity 16(1), 2–49 (2000)
2. Aldaz, M., Heintz, J., Matera, G., Montaña, J.L., Pardo, L.M.: Combinatorial hardness proofs for polynomial evaluation (extended abstract). In: Brim, L., Gruska, J., Zlatuška, J. (eds.) MFCS 1998. LNCS, vol. 1450, pp. 167–175. Springer, Heidelberg (1998)
3. Ben-Or, M.: Lower Bounds for Algebraic Computation Trees. In: STOC, pp. 80–86 (1983)
4. Blum, L., Cucker, F., Shub, M., Smale, S.: Complexity and real computation. xvi+453 pp. Springer, New York (1998)
5. Blum, L., Shub, M., Smale, S.: On a theory of computation over the real numbers: NP completeness, recursive functions and universal machines [Bull. Amer. Math. Soc (N.S.) 21 (1989), no. 1, 1–46; MR0974426 (90a:68022)]. In: Workshop on Dynamical Systems (Trieste, 1988),pp. 23–52, Pitman Res. Notes Math. Ser. 221, Longman Sci. Tech. Harlow (1990)
6. Bochnak, J., Coste, M., Roy, M.-F: Géométrie algébrique réelle (French) [Real algebraic geometry]. In: Ergebnisse der Mathematik und ihrer Grenzgebiete (3) [Results in Mathematics and Related Areas (3)], vol. 12, pp.x+373. Springer, Berlin (1987)
7. Fitchas, N., Galligo, A., Morgestern, A.: Algorithmes rapides en sequentiel et en parallele pour l'élimination de quantificateurs en géométrie élémentaire. Seminaire de Structures Algebriques Ordennes. Universite de Paris VII (1987)

8. von zur Gathen, J.: Parallel arithmetic computations, a survey. In: Math. Found. Comput. Sci. 13th Proc. (1986)
9. Goldberg, P., Jerrum, M.: Bounding the Vapnik-Chervonenkis dimension of concept classes parametrizes by real numbers. Machine Learning 18, 131–148 (1995)
10. Grigorev, D.: Complexity of deciding Tarski Algebra. J. Symp. Comp. 5, 65–108 (1988)
11. Karpinski, M., Macintyre, A.: Polynomial bounds for VC dimension of sigmoidal and general Pffafian neual networks. J. Comp. Sys. Sci. 54, 169–176 (1997)
12. Meyer Auf der Heide, F.: A polynomial linear search algorithm for the n-dimensional Knapsack problem. J. ACM 31(3), 668–676 (1984)
13. Milnor, J.: On the Betti Numbers of Real Varieties. Proc. Amer. Math. Soc. 15, 275–280 (1964)
14. Montaña, J.L., Pardo, L.M., Ramanakoraisina, R.: An extension of Warren's lower bounds for approximations. J. Pure Appl. Algebra 87(3), 251–258 (1993)
15. Montaña, J.L., Pardo, L.M.: Lower bounds for arithmetic networks. Appl. Algebra Engrg. Comm. Comput. 4(1), 1–24 (1993)
16. Montaña, J.L., Pardo, L.M., Recio, T.: The non-scalar model of complexity in Computational Geometry. In: Proc. MEGA'90. Progress in Math. 94. pp. 347–362. Birkhäuser(1991)
17. Oleinik, O.A.: Estimates of the Betti Numbers of Real Algebraic Hypersurfaces. Mat. Mat. Sbornik 70, 63–640 (1951)
18. Oleinik, O.A., Petrovsky, I.B.: On the topology of Real Algebraic Surfaces. Izv. Akad. Nauk SSSR (in Trans. of the Amer. Math. Soc.) 1, 399–417 (1962)
19. Rabin, M.: Proving simultaneous positivity of linear forms. J. Comp. System Sci. 6, 639–650 (1972)
20. Thom, R.: Sur l'Homologie des Varietes Alg ' ebriques R ' eelles. In: Differential and Combinatorial Topology (A Symposium in Honor of Marston Morse), pp. 255–265. Princeton Univ. Press (1965)
21. Vapnik, V.: Statistical learning theory. John Willey & Sons (1998)
22. Warren, H.E.: Lower Bounds for Approximation by non Linear Manifolds. Trans. A.M.S. 133, 167–178 (1968)

Parameterized Learnability of k-Juntas and Related Problems[*]

Vikraman Arvind[1], Johannes Köbler[2], and Wolfgang Lindner[3]

[1] The Institute of Mathematical Sciences, Chennai 600 113, India
arvind@imsc.res.in
[2] Institut für Informatik, Humboldt Universität zu Berlin, Germany
koebler@informatik.hu-berlin.de
[3] Sidonia Systems, Grubmühl 20, D-82131 Stockdorf, Germany
wolfgang.lindner@sidoniasystems.de

Abstract. We study the parameterized complexity of learning k-juntas and some variations of juntas. We show the hardness of learning k-juntas and subclasses of k-juntas in the PAC model by reductions from a W[2]-complete problem. On the other hand, as a consequence of a more general result we show that k-juntas are exactly learnable with improper equivalence queries and access to a W[P] oracle.

Subject Classification: Learning theory, computational complexity.

1 Introduction

Efficient machine learning in the presence of irrelevant information is an important issue in computational learning theory (see, e.g., [21]). This has motivated the fundamental problem of learning k-*juntas*: let f be an unknown boolean function defined on the domain $\{0,1\}^n$ that depends only on an unknown subset of at most k variables, where $k \ll n$. Such a boolean function f is referred to as a k-*junta*, and the problem is whether this class of functions is efficiently learnable (under different notions of learning). This is a natural parameterized learning problem that calls for techniques from parameterized complexity.

Our study is motivated by the recent exciting work by Mossel, O'Donnell and Servedio [22] and the article with open problems on k-juntas proposed by Blum [4,3], drawing to our attention the connection between the learnability of k-juntas and fixed parameter tractability. Notice that in the distribution-free PAC model, an exhaustive search algorithm can learn k-juntas in time roughly n^k. For the uniform distribution, [22] have designed an algorithm for learning k-juntas in time roughly $n^{0.7 \cdot k}$. For the smaller class of monotone k-juntas they even achieve a running time polynomial in n and 2^k (for this class an algorithm with a different running time is given in [8]). Further, for learning symmetric k-juntas, Lipton et al. [20] have provided an algorithm with running-time roughly $n^{0.1 \cdot k}$ and this bound has been subsequently improved to $O(n^{k/\log k})$ in [18].

---

[*] Work supported by a DST-DAAD project grant for exchange visits.

M. Hutter, R.A. Servedio, and E. Takimoto (Eds.): ALT 2007, LNAI 4754, pp. 120–134, 2007.
© Springer-Verlag Berlin Heidelberg 2007

Actually, natural parameters abound in the context of learning and several other learning algorithms in the literature can be seen as parameterized learning algorithms. We mention only two important further examples: Kushilevitz and Mansour [19, Theorem 5.3] give an exact learning algorithm with membership queries for boolean decision trees of depth d and n variables with $\mathbb{F}_2$-linear functions at each node with running time polynomial in n and 2^d. Blum and Rudich [5] design an exact learning algorithm with (improper) equivalence and membership queries for k-term DNFs which runs in time $n2^{O(k)}$.

Parameterized Complexity, introduced as an approach to coping with intractability by Downey and Fellows in [11], is now a flourishing area of research (see, e.g. the monographs [12,14]). Questions focussing on parameterized problems in computational learning have been first studied in [10]. Fixed parameter tractability provides a notion of feasible computation less restrictive than polynomial time. It provides a theoretical basis for the design of new algorithms that are efficient and practically useful for small parameter values. We quickly recall the rudiments of this theory relevant for the present paper. More details (especially on the levels of the W-hierarchy) will be given in the next section (see also [12,14]).

Computational problems often have inputs consisting of two or more parts where some of these parts typically take only small values. For example, an input instance of the vertex cover problem is (G, k), and the task is to determine if the graph G has a vertex cover of size k. A similar example is the k-clique problem where again an input instance is a pair (G, k) and the problem is to test if the graph G has a clique of size k. For such problems an exhaustive search will take time $O(n^k)$, where n is the number of vertices in G. However, a finer classification is possible. The vertex cover problem has an $2^k n^{O(1)}$ time algorithm, whereas no algorithm is known for the k-clique problem of running time $O(n^{o(k)})$. Thus, if the parameter k is such that $k \ll n$, then we have a faster algorithm for the k-vertex cover problem than is known for the k-clique problem.

More generally, a *parameterized decision problem* is a pair (L, κ) where $L \subseteq \{0,1\}^*$ and κ is a polynomial time computable function $\kappa : \{0,1\}^* \to \mathbb{N}$. We call $k = \kappa(x)$ the parameter value of the instance x. The problem (L, κ) is *fixed parameter tractable* $((L, \kappa) \in \text{FPT}$ for short) if L is decidable by an *fpt algorithm*, i.e., by an algorithm that runs in time $g(\kappa(x))|x|^{O(1)}$ for an arbitrary computable function g. In particular, the k-vertex cover problem has an $2^k n^{O(1)}$ time algorithm, implying that it is fixed parameter tractable. On the other hand, the k-clique problem is not known to be in FPT.

In their seminal work, Downey and Fellows [11,12] also developed a theory of intractability for parameterized problems as a tool to classify parameterized problems according to their computational hardness. The W-hierarchy consists of the levels $W[t]$, $t \geq 1$, together with the two classes W[SAT] and W[P] and we have the inclusions

$$\text{FPT} \subseteq \text{W}[1] \subseteq \text{W}[2] \subseteq \cdots \subseteq \text{W}[\text{SAT}] \subseteq \text{W}[\text{P}].$$

In this paper, we show that k-juntas and some subclasses of k-juntas are proper PAC learnable in fixed parameter time with access to an oracle in the second level W[2] of the W-hierarchy. This bound is achieved by reducing the parameterized consistency problem for k-juntas to the parameterized set cover problem. In order to achieve proper learning in fixed parameter time, the learner computes an optimal set cover with the help of a W[2] oracle. A similar approach has been used by Haussler [15] to design an efficient PAC-learning algorithm for k-monomials using $O(\varepsilon^{-1}(\log(\delta^{-1}) + k \log(n)(\log(k) + \log \log(n))))$ many examples.

As a lower bound we prove that monotone k-monomials are not even PAC learnable with k-juntas as hypotheses in randomized fixed parameter time unless W[2] has randomized FPT algorithms. The proof is an application of the well-known technique introduced by Pitt and Valiant [23] to reduce a hard problem to the consistency problem for the hypothesis class. Further, we describe a deterministic fpt algorithm that proper PAC learns k-monomials under the uniform distribution.

We next consider the question of exactly learning k-juntas with only equivalence queries. It turns out that k-juntas are learnable by a randomized fpt algorithm with improper equivalence queries and access to a W[P] oracle. As a consequence, k-juntas are also fpt PAC learnable with access to a W[P] oracle. Actually, we prove a more general result: we consider the problem of learning parameterized concept classes for which the membership of an assignment to a given concept is decidable in FPT and show that these concept classes are exactly learnable by a randomized fpt algorithm with equivalence queries and with access to a W[P] oracle, provided that the Hamming weight is used as parameter. Our learning algorithm uses a similar strategy as the algorithm designed by Bshouty et al. [7] for exactly learning boolean circuits with equivalence queries and with the help of an NP oracle.

The rest of the paper is organized as follows. In Section 2 we provide the necessary notions and concepts and fix notation. Section 3 contains our results on PAC learning and in Section 4 we prove the query-learning results.

2   Preliminaries

2.1   Parameterized Complexity

We fix the alphabet $\Sigma = \{0, 1\}$. The *Hamming weight* $w(x)$ of a string $x \in \{0, 1\}^*$ is the number of 1's in x. The cardinality of a finite set X is denoted by $\|X\|$.

The key idea in quantifying parameterized hardness is the notion of the *weft* of a boolean circuit [11]: We fix any constant $l > 2$. In a boolean circuit c we say that a gate is *large* if it has fanin at least l. The *weft* of a boolean circuit (or formula) c is the maximum number of large gates on any input to output path in c. Thus, any CNF formula is a depth 2 and weft 2 circuit, whereas k-CNF formulas (i.e. CNF formulas with at most k literals per clause) are circuits of depth 2 and weft 1.

The following parameterized problem WEIGHTED-CIRCUIT-SAT (a weighted version of the satisfiability problem for boolean circuits) is central to this theory: Given a pair (c, k), where c is a boolean circuit (or formula) and $k =$

$\kappa(c, k)$ is the parameter, the problem is to decide if there is an input of hamming weight k accepted by c. For a class C of circuits we denote the parameterized problem WEIGHTED-CIRCUIT-SAT restricted to circuits from C by WEIGHTED-CIRCUIT-SAT(C).

In order to compare the complexity of parameterized problems we use the fpt many-one and Turing reducibilities [11]. An *fpt many-one reduction* f from a parameterized problem (L, κ) to a parameterized problem (L', κ') maps an instance x for L to an equivalent instance $f(x)$ for L' (i.e., $x \in L \Leftrightarrow f(x) \in L'$), where for a computable function g, $f(x)$ can be computed in time $g(\kappa(x))|x|^{O(1)}$ and $\kappa'(f(x))$ is bounded by $g(\kappa(x))$. The notion of an *fpt Turing reduction* where the parameterized problem (L', κ') is used as an oracle is defined accordingly: An *fpt Turing reduction* from a parameterized problem (L, κ) to a parameterized problem (L', κ') is a deterministic algorithm $\mathcal{M}$ that for a computable function g, decides L with the help of oracle L' in time $g(\kappa(x))|x|^{O(1)}$ and asks only queries y with $\kappa'(y) \leq g(\kappa(x))$. Now we are ready to define the *weft hierarchy* and the class XP [12,14].

- For each $t \geq 1$, W[t] is the class of parameterized problems that for some constant d are fpt many-one reducible to the weighted satisfiability problem for boolean formulas of depth d and weft t.
- The class W[SAT] consists of parameterized problems that are fpt many-one reducible to the weighted satisfiability problem for boolean formulas.
- W[P] is the class of parameterized problems fpt many-one reducible to the weighted satisfiability problem for boolean circuits.
- For each $k \in \mathbb{N}$, the kth slice of a parameterized problem (L, κ) is the language $L_k = \{x \in L \mid \kappa(x) = k\}$. A parameterized problem (L, κ) belongs to the class XP if for any k, the kth slice L_k of (L, κ) is in P. Note that XP is a non-uniform class that even contains undecidable problems. There is also a uniform version of XP that is more suitable for our purpose. A parameterized problem (L, κ) belongs to the class uniform-XP if there is a computable function $f : \mathbb{N} \to \mathbb{N}$ and an algorithm that, given $x \in \{0, 1\}^*$, decides if $x \in L$ in at most $|x|^{f(\kappa(x))} + f(\kappa(x))$ steps.

From these definitions it is easy to see that we have the following inclusion chain:

$$\text{FPT} \subseteq \text{W}[1] \subseteq \text{W}[2] \subseteq \cdots \subseteq \text{W[SAT]} \subseteq \text{W[P]} \subseteq \text{uniform-XP} \subseteq \text{XP}.$$

2.2 Parameterized Learnability

The Boolean constants *false* and *true* are identified with 0 and 1, and B_n denotes the set of all Boolean functions $f : \{0, 1\}^n \to \{0, 1\}$. Elements x of $\{0, 1\}^n$ are called *assignments* and any pair (x, b) with $f(x) = b$ is called an *example of f*. A variable x_i is called *relevant* for f, if there is an assignment x with $f(x) \neq f(x')$, where x' is obtained from x by flipping the i-th bit.

In order to make our presentation concise, we only consider learning of *concept classes* $C \subseteq B_n$ for some fixed arity n. By abusing notation, we often identify a concept $f \in C$ with the set $\{x \in \{0, 1\}^n \mid f(x) = 1\}$.

A *representation of concepts* is a set $R \subseteq \{0,1\}^*$ of encoded pairs $\langle r, x \rangle$. A *concept name* r represents for each integer $n \geq 1$ the concept

$$R_n(r) = \{x \in \{0,1\}^n \mid \langle r, x \rangle \in R\}.$$

The concept class represented by R is $C(R) = \bigcup_{n \geq 1} C_n(R)$ where $C_n(R) = \{R_n(r) \mid r \in \{0,1\}^*\}$.

A *parameterization* of a representation R of concepts is a polynomial-time computable function $\kappa : \{0,1\}^* \to \mathbb{N}$. We call (R, κ) a *parameterized representation* of concepts and $k = \kappa(r)$ the *parameter value* of the concept description r. (R, κ) is said to be *fpt evaluable* if $\langle r, x \rangle \in R$ is decidable in time $g(\kappa(r))p(|r|, |x|)$, for some arbitrary computable function g and some polynomial p. For a pair of integers k, s we denote by $R_{k,s}$ the set $\{r \in \{0,1\}^s \mid \kappa(r) = k\}$ of all representations r of size s having parameter value k.

The concept classes we consider in the present paper are the following.

- The class $\cup_{n>0} B_n$ of all boolean functions. We usually represent these concepts by (binary encodings of) boolean circuits.
- The class $J_{k,n}$ of k-*juntas* in B_n. If we represent k-juntas by boolean circuits c, then we use the number k of input gates x_i in c having fanout at least 1 as parameter. As in [1] we can also represent concepts in $J_{k,n}$ by strings of length $n + 2^k$ having at most $k + 2^k$ ones, where the first part is of length n and contains exactly k ones (specifying the relevant variables) and the second part consists of the full value table of the k-junta. We denote this representation of k-juntas by J.
- Likewise, for the class $M_{k,n}$ of k-*monomials* consisting of all conjunctions f of at most k literals, we can represent f by a string of length $n + k$ having at most $2k$ ones, where the first n bits specify the set of relevant variables of f (exactly as for k-juntas) and the last k bits indicate which of these variables occur negated in f. Clearly, monotone k-juntas $f \in mon\text{-}J_{k,n}$ and monotone k-monomials $f \in mon\text{-}M_{k,n}$ can be represented in a similar way.

The Hamming weight $w(r)$ provides a natural parameterization of concept classes $R_n(r)$. In fact, if we use the representation J of k-juntas described above, then this parameterization is equivalent to the usual one since for every string r representing a k-junta it holds that $k \leq w(r) \leq k + 2^k$. Further, it is easy to see that all parameterized representations considered in this paper are fpt (even polynomial-time) evaluable. W.r.t. the Hamming weight parameterization, notice that $R_{k,s}$ has size $s^{O(k)}$. Furthermore, the set $R_{k,s}$ can be easily enumerated in time $s^{O(k)}$. This motivates the following definition: a parameterized representation (R, κ) is XP-*enumerable* if the set $R_{k,s}$ can be enumerated in time $s^{O(k)}$ by a uniform algorithm.

Valiant's model of probably approximately correct (PAC) learning [25] and Angluin's model of exact learning via queries [2] are two of the most well-studied models in computational learning theory. In the parameterized setting, both PAC-learning and exact learning with queries are defined in the standard way. However, the presence of the fixed parameter allows a finer complexity classification of learning problems.

To define a parameterized version of exact learning with equivalence queries, let (R, κ) and H be (parameterized) representations. An algorithm $\mathcal{A}$ *exactly learns* (R, κ) *using equivalence queries from* H, if for all $n \in \mathcal{N}$ and all concept names r,

1) $\mathcal{A}$ gets inputs n, $s = |r|$ and $k = \kappa(r)$.
2) $\mathcal{A}$ makes equivalence queries with respect to $R_n(r)$, where the query is a concept name $h \in \{0, 1\}^*$, and the answer is either "Yes" if $H_n(h) = R_n(r)$ or a counterexample x in the symmetric difference $H_n(h) \triangle R_n(r)$.
3) $\mathcal{A}$ outputs a concept name $h \in \{0, 1\}^*$ such that $H_n(h) = R_n(r)$.

We say that $\mathcal{A}$ is an *fpt EQ-learning algorithm* if for each integer $n \in \mathcal{N}$ and each target r the running time of $\mathcal{A}$ on input n, $s = |r|$ and $k = \kappa(r)$ is bounded by $g(k)p(n, s)$, for some computable function g and some polynomial p.

Next we define parameterized PAC-learning. Let (R, κ) and H be (parameterized) representations. A (possibly randomized) algorithm $\mathcal{A}$ *PAC-learns* (R, κ) *using hypotheses from* H, if for all $n \in \mathcal{N}$, all concept names r and for all $\epsilon, \delta > 0$,

1) $\mathcal{A}$ gets inputs n, $s = |r|$, $k = \kappa(r)$, ϵ and δ.
2) $\mathcal{A}$ gets random examples (x, b) of the concept $R_n(r)$, where the strings x are chosen independently according to some distribution $\mathcal{D}_n$ on $\{0, 1\}^n$.
3) With probability at least $1 - \delta$, $\mathcal{A}$ outputs a concept name $h \in \{0, 1\}^*$ such that the error

$$error(h) = \Pr_{x \in \mathcal{D}_n} [x \in R_n(r) \triangle H_n(h)]$$

of h with respect to the target r, where x is chosen according to $\mathcal{D}_n$, is at most ϵ.

$\mathcal{A}$ is an *fpt algorithm* if for each integer $n \in \mathcal{N}$, each target r and for all $\epsilon, \delta > 0$, the running time of $\mathcal{A}$ is bounded by $g(k)p(n, s, 1/\epsilon, 1/\delta)$, for an arbitrary computable function g and a polynomial p. We say that (R, κ) is *fpt PAC-learnable with hypotheses from* H, if there is an fpt algorithm $\mathcal{A}$ that PAC-learns (R, κ) using hypotheses from H.

As usual, in *distribution-free* PAC-learning, the algorithm must succeed on any unknown distribution, whereas in *distribution-specific* PAC-learning the learning algorithm only works for a fixed distribution.

3   PAC Learning of k-Juntas

By the classical algorithm due to Haussler [15] (using the modification of Warmuth as described in [17, Chapter 2]), the class of k-monomials is PAC-learnable in time $poly(n, 1/\varepsilon, \log(1/\delta))$ with $k \log(2/\varepsilon)$-monomials as hypotheses and using $O(\varepsilon^{-1}(\log(\delta^{-1}) + k \log(n) \log(\varepsilon^{-1})))$ many examples. The algorithm uses the well-known greedy heuristic to approximate the set cover problem [16,9]. By computing an optimal solution of the set cover problem, we can achieve proper learning with k-monomials as hypotheses, though at the expense of access to a

W[2] oracle. In the fixed parameterized setting, this can be extended to the class of k-juntas as well as to monotone k-monomials and to monotone k-juntas.

We first show that the parameterized consistency problem (see Definition 1) for (monotone) k-juntas and for (monotone) k-monomials is in W[2]. For this we use the parameterized version of the set cover problem defined as follows. Given a set $U = \{u_1, \ldots, u_m\}$, a family $S = \{S_1, \ldots, S_n\}$ of subsets $S_i \subseteq U$, and a positive integer k (which is the parameter), is there a subset $R \subseteq S$ of size k whose union is U. It is well-known that this problem is W[2]-complete (see for example the book [12]).

Definition 1. *The parameterized consistency problem for a concept class $C = \bigcup_{n \geq 1} C_n$, where $C_n \subseteq B_n$, is defined as follows. Given sets P and N of positive and negative examples from $\{0, 1\}^n$ and a positive integer k (which is the parameter), does C_n contain a k-junta f which is consistent with P and N (meaning that $f(x) = 1$ for all $x \in P$ and $f(x) = 0$ for all $x \in N$).*

Theorem 2. *The parameterized consistency problem is in W[2] for the following concept classes $C = \bigcup_{n \geq 1} C_n$:*

1) *for all k-juntas (i.e., $C_n = \bigcup_{k=0}^n J_{k,n} = B_n$),*
2) *for monotone k-juntas (i.e., $C_n = \bigcup_{k=0}^n$ mon-$J_{k,n}$),*
3) *for k-monomials (i.e., $C_n = \bigcup_{k=0}^n M_{k,n}$),*
4) *for monotone k-monomials (i.e., $C_n = \bigcup_{k=0}^n$ mon-$M_{k,n}$).*

Moreover, in each case, a representation for a consistent k-junta $f \in C_n$ can be constructed (if it exists) in fixed parameter time relative to a W[2] oracle.

Proof. 1) Let (P, N, k) be an instance of the consistency problem for k-juntas. We claim that there is a k-junta consistent with P and N if and only if there is an index set $I \subseteq [n]$ of size k such that

$$\forall (a, b) \in P \times N \; \exists i \in I : a_i \neq b_i. \tag{1}$$

The forward implication is immediate. For the backward implication let I be a size k index set fulfilling property (1) and consider the k-junta f defined by

$$f(x) = \begin{cases} 1, & \text{there exists an } a \in P \text{ s.t. for all indices } i \in I : x_i = a_i, \\ 0, & \text{otherwise.} \end{cases} \tag{2}$$

Then it is clear that $f(a) = 1$ for all $a \in P$. Further, since by property (1) no assignment $b \in N$ can agree with any $a \in P$ on I, it follows that f is also consistent with N. Thus we have shown that (P, N, k) is a positive instance of the consistency problem for k-juntas if and only if the weft 2 formula $\bigwedge_{(a,b) \in P \times N} \bigvee_{a_i \neq b_i} x_i$ has a satisfying assignment of weight k, implying that the consistency problem for k-juntas is in W[2].

In order to construct a consistent k-junta with the help of a W[2] oracle in time $poly(2^k, m, n)$, where $m = \|P \cup N\|$, note that there is also an easy reduction of the parameterized consistency problem to the parameterized set cover problem.

In fact, for each $i \in [n]$ consider the subset $S_i = \{(a, b) \in P \times N \mid a_i \neq b_i\}$ of $U = P \times N$. Then an index set $I \subseteq [n]$ fulfills property (1) if and only if the subfamily $R = \{S_i \mid i \in I\}$ covers U. Now observe that a set S_i is contained in a size k subfamily $R \subseteq \{S_1, \ldots, S_n\}$ covering U if and only if the set $U' = U \setminus S_i$ is covered by some size $k-1$ subfamily of $R' = \{S_1 \setminus S_i, \ldots, S_n \setminus S_i\}$. Thus, we can successively construct a cover R of size k (if it exists) by using kn oracle calls to the parameterized set cover problem. From R we immediately get an index set $I \subseteq [n]$ fulfilling property (1) and thus, a representation of the consistent k-junta f defined by Equation (2) can be computed in fixed parameter time relative to the parameterized set cover problem.

2) Similarly as above it follows that there is a monotone k-junta consistent with P and N if and only if there is an index set $I \subseteq [n]$ of size k fulfilling the property

$$\forall (a, b) \in P \times N \; \exists i \in I : a_i > b_i. \tag{3}$$

In this case, the monotone k-junta f derived from a size k index set I with property (3) has the form

$$f(x) = \begin{cases} 1, & \text{there exists an } a \in P \text{ s.t. for all indices } i \in I : x_i \geq a_i, \\ 0, & \text{otherwise.} \end{cases} \tag{4}$$

Thus, there is some monotone k-junta which is consistent with P and N if and only if the weft 2 formula $\bigwedge_{(a,b) \in P \times N} \bigvee_{a_i > b_i} x_i$ has a satisfying assignment of weight k, implying that also the consistency problem for monotone k-juntas is in W[2]. Further, a consistent monotone k-junta can be constructed by computing a size k solution for the set cover instance $(U, \{S_1, \ldots, S_n\})$, where $U = P \times N$ and $S_i = \{(a, b) \in U \mid a_i > b_i\}$ for $i = 1, \ldots, n$.

3) First observe that a monomial can only be consistent with a set P of positive assignments if it does not depend on any variable x_i such that P contains two examples a and a' with $a_i \neq a_i'$. Let $J = \{i \in [n] \mid \forall a, a' \in P : a_i = a_i'\}$ and let a be an arbitrary but fixed positive example from P. Then there is some k-monomial which is consistent with P and N if and only if there is an index set $I \subseteq J$ of size k fulfilling the property

$$\forall b \in N \; \exists i \in I : a_i \neq b_i. \tag{5}$$

Indeed, if $I \subseteq J$ has property (5), then the monomial $\bigwedge_{i \in I, a_i = 1} x_i \wedge \bigwedge_{i \in I, a_i = 0} \bar{x}_i$ is consistent with P and N. Thus, some k-monomial is consistent with P and N if and only if the weft 2 formula $\bigwedge_{b \in N} \bigvee_{i \in J, a_i \neq b_i} x_i$ has a satisfying assignment of weight k, implying that also the consistency problem for k-monomials is in W[2]. Further, a consistent k-monomial can be constructed by computing a size k solution for the set cover instance $(N, \{S_i \mid i \in J\})$, where $S_i = \{b \in N \mid a_i \neq b_i\}$ for $i = 1, \ldots, n$.

4) The reduction is very similar to the previous one. Observe that a monotone monomial can only be consistent with a set P of positive assignments if it does not depend on any variable x_i such that P contains an example a with $a_i = 0$. Let $J = \{i \in [n] \mid \forall a \in P : a_i = 1\}$. Then there is some monotone k-monomial

which is consistent with P and N if and only if there is an index set $I \subseteq J$ of size k fulfilling the property

$$\forall b \in N \ \exists i \in I : b_i = 0. \tag{6}$$

Indeed, if $I \subseteq J$ fulfills this property, then the monomial $\bigwedge_{i \in I} x_i$ is consistent with P and N. Thus, some k-monomial is consistent with P and N if and only if the weft 2 formula $\bigwedge_{b \in N} \bigvee_{i \in J} x_i$ has a satisfying assignment of weight k, implying that also the consistency problem for monotone k-monomials is in W[2]. Further, a consistent monotone k-monomial can be constructed by computing a size k solution for the set cover instance $(N, \{S_i \mid i \in J\})$, where $S_i = \{b \in N \mid b_i = 0\}$ for $i = 1, \ldots, n$. $\square$

Theorem 3. *The class of k-juntas is fpt PAC-learnable with access to a W[2] oracle and using k-juntas as hypotheses. The same holds for monotone k-juntas as well as for k-monomials and monotone k-monomials.*

Proof. We first consider the case of k-juntas and monotone k-juntas. As has been observed in [1], the set of all k-juntas has size $O(n^k 2^{2^k})$ and hence it follows from [6] that (monotone) k-juntas are proper PAC-learnable by an Occam algorithm by using $O(\varepsilon^{-1}(\log(\delta^{-1}) + 2^k + k \log(n)))$ many examples. Further, observe that using the algorithm described in the proof of Theorem 2, a (monotone) k-junta consistent with the random training sample (P, N) can be constructed with the help of a W[2] oracle in time $poly(2^k, m, n)$, where $m = \|P \cup N\|$.

For the case of (monotone) k-monomials we note that the variant of Haussler's algorithm that requests $O(\varepsilon^{-1}(\log(\delta^{-1}) + k \log(n)))$ many examples and uses the parameterized set cover problem as an oracle to determine a consistent (monotone) k-monomial learns this class in time $poly(n, 1/\varepsilon, \log(1/\delta))$. $\square$

In order to show that the W[2] oracle is indeed necessary we make use of the following hardness result that easily follows by transforming Haussler's [15] reduction of the set cover problem to the consistency problem for monotone monomials into the parameterized setting.

Lemma 4. *Let $C = \bigcup_{n \geq 1} C_n$ be a concept class where C_n contains all monotone monomials over the variables $x_1, \ldots, x_n$. Then the parameterized consistency problem for C is hard for W[2].*

Proof. Consider Haussler's [15] reduction f that maps a set cover instance $U = \{u_1, \ldots, u_m\}$, $S = \{S_1, \ldots, S_n\}$ and k to the instance $P = \{1^n\}$, $N = \{b_1, \ldots, b_m\}$ and k, where the i-th bit of the negative example b_j is 0 if and only if $u_j \in S_i$. We claim that the following statements are equivalent:

- some k-junta is consistent with P and N,
- U can be covered by a subfamily $R \subseteq S$ of size k,
- some monotone k-monomial is consistent with P and N.

Suppose that some k-junta $f \in C_n$ is consistent with the examples from P and N. Let I be the index set of the relevant variables of f. Then by the choice of $P = \{1^n\}$, each negative example b_j differs from 1^n in at least one of the k positions from I. This means that for every $j \in [m]$ there is some $i \in I$ such that the i-th bit of b_j is 0 and, hence, $u_j \in S_i$. Thus, the union of all sets S_i with $i \in I$ covers U.

Now suppose that U can be covered by a subfamily $R = \{S_i \mid i \in I\}$ for some index set $I \subseteq [n]$ of size k. Then for every $j \in [m]$ there is some index $i \in I$ such that the i-th bit of b_j is 0, implying that the monotone k-monomial $\bigwedge_{i \in I} x_i$ is false on all b_j from N and true on 1^n.

Since, by assumption, C_n contains all monotone monomials over the variables $x_1, \ldots, x_n$, this shows that f is an fpt many-one reduction of the parameterized set cover problem (which is W[2]-complete) to the parameterized consistency problem for $\mathcal{C}$. $\square$

By combining Lemma 4 with Theorem 2 we immediately get the following completeness results.

Corollary 5. *The parameterized consistency problem for the following concept classes is complete for* W[2]:

1) *all k-juntas,*
2) *monotone k-juntas,*
3) *k-monomials,*
4) *monotone k-monomials.*

Next we show that no concept class containing all monotone k-monomials is fpt PAC-learnable with boolean circuits having at most k relevant variables as hypotheses unless the second level of the W-hierarchy collapses to randomized FPT (meaning that for any problem $(L, \kappa) \in$ W[2] there is a randomized algorithm that decides L in expected time $g(\kappa(x))|x|^{O(1)}$ for a computable function g; see [13]).

Theorem 6. *Monotone k-monomials are not fpt PAC-learnable with boolean circuits having at most k relevant variables as hypotheses, unless* W[2] *is contained in randomized* FPT.

Proof. Assume that there exists a PAC-learning algorithm $\mathcal{A}$ for the set of monotone k-monomials which runs in time $g(k)poly(n, 1/\varepsilon, 1/\delta)$ and outputs boolean circuits with at most k relevant variables as hypotheses. We describe a randomized algorithm $\mathcal{M}$ which solves the parameterized set cover problem in fixed parameter time.

On input a set $U = \{u_1, \ldots, u_m\}$, a family $S = \{S_1, \ldots, S_n\}$ of subsets $S_i \subseteq U$, and a positive integer k, $\mathcal{M}$ first computes the corresponding instance $f(U, S, k) = (P, N, k)$ of the parameterized consistency problem as described in the proof of Lemma 4. Then $\mathcal{M}$ runs the PAC-learning algorithm $\mathcal{A}$ with confidence parameter $\delta = 1/4$ and error parameter $\varepsilon = 1/(\|N\| + 2)$. For each request for a random classified example, $\mathcal{M}$ randomly chooses an example from $P \cup N$ and passes it to $\mathcal{A}$ along with its classification. After at most $g(k)poly(n, m)$ steps, $\mathcal{A}$ produces a boolean circuit computing some hypothesis h. Now $\mathcal{M}$ tries to determine the relevant variables of h as follows. Observe that if h depends on at most k relevant variables, then for each relevant variable x_i and for a uniformly at random chosen assignment $x \in \{0,1\}^n$ we have $h(x) \neq h(x')$ with probability at least 2^{-k}, where x' is obtained from x by flipping the i-th bit. Thus, $\mathcal{M}$ can detect the index set I of all relevant variables of h with probability $\geq 3/4$ in time $poly(2^k, n)$, provided that h indeed depends on at most k variables (otherwise, I can be an arbitrary subset of $[n]$ and $\mathcal{M}$ might fail to find any relevant variables). Finally, $\mathcal{M}$ accepts if and only if $\|I\| \leq k$ and the monomial $\bigwedge_{i \in I} x_i$ is consistent with P and N.

Assume that (U, S, k) is a positive instance of the parameterized set cover problem. Then, by the choice of δ and ε, $\mathcal{A}$ produces with probability at least $3/4$ a k-junta h that is consistent with P and N. Now, using the properties of the instance (P, N, k) described in the proof of Lemma 4, it follows that if $\mathcal{A}$ is successful, then $\mathcal{M}$ finds with probability $\geq 3/4$ a monotone k-monomial consistent with P and N, implying that $\mathcal{M}$ accepts with probability $\geq 1/2$.

On the other hand, it is clear that $\mathcal{M}$ will never accept a negative instance (U, S, k).   □

Thus it is rather unlikely that the class of k-monomials (or any other concept class considered in Theorem 3) is proper PAC-learnable in time $g(k)poly(n)$. In contrast, in the distribution-specific setting with respect to the uniform distribution, proper PAC-learning can be achieved in fixed parameter time. For the class of monotone k-juntas, this has already been shown by Mossel et al. [22].

Theorem 7. *Under the uniform distribution, k-monomials are PAC-learnable in deterministic fixed parameter time with k-monomials as hypotheses.*

Proof. Let f be some k-monomial and for any $i \in [n]$ consider the probability $p_i = \Pr[f(x) = x_i]$ for a uniformly chosen assignment $x \in \{0,1\}^n$. If x_i does not appear in f then $p_i = 1/2$. If x_i appears unnegated in f then $p_i = 1/2 + 2^{-k}$, and if x_i appears negated in f then $p_i = 1/2 - 2^{-k}$. The probability p_i can be estimated within additive error 2^{-k-1} with high probability by using $poly(2^k)$ random examples. Thus, we can successively determine all literals of f in time $poly(2^k, n)$.   □

4   Learning k-Juntas Exactly

In this section we consider the parameterized learnability of concept classes that are evaluable in fixed parameter time. Our main result here is that any such class is randomized fpt EQ-learnable with access to an oracle in W[P], provided that the Hamming weight is used as parameter. Our learning algorithm uses a similar strategy as the exact learning algorithm of Bshouty et al. [7]. We first recall a version of the Valiant-Vazirani lemma [24] that lower bounds the probability that a randomly chosen linear function h isolates some $x \in D$ (we say that a function $h : \{0,1\}^s \to \{0,1\}^l$ *isolates* x in $D \subseteq \{0,1\}^s$, if x is the only string in D with $h(x) = 0^l$). Furthermore, it provides an upper bound on the probability that such an isolated x lies in a given small subset D' of D.

Lemma 8. *Let $D \subseteq \{0,1\}^s - \{0^s\}$ be a non-empty set of cardinality c, let $D' \subseteq D$ be of cardinality at most $c/12$, and let l be an integer such that $2^l < 3c \le 2^{l+1}$. Then, for a uniformly chosen linear function $h : \{0,1\}^s \to \{0,1\}^l$,*

- *with probability at least $2/9$, there exists exactly one element $x \in D$ such that $h(x) = 0^l$, and*
- *with probability at most $1/18$, there exists some element $x \in D'$ such that $h(x) = 0^l$.*

Theorem 9. *Any XP-enumerable representation (R, κ) is randomized fpt EQ-learnable with access to a* uniform-XP *oracle and using boolean circuits as hypotheses. Moreover, if the Hamming weight is used as parameter, then a W[P] oracle suffices.*

Proof. We give an outline of the proof. Let $\hat{r}$ be the target. We describe a randomized learning algorithm $\mathcal{A}$ that on input n, $s = |\hat{r}|$ and $k = \kappa(\hat{r})$ collects a set S of counterexamples obtained from the teacher. To build a suitable hypothesis from the current set S, $\mathcal{A}$ randomly samples a polynomial number of concept names $r_1, \ldots, r_p$ from the set

$$Cons_{k,s}(S) = \{r \in R_{k,s} \mid R_n(r) \text{ is consistent with } S\}.$$

Then $\mathcal{A}$ makes an improper equivalence query using the hypothesis

$$\text{maj}_{[r_1,\ldots,r_p]}(x) = \begin{cases} 1, & \|\{i \in \{1,\ldots,p\} \mid x \in R_n(r_i)\}\| \ge p/2, \\ 0, & \text{otherwise} \end{cases}$$

which is the majority vote on the concepts $R_n(r_1), \ldots, R_n(r_p)$. In order to do the sampling $\mathcal{A}$ will apply the hashing lemma stated above. More precisely, $\mathcal{A}$ cycles through all values $l = s, s-1, \ldots, 1$ and randomly chooses linear functions $h_i : \{0,1\}^s \to \{0,1\}^l$, $i = 1,\ldots,p$. Then $\mathcal{A}$ uses the oracle

$$B = \{(k, r, S, h, s, l) \mid \exists r' : rr' \in Cons_{k,s}(S) \text{ and } h(rr') = 0^l\},$$

where k is the parameter, to find for each function h_i a concept name r_i (if it exists) that is isolated by h_i in $Cons_{k,s}(S)$. Note that B belongs to uniform-XP as the representation (R, κ) is XP-enumerable. Now, for $i = 1,\ldots,p$ and each string $x \in \{0,1\}^n$ with the property that

$$\|\{r \in Cons_{k,s}(S) \mid x \in R_n(r) \Leftrightarrow x \in R_n(\hat{r})\}\| > (^{11}/_{12})\|Cons_{k,s}(S)\|$$

(meaning that the inclusion of the counterexample x in S discards less than a $1/12$ fraction of all representations in $Cons_{k,s}(S)$) consider the random variable

$$Z_i(x) = \begin{cases} -1, & h_i \text{ isolates an } r_i \text{ in } Cons_{k,s}(S) \text{ with } x \in R_n(r_i)\Delta R_n(\hat{r}), \\ 0, & h_i \text{ does not isolate any string in } Cons_{k,s}(S), \\ 1, & h_i \text{ isolates an } r_i \text{ in } Cons_{k,s}(S) \text{ with } x \in R_n(r_i) \Leftrightarrow x \in R_n(\hat{r}). \end{cases}$$

Then, provided that l has the right value, it follows that

$$E(Z_i(x)) \geq (2/9 - 1/18) - 1/18 = 1/9$$

and by Hoefding's inequality we get

$$Prob\left[\sum_{i=1}^{p} Z_i(x) \leq 0\right] = 2^{-\Omega(p)}.$$

Since the equivalence query $h = \mathrm{maj}_{[r_1,\ldots,r_p]}$ only disagrees with the target on the classification of x if $\sum_{i=1}^{p} Z_i(x) \leq 0$, this means that with probability $1 - 2^{n - \Omega(p)}$, h allows only counterexamples x that discard at least a $1/12$ fraction of all representations in $Cons_{k,s}(S)$.

To complete the proof outline, note that it is easy to see that if we use the Hamming weight as parameterization, then the oracle B actually belongs to W[P].   □

As an immediate consequence we get the following corollary.

Corollary 1. *Any XP-enumerable representation (R, κ) is PAC-learnable in randomized fixed parameter time with access to a* uniform-XP *oracle. Moreover, if the Hamming weight is used as parameter, then a* W[P] *oracle suffices.*

By using the representation J of k-juntas described in Section 2.2, we immediately get the following positive learning result for k-juntas.

Corollary 2. *k-juntas are randomized fpt EQ-learnable with access to a* W[P] *oracle.*

Note that the hypotheses used by the query-learning algorithm can have up to n relevant variables. It is not hard to verify that this is essentially optimal for any algorithm with fixed parameter running time. To see this, suppose that $\mathcal{A}$ learns k-juntas with $g(k)n^c$ equivalence queries using circuits having at most l relevant variables as hypotheses. Consider the subclass D consisting of all monotone monomials with exactly k variables.

If $\mathcal{A}$ asks the constant $h \equiv 0$ function as an equivalence query, then no monotone monomial from D agrees with h on the counterexample $a = 1^n$. Otherwise let a be a counterexample such that $h(a) = 1$ where $a_i = 0$ on all positions i for which h does not depend on x_i. The number of hypotheses from D that agree with h on a is at most $\binom{l}{k}$. Hence, for every equivalence query h there is some counterexample a such that the algorithm $\mathcal{A}$ can discard at most $\binom{l}{k}$ hypotheses from D. By a simple counting argument it follows that

$$g(k)n^c \binom{l}{k} \geq \binom{n}{k} - 1,$$

implying that $l = \Omega(n^{1-c/k}/g(k)^{1/k})$. Thus it follows for all ε and for sufficiently large k and n that $l \geq n^{1-\varepsilon}/g(k)$.

We conclude this section with a remark on exactly learning a generalization of juntas with membership queries. Consider the natural generalization of k-juntas where the target f is a boolean function of k linear forms on the n variables over the field $\mathbb{F}_2$. More precisely, $f(x_1, \ldots, x_n) = g(a_1(x), \ldots, a_k(x))$, where each $a_i(x)$ is defined as a linear function $\sum_{j=1}^{n} a_{ij} x_j$ over $\mathbb{F}_2$, where $a_{ij} \in \{0, 1\}$. Using membership queries such "generalized" k-juntas are exactly learnable in time $2^{O(k)} n^{O(1)}$ by a direct application of the learning algorithm of Kushilevitz and Mansour [19, Theorem 5.3]. According to this result, a boolean decision tree of depth d and n variables with $\mathbb{F}_2$-linear functions at each node can be exactly learned with membership queries in deterministic time polynomial in n and 2^d. Now it suffices to observe that a generalized k-junta can be transformed into a decision tree of depth k with a linear function at each node.

5 Discussion and Open Problems

We have examined the parameterized complexity of learning k-juntas, with our notion of efficient learning as fixed parameter tractable learnability. Our main results are about the hardness of learning k-juntas and subclasses of k-juntas in the PAC model by reductions ¿from a W[2]-complete problem. On the other hand, as a consequence of a more general result we show that k-juntas are exactly learnable with improper equivalence queries and access to a W[P] oracle. Some interesting open questions remain.

The main open question is whether the learning result of [22] for k-juntas can be improved to show that k-juntas are fpt PAC-learnable with boolean circuits as hypotheses. A more modest question is whether (monotone) k-monomials are fpt PAC-learnable with boolean circuits as hypotheses having k' relevant variables, where $k' = g(k)$ only depends on k. From Theorem 6 we only know that if we choose for g the identity function, then this is not possible unless W[2] collapses. On the other hand, Warmuth's modification of Haussler's algorithm achieves PAC learning of k-monomials in polynomial time with $k \log(2/\varepsilon)$-monomials as hypotheses.

Acknowledgments

We thank the anonymous referees for their valuable comments.

References

1. Almuallim, H., Dietterich, T.G.: Learning boolean concepts in the presence of many irrelevant features. Artificial Intelligence 69(1-2), 279–305 (1994)
2. Angluin, D.: Learning regular sets from queries and counterexamples. Information and Control 75, 87–106 (1987)
3. Blum, A.: My favorite open problems (and a few results). In: Talk given at 2001 NeuroCOLT Alpine Workshop on Computational Complexity Aspects of Learning, March 26-29,2001 Sestriere, Italy (2001)

4. Blum, A.: Learning a function of r relevant variables. In: COLT, pp. 731–733 (2003)
5. Blum, A., Rudich, S.: Fast learning of k-term DNF formulas with queries. Journal of Computer and System Sciences 51(3), 367–373 (1995)
6. Blumer, A., Ehrenfeucht, A., Haussler, D., Warmuth, M.K.: Occam's razor. Information Processing Letters 24(6), 377–380 (1987)
7. Bshouty, N., Cleve, R., Gavaldà, R., Kannan, S., Tamon, C.: Oracles and queries that are sufficient for exact learning. Journal of Computer and System Sciences 52, 421–433 (1996)
8. Bshouty, N., Tamon, C.: On the fourier spectrum of monotone functions. Journal of the ACM 43(4), 747–770 (1996)
9. Chvatal, V.: A greedy heuristic for the set covering problem. Mathematics of Operations Research 4(3), 233–235 (1979)
10. Downey, R.G., Evans, P.A., Fellows, M.R.: Parameterized learning complexity. In: Proc. 6th Annual ACM Conference on Computational Learning Theory, pp. 51–57. ACM Press, New York (1993)
11. Downey, R.G., Fellows, M.R.: Fixed-parameter tractability and completeness I: Basic results. SIAM Journal on Computing 24(4), 873–921 (1995)
12. Downey, R.G., Fellows, M.R.: Parameterized Complexity. Springer, Heidelberg (1999)
13. Fellows, M., Koblitz, N.: Fixed-parameter complexity and cryptography. In: Moreno, O., Cohen, G., Mora, T. (eds.) Proc. Tenth International Symposium on Applied Algebra, Algebraic Algorithms, and Error Correcting Codes. LNCS, vol. 673, pp. 121–131. Springer, Heidelberg (1993)
14. Flum, J., Grohe, M.: Parameterized Complexity Theory. Springer, Heidelberg (2006)
15. Haussler, D.: Quantifying inductive bias: AI learning algorithms and Valiant's learning framework. Artificial Intelligence 36, 177–221 (1988)
16. Johnson, D.S.: Approximation algorithms for combinatorial problems. Journal of Computer and System Sciences 9, 256–278 (1974)
17. Kearns, M.J., Vazirani, U.V.: An Introduction to Computational Learning Theory. MIT Press, Cambridge (1994)
18. Kolountzakis, M., Markakis, E., Mehta, A.: Learning symmetric juntas in time $n^{o(k)}$. Interface between Harmonic Analysis and Number Theory (2005)
19. Kushilevitz, E., Mansour, Y.: Learning decision trees using the fourier spectrum. SIAM Journal on Computing 22(6), 1331–1348 (1993)
20. Lipton, R., Markakis, E., Mehta, A., Vishnoi, N.: On the fourier spectrum of symmetric boolean functions with applications to learning symmetric juntas. In: Twentieth Annual IEEE Conference on Computational Complexity, pp. 112–119 (2005)
21. Littlestone, N.: Learning quickly when irrelevant attributes abound: A new linear-threshold algorithm. Machine Learning 2, 285–318 (1988)
22. Mossel, E., O'Donnell, R., Servedio, R.P.: Learning juntas. In: Proc. 35th ACM Symposium on Theory of Computing, pp. 206–212. ACM Press, New York (2003)
23. Pitt, L., Valiant, L.G.: Computational limitations on learning from examples. Journal of the ACM 35(4), 965–984 (1988)
24. Valiant, L., Vazirani, V.: NP is as easy as detecting unique solutions. Theoretical Computer Science 47, 85–93 (1986)
25. Valiant, L.G.: A theory of the learnable. Communications of the ACM 27(11), 1134–1142 (1984)

On Universal Transfer Learning

M.M. Hassan Mahmud

Department of Computer Science
University of Illinois at Urbana Champaign
201 N. Goodwin Avenue, Urbana, IL 61801, USA
mmmahmud@uiuc.edu

Abstract. In transfer learning the aim is to solve new learning tasks using fewer examples by using information gained from solving related tasks. Existing transfer learning methods have been used successfully in practice and PAC analysis of these methods have been developed. But the key notion of relatedness between tasks has not yet been defined clearly, which makes it difficult to understand, let alone answer, questions that naturally arise in the context of transfer, such as, how much information to transfer, whether to transfer information, and how to transfer information across tasks. In this paper we look at transfer learning from the perspective of Algorithmic Information Theory, and formally solve these problems in the same sense Solomonoff Induction solves the problem of inductive inference. We define universal measures of relatedness between tasks, and use these measures to develop universally optimal Bayesian transfer learning methods.

1   Introduction

In Transfer Learning (TL) (e.g. [1], [2]) we are concerned with reducing sample complexity required to learn a particular task by using information from solving *related tasks* (see [3] for a review). Each task in TL corresponds to a particular probability measure generating the data for the task. Transfer learning has in general been inspired by noting that to solve a problem at hand, people almost always use knowledge from solving related problems previously. This motivation has been borne out by practical successes; TL was used to recognize related parts of a visual scene in robot navigation tasks, predict rewards in related regions in reinforcement learning based robot navigation problems, and predict results of related medical tests for the same group of patients etc. A key concept in transfer learning, then, is this notion of relatedness between tasks. As we will see, it is not yet clear what the proper way to define this notion is (see also [4]), and in addition to being conceptually troubling, this problem has also hampered development of even more powerful and principled transfer algorithms.

Many current TL methods are in essence based on the method developed in [1]. The basic idea is to learn m related tasks in *parallel* using neural networks, with all the tasks defined on the same input space (Fig. 1). Different tasks are related by virtue of requiring the same set of good 'high level features' encoded by

M. Hutter, R.A. Servedio, and E. Takimoto (Eds.): ALT 2007, LNAI 4754, pp. 135–149, 2007.

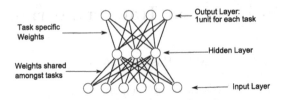

Fig. 1. A typical transfer learning method

the hidden units. The hope is that by training with alternating training samples from different tasks, these common high level features will be learned quicker. The same idea has been used for sequential transfer - i.e. input-to-hidden layer weights from previously learned related tasks were used to speed up learning of new tasks. So tasks are considered related if they can be learned faster together than individually - i.e. if they have a *common near-optimal inductive bias* with respect to a given hypothesis space (e.g. the common hidden units in Fig. 1).

This case was analyzed extensively in a PAC setting in [5]. Here a probability distribution P was assumed over the space of tasks, and bounds were derived on the sample complexity required to estimate the expected error (with respect to P) of the m tasks when the tasks were learned using a sub-space of the hypothesis space. That is bounds were derived for sample complexity for estimating fitness of inductive biases. Most work done on TL is subsumed by this analysis, and they all begin with the assumption that tasks have a common, near optimal inductive bias. So no actual measure of similarity between tasks is prescribed, and hence it becomes difficult to understand, let alone answer, questions such as 'how and when should we transfer information between tasks?' and 'how much information should we transfer?'.

There has been some work which attempts to solve these problems. [4] gives a more explicit measure of relatedness in which two tasks P and Q are said to be similar with respect to a given set of functions F if $\exists f \in F$ such that $P(a) = Q(f(a))$ for all events a. Using F, the authors derive PAC sample complexity bounds for the error of each task (as opposed to expected error in [5]), which can be smaller than single task bounds under certain conditions. More interesting is the approach in [6] (see Sects. 4.3, 6) which gives PAC bounds in the setting of [5]. Here, the sample complexity is proportional to the joint Kolmogorov complexity of the m hypotheses, and so the Kolmogorov complexity measures task relatedness. However, the bounds hold only for ≥ 8192 tasks (Theorem 3).

In this paper we address the problems with transfer learning mentioned above in the framework of Algorithmic Information Theory. Our aim will be to look at what is the best we can do (most amount of similarity between tasks we can uncover, most amount of information we can transfer etc.) given unlimited amount of computational time and space. We use and extend the theory of Information Distance [7] to measure relatedness between tasks, transfer the right amount of information etc. For our task space we restrict ourselves to probability measures that are lower semi-computable, which is reasonable as it covers all situations where we learn using computers. In this space the Information Distance is a universal measure of relatedness between tasks. We give a sharp characterization of Information Distance by showing it is, upto a constant, equal to the Cognitive Distance (Theorems 3.1 and 3.2, which are quite interesting results in and of themselves). Based on this distance we develop universally optimal Bayesian transfer learning methods for doing sequential transfer (Theorem 4.1). We show that sequential transfer is always justified from a formal perspective (Theorem 4.2). We also show that, while universally optimal parallel transfer/multitask learning methods exist (Theorem 4.3), in contrast to sequential transfer methods, it is not clear that these methods are transfer learning methods or are justified when we do not know a-priori that parallel transfer will be useful.

2   Preliminaries

We use $a := b$ to mean expression a is defined by expression b. For any finite alphabet A, let A^*, A^n, A^∞ be the set of all finite strings, length n strings and infinite sequences in A respectively. Let ε be the empty string. For $x, y \in A^*$, xy denotes y concatenated to the end of x. Let $l(x)$ denote the length of a finite string x. We use $\langle \cdot, \cdot \rangle$ to denote a standard bijective mapping from $A^* \times A^* \to A^*$. $\langle \rangle^m$ denotes the m-arity version of this, and $\rangle\langle_i^m$ denotes the ith component of the inverse of $\langle \rangle^m$. We assume the standard 'lexicographical' correspondence between A^* and $\mathbb{N}$ – e.g. for $A := \{0,1\}$, this is $(\varepsilon, 0), (0, 1), (1, 2), (00, 3), (01, 4), \cdots$. Depending on the context, elements of each pair will be used interchangeably (so 01 (and 4) may mean either 01 or 4). A rational number a/b is represented by $\langle a, b \rangle$. We use $\overset{+}{\leq}$ to denote $\leq$ upto an additive constant independent of the variables in the relation i.e. $f(x) \overset{+}{\leq} g(x) \equiv f(x) \leq g(x) + c$. We use the same convention for all the usual binary inequality relations. Let $2^{-\infty} := 0$, $\log := \log_2$ and $\bar{m}$ the self delimiting encoding of $m \in \mathbb{N}$ using $l(m) + 2l(l(m)) + 1$ bits [8].

   We fix a reference prefix universal Turing machine $U : \mathcal{B}^* \times \mathcal{A}^* \to \mathcal{A}^*$, where $\mathcal{B} := \{0,1\}$ is the alphabet for programs, and $\mathcal{A}$, $\mathcal{A} \supset \mathcal{B}$, is an arbitrary alphabet for inputs and outputs. Fixing U as a reference machine is fine because of the *Invariance Theorem* – see [8]. The prefix property means that programs are self-delimiting and the lengths of programs satisfy the Kraft inequality: $\sum_p 2^{-l(p)} \leq 1$. $U(p, x)$ denotes running the program p on input x. When it is clear from the context that p is a program, we will denote $U(p, x)$ simply by $p(x)$. A real function f is *upper semi-computable* if there is a program p such that for $x, t \in \mathbb{N}$, 1) $p(\langle x, t \rangle)$ halts in finite time 2) $p(\langle x, t \rangle) \geq p(\langle x, t+1 \rangle)$ 3) $\lim_{t \to \infty} p(\langle x, t \rangle) = f(x)$. A real function f is *lower semi-computable* if $-f$ is upper semi-computable. A function f is *computable/recursive* if there is a p such that for $n, x \in \mathbb{N}$, $|p(\langle x, n \rangle) - f(x)| < 2^{-n}$, and $p(\langle x, n \rangle)$ halts in finite time.

3   Universal Transfer Learning Distances

In this section we will first describe our task space and the learning problem we consider. Then we will discuss our universal transfer learning distances.

3.1   Task Space $\mathcal{V}$ and the Learning Problem

We consider as our task space a particular subset of the set of all *semi-measures*.

Definition 3.1. *A* **semi-measure** *is a function* $f : \mathcal{A}^* \to [0,1]$ *such that* $\forall x \in \mathcal{A}^*, f(x) \geq \sum_{a \in \mathcal{A}} f(xa)$.

$f(x)$ is the 'defective probability' that a particular infinite sequence starts with the prefix string x (f is a probability measure if $f(\varepsilon) = 1$ and the inequality is an equality). So f is equivalent to a probability measure p defined on $[0,1]$ such that $f(x) = p([0.x, 0.x + |\mathcal{A}|^{-l(x)}))$ where $0.x$ is in base $|\mathcal{A}|$. The conditional probability of the next letter being a given the string x observed so far is $f(a|x) := f(xa)/f(x)$.

[9] showed that the set of all lower semi-computable semi-measures is recursively enumerable. That is, there is a Turing machine T such that $T(\langle i, \cdot \rangle)$ lower semi-computes $f_i(\cdot)$, the ith semi-measure in this effective enumeration. Since U is universal, for each $i \in \mathbb{N}$, there is a program p_i such that $p_i(x) = T(\langle i, x \rangle)$. Let $\mathcal{V}$ be the enumeration of these programs - i.e. $p_i \in \mathcal{V}$ lower semi-computes f_i, and each lower semi-computable semi-measure f is computed by at least one $p_j \in \mathcal{V}$. We will consider enumerable subsets $\mathcal{V}'$ of $\mathcal{V}$ as our task space, as any probability measure that we may expect to be able to learn must either belong to the set of computable measures, or have a reasonable approximation (however it may be defined) that does. $\mathcal{V}$ is the largest superset of this that contains any Bayes mixture of its own elements, which is important in Sect. 4 (see also [10, Sect. 2.6] and [8]).

The learning problem we consider is the online learning setting. When learning task μ, at each step t, $a \in \mathcal{A}$ is generated according to $\mu(.|x)$, where x is the sequence of length $t-1$ generated by μ in the previous $t-1$ steps. The learning problem is to predict the letter a at each step (see e.g. [10, Sect. 6.2] for how i.i.d. learning problems are a special case of this setting).

3.2   Universal Transfer Learning Distance

We want our transfer learning distance to measure the amount of constructive information $\mu, \varphi \in \mathcal{V}$ contain about each other. Elements of $\mathcal{V}$ are strings, and the following defines amount of constructive information any string y contains about another string x.

Definition 3.2. *The* **conditional Kolmogorov complexity** *of x given y, $x, y \in \mathcal{A}^*$ is the length of the shortest program that outputs x given y:*

$$K(x|y) := \min_{p}\{l(p) : p(y) = x\}.$$

Conditional Kolmogorov complexity measures absolute information content of individual objects, and is a sharper version of information-theoretic entropy which measures information content of ensemble of objects relative to a distribution over them. When $y = \varepsilon$, the above is just called Kolmogorov complexity and denoted by $K(x)$. For m strings we use $\langle\rangle^m$ - e.g. $K(x, y|z, w) :=$

$K(\langle x, y\rangle | \langle z, w\rangle)$ etc. We will use the following minimality property of $K(x|y)$ - for any partial, non-negative, upper semi-computable function $f : \mathcal{A}^* \times \mathcal{A}^* \to \mathbb{R}$, if $\sum_x 2^{-f(x,y)} \leq 1$ (taking $f(x, y) = \infty$ when it is undefined)

$$K(x|y) \overset{+}{\leq} f(x, y). \tag{1}$$

Kolmogorov complexity is upper semi-computable, which is in agreement with our desire to investigate information transfer in principle. See [8] for the above results and a comprehensive introduction to AIT.

To measure the amount of information two strings contain about each other in [7] the authors defined the following upper semi-computable function:

Definition 3.3. *The **Information Distance** between $x, y \in \mathcal{A}^*$ is the length of the shortest program that given x outputs y, and vice versa:*

$$E_0(x, y) := \min_p \{l(p) : p(x) = y, p(y) = x\}.$$

So for $\mu, \varphi \in \mathcal{V}$, $E_0(\mu, \varphi)$ measures the amount of information μ and φ contain about each other. Hence E_0 is the natural candidate for a transfer learning distance. We will however use a sharper characterization of E_0:

Definition 3.4. *The **Cognitive Distance** between $x, y \in \mathcal{A}^*$ is given by*

$$E_1(x, y) := \max\{K(x|y), K(y|x)\}.$$

E_1 is upper semi-computable - we simply upper semi-compute in 'parallel' (by dovetailing) each term in the definition of E_1. In [7] it was proved:

$$E_0(x, y) = E_1(x, y) + O[\log(E_1(x, y))]. \tag{2}$$

We will actually prove a sharper version of the above where the log term is replaced by a constant. Now, we need:

Definition 3.5. *An **admissible distance** D is a partial, upper semi-computable, non-negative, symmetric function on $\mathcal{A}^* \times \mathcal{A}^*$ with $\forall y \sum_x 2^{-D(x,y)} \leq 1$ (we will assume $D(x, y) = \infty$ when it is undefined). Let $\mathcal{D}$ be the set of admissible distances.*

$A\ D \in \mathcal{D}$ *is **universal** in $\mathcal{D}$ if $\forall D' \in \mathcal{D}, \forall x, y \in \mathcal{A}^*, D(x, y) \overset{+}{\leq} D'(x, y)$.*

In [7] it was shown that $\forall D \in \mathcal{D}, \forall x, y \in \mathcal{A}^*$

$$E_1(x, y) \overset{+}{\leq} D(x, y). \tag{3}$$

That is, E_1 is universal in $\mathcal{D}$ (this was proven via (1) with $f = D$, as D satisfies the requisite conditions due to its admissibility). Note that [7] showed that the above holds for admissible *metrics*, but as pointed out in [11] this holds for admissible distances as well. Admissible distances include admissible versions of Hamming, Edit, Euclidean, Lempel-Ziv etc. distances [7,11,12]. See [7] for an eloquent account of why admissible distances (and distances satisfying the Kraft Inequality) are interesting for strings. Normalized, practical versions of E_1 has been applied very successfully in various clustering tasks - see [11] and especially [12]. We now state a sharper version of (2) (the proof is in the Appendix).

Theorem 3.1
$$E_0(x, y) \overset{+}{=} E_1(x, y).$$

Given Theorem 3.1, we now define:

Definition 3.6. *The* **transfer learning distance** *between two tasks* $\mu, \varphi \in \mathcal{V}$ *is defined as* $E_1(\mu, \varphi)$.

So from the above, we immediately get that transfer learning distance is universal in the class of admissible distances that may be used for measuring task similarity. This formally solves the conceptual problem of how one measures task similarity. We will use this distance function in Sect. 4 to formally solve other problems in transfer learning mentioned in the Introduction and give more reasons why it is sufficient to consider only admissible distances (see discussion following the proof of Theorem 4.1). E_1 and K are sufficient for sequential transfer (Sect. 4.2), however, for parallel transfer/multitask learning (Sect. 4.3), we do not even need this, as it is not clear that these are transfer methods.

3.3   Universal Transfer Learning Distance for m Tasks

The material in this section may be skipped as it is not used below, but we include it here for the sake of completeness and because the results are interesting in and of themselves. We also hope that the functions here will find application in task clustering problems which are important for designing 'Long Lived' transfer learning agents [3], and in clustering problems in general, as in [12]. The distance functions in this section apply to arbitrary strings in addition to elements of $\mathcal{V}$.

   Let $X := \{x_1, x_2, \cdots, x_m\}, x_j \in \mathcal{A}^*$, $X_i^{m_1}$ the ith subset of X of size m_1, $0 < m_1 < m$, $0 < i < \binom{m}{m_1}$. Let $\sigma(X_i^{m_1})$ be the set of permutations of elements of $X_i^{m_1}$. Then, to generalize E_0 to measure how much each group of m_1 x_js, $0 < m_1 < m$, contain about the other $m - m_1$ x_js, we define:

Definition 3.7. *The* m **fold information distance** $E_0^m(x_1, x_2, \cdots, x_m)$ *between* $x_1, x_2, \cdots, x_m \in \mathcal{A}^*$ *is the length of the shortest program that given any permutation of* m_1 x_js, $1 < m_1 < m$, *outputs a permutation of the other* $m - m_1$ x_js. *That is:*

$$E_0^m(x_1, x_2, \cdots, x_m) := \min_p \{l(p) : \forall m_1, i, x, 0 < m_1 < m, 1 \le i \le \binom{m}{m_1},$$
$$x \in \sigma(X_i^{m_1}), p(\langle \langle x \rangle^{m_1}, m_1 \rangle) = \langle y \rangle^{m-m_1}, \text{ where } y \in \sigma(X \setminus X_i^{m_1})\}.$$

In contrast to E_0 the additional information m_1 is included in the definition for E_0^m to determine how to interpret the input, – i.e. which $\rangle\langle^{m_1}$ to use to decode the input. E_0^m is upper semi-computable by the same reasoning E_0 is [7]. To give a sharper characterization of E_0^m, we define:

Definition 3.8. *The* **Cognitive Distance** *for* m *strings in* $\mathcal{A}^*$ *is:*

$$E_1^m(x_1, x_2, \cdots, x_m) := \max_{x_i} \max_{y \in \sigma(X \setminus \{x_i\})} E_1(x_i, \langle y \rangle^{m-1}).$$

E_1^m is upper semi-computable by the same reasoning E_1 is. We can now state the analogue of Theorem 3.1 for m strings (the proof is in the Appendix):

Theorem 3.2. $E_0^m(x_1, x_2, \cdots, x_m) \overset{+}{=} E_1^m(x_1, x_2, \cdots, x_m)$.

Definition 3.9. *The m-fold transfer learning distance between m tasks $\mu_1, \mu_2, \cdots, \mu_m \in \mathcal{V}$ is defined as $E_1^m(\mu_1, \mu_2, \cdots, \mu_m)$.*

Theorem 3.3. *The following are true:*

1. E_1^m *is universal in the class of admissible distances for m strings - i.e. functions $D_m : \times_m \mathcal{A}^* \to \mathbb{R}$ that are non-negative, upper semi-computable, m-wise symmetric, and satisfies the following version of the Kraft inequality: $\forall x, y_1, y_2, \cdots, y_{m-1} \in \mathcal{A}^*$, $\sum_{z_1, \cdots, z_{m-1} \in \mathcal{A}^*} 2^{-D_m(x, z_1, \cdots, z_{m-1})} \leq 1$ and $\sum_{w \in \mathcal{A}^*} 2^{-D_m(w, y_1, \cdots, y_{m-1})} \leq 1.$*
2. E_1^m *satisfies the above version of the Kraft inequality.*

Proof. Let $x, y_1, y_2, \cdots, y_{m-1} \in \mathcal{A}^*$. For part 1, by (1) and admissibility of D_m, $K(x|y_1, y_2, \cdots, y_{m-1}), K(y_1, y_2, \cdots, y_{m-1}|x) \overset{+}{\leq} D_m(x, y_1, y_2, \cdots, y_{m-1})$. The desired result now follows from the definition of E_1^m. Part 2 follows because by definition $E_1(x, \langle y_1, y_2, \cdots, y_{m-1} \rangle^{m-1}) \leq E_1^m(x, y_1, y_2, \cdots, y_{m-1})$, and $E_1(x, \langle y_1, y_2, \cdots, y_{m-1} \rangle^{m-1})$ satisfies the Kraft inequality in part 1 (see Sect. 4.2). $\square$

4   Universal Bayesian Transfer Learning

In this section we will discuss how to do transfer learning in Bayes mixtures over enumerable subsets $\mathcal{V}'$ of $\mathcal{V}$, which we consider as our task spaces. That is we will present a transfer learning analogue of Solomonoff Induction [13]. First we will discuss relevant error bounds for Bayesian sequence prediction, and then we will present our transfer learning methods.

4.1   Bayesian Convergence Results

A Bayes Mixture $\mathbf{M}$ over $\mathcal{V}'$ is defined by:

$$\mathbf{M}(x) := \sum_{\mu_i \in \mathcal{V}'} \mu_i(x) W(\mu_i). \tag{4}$$

where W is a prior with $W(\mu_i) \geq 0$ for each μ_i and $\sum_{\mu_i \in \mathcal{V}'} W(\mu_i) \leq 1$. Then the following well-known extraordinary result holds true $\forall \mu_j \in \mathcal{V}'$:

$$\sum_{t=0}^{\infty} \sum_{x \in \mathcal{A}^t} \mu_j(x) \left(\sum_{a \in \mathcal{A}} [\mathbf{M}(a|x) - \mu_j(a|x)]^2 \right) \leq -\ln W(\mu_j). \tag{5}$$

Note that, for finite $-\ln W(\mu_j)$, convergence is rapid; the expected number of times $t\ |\mathbf{M}(a|x) - \mu_j(a|x)| > \epsilon$ is $\leq -\ln W(\mu_j)/\epsilon^2$, and the probability that the number of ϵ deviations $> -\ln W(\mu_j)/\epsilon^2 \delta$ is $< \delta$. Now define:

Definition 4.1. *For a prior W, the* **error bound** *under (5) is defined as* $\mathsf{Eb}_W(\mu) := -\ln W(\mu)$. *A prior W is said to be* **universally optimal** *in some class C if for all priors $W' \in C$, $\forall \mu \in \mathcal{V}'$, $\mathsf{Eb}_W(\mu) \overset{+}{\leq} \mathsf{Eb}_{W'}(\mu)$.*

As we wish to investigate transfer in the limit, we will only consider lower semi-computable priors. Of particular interest is the Solomonoff-Levin prior: $2^{-K(\mu_i)}$. In this case, the error bound is $K(\mu_j)\ln 2$. This is intuitively appealing because it shows the smaller the code for μ_j, the smaller the bound, which is a instantiation of Occam's razor. In addition, for any other lower semi-computable prior W, the error bound $-\ln W(\mu_j)$ is upper semi-computable, and $-\ln W/\ln 2$ satisfies the conditions for (1) (with $y = \varepsilon$ and $W(x)$ undefined if $x \notin \mathcal{V}'$), so:

$$K(\mu_j)\ln 2 \overset{+}{\leq} -\ln W(\mu_j). \tag{6}$$

i.e. the Solomonoff-Levin prior is universally optimal in the class of lower semi-computable priors. Equation (5) was first proved in [13] for $\mathcal{V}' = \mathcal{V}$ and $\mathcal{A} = \mathcal{B}$, and was then extended to arbitrary finite alphabets, $\mathcal{V}$'s and bounded loss functions in [10], [14]. In [10] Hutter has also shown that Bayes mixtures are Pareto optimal, and that if $\mu_j \notin \mathcal{V}'$, but there is a $\rho \in \mathcal{V}'$ such that $\forall t \in \mathbb{N}$, the tth order KL divergence between ρ and $\mu_j \leq k$, then $\mathsf{Eb}_W(\mu_j) = -\ln W(\rho) + k$.

4.2   Universal Sequential Transfer Learning

We assume that we are given tasks $\varphi_1, \varphi_2, \cdots, \varphi_{m-1}$, $\varphi_i \in \mathcal{V}$, as previously learned tasks. We do not care about how these were learned - for instance each φ_i may be a weighted sum of elements of $\mathcal{V}'$ after having observed a finite sequence $x^{(i)}$ [10, Sect. 2.4] or each φ_i may be given by the user. Let $\varphi := \langle \varphi_1, \varphi_2, \ldots, \varphi_{m-1} \rangle^{m-1}$. The aim of transfer learning is to use φ as prior knowledge when predicting for the mth task with some unknown generating semi-measure $\mu \in \mathcal{V}'$. Given this, a transfer learning scheme is just a conditional prior over $\mathcal{V}'$, and it may or may not be based on a distance function. So,

Definition 4.2. *A transfer learning scheme is a lower semi-computable prior $W(\mu_i|\varphi)$ with $\sum_{\mu_i \in \mathcal{V}'} W(\mu_i|\varphi) \leq 1$, and $W(x|\varphi)$ undefined for $x \notin \mathcal{V}'$. A* **symmetric distance** D **based transfer learning scheme** *is a transfer learning scheme $W_D(\mu_i|\varphi)$ with $W_D(\mu_i|\varphi) := g(D(\mu_i, \varphi))$ for a symmetric function $D : \mathcal{A}^* \times \mathcal{A}^* \to \mathbb{R}$ and $g : \mathbb{R} \to [0,1]$.*

W_D is defined in terms of g because we do not want to put restrictions on how the distance function D may be used to induce a prior, or even what constraints D must satisfy other than being symmetric.

Definition 4.3. *Our* **universal transfer learning scheme** *is the prior $\xi_{\mathbf{TL}}$ $(\mu_i|\varphi) := 2^{-K(\mu_i|\varphi)}$. Our* **TL distance based universal transfer learning scheme** *for Bayes mixtures over $\mathcal{V}'$ is the prior $\xi_{\mathbf{DTL}}(\mu_i|\varphi) := 2^{-E_1(\mu_i, \varphi)}$.*

For $\xi_{\mathbf{DTL}}$ we use E_1 instead of E_1^m because E_1 measures amount of information between the mth task and previous $m-1$ tasks, which is what we want, whereas E_1^m measures amount of information between all possible disjoint groupings of tasks, and hence it measures more information than we are interested in. $\xi_{\mathbf{DTL}}$ is a prior since $\sum_{\mu_i \in \mathcal{V}'} 2^{-E_1(\mu_i, \varphi)} \leq \sum_{\mu_i \in \mathcal{V}'} 2^{-K(\mu_i|\varphi)} \leq 1$ ($K(\mu_i|\varphi)$ being lengths of programs). As $E_1(\cdot, \varphi)$ and $K(\cdot|\varphi)$ are upper semi-computable, $\xi_{\mathbf{DTL}}$ and $\xi_{\mathbf{TL}}$ are lower semi-computable.

So in the Bayesian framework $\xi_{\mathbf{DTL}}$ automatically transfers the right amount of information from previous tasks to a potential new task by weighing it according to how related it is to older tasks. $\xi_{\mathbf{TL}}$ is less conceptually pleasing as $K(\mu_i|\varphi)$ is not a distance, and a goal of TL has been to define transfer learning scheme using TL distance functions. But as we see below, $\xi_{\mathbf{TL}}$ is actually more generally applicable for sequential transfer.

Theorem 4.1. $\xi_{\mathbf{TL}}$ *and* $\xi_{\mathbf{DTL}}$ *are universally optimal in the class of transfer learning schemes and distance based transfer learning schemes respectively.*

Proof. Let W be a transfer learning scheme. $\mathsf{Eb}_{\xi_{\mathbf{TL}}}(\mu) = K(\mu|\varphi)\ln 2$ and $\mathsf{Eb}_W(\mu) = -\ln W(\mu|\varphi)$. W is lower semi-computable, which implies $-\ln W$ is upper semi-computable; $-\ln W/\ln 2$, restricted to $\mathcal{V}'$, satisfies the requisite conditions for (1) with $y = \varphi$, and so $\mathsf{Eb}_{\xi_{\mathbf{TL}}}(\mu) \overset{+}{\leq} \mathsf{Eb}_W(\mu)$.

Let W_D be a distance based transfer learning scheme. $\mathsf{Eb}_{\xi_{\mathbf{DTL}}}(\mu) = E_1(\mu, \varphi) \ln 2$ and $\mathsf{Eb}_{W_D}(\mu) = -\ln W_D(\mu|\varphi)$. $-\ln W_D$ is upper semi-computable as W_D is lower semi-computable; $-\ln W_D$ is symmetric, and restricted to $\mathcal{V}'$, $-\ln W_D/\ln 2$ satisfies the Kraft inequality condition in Definition 3.5; therefore $-\ln W_D/\ln 2 \in \mathcal{D}$. Now by (3) $\mathsf{Eb}_{\xi_{\mathbf{DTL}}}(\mu) \overset{+}{\leq} \mathsf{Eb}_{W_D}(\mu)$. $\square$

Note that for W_D the error bound is given by $-\ln W_D /\ln 2$ which is $\in \mathcal{D}$, and so whether D itself is admissible or not is irrelevant. This further justifies considering only admissible distances. So from the theorem and discussion above, our method formally solves the problem of sequential transfer. It is universally optimal, and it automatically determines how much information to transfer. Additionally, $\xi_{\mathbf{TL}}$ does not transfer information when the tasks are not related in the following sense. By (6), the non-transfer universally optimal prior is $2^{-K(.)}$, with error bound $K(\mu)\ln 2$. As $K(\mu|\varphi) \overset{+}{\leq} K(\mu)$ (by definition), we have

Theorem 4.2. $\xi_{\mathbf{TL}}$ *is universally optimal in the class of non-transfer priors.*

The above implies, that, from a *formal* perspective, sequential transfer is always justified - i.e. it never hurts to transfer (see last paragraph of Sect. 4.3).

4.3   Universal Parallel Transfer Learning

Multitask learning methods are considered to be 'parallel transfer' methods, but as we will see in this section, it is not entirely clear if this is true. In parallel transfer we learn m related tasks in parallel. There are m generating semi-measures $\mu_1, \mu_2, \cdots, \mu_m \in \mathcal{V}$ generating sequences $x^{(1)}, x^{(2)}, \cdots, x^{(m)}$ respectively. At step

t, μ_i generates the tth bit of sequence $x^{(i)}$ in the usual way. To apply (5) in this scenario, we assume that our semi-measures are defined over an alphabet $\mathcal{A}_m$ of size $|\mathcal{A}|^m$, i.e. we use an m vector of $\mathcal{A}$ to represent each element of $\mathcal{A}_m$. So given a sequence x of elements of $\mathcal{A}_m$, i.e. $x \in \mathcal{A}_m^*$, $x^{(i)}$ will be the ith components of vectors in x, for $1 \leq i \leq m$. A semi-measure over $\mathcal{A}_m$ is now defined as in Definition 3.1. Our task space $\mathcal{V}_m$ is now defined by:

$$\mathcal{V}_m := \{\rho : \forall x \in \mathcal{A}_m^*, \rho(x) = \prod_i \rho_i^m(x^{(i)}) \text{ where } \rho_i^m \in \mathcal{V}\}.$$

We denote the m different components of $\rho \in \mathcal{V}_m$ by ρ_i^m. It is easy to see that $\mathcal{V}_m$ is enumerable: as we enumerate $\mathcal{V}$, we use the $\langle\rangle^m$ map to determine the elements of $\mathcal{V}$ that will be the components of a particular $\rho \in \mathcal{V}_m$. We will consider as our task spaces enumerable subsets $\mathcal{V}_m'$ of $\mathcal{V}_m$. A Bayes mixture is given, as in (4), by $\mathbf{M}_m(x) := \sum_{\mu_i \in \mathcal{V}_m'} \mu_i(x) W(\mu_i)$. As before we define:

Definition 4.4. *A* **parallel transfer learning scheme** W_m *is a lower semi-computable prior over* $\mathcal{V}_m'$ *($W_m(x)$ undefined for $x \notin \mathcal{V}_m'$):*

$$W_m(\rho) := W_m(\rho_1^m, \rho_2^m, \cdots, \rho_m^m) \text{ with } \sum_{\rho \in \mathcal{V}_m'} W_m(\rho) \leq 1.$$

The **universal parallel transfer learning scheme** *is defined as the prior:*

$$\xi_{\mathbf{PTL}}(\rho) := \xi_{\mathbf{PTL}}(\rho_1^m, \rho_2^m, \cdots, \rho_m^m) := 2^{-K(\rho_1^m, \rho_2^m, \cdots, \rho_m^m)}.$$

Theorem 4.3. $\xi_{\mathbf{PTL}}$ *is universally optimal in the class of parallel transfer learning schemes.*

Proof. Let the generating semi-measure be $\mu = \mu_1^m, \mu_2^m, \cdots, \mu_m^m$. $\mathsf{Eb}_{\xi_{\mathbf{PTL}}}(\mu) = K(\mu_1^m, \mu_2^m, \cdots, \mu_m^m) \ln 2$ while for any W_m, $\mathsf{Eb}_{W_m}(\mu) = -\ln W_m(\mu_1^m, \mu_2^m, \cdots, \mu_m^m)$. By minimality (1), $K(\mu_1^m, \mu_2^m, \cdots, \mu_m^m) \ln 2 \overset{+}{\leq} -\ln W_m(\mu_1^m, \mu_2^m, \cdots, \mu_m^m)$. Hence the prior $\xi_{\mathbf{PTL}}$ is universally optimal. $\qquad\square$

Indeed, $K(\rho_1^m, \rho_2^m, \cdots, \rho_m^m)$ is the measure of similarity that was used in [6] to analyze multitask learning in a PAC setting (as mentioned in the Introduction). However, $\xi_{\mathbf{PTL}}$ is also the non-transfer Solomonoff-Levin prior for the space $\mathcal{V}_m$. Therefore, it seems that in multitask transfer, in contrast to sequential transfer, no actual transfer of information is occurring. Plain single task learning is taking place, but in a product space. The benefit of this is not clear from a formal perspective as $K(x) \overset{+}{\leq} K(y_1, y_2, x, \cdots, y_{m-1})$, and so this type of 'transfer', in general, should not help learning. Note that, for other performance measures (e.g. sum of the errors for the m tasks) where the error bound is $-\ln W_m$, via similar arguments as above $2^{-K(.)}$ etc. will still be the universally optimal prior.

However: In the majority of multitask learning methods used in practice, each $x^{(i)}$ corresponds to training samples for task i. In a Bayesian setting, for each task i, $x^{(j)}, j \neq i$ now function as prior knowledge, and we have priors of the form : $W(\mu|x^{(j)}, 1 \leq j \leq m, j \neq i)$. So current multitask learning methods seem to be performing m sequential transfers in parallel, rather than 'pure' parallel transfer. However, it has been observed that transferring from unrelated tasks hurts generalization [1], which, given Theorem 4.2, seems to contradict the above conclusion. Nonetheless, our own empirical investigations [15] lead us to believe that this is not because of parallel transfer but use of improper algorithms.

5 Kolmogorov Complexity of Functions

One natural definition of Kolmogorov complexity of a function f given string q is $K'(f|q)$, the length of the shortest program that computes f given q as extra information [16, Sect. 7], [17, Sect 2.2.3] . So one objection to the definitions in this paper may be that, since we are interested in $\mu \in \mathcal{V}$ as semi-measures (i.e. functions), perhaps we should define the complexity of $\mu \in \mathcal{V}$ as $K'(\mu|q)$. However K' is not computable in the limit, so to address this concern, we adapt the definition in [16], and define $K''(\mu|q)$ (which is upper semi-computable):

Definition 5.1. *For* $\mu \in \mathcal{V}, q \in \mathcal{A}^*$,

$$K''(\mu|q) := \min_r \{l(r) : r(q) = \alpha \ and \ \exists \ proof \ \S \ \forall x : \mu(x) \Uparrow \alpha(x) \S\}.$$

The above definition means $K''(\mu|q)$ is the length of the shortest program that given q as input, outputs a program α that *provably* lower semi-computes (denoted by $\Uparrow$) the same semi-measure as μ. The proof is in a formal system $\mathcal{F}$, in which we can formalize equality of programs in the sense of $\Uparrow$. Formulas in $\mathcal{F}$ are enclosed in $\S\,\S$ - so $\S \forall x : \mu(x) \Uparrow \alpha(x) \S$ is true if and only if $\forall x, U(\mu, x) \Uparrow U(\alpha, x)$. Another property of $\mathcal{F}$ we use is that the set of correct proofs is enumerable (see [16] for more details). The following is true:

Lemma 5.1. *Let* $\arg K''(\mu|q)$ *denote the* α *that is the witness in Definition 5.1. Then, 1)* $\forall \mu \in \mathcal{V}, q \in \mathcal{A}^*, K''(\mu|q) \leq K(\mu|q)$. *2)* $\forall n \in \mathbb{N}, q \in \mathcal{A}^* \ \exists \mu \in \mathcal{V}$ *such that* $K(\mu|q) - K''(\mu|q) \overset{+}{\geq} n$. *3)* $\forall \mu \in \mathcal{V}, q \in \mathcal{A}^*, K(\arg K''(\mu|q)) \overset{+}{=} K''(\arg K''(\mu|q))$. *4)* $\forall q \in \mathcal{A}^*, \sum_{\mu \in \mathcal{V}} 2^{-K''(\mu|q)} = \infty$.

Proof. The results are mostly self-evident. Part 1 follows from definition since each $\mu \in \mathcal{V}$ provably computes the same function as itself. For part 2, fix $q \in \mathcal{A}^*, \varphi \in \mathcal{V}$, and $n \in \mathbb{N}$. Now by the theory of random strings (see [8]), there exists infinitely many incompressible strings - i.e. strings s such that $K(s|\varphi, K(\varphi|q), q) \geq l(s)$. Let $l(s) = n$, and construct a $\mu \in \mathcal{V}$ which is just φ with s encoded in it at a fixed position t. Now $K(\mu|q) \overset{+}{=} K(s, \varphi|q)$, since, using t, given a program to generate μ given q, we can recover φ and s from it, and given a program to generate $\langle s, \varphi \rangle$ given q, we can construct μ. A fundamental and deep result in AIT, due to Gacs and Chaitin, gives us $K(s, \varphi|q) \overset{+}{=} K(s|\varphi, K(\varphi|q), q) + K(\varphi|q)$. By definition $K''(\mu|q) \leq K(\varphi|q)$, so we get, $K(\mu|q) - K''(\mu|q) \overset{+}{=} K(\varphi, s|q) - K''(\mu|q) \overset{+}{=} K(s|\varphi, K(\varphi|q), q) + K(\varphi|q) - K''(\mu|q) \geq n + K(\varphi|q) - K''(\mu|q) \geq n + K(\varphi|q) - K(\varphi|q) = n$.

Part 3 follows from definition. For part 4, for each $\varphi \in \mathcal{V}$, by the method in part 2, there are infinitely many $\mu \in \mathcal{V}$ such that $\forall x, \varphi(x) \Uparrow \mu(x)$ provably . So $\sum_{\mu_i \in \mathcal{V}} 2^{-K''(\mu_i|q)} = \infty$, as infinitely many μ_is have the same $K''(\mu_i|q)$ value. $\quad\square$

Parts 1 and 2 in the lemma show that the K''s can uncover much more similarity between tasks than K. However, there is no advantage to using K'' for Bayesian transfer learning, as for any enumerable set $\mathcal{V}'$, the set of programs $\mathcal{V}'_{proof}$ that are provably equal to the elements of $\mathcal{V}'$ is also enumerable (because the set of correct proofs in $\mathcal{F}$ are enumerable). Therefore we get that for any $\mu \in \mathcal{V}'$, $\arg K''(\mu|q)$ is in $\mathcal{V}'_{proof}$. Since the error bound in Bayes mixtures depends only on the weight assigned to the generating semi-measure , from part 3 of the above lemma, substituting $\mathcal{V}'$ with $\mathcal{V}'_{proof}$ counteracts the benefit of using K''. However, part 2 in the lemma shows that K'' deserves further study.

6   Discussion

In this paper we formally solved some of the key problems of transfer learning in the same sense that Solomonoff Induction solves the problem of inductive inference. We defined universal transfer learning distances and showed how these may be used to automatically transfer the right amount of information in our universally optimal Bayesian sequential transfer method. We also showed that from a formal perspective sequential transfer is always justified, and while optimal parallel transfer method exists, it is not clear that it is a transfer method; so this issue needs further investigation. We also showed that practical parallel transfer methods (i.e. [1]) may in fact be sequential transfer methods in disguise. Practical approximations to our methods to transfer information across arbitrary databases from the UCI ML repository are described in [15]. We note that the results and discussion in Sects. 3 and 4 (except the results of Sect. 4.3) also apply when instead of previous tasks we use arbitrary strings. So our methods are also universally optimal Bayesian methods for using prior knowledge.

We will conclude with a brief comparison of our method to [6]. [6] deals only with finite sample spaces, and computable tasks and hypothesis spaces, and gives PAC bounds, where the sample complexity required to bound the error for given ϵ, δ is proportional to Kolmogorov Complexity of the m hypothesis being considered. The number of tasks required for the bounds to hold is ≥ 8192 (Theorem 3). In contrast, our results are elegant and far more general. They are incomputable, but serve as explicit guidelines on how one may approximate them to design principled and powerful practical algorithms [12,15].

Acknowledgements. We would like to thank Samarth Swarup, Sylvian Ray and Kiran Lakkaraju for their comments. The comments from three anonymous referees and Dr. Marcus Hutter were also invaluable in improving the paper.

References

1. Caruana, R.: Multitask learning. Machine Learning 28, 41–75 (1997)
2. Thrun, S., Mitchell, T.: Lifelong robot learning. Robotics and Autonomous Systems 15, 25–46 (1995)
3. Thrun, S., Pratt, L.Y. (eds.): Learning To Learn. Kluwer Academic Publishers, Boston (1998)
4. Ben-David, S., Schuller, R.: Exploiting task relatedness for learning multiple tasks. In: Proceedings of the 16th Annual Conference on Learning Theory (2003)
5. Baxter, J.: A model of inductive bias learning. Journal of Artificial Intelligence Research 12, 149–198 (2000)
6. Juba, B.: Estimating relatedness via data compression. In: Proceedings of the 23rd International Conference on Machine Learning (2006)
7. Bennett, C., Gacs, P., Li, M., Vitanyi, P., Zurek, W.: Information distance. IEEE Transactions on Information Theory 44(4), 1407–1423 (1998)
8. Li, M., Vitanyi, P.: An Introduction to Kolmogorov Complexity and its Applications, 2nd edn. Springer, New York (1997)
9. Zvonkin, A.K., Levin, L.A.: The complexity of finite objects and the development of the concepts of information and randomness by means of the theory of algorithms. Russian Math. Surveys 25(6), 83–124 (1970)
10. Hutter, M.: Optimality of Bayesian universal prediction for general loss and alphabet. Journal of Machine Learning Research 4, 971–1000 (2003)
11. Li, M., Chen, X., Ma, B., Vitanyi, P.: The similarity metric. IEEE Transactions on Information Theory 50(12), 3250–3264 (2004)
12. Cilibrasi, R., Vitanyi, P.: Clustering by compression. IEEE Transactions on Information theory 51(4), 1523–1545 (2004)
13. Solomonoff, R.J.: Complexity-based induction systems: comparisons and convergence theorems. IEEE Transactions on Information Theory 24(4), 422–432 (1978)
14. Hutter, M.: Universal Artificial Intelligence: Sequential Decisions Based on Algorithmic Probability. Springer, Berlin (2004)
15. Mahmud, M.M.H., Ray, S.: Transfer learning using Kolmogorov complexity:basic theory and empirical evaluations. Technical Report UIUC-DCS-R-2007-2875, Department of Computer Science, University of Illinois at Urbana-Champaign (2007)
16. Hutter, M.: The fastest and shortest algorithm for all well defined problems. International Journal of Foundations of Computer Science 13(3), 431–443 (2002)
17. Grunwald, P., Vitanyi, P.: Shannon information and Kolmogorov complexity. Submitted to IEEE Transactions on Information Theory (2004)

Appendix: Proofs

Proof of Theorem 3.1. $E_0(x,y) \overset{+}{=} E_1(x,y)$.

Proof. Let p be a program such that $p(x) = y$ and $p(y) = x$. So by definition $E_1(x,y) \leq l(p)$ for all such p. Since $\arg E_0(x,y)$ is a such a p, we have $E_1(x,y) \overset{+}{\leq} E_0(x,y)$. Now we prove the inequality in the other direction. Fix any two strings α, β and set $E_1(\alpha, \beta) = E1$. Now we will derive a program q_{E1} with $l(q_{E1}) \overset{+}{=} E1$ which given α outputs β and given β outputs α. We will do so by constructing a graph G that assigns a unique color/code of length $\leq E1 + 1$ to each pair of strings x, y with $E_1(x,y) \leq E1$, and the code will turn out to be more or less the program q_{E1} we need to convert α to β and vice versa. We note that the proof of (2) also uses a similar graph construction method. Define $G := (V, E)$ with vertices V and undirected edges E:

$$V := \{x : x \in A\} \text{ and } E := \{\{x, y\} : x \in A, y \in A_x\}, \text{ where,}$$
$$A := \{x : \exists y, E_1(x, y) \le E1\} \text{ and } \forall x \in A, A_x := \{y : E_1(x, y) \le E1\}.$$

The degree of $x \in V$ is $|A_x|$ by construction. Hence the maximum degree of G is $\Delta_G = \max_{x \in A} |A_x|$. We define the set of colors/code $\mathcal{C}_{E1}$ as:

$$\mathcal{C}_{E1} := \{p0 : p \in B\} \cup \{p1 : p \in B\}, \text{ where,}$$
$$B := \{p : p(x) = y, x \in A, y \in A_x, l(p) \le E1\}.$$

q_{E1} will need to dynamically construct G and $\mathcal{C}_{E1}$, and assign a *valid* coloring to the edges in G using $\mathcal{C}_{E1}$. For this, all we need is $E1$. We run all programs p with $l(p) \le E1$ on all $x \in \mathcal{A}^*$ in 'parallel' by dovetailing and record triples (p, x, y) such that $p(x) = y$. Whenever we record (p, x, y) we check to see if we have previously recorded (q, y, x). If so, we add $p0, p1, q0, q1$ to $\mathcal{C}_{E1}$, x, y to V and $\{x, y\}$ to E. Of course, if any of these already exist in the respective sets, we do not add it again. We color a newly added edge $\{x, y\}$ using a color from $\mathcal{C}_{E1}$ using the First-Fit algorithm - i.e. the first color that has not been assigned to any other $\{x, w\}$ or $\{y, z\}$. So, by dynamically reconstructing G, given x (y) and the color for $\{x, y\}$, q_{E1} can use the color to recognize and output y (x).

That $\mathcal{C}_{E1}$ has sufficient colors to allow valid coloring can be seen as follows. $p \in B$ iff $l(p) \le E1$ and for some A_x, $y \in A_x, p(x) = y$. So for each A_x, for each $y \in A_x$, $\exists p_y \in B$, and $p_y \neq p_{y'}$ $\forall y' \in A_x, y' \neq y$ since $p_y(x) \neq y'$. This means, for each A_x, $|\mathcal{C}_{E1}| \ge 2|A_x|$, or $|\mathcal{C}_{E1}| \ge 2\Delta_G$. By the same reasoning and the construction procedure above, as we dynamically construct G and $\mathcal{C}_{E1}$, the estimates $\mathcal{C}_{E1}^t$ and Δ_G^t at step t of the construction process also satisfies $|\mathcal{C}_{E1}^t| \ge 2\Delta_G^t$. Now at step t First-Fit requires at most $2\Delta_G^t - 1$ colors to assign a valid color, as two vertices could have exhausted at most $2\Delta_G^t - 2$ colors between them. Therefore First-Fit always has sufficient colors to assign a valid coloring

Each color/code in $\mathcal{C}_{E1}$ is at most $E1 + 1$ in length by construction. So, as we construct G, α and β shows up in the graph at some point with code/color (say) γ, and $l(\gamma) \le E1 + 1$. From construction of $\mathcal{C}_{E1}$, γ is a self-delimiting string p, followed by 0 or 1. γ and $E1$ can be encoded by a string $pa0^{E1-l(p)}1$, where a is 0 if $\gamma = p0$, or 1 if $\gamma = p1$, and $0^{E1-l(p)}$ is 0 repeated $E1 - l(p)$ times.

The desired program q_{E1} has encoded in it the string $pa0^{E1-l(p)}1$ at some fixed position, and $q_{E1}(z)$ works as follows. q_{E1} decodes p (which is possible as it is self-delimiting) and then reads the next bit, which is a, to get γ. It computes $E1$ from counting the number of 0s after a and $l(p)$. When $a = 0$, it is not confused with the 0s following it because it is the bit that appears immediately after p, and p can be decoded by itself. q_{E1} then reconstructs G using $E1$, and finds the edge $\{z, w\}$ with color γ, and outputs w. By construction, if $z = \alpha$ then $w = \beta$ and if $z = \beta$ then $w = \alpha$. Since $l(q_{E1}) \stackrel{+}{=} E1$ (the constant being for the extra bits in $pa0^{E1-l(p)}1$ and other program code in q), we have $E_0(\alpha, \beta) \le l(q_{E1}) \stackrel{+}{=} E_1(\alpha, \beta)$, and therefore $E_0(\alpha, \beta) \stackrel{+}{=} E_1(\alpha, \beta)$. $\qquad \square$

Proof of Theorem 3.2. $E_0^m(x_1, x_2, \cdots, x_m) \stackrel{+}{=} E_1^m(x_1, x_2, \cdots, x_m)$.

Proof. (Sketch) We assume notation of Sect. 3.3, and use $x_{1,m}$ as a shorthand for $x_1, x_2, \cdots, x_m$. The proof is similar to the proof of Theorem 3.1. Fix $\Lambda := \{\lambda_{1,m}\}$. $E_1^m(\lambda_{1,m}) \stackrel{+}{\leq} E_0^m(\lambda_{1,m})$ follows using the method used in proving the first inequality in Theorem 3.1. But now we have to modify arg $E_0^m(\lambda_{1,m})$ to give it the extra information of length $\log m$ and $m \log m$ to interpret its input and order the output strings, respectively, so that its output for the relevant elements of Λ is like that of arg $E_1^m(\lambda_{1,m})$. So with m treated as a constant $E_1^m(\lambda_{1,m}) \stackrel{+}{\leq} E_0^m(\lambda_{1,m})$, and otherwise $E_1^m(\lambda_{1,m}) \leq E_0^m(\lambda_{1,m}) + (m+1) \log m$. For the inequality in the other direction let $E1m = E_1^m(\lambda_{1,m})$. We will construct a program q_{E1m} with $l(q_{E1m}) \stackrel{+}{=} E1m$ that will have the same outputs as arg $E_0^m(\lambda_{1,m})$ on $\langle \langle y \rangle^{m_1}, m_1 \rangle$, $y \in \sigma(\Lambda_i^{m_1}), 0 < m_1 < m, 1 \leq i \leq \binom{m}{m_1}$. For this, we need the set L, and colors $\mathcal{C}_{E1m}$ – additionally we define sets A_xs and B to make elucidation of the proof easier:

$$L := \{\{x_{1,m}\} : E_1^m(x_{1,m}) \leq E1m\} \text{ and } A_x := \{\{z_{1,m-1}\} : \{x, z_{1,m-1}\} \in L\}$$

$$\mathcal{C}_{E1m} := \{pj : p \in B, j \leq m\}, \text{ where,}$$

$$B := \{p : p(x) = \langle y_{1,m-1} \rangle^{m-1}, \{y_{1,m-1}\} \in A_x, l(p) \leq E1m\}.$$

By using $E1m$ and m, q_{E1m} will construct L dynamically and color each element of L. To do so, we run all programs p with $l(p) \leq E1m$ on all $x \in \mathcal{A}^*$ in parallel. If we find $p(x) = y$, we record the tuples $(p, (w_{1,m-1}), y)$ and $(p, x, (z_{1,m-1}))$, where $x = \langle w_{1,m-1} \rangle^{m-1}$ and $y = \langle z_{1,m-1} \rangle^{m-1}$. If we find a $x_{1,m}$ such that we have recorded $(p_{x_i,y}, x_i, y)$ and (p_{y,x_i}, y, x_i) for each x_i and $\forall y \in \sigma(\{x_{1,m}\} \backslash \{x_i\})$, then we add each of the $p_{x_i,y}, p_{y,x_i}$s to B and add the corresponding colors to $\mathcal{C}_{E1m}$. We add $X := \{x_{1,m}\}$ to L and color it using a variation of First-Fit in Theorem 3.1 as follows. Denote by $\mathcal{C}(X)$ the color assigned to X. Then $\mathcal{C}(X)$ is set to the first $\gamma \in \mathcal{C}_{E1m}$ such that $\forall x \in X$, if $x \in X'$, $X' \in L$, then $\gamma \neq \mathcal{C}(X')$. So given any $x \in X$, and $\mathcal{C}(X)$, q_{E1m} can reconstruct and color L as above and hence find X. To see that $\mathcal{C}_{E1m}$ has enough colors: Let $\Delta_L := \max_x |A_x|$. For each $\kappa \in A_x$, $\exists p_\kappa \in B$, $p_\kappa(x) = \langle y \rangle^{m-1}, y \in \sigma(\kappa)$ and $p_{\kappa'} \neq p_\kappa \forall \kappa' \in A_x, \kappa' \neq \kappa$. Therefore $|\mathcal{C}_{E1m}| \geq m\Delta_L$. Also, from the construction method for L above, $|\mathcal{C}_{E1m}^t| \geq m\Delta_L^t$ for the estimates at each step t of the construction process. When coloring X at step t, each $x \in X$ has used $\leq \Delta_L^t - 1$ colors previously. So, as $|X| = m$, First-Fit will require at most $m(\Delta_L^t - 1) + 1$ colors to assign a valid color to X.

Now $\max_{\gamma \in \mathcal{C}_{E1m}} l(\gamma) \leq E1m + l(m)$ $(l(m) = \lfloor \log(m+1) \rfloor$ [8]), and with m as a constant, this becomes $E1m + c$. Like q_{E1} from Theorem 3.1, q_{E1m} can encode $E1m$, m, and the color $\gamma_\Lambda = pj$ for Λ in itself as $p\bar{j}\bar{m}0^{E1m-l(p)}1$ (for definition of $\bar{j}$ and $\bar{m}$ see Sect. 2). Using this, q_{E1m} can dynamically construct L, $\mathcal{C}_{E1m}$ and color L. For input $\langle x, m_1 \rangle$, $0 < m_1 < m$, q_{E1m} decodes x with $\beta_j := \rangle x \langle_j^{m_1}$, $0 < j < m_1$. By construction of L, using any β_j and γ_Λ, q_{E1m} can find Λ in L, and output $\langle y \rangle^{m-m_1}$, $y \in \sigma(\Lambda \backslash \{\beta_{1,m_1}\})$, which is what is required. This proves, with m as a constant $E_0^m(\lambda_{1,m}) \stackrel{+}{\leq} E_1^m(\lambda_{1,m})$ and $E_0^m(\lambda_{1,m}) \stackrel{+}{\leq} E_1^m(\lambda_{1,m}) + 3\lceil \log m \rceil$ otherwise. This and the first inequality completes the proof. $\square$

Tuning Bandit Algorithms in Stochastic Environments

Jean-Yves Audibert[1], Rémi Munos[2], and Csaba Szepesvári[3]

[1] CERTIS - Ecole des Ponts
19, rue Alfred Nobel - Cité Descartes
77455 Marne-la-Vallée - France
audibert@certis.enpc.fr
[2] INRIA Futurs Lille, SequeL project,
40 avenue Halley, 59650 Villeneuve d'Ascq, France
remi.munos@inria.fr
[3] University of Alberta, Edmonton T6G 2E8, Canada
szepesva@cs.ualberta.ca

Abstract. Algorithms based on upper-confidence bounds for balancing exploration and exploitation are gaining popularity since they are easy to implement, efficient and effective. In this paper we consider a variant of the basic algorithm for the stochastic, multi-armed bandit problem that takes into account the empirical variance of the different arms. In earlier experimental works, such algorithms were found to outperform the competing algorithms. The purpose of this paper is to provide a theoretical explanation of these findings and provide theoretical guidelines for the tuning of the parameters of these algorithms. For this we analyze the expected regret and for the first time the concentration of the regret. The analysis of the expected regret shows that variance estimates can be especially advantageous when the payoffs of suboptimal arms have low variance. The risk analysis, rather unexpectedly, reveals that except for some very special bandit problems, the regret, for upper confidence bounds based algorithms with standard bias sequences, concentrates only at a polynomial rate. Hence, although these algorithms achieve logarithmic expected regret rates, they seem less attractive when the risk of suffering much worse than logarithmic regret is also taken into account.

1 Introduction and Notations

In this paper we consider *stochastic multi-armed bandit problems*. The original motivation of bandit problems comes from the desire to optimize efficiency in clinical trials when the decision maker can choose between treatments but initially does not know which of the treatments is the most effective one [9]. Multi-armed bandit problems became popular with the seminal paper of Robbins [8], after which they have found applications in diverse fields, such as control, economics, statistics, or learning theory.

Formally, a K-armed bandit problem ($K \geq 2$) is defined by K distributions, $\nu_1, \ldots, \nu_K$, one for each "arm" of the bandit. Imagine a gambler playing with

M. Hutter, R.A. Servedio, and E. Takimoto (Eds.): ALT 2007, LNAI 4754, pp. 150–165, 2007.

these K slot machines. The gambler can pull the arm of any of the machines. Successive plays of arm k yield a sequence of independent and identically distributed (i.i.d.) real-valued random variables $X_{k,1}, X_{k,2}, \ldots$, coming from the distribution ν_k. The random variable $X_{k,t}$ is the payoff (or reward) of the k-th arm when this arm is pulled the t-th time. Independence also holds for rewards across the different arms. The gambler facing the bandit problem wants to pull the arms so as to maximize his cumulative payoff.

The problem is made challenging by assuming that the payoff distributions are initially unknown. Thus the gambler must use exploratory actions in order to learn the utility of the individual arms, making his decisions based on the available past information. However, exploration has to be carefully controlled since excessive exploration may lead to unnecessary losses. Hence, efficient online algorithms must find the right balance between *exploration and exploitation.*

Since the gambler cannot use the distributions of the arms (which are not available to him) he must follow a *policy*, which is a mapping from the space of possible histories, $\cup_{t \in \mathbb{N}^+} \{1, \ldots, K\}^t \times \mathbb{R}^t$, into the set $\{1, \ldots, K\}$, which indexes the arms. Let $\mu_k = \mathbb{E}[X_{k,1}]$ denote the expected reward of arm k.[1] By definition, an *optimal arm* is an arm having the largest expected reward. We will use k^* to denote the index of such an arm. Let the optimal expected reward be $\mu^* = \max_{1 \leq k \leq K} \mu_k$.

Further, let $T_k(t)$ denote the number of times arm k is chosen by the policy during the first t plays and let I_t denote the arm played at time t. The *(cumulative) regret at time n* is defined by

$$\hat{R}_n \triangleq \sum_{t=1}^{n} X_{k^*,t} - \sum_{t=1}^{n} X_{I_t, T_{I_t}(t)}.$$

Oftentimes, the goal is to minimize the *expected (cumulative) regret of the policy*, $\mathbb{E}[\hat{R}_n]$. Clearly, this is equivalent to maximizing the total expected reward achieved up to time n. It turns out that the expected regret satisfies

$$\mathbb{E}[\hat{R}_n] = \sum_{k=1}^{K} \mathbb{E}[T_k(n)]\Delta_k,$$

where $\Delta_k \triangleq \mu^* - \mu_k$ is the expected loss of playing arm k. Hence, an algorithm that aims at minimizing the expected regret should minimize the expected sampling times of suboptimal arms.

Early papers studied stochastic bandit problems under Bayesian assumptions (e.g., [5]). Lai and Robbins [6] studied bandit problems with parametric uncertainties. They introduced an algorithm that follows what is now called the "optimism in the face of uncertainty". Their algorithm computes *upper confidence bounds* for all the arms by maximizing the expected payoff when the parameters are varied within appropriate confidence sets derived for the parameters. Then the algorithm chooses the arm with the highest such bound. They

---

[1] $\mathbb{N}$ denotes the set of natural numbers, including zero and $\mathbb{N}^+$ denotes the set of strictly positive integers.

show that the expected regret increases logarithmically only with the number of trials and prove that the regret is asymptotically the smallest possible up to a sublogarithmic factor for the considered family of distributions. Agrawal [1] has shown how to construct such optimal policies starting from the sample-mean of the arms. More recently, Auer et al. [3] considered the case when the rewards come from a bounded support, say $[0, b]$, but otherwise the reward distributions are unconstrained. They have studied several policies, most notably UCB1 which constructs the Upper Confidence Bounds (UCB) for arm k at time t by adding the *bias factor*

$$\sqrt{\frac{2b^2 \log t}{T_k(t-1)}}$$

to its sample-mean. They have proven that the expected regret of this algorithm satisfies

$$\mathbb{E}[\hat{R}_n] \leq 8 \left(\sum_{k:\mu_k < \mu^*} \frac{b^2}{\Delta_k} \right) \log(n) + O(1). \tag{1}$$

In the same paper they propose UCB1-NORMAL, that is designed to work with normally distributed rewards only. This algorithm estimates the variance of the arms and uses these estimates to refine the bias factor. They show that for this algorithm when the rewards are indeed normally distributed with means μ_k and variances σ_k^2,

$$\mathbb{E}[\hat{R}_n] \leq 8 \sum_{k:\mu_k < \mu^*} \left(\frac{32\sigma_k^2}{\Delta_k} + \Delta_k \right) \log(n) + O(1). \tag{2}$$

Note that one major difference of this result and the previous one is that the regret-bound for UCB1 scales with b^2, while the regret bound for UCB1-NORMAL scales with the variances of the arms. First, let us note that it can be proven that the scaling behavior of the regret-bound with b is not a proof artifact: The expected regret indeed scales with $\Omega(b^2)$. Since b is typically just an *a priori* guess on the size of the interval containing the rewards, which might be overly conservative, it is desirable to lessen the dependence on it.

Auer et al. introduced another algorithm, UCB1-Tuned, in the experimental section of their paper. This algorithm, similarly to UCB1-NORMAL uses the empirical estimates of the variance in the bias sequence. Although no theoretical guarantees were derived for UCB1-Tuned, this algorithm has been shown to outperform the other algorithms considered in the paper in essentially all the experiments. The superiority of this algorithm has been reconfirmed recently in the latest Pascal Challenge [4]. Intuitively, algorithms using variance estimates should work better than UCB1 when the variance of some suboptimal arms is much smaller than b^2, since these arms will be less often drawn: suboptimal arms are more easily spotted by algorithms using variance estimates.

In this paper we study the regret of *UCB-V*, which is a generic UCB algorithm that use variance estimates in the bias sequence. In particular, the bias sequences of UCB-V take the form

$$\sqrt{\frac{2V_{k,T_k(t-1)}\mathcal{E}_{T_k(t-1),t}}{T_k(t-1)}} + c\frac{3b\mathcal{E}_{T_k(t-1),t}}{T_k(t-1)},$$

where $V_{k,s}$ is the empirical variance estimate for arm k based on s samples, $\mathcal{E}$ (viewed as a function of (s,t)) is a so-called *exploration function* for which a typical choice is $\mathcal{E}_{s,t} = \zeta \log(t)$ (thus in this case, $\mathcal{E}$ independent of s). Here $\zeta, c > 0$ are tuning parameters that can be used to control the behavior of the algorithm.

One major result of the paper (Corollary 1) is a bound on the expected regret that scales in an improved fashion with b. In particular, we show that for a particular settings of the parameters of the algorithm,

$$\mathbb{E}[\hat{R}_n] \le 10 \sum_{k:\mu_k<\mu^*} \left(\frac{\sigma_k^2}{\Delta_k} + 2b \right) \log(n).$$

The main difference to the bound (1) is that b^2 is replaced by σ_k^2, though b still appears in the bound. This is indeed the major difference to the bound (2).[2] In order to prove this result we will prove a novel tail bound on the sample average of i.i.d. random variables with bounded support that, unlike previous similar bounds, involves the empirical variance and which may be of independent interest (Theorem 1). Otherwise, the proof of the regret bound involves the analysis of the sampling times of suboptimal arms (Theorem 2), which contains significant advances compared with the one in [3]. This way we obtain results on the expected regret for a wide class of exploration functions (Theorem 3). For the "standard" logarithmic sequence we will give lower limits on the tuning parameters: If the tuning parameters are below these limits the loss goes up considerably (Theorems 4,5).

The second major contribution of the paper is the analysis of the risk that the studied upper confidence based policies have a regret much higher than its expected value. To our best knowledge no such analysis existed for this class of algorithms so far. In order to analyze this risk, we define the *(cumulative) pseudo-regret* at time n via

$$R_n = \sum_{k=1}^{K} T_k(n)\Delta_k.$$

Note that the expectation of the pseudo-regret and the regret are the same: $\mathbb{E}[R_n] = \mathbb{E}[\hat{R}_n]$. The difference of the regret and the pseudo-regret comes from the randomness of the rewards. Sections 4 and 5 develop high probability bounds for the pseudo-regret . The same kind of formulae can be obtained for the cumulative regret (see Remark 2 p.162).

Interestingly, our analysis revealed some tradeoff that we did not expect: As it turns out, if one aims for logarithmic expected regret (or, more generally, for subpolynomial regret) then the regret does not necessarily concentrate exponentially fast around its mean (Theorem 7). In fact, this is the case when with positive probability the optimal arm yields a reward smaller than the expected

---

[2] Although this is unfortunate, it is possible to show that the dependence on b is unavoidable.

reward of some suboptimal arm. Take for example two arms satisfying this condition and with $\mu_1 > \mu_2$: the first arm is the optimal arm and $\Delta_2 = \mu_1 - \mu_2 > 0$. Then the distribution of the pseudo-regret at time n will have two modes, one at $\Omega(\log n)$ and the other at $\Omega(\Delta_2 n)$. The probability mass associated with this second mass will decay polynomially with n where the rate of decay depends on Δ_2. Above the second mode the distribution decays exponentially. By increasing the exploration rate the situation can be improved. Our risk tail bound (Theorem 6) makes this dependence explicit. Of course, increasing exploration rate increases the expected regret. This illustrates the tradeoff between the expected regret and the risk of achieving much worse than the expected regret. One lesson is thus that if in an application risk is important then it might be better to increase the exploration rate.

In Section 5, we study a variant of the algorithm obtained by $\mathcal{E}_{s,t} = \mathcal{E}_s$. In particular, we show that with an appropriate choice of $\mathcal{E}_s = \mathcal{E}_s(\beta)$, for any $0 < \beta < 1$, for an infinite number of plays, the algorithm achieves *finite* cumulative regret with probability $1 - \beta$ (Theorem 8). Hence, we name this variant PAC-UCB ("Probably approximately correct UCB"). Besides, for a finite time-horizon n, choosing $\beta = 1/n$ then yields a logarithmic bound on the regret that fails with probability $O(1/n)$ only. This should be compared with the bound $O(1/\log(n)^a)$, $a > 0$ obtained for the standard choice $\mathcal{E}_{s,t} = \zeta \log t$ in Corollary 2. Hence, we conjecture that knowing the time horizon might represent a significant advantage.

Due to limited space, some of the proofs are absent from this paper. All the proofs can be found in the extended version [2].

2   The UCB-V Algorithm

For any $k \in \{1, \ldots, K\}$ and $t \in \mathbb{N}$, let $\overline{X}_{k,t}$ and $V_{k,t}$ be the empirical estimates of the mean payoff and variance of arm k:

$$\overline{X}_{k,t} \triangleq \frac{1}{t} \sum_{i=1}^{t} X_{k,i} \quad \text{and} \quad V_{k,t} \triangleq \frac{1}{t} \sum_{i=1}^{t} (X_{k,i} - \overline{X}_{k,t})^2,$$

where by convention $\overline{X}_{k,0} \triangleq 0$ and $V_{k,0} \triangleq 0$. We recall that an *optimal arm* is an arm having the best expected reward $k^* \in \operatorname{argmax}_{k \in \{1,\ldots,K\}} \mu_k$. We denote quantities related to the optimal arm by putting $*$ in the upper index.

In the following, we assume that the rewards are bounded. Without loss of generality, we may assume that all the rewards are almost surely in $[0, b]$, with $b > 0$. We summarize our assumptions on the reward sequence here:

Assumptions: Let $K > 2$, $\nu_1, \ldots, \nu_K$ distributions over reals with support $[0, b]$. For $1 \leq k \leq K$, let $\{X_{k,t}\} \sim \nu_k$ be an i.i.d. sequence of random variables specifying the rewards for arm k.[3] Assume that the rewards of different arms are independent of each other, i.e., for any k, k', $1 \leq k < k' \leq K$, $t \in \mathbb{N}^+$,

---

[3] The i.i.d. assumption can be relaxed, see e.g., [7].

the collection of random variables, $(X_{k,1}, \ldots, X_{k,t})$ and $(X_{k',1}, \ldots, X_{k',t})$, are independent of each other.

2.1 The Algorithm

Let $c \geq 0$. Let $\mathcal{E} = (\mathcal{E}_{s,t})_{s \geq 0, t \geq 0}$ be nonnegative real numbers such that for any $s \geq 0$, the function $t \mapsto \mathcal{E}_{s,t}$ is nondecreasing. We shall call $\mathcal{E}$ (viewed as a function of (s, t)) the exploration function. For any arm k and any nonnegative integers s, t, introduce

$$B_{k,s,t} \triangleq \overline{X}_{k,s} + \sqrt{\frac{2V_{k,s}\mathcal{E}_{s,t}}{s}} + c\frac{3b\mathcal{E}_{s,t}}{s} \tag{3}$$

with the convention $1/0 = +\infty$.

> **UCB-V policy:**
> At time t, play an arm maximizing $B_{k,T_k(t-1),t}$.

Let us roughly describe the behavior of the algorithm. At the beginning (i.e., for small t), every arm that has not been drawn is associated with an infinite bound which will become finite as soon as the arm is drawn. The more an arm k is drawn, the closer the bound (3) gets close to its first term, and thus, from the law of large numbers, to the expected reward μ_k. So the procedure will hopefully tend to draw more often arms having greatest expected rewards.

Nevertheless, since the obtained rewards are stochastic it might happen that during the first draws the (unknown) optimal arm always gives low rewards. Fortunately, if the optimal arm has not been drawn too often (i.e., small $T_{k^*}(t-1)$), for appropriate choices of $\mathcal{E}$ (when $\mathcal{E}_{s,t}$ increases without bounds in t for any fixed s), after a while the last term of (3) will start to dominate the two other terms and will also dominate the bound associated with the arms drawn very often. Thus the optimal arm will be drawn even if the empirical mean of the obtained rewards, $\overline{X}_{k^*,T_{k^*}(t-1)}$, is small. More generally, such choices of $\mathcal{E}$ lead to the exploration of arms with inferior empirical mean. This is why $\mathcal{E}$ is referred to as the exploration function. Naturally, a high-valued exploration function also leads to draw often suboptimal arms. Therefore the choice of $\mathcal{E}$ is crucial in order to explore possibly suboptimal arms while keeping exploiting (what looks like to be) the optimal arm.

The actual form of $B_{k,s,t}$ comes from the following novel tail bound on the sample average of i.i.d. random variables with bounded support that, unlike previous similar bounds (Bennett's and Bernstein's inequalities), involves the empirical variance.

Theorem 1. *Let $X_1, \ldots, X_t$ be i.i.d. random variables taking their values in $[0, b]$. Let $\mu = \mathbb{E}[X_1]$ be their common expected value. Consider the empirical expectation $\overline{X}_t$ and variance V_t defined respectively by*

$$\overline{X}_t = \frac{\sum_{i=1}^{t} X_i}{t} \quad and \quad V_t = \frac{\sum_{i=1}^{t}(X_i - \overline{X}_t)^2}{t}.$$

Then for any $t \in \mathbb{N}$ and $x > 0$, with probability at least $1 - 3e^{-x}$,

$$|\overline{X}_t - \mu| \leq \sqrt{\frac{2V_t x}{t}} + \frac{3bx}{t}. \tag{4}$$

Furthermore, introducing

$$\beta(x,t) = 3 \inf_{1 < \alpha \leq 3} \left(\frac{\log t}{\log \alpha} \wedge t \right) e^{-x/\alpha}, \tag{5}$$

we have for any $t \in \mathbb{N}$ and $x > 0$, with probability at least $1 - \beta(x,t)$

$$|\overline{X}_s - \mu| \leq \sqrt{\frac{2V_s x}{s}} + \frac{3bx}{s} \tag{6}$$

hold simultaneously for $s \in \{1, 2, \ldots, t\}$.

Remark 1. The uniformity in time is the only difference between the two assertions of the previous theorem. When we use (6), the values of x and t will be such that $\beta(x,t)$ is of order of $3e^{-x}$, hence there will be no real price to pay for writing a version of (4) that is uniform in time. In particular, this means that if $1 \leq S \leq t$ is a random variable then (6) still holds with probability at least $1 - \beta(x,t)$ and when s is replaced with S.

Note that (4) is useless for $t \leq 3$ since its r.h.s. is larger than b. For any arm k, time t and integer $1 \leq s \leq t$ we may apply Theorem 1 to the rewards $X_{k,1}, \ldots, X_{k,s}$, and obtain that with probability at least $1 - 3\sum_{s=4}^{\infty} e^{-(c \wedge 1)\mathcal{E}_{s,t}}$, we have $\mu_k \leq B_{k,s,t}$. Hence, by our previous remark at time t with high probability (for a high-valued exploration function $\mathcal{E}$) the expected reward of arm k is upper bounded by $B_{k,T_k(t-1),t}$. The user of the generic UCB-V policy has two parameters to tune: the exploration function $\mathcal{E}$ and the positive real number c.

A cumbersome technical analysis (not reproduced here) shows that there are essentially two interesting types of exploration functions:

- the ones in which $\mathcal{E}_{s,t}$ depends only on t (see Sections 3 and 4).
- the ones in which $\mathcal{E}_{s,t}$ depends only on s (see Section 5).

2.2   Bounds for the Sampling Times of Suboptimal Arms

The natural way of bounding the regret of UCB policies is to bound the number of times suboptimal arms are drawn (or the inferior sampling times). The bounds presented here significantly improve the ones used in [3]. This improvement is necessary to get tight bounds for the interesting case where the exploration function is logarithmic. The idea of the bounds is that the inferior sampling time of an arm can be bounded in terms of the number of times the UCB for the arm considered is over a some threshold value (τ in the statement below) and the number of times the UCB for an optimal arm is below the same threshold. Note that even though the above statements hold for any arm, they will be only useful for suboptimal arms. In particular, for a suboptimal arm the threshold can be chosen to lie between the payoff of an optimal arm and the payoff of the arm considered.

Theorem 2. *Consider UCB-V. Then, after K plays, each arm has been pulled once. Further, the following holds: Let arm k and time $n \in \mathbb{N}^+$ be fixed. For any $\tau \in \mathbb{R}$ and any integer $u > 1$, we have*

$$T_k(n) \leq u + \sum_{t=u+K-1}^{n} \left(\mathbb{1}_{\{\exists s: u \leq s \leq t-1 \text{ s.t. } B_{k,s,t} > \tau\}} + \mathbb{1}_{\{\exists s^*: 1 \leq s^* \leq t-1 \text{ s.t. } \tau \geq B_{k^*,s^*,t}\}} \right), \tag{7}$$

hence

$$\mathbb{E}\left[T_k(n)\right] \leq u + \sum_{t=u+K-1}^{n} \sum_{s=u}^{t-1} \mathbb{P}\left(B_{k,s,t} > \tau\right) + \sum_{t=u+K-1}^{n} \mathbb{P}\left(\exists s: 1 \leq s \leq t-1 \text{ s.t. } B_{k^*,s,t} \leq \tau\right). \tag{8}$$

Besides we have

$$\mathbb{P}\left(T_k(n) > u\right) \leq \sum_{t=3}^{n} \mathbb{P}\left(B_{k,u,t} > \tau\right) + \mathbb{P}\left(\exists s: 1 \leq s \leq n-u \text{ s.t. } B_{k^*,s,u+s} \leq \tau\right). \tag{9}$$

Proof. The first assertion is trivial since at the beginning all arms has an infinite UCB, which becomes finite as soon as the arm has been played once. To obtain (7), we note that

$$T_k(n) - u \leq \sum_{t=u+K-1}^{n} \mathbb{1}_{\{I_t = k; T_k(t) > u\}} = \sum_{t=u+K-1}^{n} Z_{k,t,u},$$

where

$$Z_{k,t,u} = \mathbb{1}_{\{I_t = k; \, u \leq T_k(t-1); \, 1 \leq T_{k^*}(t-1); \, B_{k,T_k(t-1),t} \geq B_{k^*,T_{k^*}(t-1),t}\}}$$
$$\leq \mathbb{1}_{\{\exists s: u \leq s \leq t-1 \text{ s.t. } B_{k,s,t} > \tau\}} + \mathbb{1}_{\{\exists s^*: 1 \leq s^* \leq t-1 \text{ s.t. } \tau \geq B_{k^*,s^*,t}\}}$$

Taking the expectation on both sides of (7) and using the probability union bound, we obtain (8). Finally, (9) comes from a more direct argument that uses the fact that the exploration function $\xi_{s,t}$ is a nondecreasing function with respect to t. Consider an event such that the following statements hold:

$$\begin{cases} \forall t: 3 \leq t \leq n \text{ s.t. } B_{k,u,t} \leq \tau, \\ \forall s: 1 \leq s \leq n-u \text{ s.t. } B_{k^*,s,u+s} > \tau. \end{cases}$$

Then for any $1 \leq s \leq n-u$ and $u+s \leq t \leq n$,

$$B_{k^*,s,t} \geq B_{k^*,s,u+s} > \tau \geq B_{k,u,t}.$$

This implies that arm k will not be pulled a $(u+1)$-th time. Therefore we have proved by contradiction that

$$\{T_k(n) > u\} \subset \left(\{\exists t: 3 \leq t \leq n \text{ s.t. } B_{k,u,t} > \tau\} \right. $$
$$\left. \cup \{\exists s: 1 \leq s \leq n-u \text{ s.t. } B_{k^*,s,u+s} \leq \tau\} \right), \tag{10}$$

which by taking probabilities of both sides gives the announced result.

3   Expected Regret of UCB-V

In this section, we consider that the exploration function does not depend on s (yet, $\mathcal{E} = (\mathcal{E}_t)_{t\geq 0}$ is still nondecreasing with t). We will see that as far as the expected regret is concerned, a natural choice of $\mathcal{E}_t$ is the logarithmic function and that c should not be taken too small if one does not want to suffer polynomial regret instead of logarithmic one. We derive bounds on the expected regret and conclude by specifying natural constraints on c and $\mathcal{E}_t$.

Theorem 3. *We have*

$$\mathbb{P}(B_{k,s,t} > \mu^*) \leq 2e^{-s\Delta_k^2/(8\sigma_k^2+4b\Delta_k/3)}, \tag{11}$$

and

$$\mathbb{E}[R_n] \leq \sum_{k:\Delta_k>0} \left\{ 1 + 8(c \vee 1)\mathcal{E}_n \left(\frac{\sigma_k^2}{\Delta_k^2} + \frac{2b}{\Delta_k} \right) \right.$$
$$\left. + ne^{-\mathcal{E}_n} \left(\frac{24\sigma_k^2}{\Delta_k^2} + \frac{4b}{\Delta_k} \right) + \sum_{t=16\mathcal{E}_n}^{n} \beta\big((c \wedge 1)\mathcal{E}_t, t\big) \right\} \Delta_k, \tag{12}$$

where we recall that $\beta\big((c \wedge 1)\mathcal{E}_t, t\big)$ is essentially of order $e^{-(c\wedge 1)\mathcal{E}_t}$ (see (5) and Remark 1).

Proof. Let $\mathcal{E}_n' = (c \vee 1)\mathcal{E}_n$. We use (8) with u being the smallest integer larger than $8\big(\frac{\sigma_k^2}{\Delta_k^2} + \frac{2b}{\Delta_k}\big)\mathcal{E}_n'$ and $\tau = \mu^*$. For any $s \geq u$ and $t \geq 2$, we have

$$\mathbb{P}(B_{k,s,t} > \mu^*) = \mathbb{P}\Big(\overline{X}_{k,s} + \sqrt{\tfrac{2V_{k,s}\mathcal{E}_t}{s}} + 3bc\tfrac{\mathcal{E}_t}{s} > \mu_k + \Delta_k\Big)$$
$$\leq \mathbb{P}\Big(\overline{X}_{k,s} + \sqrt{\tfrac{2[\sigma_k^2+b\Delta_k/2]\mathcal{E}_t}{s}} + 3bc\tfrac{\mathcal{E}_t}{s} > \mu_k + \Delta_k\Big) + \mathbb{P}\big(V_{k,s} \geq \sigma_k^2 + b\Delta_k/2\big)$$
$$\leq \mathbb{P}\big(\overline{X}_{k,s} - \mu_k > \Delta_k/2\big) + \mathbb{P}\Big(\tfrac{\sum_{j=1}^{s}(X_{k,j}-\mu_k)^2}{s} - \sigma_k^2 \geq b\Delta_k/2\Big)$$
$$\leq 2e^{-s\Delta_k^2/(8\sigma_k^2+4b\Delta_k/3)}, \tag{13}$$

proving (11). Here in the last step we used Bernstein's inequality twice and in the second inequality we used that the choice of u guarantees that for any $u \leq s < t$ and $t \geq 2$,

$$\sqrt{\tfrac{2[\sigma_k^2+b\Delta_k/2]\mathcal{E}_t}{s}} + 3bc\tfrac{\mathcal{E}_t}{s} \leq \sqrt{\tfrac{[2\sigma_k^2+b\Delta_k]\mathcal{E}_n'}{u}} + 3b\tfrac{\mathcal{E}_n'}{u} \leq \sqrt{\tfrac{[2\sigma_k^2+b\Delta_k]\Delta_k^2}{8[\sigma_k^2+2b\Delta_k]}} + \tfrac{3b\Delta_k^2}{8[\sigma_k^2+2b\Delta_k]}$$
$$= \tfrac{\Delta_k}{2}\left[\sqrt{\tfrac{2\sigma_k^2+b\Delta_k}{2\sigma_k^2+4b\Delta_k}} + \tfrac{3b\Delta_k}{4\sigma_k^2+8b\Delta_k}\right] \leq \tfrac{\Delta_k}{2}, \tag{14}$$

since the last inequality is equivalent to $(x-1)^2 \geq 0$ with $x = \sqrt{\tfrac{2\sigma_k^2+b\Delta_k}{2\sigma_k^2+4b\Delta_k}}$.
Summing up the probabilities in Equation (13) we obtain

$$\sum_{s=u}^{t-1} \mathbb{P}(B_{k,s,t} > \mu^*) \le 2 \sum_{s=u}^{\infty} e^{-s\Delta_k^2/(8\sigma_k^2 + 4b\Delta_k/3)} = 2\frac{e^{-u\Delta_k^2/(8\sigma_k^2 + 4b\Delta_k/3)}}{1 - e^{-\Delta_k^2/(8\sigma_k^2 + 4b\Delta_k/3)}}$$

$$\le \left(\frac{24\sigma_k^2}{\Delta_k^2} + \frac{4b}{\Delta_k}\right) e^{-u\Delta_k^2/(8\sigma_k^2 + 4b\Delta_k/3)} \le \left(\frac{24\sigma_k^2}{\Delta_k^2} + \frac{4b}{\Delta_k}\right) e^{-\mathcal{E}'_n}, \quad (15)$$

where we have used that $1 - e^{-x} \ge 2x/3$ for $0 \le x \le 3/4$. By using (6) of Theorem 1 to bound the other probability in (8), we obtain that

$$\mathbb{E}[T_k(n)] \le 1 + 8\mathcal{E}'_n \left(\frac{\sigma_k^2}{\Delta_k^2} + \frac{2b}{\Delta_k}\right) + ne^{-\mathcal{E}'_n} \left(\frac{24\sigma_k^2}{\Delta_k^2} + \frac{4b}{\Delta_k}\right) + \sum_{t=u+1}^{n} \beta((c \wedge 1)\mathcal{E}_t, t),$$

which by $u \ge 16\mathcal{E}_n$ gives the announced result.

In order to balance the terms in (12) the exploration function should be chosen to be proportional to $\log t$. For this choice, the following corollary gives an explicit bound on the expected regret:

Corollary 1. If $c = 1$ and $\mathcal{E}_t = \zeta \log t$ for $\zeta > 1$, then there exists a constant c_ζ depending only on ζ such that for $n \ge 2$

$$\mathbb{E}[R_n] \le c_\zeta \sum_{k:\Delta_k>0} \left(\frac{\sigma_k^2}{\Delta_k} + 2b\right) \log n. \quad (16)$$

For instance, for $\zeta = 1.2$, the result holds for $c_\zeta = 10$.

Proof (Sketch of the proof). The first part, (16), follows directly from Theorem 3. Let us thus turn to the numerical result. For $n \ge K$, we have $R_n \le b(n - 1)$ (since in the first K rounds, the optimal arm is chosen at least once). As a consequence, the numerical bound is nontrivial only for $20 \log n < n - 1$, so we only need to check the result for $n > 91$. For $n > 91$, we bound the constant term using $1 \le \frac{\log n}{\log 91} \le a_1 \frac{2b}{\Delta_k}(\log n)$, with $a_1 = 1/(2\log 91) \approx 0.11$. The second term between the brackets in (12) is bounded by $a_2(\frac{\sigma_k^2}{\Delta_k^2} + \frac{2b}{\Delta_k}) \log n$, with $a_2 = 8 \times 1.2 = 9.6$. For the third term, we use that for $n > 91$, we have $24n^{-0.2} < a_3 \log n$, with $a_3 = \frac{24}{91^{0.2} \times \log 91} \approx 0.21$. By tedious computations, the fourth term can be bounded by $a_4 \frac{2b}{\Delta_k}(\log n)$, with $a_4 \approx 0.07$. This gives the desired result since $a_1 + a_2 + a_3 + a_4 \le 10$.

As promised, Corollary 1 gives a logarithmic bound on the expected regret that has a linear dependence on the range of the reward, contrary to bounds for algorithms that do not take into account the empirical variance of the rewards (see e.g. the bound (1) that holds for UCB1).

The previous corollary is well completed by the following result, which essentially says that we should not use $\mathcal{E}_t = \zeta \log t$ with $\zeta < 1$.

Theorem 4. *Consider $\mathcal{E}_t = \zeta \log t$ and let n denote the total number of draws. Whatever c is, if $\zeta < 1$, then there exist some reward distributions (depending on n) such that*

- *the expected number of draws of suboptimal arms using the UCB-V algorithm is polynomial in the total number of draws*
- *the UCB-V algorithm suffers a polynomial loss.*

So far we have seen that for $c = 1$ and $\zeta > 1$ we obtain a logarithmic regret, and that the constant ζ should not be taken below 1 (whatever c is) without risking to suffer polynomial regret. Now we consider the last term in $B_{k,s,t}$, which is linear in the ratio $\mathcal{E}_t/s$, and show that this term is also necessary to obtain a logarithmic regret, since we have:

Theorem 5. *Consider $\mathcal{E}_t = \zeta \log t$. Whatever ζ is, if $c\zeta < 1/6$, there exist probability distributions of the rewards such that the UCB-V algorithm suffers a polynomial loss.*

To conclude the above analysis, natural values for the constants appearing in the bound are the following ones

$$B_{k,s,t} \triangleq \overline{X}_{k,s} + \sqrt{\frac{2V_{k,s}\log t}{s}} + \frac{b\log t}{2s}.$$

This choice corresponds to the critical exploration function $\mathcal{E}_t = \log t$ and to $c = 1/6$, that is, the minimal associated value of c in view of the previous theorem. In practice, it would be unwise (or risk seeking) to use smaller constants in front of the last two terms.

4   Concentration of the Regret

In real life, people are not only interested in the expected rewards that they can obtain by some policy. They also want to estimate probabilities of obtaining much less rewards than expected, hence they are interested in the concentration of the regret. This section starts with the study of the concentration of the pseudo-regret, since, as we will see in Remark 2 p.162, the concentration properties of the regret follow from the concentration properties of the pseudo-regret.

We still assume that the exploration function does not depend on s and that $\mathcal{E} = (\mathcal{E}_t)_{t \geq 0}$ is nondecreasing. Introduce

$$\tilde{\beta}_n(t) \triangleq 3 \min_{\substack{\alpha \geq 1 \\ s_0 = 0 < s_1 < \cdots < s_M = n \\ \text{s.t. } s_{j+1} \leq \alpha(s_j+1)}}^{M \in \mathbb{N}} \sum_{j=0}^{M-1} e^{-\frac{(c \wedge 1)\mathcal{E}_{s_j+t+1}}{\alpha}}. \tag{17}$$

We have seen in the previous section that in order to obtain logarithmic expected regret, it is natural to take a logarithmic exploration function. In this case, and also when the exploration function goes to infinity faster than the logarithmic function, the complicated sum in (17), up to second order logarithmic terms, is of the order $e^{-(c \wedge 1)\mathcal{E}_t}$. This can be seen by considering (disregarding rounding issues) the geometric grid $s_j = \alpha^j$ with α close to 1. The next theorem provides a bound for the tails of the pseudo-regret.

Theorem 6. *Let*

$$v_k \triangleq 8(c \vee 1)\left(\frac{\sigma_k^2}{\Delta_k^2} + \frac{4b}{3\Delta_k}\right), \qquad r_0 \triangleq \sum_{k:\Delta_k>0} \Delta_k(1 + v_k\mathcal{E}_n).$$

Then, for any $x \geq 1$, we have

$$\mathbb{P}(R_n > r_0 x) \leq \sum_{k:\Delta_k>0} \left\{ 2ne^{-(c\vee 1)\mathcal{E}_n x} + \tilde{\beta}_n(\lfloor v_k\mathcal{E}_n x\rfloor) \right\}, \qquad (18)$$

where we recall that $\tilde{\beta}_n(t)$ is essentially of order $e^{-(c\wedge 1)\mathcal{E}_t}$ (see the above discussion).[4]

Proof (sketch of the proof). First note that

$$\mathbb{P}(R_n > r_0 x) = \mathbb{P}\left\{ \sum_{k:\Delta_k>0} \Delta_k T_k(n) > \sum_{k:\Delta_k>0} \Delta_k(1 + v_k\mathcal{E}_n)x \right\}$$

$$\leq \sum_{k:\Delta_k>0} \mathbb{P}\left\{ T_k(n) > (1 + v_k\mathcal{E}_n)x \right\}.$$

Let $\mathcal{E}'_n = (c\vee 1)\mathcal{E}_n$. We use (9) with $\tau = \mu^*$ and $u = \lfloor(1 + v_k\mathcal{E}_n)x\rfloor \geq v_k\mathcal{E}_n x$. From (11) of Theorem 3, we have $\mathbb{P}(B_{k,u,t} > \mu^*) \leq 2e^{-u\Delta_k^2/(8\sigma_k^2+4b\Delta_k/3)} \leq 2e^{-\mathcal{E}'_n x}$. To bound the other probability in (9), we use $\alpha \geq 1$ and the grid $s_0, \ldots, s_M$ realizing the minimum of (17) when $t = u$. Let $I_j = \{s_j + 1, \ldots, s_{j+1}\}$. Then

$$\mathbb{P}(\exists s : 1 \leq s \leq n - u \text{ s.t. } B_{k^*,s,u+s} \leq \mu^*) \leq \sum_{j=0}^{M-1} \mathbb{P}(\exists s \in I_j \text{ s.t. } B_{k^*,s,s_j+u+1} \leq \mu^*)$$

$$\leq \sum_{j=0}^{M-1} \mathbb{P}(\exists s \in I_j \text{ s.t. } s(\overline{X}_{k^*,s} - \mu^*) + \sqrt{2sV_{k^*,s}\mathcal{E}_{s_j+u+1}} + 3bc\mathcal{E}_{s_j+u+1} \leq 0)$$

$$\leq 3 \sum_{j=0}^{M-1} e^{-\frac{(c\wedge 1)\mathcal{E}_{s_j+u+1}}{\alpha}} = \tilde{\beta}_n(u) \leq \tilde{\beta}_n(\lfloor v_k\mathcal{E}_n x\rfloor),$$

where the second to last inequality comes from an appropriate union bound argument (see [2] for details).

When $\mathcal{E}_n \geq \log n$, the last term is the leading term. In particular, when $c = 1$ and $\mathcal{E}_t = \zeta \log t$ with $\zeta > 1$, Theorem 6 leads to the following corollary, which essentially says that for any $z > \gamma \log n$ with γ large enough,

$$\mathbb{P}(R_n > z) \leq C z^{-\zeta},$$

for some constant $C > 0$:

---

[4] Here $\lfloor x\rfloor$ denotes the largest integer smaller or equal to x.

Corollary 2. *When $c = 1$ and $\mathcal{E}_t = \zeta \log t$ with $\zeta > 1$, there exist $\kappa_1 > 0$ and $\kappa_2 > 0$ depending only on b, K, $(\sigma_k)_{k \in \{1,\dots,K\}}$, $(\Delta_k)_{k \in \{1,\dots,K\}}$ satisfying that for any $\varepsilon > 0$ there exists $\Gamma_\varepsilon > 0$ (tending to infinity when ε goes to 0) such that for any $n \geq 2$ and any $z > \kappa_1 \log n$*

$$\mathbb{P}(R_n > z) \leq \kappa_2 \frac{\Gamma_\varepsilon \log z}{z^{\zeta(1-\varepsilon)}}$$

Since the regret is expected to be of order $\log n$ the condition $z = \Omega(\log n)$ is not an essential restriction. Further, the regret concentration, although increasing with ζ, is pretty slow. For comparison, remember that a zero-mean martingale M_n with increments bounded by 1 would satisfy $\mathbb{P}(M_n > z) \leq \exp(-2z^2/n)$. The slow concentration for UCB-V happens because the first $\Omega(\log(t))$ choices of the optimal arm can be unlucky, in which case the optimal arm will not be selected any more during the first t steps. Hence, the distribution of the regret will be of a mixture form with a mode whose position scales linearly with time and which decays only at a polynomial rate, which is controlled by ζ.[5] This reasoning relies crucially on that the choices of the optimal arm can be unlucky. Hence, we have the following result:

Theorem 7. *Consider $\mathcal{E}_t = \zeta \log t$ with $c\zeta > 1$. Let $\tilde{k}$ denote the second optimal arm. If the essential infimum of the optimal arm is strictly larger than $\mu_{\tilde{k}}$, then the pseudo-regret has exponentially small tails. Inversely, if the essential infimum of the optimal arm is strictly smaller than μ_k, then the pseudo-regret has only polynomial tail.*

Remark 2. In Theorem 6 and Corollary 2, we have considered the pseudo-regret: $R_n = \sum_{k=1}^{K} T_k(n) \Delta_k$ instead of the regret $\hat{R}_n \triangleq \sum_{t=1}^{n} X_{k^*,t} - \sum_{t=1}^{n} X_{I_t, T_{I_t}(t)}$. Our main motivation for this was to provide as simple as possible formulae and assumptions. The following computations explains that when the optimal arm is unique, one can obtain similar contraction bounds for the regret. Consider the interesting case when $c = 1$ and $\mathcal{E}_t = \zeta \log t$ with $\zeta > 1$. By modifying the analysis slightly in Corollary 2, one can get that there exists $\kappa_1 > 0$ such that for any $z > \kappa_1 \log n$, with probability at least $1 - z^{-1}$, the number of draws of suboptimal arms is bounded by $C z$ for some $C > 0$. This means that the algorithm draws an optimal arm at least $n - C z$ times. Now if the optimal arm is unique, this means that $n - Cz$ terms cancel out in the summations of the definition of the regret. For the Cz terms which remain, one can use standard Bernstein inequalities and union bounds to prove that with probability $1 - Cz^{-1}$, we have $\hat{R}_n \leq R_n + C'\sqrt{z}$. Since the bound on the pseudo-regret is of order z (Corollary 2), a similar bound holds for the regret.

5   PAC-UCB

In this section, we consider that the exploration function does not depend on t: $\mathcal{E}_{s,t} = \mathcal{E}_s$. We show that for an appropriate sequence $(\mathcal{E}_s)_{s \geq 0}$, this leads to

---

[5] Note that entirely analogous results hold for UCB1.

an UCB algorithm which has nice properties with high probability (Probably Approximately Correct), hence the name of it. Note that in this setting, the quantity $B_{k,s,t}$ does not depend on the time t so we will simply write it $B_{k,s}$. Besides, in order to simplify the discussion, we take $c = 1$.

Theorem 8. *Let $\beta \in (0,1)$. Consider a sequence $(\mathcal{E}_s)_{s \geq 0}$ satisfying $\mathcal{E}_s \geq 2$ and*

$$4K \sum_{s \geq 7} e^{-\mathcal{E}_s} \leq \beta. \tag{19}$$

Consider u_k the smallest integer such that

$$\frac{u_k}{\mathcal{E}_{u_k}} > \frac{8\sigma_k^2}{\Delta_k^2} + \frac{26b}{3\Delta_k}. \tag{20}$$

With probability at least $1 - \beta$, the PAC-UCB policy plays any suboptimal arm k at most u_k times.

Let $q > 1$ be a fixed parameter. A typical choice for $\mathcal{E}_s$ is

$$\mathcal{E}_s = \log(Ks^q \beta^{-1}) \vee 2, \tag{21}$$

up to some additive constant ensuring that (19) holds. For this choice, Theorem 8 implies that for some positive constant κ, with probability at least $1 - \beta$, for any suboptimal arm k (i.e., $\Delta_k > 0$), its number of play is bounded by

$$T_{k,\beta} \triangleq \kappa \Big(\frac{\sigma_k^2}{\Delta_k^2} + \frac{1}{\Delta_k} \Big) \log \Big[K \Big(\frac{\sigma_k^2}{\Delta_k^2} + \frac{b}{\Delta_k} \Big) \beta^{-1} \Big],$$

which is independent of the total number of plays! This directly leads to the following upper bound on the regret of the policy at time n

$$\sum_{k=1}^{K} T_k(n) \Delta_k \leq \sum_{k:\Delta_k > 0} T_{k,\beta} \Delta_k. \tag{22}$$

One should notice that the previous bound holds with probability at least $1 - \beta$ and on the complement set no small upper bound is possible: one can find a situation in which with probability of order β, the regret is of order n (even if (22) holds with probability greater than $1-\beta$). More formally, this means that the following bound cannot be essentially improved (unless additional assumptions are imposed):

$$\mathbb{E}[R_n] = \sum_{k=1}^{K} \mathbb{E}[T_k(n)] \Delta_k \leq (1 - \beta) \sum_{k:\Delta_k > 0} T_{k,\beta} \Delta_k + \beta n$$

As a consequence, if one is interested in having a bound on the expected regret at some fixed time n, one should take β of order $1/n$ (up to a logarithmic factor):

Theorem 9. *Let $n \geq 7$ be fixed. Consider the sequence $\mathcal{E}_s = \log[Kn(s+1)]$. For this sequence, the PAC-UCB policy satisfies*

- *with probability at least $1 - \frac{4\log(n/7)}{n}$, for any $k : \Delta_k > 0$, the number of plays of arm k up to time n is bounded by $1 + \Big(\frac{8\sigma_k^2}{\Delta_k^2} + \frac{26b}{3\Delta_k} \Big) \log(Kn^2)$.*
- *the expected regret at time n satisfies*

$$\mathbb{E}[R_n] \leq \sum_{k:\Delta_k > 0} \Big(\frac{24\sigma_k^2}{\Delta_k} + 30b \Big) \log(n/3). \tag{23}$$

6   Open Problem

When the horizon time n is known, one may want to choose the exploration function $\mathcal{E}$ depending on the value of n. For instance, in view of Theorems 3 and 6, one may want to take $c = 1$ and a constant exploration function $\mathcal{E} \equiv 3 \log n$. This choice ensures logarithmic expected regret and a nice concentration property:

$$\mathbb{P}\left\{ R_n > 24 \sum_{k:\Delta_k>0} \left(\frac{\sigma_k^2}{\Delta_k} + 2b \right) \log n \right\} \le \frac{C}{n}. \tag{24}$$

This algorithm does not behave as the one which simply takes $\mathcal{E}_{s,t} = 3 \log t$. Indeed the algorithm with constant exploration function $\mathcal{E}_{s,t} = 3 \log n$ concentrates its exploration phase at the beginning of the plays, and then switches to exploitation mode. On the contrary, the algorithm which adapts to the time horizon explores and exploits during all the time interval $[0; n]$. However, in view of Theorem 7, it satisfies only

$$\mathbb{P}\left\{ R_n > 24 \sum_{k:\Delta_k>0} \left(\frac{\sigma_k^2}{\Delta_k} + 2b \right) \log n \right\} \le \frac{C}{(\log n)^C}.$$

which is significantly worse than (24). The open question is: is there an algorithm that adapts to time horizon which has a logarithmic expected regret and a concentration property similar to (24)? We conjecture that the answer is no.

Acknowledgements. Csaba Szepesvári greatly acknowledges the support received through the Alberta Ingenuity Center for Machine Learning (AICML) and the Computer and Automation Research Institute of the Hungarian Academy of Sciences, and the PASCAL pump priming project "Sequential Forecasting". The authors thank Yizao Wang for the careful reading of an earlier version of this paper.

References

1. Agrawal, R.: Sample mean based index policies with $O(\log n)$ regret for the multi-armed bandit problem. Advances in Applied Probability 27, 1054–1078 (1995)
2. Audibert, J.-Y., Munos, R., Szepesvári,Cs.: Variance estimates and exploration function in multi-armed bandit. Research report 07-31, Certis - Ecole des Ponts (2007), http://cermics.enpc.fr/~audibert/RR0731.pdf
3. Auer, P., Cesa-Bianchi, N., Fischer, P.: Finite time analysis of the multiarmed bandit problem. Machine Learning 47(2-3), 235–256 (2002)
4. Auer, P., Cesa-Bianchi, N., Shawe-Taylor, J.: Exploration versus exploitation challenge. In: 2nd PASCAL Challenges Workshop. Pascal Network (2006)
5. Gittins, J.C.: Multi-armed Bandit Allocation Indices. In: Wiley-Interscience series in systems and optimization. Wiley, Chichester (1989)
6. Lai, T.L., Robbins, H.: Asymptotically efficient adaptive allocation rules. Advances in Applied Mathematics 6, 4–22 (1985)

7. Lai, T.L., Yakowitz, S.: Machine learning and nonparametric bandit theory. IEEE Transactions on Automatic Control 40, 1199–1209 (1995)
8. Robbins, H.: Some aspects of the sequential design of experiments. Bulletin of the American Mathematics Society 58, 527–535 (1952)
9. Thompson, W.R.: On the likelihood that one unknown probability exceeds another in view of the evidence of two samples. Biometrika 25, 285–294 (1933)

Following the Perturbed Leader to Gamble at Multi-armed Bandits

Jussi Kujala and Tapio Elomaa

Institute of Software Systems
Tampere University of Technology
P. O. Box 553, FI-33101 Tampere, Finland
jussi.kujala@tut.fi, elomaa@cs.tut.fi

Abstract. Following the perturbed leader (FPL) is a powerful technique for solving online decision problems. Kalai and Vempala [1] rediscovered this algorithm recently. A traditional model for online decision problems is the multi-armed bandit. In it a gambler has to choose at each round one of the k levers to pull with the intention to minimize the cumulated cost. There are four versions of the nonstochastic optimization setting out of which the most demanding one is a game played against an adaptive adversary in the bandit setting. An adaptive adversary may alter its game strategy of assigning costs to decisions depending on the decisions chosen by the gambler in the past. In the bandit setting the gambler only gets to know the cost of the choice he made, rather than the costs of all available alternatives. In this work we show that the very straightforward and easy to implement algorithm Adaptive Bandit FPL can attain a regret of $O(\sqrt{T \ln T})$ against an adaptive adversary. This regret holds with respect to the best lever in hindsight and matches the previous best regret bounds of $O(\sqrt{T \ln T})$.

1 Introduction

Following the perturbed leader (FPL) is a natural and straightforward approach that can be applied in a wide variety of online decision problems, where costs associated with future decisions are not known [1,2]. In FPL one simply chooses the decision that has fared the best in the earlier rounds (the leader). In order to cope with an adversary, the necessary randomization is implemented by adding a perturbation to the total costs prior to selecting the leader. We have previously shown that FPL will attain an expected regret $O(\sqrt{T})$ in T rounds against an oblivious (non-adaptive) adversary in a bandit version of the expert setting [3].[1]

In this paper we study the performance of the FPL algorithm when faced with an adaptive adversary in the multi-armed bandit model. Dani and Hayes [4] give an example where the EXP3 algorithm of Auer et al. [5] can attain, with a constant probability, a regret of $\Omega(T^{2/3})$ against an adaptive adversary. Their

---

[1] In this paper, instead of the bandit version of the expert setting, we talk about the multi-armed bandit model. However, we do not distinguish the two settings in any way.

M. Hutter, R.A. Servedio, and E. Takimoto (Eds.): ALT 2007, LNAI 4754, pp. 166–180, 2007.

argument applies as such to our earlier FPL algorithm [3]. However, a small modification to the algorithm lets us circumnavigate this lower bound.

In this paper we introduce a new FPL algorithm that matches the best regret bound $O(\sqrt{T \ln T})$ of the best algorithms for this problem [6,4]. The new algorithm is the same as our earlier one, except that instead of using an unbiased estimate of the true cost vector, it uses one which guarantees that we almost never estimate the cost of a lever too high. Thus, it is difficult to hide the actual cost under the variance of the estimate of the cost. This method does not produce too much perturbation to the estimated costs, leading to a good regret bound. This technique resembles that of Auer [6], who modified EXP3 algorithm to have $O(\sqrt{T \ln T})$ regret against an adaptive adversary.

In short, the contribution of our algorithm is that it is the first FPL algorithm in the multi-armed bandit model that is $O(\sqrt{T \ln T})$ competitive against an adaptive adversary. In addition, the terms in the regret bound are somewhat lower than the previous expected regrets of the algorithms by Auer [6] and Dani and Hayes [4].

In the following section we briefly recapitulate the multi-armed bandit model. In Section 3 we describe the FPL framework and the online geometric optimization problem in more detail. We also review related work. Section 4, then, introduces the Adaptive Bandit FPL algorithm. Its regret bound is analyzed in Section 5. Some proofs are deferred to Appendix A. Finally, Section 6 concludes this paper.

2 Multi-armed Bandit Model

Multi-armed bandit [7] is a formulation for the exploration-exploitation dilemma: Whether a gambler should stick with the lever of a slot machine that has paid off best so far or should he go looking for a a better one among those that were not tried yet. The gambler's trade-off is in wasting effort on exploring unprofitable alternatives versus risking to miss a superior lever. This online optimization problem has been under scrutiny for over fifty years [7,2] and is finally starting to be well understood [8,5,4,9]. The early research on the multi-armed bandit model concentrated on studying stochastic and probabilistic slot machines, leaving the gambler to learn a good play strategy (for a review and references see [5]). The recent research, on the other hand, has deemed such machines inadequate to model all situations of interest [5], turning instead to study adversarial slot machines and randomized algorithms to play against them.

Table 1. The four variations of the nonstochastic optimization problem

	Oblivious Adversary	Adaptive Adversary
Full Information	OFI	AFI
Bandit Setting	OBS	ABS

There are four different variations of this iterated zero-sum two-player game between the gambler and the k-armed slot machine (see Table 1). The gambler may be told the payoff of each lever in each round or, more realistically, he may know only the result of the lever that he chose to play. The former model is the *full information* game, while the latter one is known as the *bandit setting*. The slot machine's payoff may be insensitive to the gambler's actions (i.e., it can as well be chosen in advance), in which case we have an *oblivious (non-adaptive) adversary*, or it may depend on the gambler's earlier choices and then the slot machine is an *adaptive adversary*. In this paper we are concerned with the most demanding setting out of these alternatives — ABS.

Gambling at a multi-armed bandit is a zero-sum game because the gambler's losses equal the winnings of the casino. If the duration of the game is not restricted, the gambler's total loss clearly has no bounds. Therefore, his performance is usually measured as the *regret*; the difference between his actual loss on the T rounds played and the single best lever in hindsight. With respect to this performance measure the gambler can do almost as well as the best static decision. To formulate, let $\ell_t \in \{1, \ldots, k\}$ be the lever chosen by the gambler at round (or turn) $t \in \{1, \ldots, T\}$. Simultaneously with the gambler's decision the adversary assigns costs $c_t = \langle c_{1,t}, \ldots, c_{k,t} \rangle$ to the levers. Of course, the costs need to be restricted somehow, because otherwise it is possible to suffer an unbounded loss on one lever pull. Let us assume that the costs are normalized to the range $[0, 1]$.

After both parties are ready, the gambler's choice is communicated to the adversary and (in the bandit setting) the gambler learns the cost $c_{\ell_t,t}$ of the lever he played. Eventually, when all T rounds have been played, the gambler has suffered a total *loss* $\sum_{t=1}^{T} c_{\ell_t,t}$. The best lever $j \in \{1, \ldots, k\}$ in hindsight is the one that minimizes $\sum_{t=1}^{T} c_{j,t}$. Hence, the performance measure regret is

$$\sum_{t=1}^{T} c_{\ell_t,t} - \min_{j \in \{1,\ldots,k\}} \sum_{t=1}^{T} c_{j,t}.$$

In the OFI case the optimal expected regret bound $\Theta(\sqrt{T \ln k})$ can be obtained by Freund and Schapire's [8] HEDGE algorithm, a variant of the WEIGHTED MAJORITY algorithm [10]. The same bound directly also holds in the AFI case [11,12]. In the bandit setting the situation is different; an adaptive adversary is strictly more powerful than an oblivious one [11].

For the oblivious bandit setting Auer et al. [5] put forward the EXP3 algorithm which is similar to the HEDGE algorithm. Dani and Hayes [4] note that while EXP3 guarantees that

$$\mathbf{E}(\text{regret vs. a lever } i) = O(\sqrt{kT \ln k}),$$

for a certain adaptive adversary it unfortunately holds that

$$\mathbf{E}(\text{regret vs. the best lever}) = \Omega(T^{2/3}).$$

Auer [6] has demonstrated how to modify the EXP3 algorithm to obtain a regret of $O(\sqrt{kT \ln T})$ against an adaptive adversary. Dani and Hayes also give an algorithm called ACCOUNTS that has a regret of $O(\sqrt{kT \ln T})$ against an adaptive adversary, although, with high constant terms.

3   Online Geometric Optimization and Following the Perturbed Leader

In the generalization of the problem — online geometric optimization [1] — an algorithm has to choose, in each round $t \in \{1, \ldots, T\}$, a vector $\boldsymbol{x}_t$ from a fixed decision set $S_D \subset \mathbb{R}^d$. The adversary simultaneously chooses a cost vector $\boldsymbol{c}_t$ from a fixed cost set $S_C \subset \mathbb{R}^d$. The loss of the algorithm at round t is the inner product $\boldsymbol{x}_t \cdot \boldsymbol{c}_t$ and its goal is to minimize regret

$$\sum_{t=1}^{T} \boldsymbol{x}_t \cdot \boldsymbol{c}_t - \min_{\boldsymbol{x}_* \in S_D} \sum_{t=1}^{T} \boldsymbol{x}_* \cdot \boldsymbol{c}_t. \tag{1}$$

In order to formulate the multi-armed bandit model in the same inner product loss setting, we can present the gambler's choice of a lever as an indicator vector; one in which all elements are zero except the one corresponding to the chosen lever.

There are a number of online problems that fit the geometric setting; e.g., *online routing* [1], *online set cover* [13], and *online traveling salesman problem* [13]. The usefulness of this model stems from the fact that there are algorithms with regret bounds depending on the dimension of the vector space of the decisions. In the above-mentioned problems the costs between different decisions can be linearly related and, hence, the dimension of the decision space is lower than the number of decisions.

Kalai and Vempala [1] showed that online geometric decision problems have efficient solutions, given an oracle for the offline version of the problem. They reintroduced the FPL algorithm, which selects a decision based on past cost vectors. Originally this method was proposed by Hannan [2] already fifty years ago. In short, the algorithm selects the best decision for the past cost vectors which are perturbed by a random vector. Zinkevich [14] has proposed another approach for solving the online geometric setting, but FPL is more relevant when the space of the decision vectors is not convex.

The regret bound of FPL in the full information setting depends on the structure of the decision vector space S_D and the cost vector space S_C. The following restrictions apply. There is a bound on

- the diameter of the decision set: $\|\boldsymbol{x}_t - \boldsymbol{x}_t'\|_1 \leq D$ for all $\boldsymbol{x}_t, \boldsymbol{x}_t' \in S_D$,
- the cost set: $\|\boldsymbol{c}_t\|_1 \leq A$ for any $\boldsymbol{c}_t \in S_C$, and
- the maximum cost: $|\boldsymbol{c}_t \cdot \boldsymbol{x}_t| \leq R$.

Let $\mathrm{opt}(\boldsymbol{c})$ be the optimal decision for a cost vector $\boldsymbol{c}$. We abbreviate the sum of cost vectors up to the time step t by $\boldsymbol{c}_{1:t} = \sum_{i=1}^{t} \boldsymbol{c}_i$. Hence, if T is the total number of rounds, then $\mathrm{opt}(\boldsymbol{c}_{1:T})$ is the best static decision and

Table 2. Multiplicative FPL* [1]

On turn t:

1. Choose a random perturbation vector $\mu_t \propto \exp(-\epsilon_t \|x\|_1)$ drawn from the exponential distribution, in which $\epsilon_t = O(1/\sqrt{t})$.
2. Pick decision $\text{opt}(c_{1:t-1} + \mu_t)$.

$\text{opt-cost}(c_{1:T}) = \text{opt}(c_{1:T}) \cdot c_{1:T}$ is its cost. With these restrictions the FPL algorithm (see Table 2) achieves a cost of

$$(1 + 2\epsilon A)\,\text{opt-cost}(c_{1:T}) + \frac{D \ln d}{\epsilon}$$

for a parameter $\epsilon < 1$ [1]. Thus, the regret of the FPL algorithm is $O(\sqrt{DRAT})$ when a suitable ϵ is plugged into the above formula.

The regret bound of FPL does not depend on the number of different decisions, whereas the regret bound of WEIGHTED MAJORITY depends on the number of experts. If the decisions are independent — i.e., experts of WEIGHTED MAJORITY — the difference between FPL and WEIGHTED MAJORITY is of implementation. Although the regret bound of FPL [1] is a constant factor $\sqrt{2}$ worse than that of HEDGE, Kalai has observed that we can choose the perturbation vector so that the decisions of FPL and HEDGE are identical [4]. Unfortunately, to the best of our knowledge there is no proof that this method works in the more general online geometric optimization.

Hutter and Poland have shown that many of the results obtained in the expert setting can be elegantly proved also in the FPL setting, albeit with worse constants [12].

3.1 Bandit FPL

FPL solves a full information problem (OFI or AFI). For the bandit setting, where only the cost of one decision per time step is revealed, there are several variations of FPL [15,16]. Bandit FPL algorithms typically choose at each time step between exploration and exploitation, like algorithms based on WEIGHTED MAJORITY. The exploration rate $\gamma_t \in [0, 1]$ determines which option to choose at step t. With probability γ_t one explores a representative sample from the decisions and with probability $1 - \gamma_t$ one exploits by choosing the decision that appears best in light of the (estimated) past costs.

Awerbuch and Kleinberg [15] considered the oblivious bandit setting and their algorithm has a regret of the order $O(T^{2/3})$. They also give an algorithm with the same regret bound designed specifically for the shortest online routing problem in the ABS case. McMahan and Blum [16] obtained a regret of $O(T^{3/4}\sqrt{\ln T})$ for the bandit FPL in the general ABS case. Dani and Hayes [11] improved the regot of McMahan and Blum's algorithm in the ABS case to $O(T^{2/3})$. Note that the above bounds omit dependences on other factors than T, which are typically quite complicated.

These algorithms for the geometric bandit setting, in general, estimate the cost vector only during exploration turns and ignore the costs on exploitation turns. If we want to get below the $O(T^{2/3})$ regret, it is necessary to use also the information obtained during exploitation turns [11]. Currently no general algorithm for the online geometric bandit setting exists with $O(\sqrt{T})$ regret, but there are customized algorithms for particular problems, like for the online shortest routing, which achieve a regret close to $O(\sqrt{T})$ [17].

In [3] we showed that a FPL variant — Bandit FPL (B-FPL) — can attain $O(\sqrt{T})$ expected regret in the OBS setting when restricted to the expert framework. In this paper we will turn to the ABS case, this time restricting our considerations to the multi-armed bandit model. We will prove that a FPL algorithm — Adaptive Bandit FPL, AB-FPL — attains an expected regret $O(\sqrt{kT\ln T})$. Thus, we match the asymptotic bound $O(\sqrt{kT\ln T})$ of the ACCOUNTS algorithm of Dani and Hayes [4] and of the algorithm given by Auer [6], but our hidden terms are somewhat lower. Although AB-FPL is currently restricted to multi-armed bandits, we hope that the idea in AB-FPL is general enough to be implemented in a B-FPL algorithm that has $O(\sqrt{T})$ regret in the online geometric bandit setting against a non-adaptive adversary.

4   Adaptive Bandit FPL

In this and the following section we show how FPL can be applied against an adaptive adversary in the multi-armed bandit setting. An argument by Dani and Hayes [4] shows that the B-FPL [3] algorithm cannot attain a regret of the order $O(\sqrt{T})$ in the ABS case and that its worst-case regret is $\Omega(T^{2/3})$. We now demonstrate that it is possible to modify the B-FPL algorithm so that it is competitive against an adaptive adversary. We call the resulting algorithm Adaptive Bandit FPL (AB-FPL). The modified algorithm AB-FPL is given in Table 3.

In short, AB-FPL is the same as B-FPL, except that instead of using an unbiased estimate $\hat{c}_{1:t}$ of the real vector $c_{1:t}$ it uses $\hat{c}'_{1:t} = \hat{c}_{1:t} - \lambda \boldsymbol{\sigma}_t \sqrt{\ln t}$, where λ is a small constant and $\boldsymbol{\sigma}_t$ is a vector that consists of upper bounds to conditional standard deviations of items in $\hat{c}_{1:t}$. Intuitively, this guarantees that we almost never estimate the cost of a lever too high and, thus, it is difficult to hide the actual cost under the variance of the estimate of the cost. Fortunately this method does not produce a too great perturbation to the estimated costs, making it possible to have a reasonable regret bound as we will show below. Similar work has been done by Auer [6] with WEIGHTED MAJORITY style algorithms. Let us first state the regret bound explicitly.

Theorem 1. *The expected regret of AB-FPL against the best lever in hindsight is at most $O(\sqrt{T\ln T})$ against an adaptive adversary. Asymptotically the most significant term is $5\sqrt{kT\ln T}$.*

Table 3. Adaptive Bandit FPL (AB-FPL)

- Let there be k levers $\{1, \ldots k\}$ and let $p_{i,t}$ be the probability of choosing lever i at turn t. We will describe how to compute these probabilities in Section 5.3.
- Let $c_{i,t}$ be the cost of lever i at turn t and let vector c_t consists of the costs for all levers.
- Moreover, let $c_{1:t}$ be the sum of cost vectors up to turn t.
- Let $\mathrm{opt}(c)$ be the indicator vector of some minimal value in c; i.e., $\mathrm{opt}(c) = i$ if i is a minimal value in c.
- Define two parameters:
  1. ϵ_t is intuitively the width of the perturbation vector μ_t used at an exploitation turn. We use $\epsilon_t = \epsilon_T = \sqrt{\ln T}/(3\sqrt{k\,T})$.
  2. $\gamma_t = \min(1, k\epsilon_t)$ is the probability of sampling a lever uniformly at random.
- Maintain two variables:
  1. Vector $\hat{c}_{1:t}$ that is an estimate of $c_{1:t}$. Estimated in the usual way as in EXP3 and B-FPL

$$\hat{c}_{i,t} = \begin{cases} c_{i,t}/p_{i,t} & \text{if } i \text{ was chosen on turn } t; \\ 0 & \text{otherwise.} \end{cases}$$

  2. Vector σ_t^2 that contains upper bounds of the conditional variances of random variables $\hat{c}_{i,1:t}$. It is defined as

$$\sigma_{i,t}^2 = \sum_{\tau=1}^{t} \frac{1}{p_{i,\tau}} \geq \sum_{\tau=1}^{t} \frac{c_{i,\tau}^2}{p_{i,\tau}} - c_{i,\tau}^2 = \sum_{\tau=1}^{t} \mathbf{Var}\left(\hat{c}_{i,\tau} \mid \tau - 1, \ldots, 1\right).$$

- Define $\hat{c}_t' = \hat{c}_t - \lambda \sigma_t \sqrt{\ln(t+1)}$, where $\lambda = \sqrt{1 + \sqrt{2}/\sqrt{k}}$.

On turn t:

Exploration step: With probability γ_t choose a lever uniformly at random.
Exploitation step: Otherwise choose a lever given by

$$\mathrm{opt}\left(\hat{c}_{1:t-1}' + \mu_t\right),$$

where $\mu_t \propto \exp(-\epsilon_t \|x\|_1)$ is a random perturbation vector in which the elements are drawn from the two-sided exponential distribution.

---

5   Proof for the Regret Bound

We first outline the proof of Theorem 1. First we observe that the cumulated cost of AB-FPL during the *exploitation steps* is upper bounded by

$$\mathbf{E}\left(\sum_{t=1}^{T} \mathrm{opt}\left(\hat{c}_{1:t-1}' + \mu_t\right) \cdot c_t\right).$$

We further bound this with
$$\text{opt-cost}(c_{1:T}) + O(\sqrt{T \ln T}).$$

The expected cumulative cost during the *exploration steps* in AB-FPL is upper bounded by $\sum_{t=1}^{T} \gamma_t = O(\sqrt{T \ln T})$, and hence for all steps we get a regret bound that is of the order $O(\sqrt{T \ln T})$.

We go through the following steps:

1. FPL run on *unbiased* estimates $\hat{c}_t$ of AB-FPL has the same expected regret as AB-FPL on actual costs. Several authors have made this observation [12,11]. It can be proved using a simple calculation of expected value.

$$\mathbf{E}_{\text{act}}\left(\sum_{t=1}^{T} \text{opt}(\hat{c}'_{1:t-1} + \mu_t) \cdot c_t\right) = \mathbf{E}_{\text{act}}\mathbf{E}_{\text{sim}}\left(\sum_{t=1}^{T} \text{opt}(\hat{c}'_{1:t-1} + r_t) \cdot \hat{c}_t\right)$$

From now on we will refer by simulated FPL to a run of the FPL algorithm with cost vectors $\hat{c}'_t$ as the input. These are available after an actual run of AB-FPL. Vectors r_t are perturbations chosen by a simulated run of FPL on vectors $\hat{c}'_t$; thus, they are independently chosen from the same distribution as perturbations μ_t. By $\mathbf{E}_{\text{act}}$ we denote expectation with regards to the actual choices made by AB-FPL and the adaptive adversary. $\mathbf{E}_{\text{sim}}$, on the other hand, refers to expectation with regards to choices made by a simulated FPL using cost vectors $\hat{c}'_t$. Hence, $\mathbf{E}_{\text{act}}$ is taken over the randomness of the joint distribution on vectors $\mu_1, \ldots, \mu_t$ and $c_1, \ldots, c_t$, while $\mathbf{E}_{\text{sim}}$ is taken over the randomness of vectors $r_1, \ldots, r_t$ given random choices that happen under $\mathbf{E}_{\text{act}}$. Plain $\mathbf{E}$ always refers to all randomness within the scope of the expectation operator.

When compared to the right-hand side of the above equation, in AB-FPL we have an additional term that results from it having extra vectors $\sigma_t \sqrt{\ln t} - \sigma_{t-1}\sqrt{\ln(t-1)}$ in the estimate $\hat{c}'_{1:t}$ which is used. In equations we get

$$\mathbf{E}\left(\sum_{t=1}^{T} \text{opt}(\hat{c}'_{1:t-1} + \mu_t) \cdot c_t\right)$$

$$= \underbrace{\mathbf{E}_{\text{act}}\mathbf{E}_{\text{sim}}\left(\sum_{t=1}^{T} \text{opt}(\hat{c}'_{1:t-1} + r_t) \cdot \hat{c}'_t\right)}_{A}$$

$$+ \underbrace{\lambda \mathbf{E}_{\text{act}}\mathbf{E}_{\text{sim}}\left(\sum_{t=1}^{T} \text{opt}(\hat{c}'_{1:t-1} + r_t) \cdot \left(\sigma_t\sqrt{\ln t} - \sigma_{t-1}\sqrt{\ln(t-1)}\right)\right)}_{B}.$$

2. In Section 5.1 we show that $B \leq 2\sqrt{kT \ln T}$.
3. In Appendix A we bound the term A with

$$\mathbf{E}_{\text{act}}\left(\text{opt-cost}(\hat{c}'_{1:T})\right) + e(e-1)\left(k\sum_{t=1}^{T} \epsilon_t + \frac{2\lambda\sqrt{T \ln T}}{\sqrt{k}}\right) + \frac{2(1 + \ln k)}{\epsilon T}.$$

This bound is quite standard and uses techniques introduced earlier [1,3].

4. Finally, by definition it holds that

$$\text{opt-cost}(\hat{c}'_{1:t}) \leq \text{opt}(c_{1:t}) \cdot \hat{c}'_{1:t}.$$

Using a martingale concentration inequality we can show that this is, with a high probability, less than the cost of the best static decision $\text{opt-cost}(c_{1:t})$. We do this in Section 5.2.

These steps bound the cost on the exploitation steps, as argued above. In Table 3 we set the parameters $\gamma_t/k = \epsilon_t = \epsilon_T = \sqrt{\ln T}/(3\sqrt{kT})$. Counting the cost during the sampling turns we obtain a final regret bound where asymptotically the most significant term is

$$5\sqrt{kT \ln T}.$$

The remaining terms are bounded by

$$2\sqrt{2}\frac{\sqrt{T \ln T}}{\sqrt{k}} + \frac{6(1 + \ln k)\sqrt{kT}}{\sqrt{\ln T}}.$$

Using the concentration inequality yields an additional term, but it is of the order $o(\sqrt{T})$ with a small constant factor.

5.1   Bound on the Term B

We now proceed to bound the term

$$B = \mathbf{E}_{\text{act}}\mathbf{E}_{\text{sim}}\left(\sum_{t=1}^{T} \text{opt}(\hat{c}'_{1:t-1} + r_t) \cdot \left(\sigma_t\sqrt{\ln t} - \sigma_{t-1}\sqrt{\ln(t-1)}\right)\right).$$

First, because $\sigma_{i,t}$ grows faster than $\sqrt{\ln t}$, the inequality

$$\sigma_{i,t}\sqrt{\ln t} - \sigma_{i,t-1}\sqrt{\ln(t-1)} \leq (\sigma_{i,t} - \sigma_{i,t-1})\left(\sqrt{\ln t} + \sqrt{\ln(t-1)}\right)$$

holds. Recall that $\sqrt{x}$ is concave and hence, by linearization, $\sqrt{x+h} \leq \sqrt{x} + h/(2\sqrt{x})$ for any $x, h > 0$. Direct application of this inequality shows that for each $\sigma_{i,t}$

$$\sigma_{i,t} - \sigma_{i,t-1} \leq \frac{1}{2p_{i,t}\,\sigma_{i,t-1}}.$$

Hence, the term B is upper bounded by $\mathbf{E}_{\text{act}}\left(\sqrt{\ln T}\sum_{t=1}^{T}\sum_{i=1}^{k} 1/\sigma_{i,t}\right)$ when the expectation $\mathbf{E}_{\text{sim}}$ is calculated. Since the function $1/\sqrt{x}$ is both convex and decreasing, it is not difficult to see that to maximize individually each $\sum_{i=1}^{k} 1/\sigma_{i,t}$ all probabilities $p_{i,\tau}$ in $\sigma_{i,t} = \sqrt{\sum_{\tau=1}^{t} 1/p_{i,\tau}}$ should be set to $1/k$. We can formally prove this using an involved manipulation of Lagrange multipliers to obtain the inequality:

$$\mathbf{E}_{\text{act}}\left(\sum_{t=1}^{T}\sum_{i=1}^{k} 1/\sigma_{it}\right) \leq \sum_{t=1}^{T}\sum_{i=1}^{k}\frac{1}{\sqrt{kt}} \leq 2\sqrt{kT}. \tag{2}$$

Thus, we obtain the desired bound

$$B \le 2\sqrt{kT \ln T}.$$

5.2   Bounding the Estimated Cost

We claim that with a high enough probability it holds that

$$\mathrm{opt}(\mathbf{c}_{1:T}) \cdot \hat{\mathbf{c}}'_{1:T} \le \mathrm{opt\text{-}cost}(\mathbf{c}_{1:T}).$$

Intuitively, $\hat{\mathbf{c}}_{1:T}$ is a random variable that is concentrated to a certain degree around its expected value and we take advantage of this in lower bounding the vector $\hat{\mathbf{c}}_t$ by $\hat{\mathbf{c}}'_t$ with high probability. Formally we use the following McDiarmid's [18] martingale concentration inequality. Note that this version of the inequality allows us to use the conditional variances.

Theorem 2 (Theorem 3.15, p. 224, in [18]). *Let $X_1, \ldots, X_T$ be a martingale difference sequence and V the sum of conditional variances*

$$V = \sum_{t=1}^{T} \mathrm{Var}\left(X_t \mid X_1, \ldots, X_{t-1}\right),$$

then for every $x, v \ge 0$

$$\mathbf{P}\left(\sum_t X_t \ge x \text{ and } V \le v\right) \le \exp\left(\frac{-x^2}{2v + 2xu/3}\right),$$

where u is a uniform upper bound for X_t.

We apply this inequality and set

- $X_{i,t} = \hat{c}_{i,t} - c_{i,t}$,
- $x = \lambda \sigma_{i,T} \ln T$,
- $u = 1/\gamma_T$, and
- $v = \sigma_{i,T}^2$.

Note that $V \le \sigma_{i,T}^2$ holds always because $\sigma_{i,T}^2$ is defined to be a upper bound of the variance of $\sum_{t=1}^{T} X_{i,t}$. Now

$$\mathbf{P}\left(\sum_{t=1}^{T} \hat{c}_{i,t} - c_{i,t} \ge \lambda \sigma_{i,T}\sqrt{\ln T}\right) \le \exp\left(\frac{-\lambda^2 \sigma_{i,T}^2 \ln T}{2\sigma_{i,T}^2 + 2\lambda \sigma_{i,T}\sqrt{\ln T}/(3\gamma_T)}\right)$$

$$\le \exp\left(\frac{-\lambda^2 \ln T}{2 + 2\lambda\sqrt{\ln T}/(3k\epsilon_T\sqrt{T})}\right) \qquad (3)$$

$$\le \exp\left(\frac{-\lambda^2 \ln T}{2 + 2\lambda/\sqrt{k}}\right) \qquad (4)$$

$$\le \exp\left(\frac{-\lambda^2 \ln \sqrt{T}}{1 + \sqrt{2}/\sqrt{k}}\right)$$

$$= \frac{1}{\sqrt{T}}. \qquad (5)$$

In Inequality 3 we use the definition $\gamma_T = k\epsilon_T$ and the fact that $\sigma_{i,T} \geq \sqrt{T}$. In Inequality 4 we write out $\epsilon_T = \sqrt{\ln T}/(3\sqrt{kT})$. Inequality 5 follows from $\lambda^2 = 1 + \sqrt{2}/\sqrt{k}$. When this bounding is done for each of the k levers separately, the bad case happens with probability at most $k/\sqrt{T}$. Even the bad case is of the order $o(T)$ with extremely high probability, because $\sigma_{i,T} \leq T^{2/3}$, and we can again use the same concentration inequality. Hence,

$$\mathbf{E}\big(\mathrm{opt}(\boldsymbol{c}_{1:T}) \cdot \hat{\boldsymbol{c}}'_{1:T} - \mathrm{opt\text{-}cost}(\boldsymbol{c}_{1:T})\big) \leq o(\sqrt{T}).$$

We do not include this term in our regret bound, because asymptotically it is not significant.

5.3   Computing the Probabilities in AB-FPL

We need to obtain the probability $\boldsymbol{p}_{i,t}$ of selecting lever i at turn t. On an exploitation turn the probability of selecting lever i is the probability that $\hat{c}_{i,1:t-1}$ plus perturbation is smaller than the corresponding value for the other levers. Each of these values is distributed according to a two-sided exponential distribution with the expected value at $\hat{c}'_{j,1:t-1}$ for lever j. The probability that the lever i has the lowest value is $\int_{-\infty}^{\infty} p_1(x)p_2(x)dx$, where $p_1(x)$ is the density of the value of lever i, and $p_2(x)$ is the probability that the values for the other levers are higher. $p_2(x)$ is a product of cumulative distributions for the exponential distribution. Hence the integral can be calculated piecewise, where the pieces are all intervals between successive expectations $\hat{c}'_{j,1:t-1}$. In AB-FPL we must also account for the sampling that is done.

6   Conclusion and Discussion

We have demonstrated that the FPL algorithm can be successfully applied in the multi-armed bandit model against an adaptive adversary with a regret of order $O(\sqrt{T \ln T})$. In the general bandit geometric setting, however, the best known regret is $O(T^{2/3})$ even against an oblivious adversary [16]. The method we have applied is plausibly general enough to be applied to a hypothetical geometric bandit FPL algorithm that has been designed against an oblivious adversary with a regret of $O(\sqrt{T})$.

Interesting future direction is to examine more closely what are the exact constant factors that are achievable in the regret bounds. Constant factors are important, because even halving the regret in an application where the cost is expensive is a significant improvement. In Vermorel and Mohri's [19] experiments — on non-adversarial stochastically generated data — simple algorithms outperformed those that were designed against an oblivious adversary. It is not surprising that theoretical worst-case regret bounds are higher for more powerful adversaries. It would, nevertheless, be interesting to know if a better regret could be simultaneously obtained in a case that is not worst-case.

Acknowledgments

We thank the anonymous reviewers for valuable comments that helped us to improve the final version of this paper. This work was supported by Academy of Finland projects "INTENTS: Intelligent Online Data Structures" and "ALEA: Approximation and Learning Algorithms." Moreover, the work of J. Kujala is financially supported by Tampere Graduate School in Information Science and Engineering (TISE) and Nokia Foundation.

References

1. Kalai, A.T., Vempala, S.: Efficient algorithms for online decision problems. Journal of Computer and System Sciences 71(3), 26–40 (2005)
2. Hannan, J.: Approximation to Bayes risk in repeated plays. In: Dresher, M., Tucker, A., Wolfe, P. (eds.) Contributions to the Theory of Games, vol. 3, pp. 97–139. Princeton University Press, Princeton (1957)
3. Kujala, J., Elomaa, T.: On following the perturbed leader in the bandit setting. In: Jain, S., Simon, H.U., Tomita, E. (eds.) ALT 2005. LNCS (LNAI), vol. 3734, pp. 371–385. Springer, Heidelberg (2005)
4. Dani, V., Hayes, T.P.: How to beat the adaptive multi-armed bandit. Technical report, Cornell University (2006), http://arxiv.org/cs.DS/602053
5. Auer, P., Cesa-Bianchi, N., Freund, Y., Schapire, R.E.: The non-stochastic multi-armed bandit problem. SIAM Journal on Computing 32(1), 48–77 (2002)
6. Auer, P.: Using upper confidence bounds for online learning. In: Proceedings of the 41st Annual Symposium on Foundations of Computer Science, pp. 270–279. IEEE Computer Society Press, Los Alamitos (2000)
7. Robbins, H.: Some aspects of the sequential design of experiments. Bulletin of the American Mathematical Society 58, 527–535 (1952)
8. Freund, Y., Schapire, R.E.: A decision-theoretic generalization of on-line learning and an application to boosting. Journal of Computer and System Sciences 55(1), 119–139 (1997)
9. Cesa-Bianchi, N., Lugosi, G.: Prediction, Learning, and Games. Cambridge University Press, Cambridge (2006)
10. Littlestone, N., Warmuth, M.K.: The weighted majority algorithm. Information and Computation 108(2), 212–261 (1994)
11. Dani, V., Hayes, T.P.: Robbing the bandit: Less regret in online geometric optimization against an adaptive adversary. In: Proceeding of the 17th Annual ACM-SIAM Symposium on Discrete Algorithms, pp. 937–943. ACM Press, New York (2006)
12. Hutter, M., Poland, J.: Adaptive online prediction by following the perturbed leader. Journal of Machine Learning Research 6, 639–660 (2005)
13. Kakade, S., Kalai, A.T., Ligett, K.: Approximation algorithms going online. Technical Report CMU-CS-07-102, Carnegie Mellon University, reports-archive.adm.cs.cmu.edu/anon/2007/CMU-CS-07-102.pdf (2007)
14. Zinkevich, M.: Online convex programming and generalized infinitesimal gradient ascent. In: Fawcett, T., Mishra, N. (eds.) Proceeding of the 20th International Conference on Machine Learning, Menlo Park, pp. 928–936. AAAI Press (2003)
15. Awerbuch, B., Kleinberg, R.: Near-optimal adaptive routing: Shortest paths and geometric generalizations. In: Proceeding of the 36th Annual ACM Symposium on Theory of Computing, pp. 45–53. ACM Press, New York (2004)

16. McMahan, H.B., Blum, A.: Geometric optimization in the bandit setting against an adaptive adversary. In: Shawe-Taylor, J., Singer, Y. (eds.) COLT 2004. LNCS (LNAI), vol. 3120, pp. 109–123. Springer, Heidelberg (2004)
17. György, A., Linder, T., Lugosi, G.: Tracking the best of many experts. In: Auer, P., Meir, R. (eds.) COLT 2005. LNCS (LNAI), vol. 3559, pp. 204–216. Springer, Heidelberg (2005)
18. McDiarmid, C.: Concentration. In: Habib, M., McDiarmid, C., Ramirez-Alfonsin, J., Reed, B. (eds.) Probabilistic Methods for Algorithmic Discrete Mathematics, pp. 195–248. Springer, Heidelberg (1998)
19. Vermorel, J., Mohri, M.: Multi-armed bandit algorithms and empirical evaluation. In: Gama, J., Camacho, R., Brazdil, P.B., Jorge, A.M., Torgo, L. (eds.) ECML 2005. LNCS (LNAI), vol. 3720, pp. 437–448. Springer, Heidelberg (2005)

A   Bound on the Term A

In this appendix we bound the term

$$A = \mathbf{E}_{\mathrm{act}}\mathbf{E}_{\mathrm{sim}}\left(\sum_{t=1}^{T}\mathrm{opt}(\hat{c}'_{1:t-1} + r_t)\cdot\hat{c}'_t\right).$$

Recall that $\mathbf{E}_{\mathrm{sim}}$ refers to the expectation on randomness originating from vectors r_t, i.e., a simulated run of FPL on vectors $\hat{c}'_{1:t}$, and the adversary for the *simulated run* is an oblivious one. The bound for the simulated run of FPL is similar to what Kalai and Vempala [1] have derived against an oblivious adversary. The difference is that in the simulated run of FPL on cost vectors $\hat{c}'_t$ the worst-case cost on a single step is too large, although under expectation the worst-case cost is small enough. The following step is different from their proof.

Lemma 1

$$\mathbf{E}\left(\sum_{t=1}^{T}\mathrm{opt}(\hat{c}'_{1:t-1} + r_t)\cdot\hat{c}'_t\right) \leq \mathbf{E}\left(\sum_{t=1}^{T}\mathrm{opt}(\hat{c}'_{1:t} + r_t)\cdot\hat{c}'_t\right)$$
$$+ e\,(e-1)\left(k\sum_{t=1}^{T}\epsilon_t + \frac{2\lambda\sqrt{T\ln T}}{\sqrt{k}}\right).$$

We also need the following lemma to complete the desired bound on A.

Lemma 2 (Lemmas 1 and 3 in [1]).

$$\mathbf{E}_{\mathrm{sim}}\left(\sum_{t=1}^{T}\mathrm{opt}(\hat{c}'_{1:t} + r_t)\cdot\hat{c}'_t\right) \leq \mathrm{opt\text{-}cost}(\hat{c}'_{1:T}) + \frac{2\,(1+\ln k)}{\epsilon_T}.$$

Applying both of these lemmas gives us the final bound on A:

$$\mathbf{E}_{\mathrm{act}}\big(\mathrm{opt\text{-}cost}(\hat{c}'_{1:T})\big) + e\,(e-1)\left(k\sum_{t=1}^{T}\epsilon_t + \frac{2\lambda\sqrt{T\ln T}}{\sqrt{k}}\right) + \frac{2(1+\ln k)}{\epsilon_T}.$$

Proof (of Lemma 1). As observed in [1,3] when the perturbation vector is chosen from the exponential distribution, the probability of selecting a lever in AB-FPL changes in a well defined way when the cost vector $\hat{c}_{1:t}$ changes:

$$e^{-\epsilon_t \delta} \leq p_{\text{new}}/p_{\text{old}} \leq e^{\epsilon_t \delta}, \tag{6}$$

where δ is the change in the 1-norm of the cost vector, p_{old} was the probability of selecting the lever before the cost vector changed, and p_{new} is the probability after the cost vector changed.

Hence, if $p_{i,t}$ is the probability of selecting lever i at step t,

$$\mathbf{E}_{\text{sim}}\left(\text{opt}\left(\hat{c}'_{1:t-1} + r_t\right) \cdot \hat{c}'_t\right)$$

$$= \sum_{i=1}^{k} \hat{c}'_{i,t}\, p_{i,t} \tag{7}$$

$$\leq \sum_{i=1}^{k} \hat{c}'_{i,t}\, p_{i,t+1}\, e^{\epsilon_t \left\|\hat{c}'_t\right\|_1} \tag{8}$$

$$\leq \sum_{i=1}^{k} \hat{c}'_{i,t}\, p_{i,t+1}\left((1 + (e-1)\,\epsilon_t \left\|\hat{c}'_t\right\|_1)\right) \tag{9}$$

$$\leq \mathbf{E}_{\text{sim}}\left(\text{opt}\left(\hat{c}'_{1:t} + r_t\right) \cdot \hat{c}'_t\right) + (e-1)\epsilon_t \sum_{i=1}^{k} \hat{c}_{i,t}\left\|\hat{c}'_t\right\|_1\, p_{i,t+1}. \tag{10}$$

Equality 7 follows by definition, Inequality 8 is an application of Inequality 6, and Inequality 9 follows from the behavior of the exponent function e^x when $x < 1$. We can assume that $x < 1$ because for all $i \in \{1, \ldots, k\}$ and $t \in \{1, \ldots, T\}$ the probability $p_{i,t} > \gamma_t/k = \epsilon_t$.

Next we bound the remaining regret term on the right in Inequality 10 under $\mathbf{E}_{\text{act}}$:

$$\mathbf{E}_{\text{act}}\left(\sum_{i=1}^{k} \hat{c}_{i,t}\left\|\hat{c}'_t\right\|_1\, p_{i,t+1}\right)$$

$$= \mathbf{E}_{\text{act}}\left(c_{j,t}\, \frac{p_{j,t+1}}{p_{j,t}}\left\|\hat{c}'_t\right\|_1\right) \tag{11}$$

$$\leq e\, \mathbf{E}_{\text{act}}\left(\left\|\hat{c}_t\right\|_1 + \lambda\left\|\sigma_t\sqrt{\ln t} - \sigma_{t-1}\sqrt{\ln(t-1)}\right\|_1\right) \tag{12}$$

$$\leq e\left(k + \frac{\lambda\sqrt{\ln t}}{k\epsilon_t}\mathbf{E}_{\text{act}}\left(\sum_{i=1}^{k} \frac{1}{\sigma_{i,t}}\right)\right). \tag{13}$$

In Equality 11 we write out the sum and j refers to the lever that was chosen by AB-FPL. In Inequality 12 we use Inequality 6 and the fact that $\delta < 1$ as well as the triangle inequality $\left\|\hat{c}'_t\right\|_1 \leq \left\|\hat{c}_t\right\|_1 + \lambda\left\|\sigma_t\sqrt{\ln t} - \sigma_{t-1}\sqrt{\ln(t-1)}\right\|_1$.

Inequality 13 follows from calculation of the expectation $\mathbf{E}_{\text{act}}(\|\hat{c}_t\|_1 \mid t-1, \ldots, 1)$ and from the following inequality, which was justified in Section 5.1,

$$\left\|\sigma_t\sqrt{\ln t} - \sigma_{t-1}\sqrt{\ln(t-1)}\right\|_1 \leq \sqrt{\ln t}\sum_{i=1}^{k}\frac{1}{p_{i,t}\sigma_{i,t-1}}.$$

Putting Inequalities 10 and 13 together and summing them over all turns t yields the final regret

$$e(e-1)\left(k\sum_{t=1}^{T}\epsilon_t + \frac{\lambda\sqrt{\ln T}}{k}\mathbf{E}_{\text{act}}\left(\sum_{t=1}^{T}\sum_{i=1}^{k}\frac{1}{\sigma_{i,t}}\right)\right)$$

$$\leq e(e-1)\left(k\sum_{t=1}^{T}\epsilon_t + \frac{2\lambda\sqrt{T\ln T}}{\sqrt{k}}\right).$$

The term under $\mathbf{E}_{\text{act}}$ was bounded in Section 5.1 by $2\sqrt{kT}$ in Inequality 2.

Online Regression Competitive with Changing Predictors

Steven Busuttil and Yuri Kalnishkan

Computer Learning Research Centre and Department of Computer Science,
Royal Holloway, University of London,
Egham, Surrey, TW20 0EX, United Kingdom
{steven,yura}@cs.rhul.ac.uk

Abstract. This paper deals with the problem of making predictions in the online mode of learning where the dependence of the outcome y_t on the signal $\mathbf{x}_t$ can change with time. The Aggregating Algorithm (AA) is a technique that optimally merges experts from a pool, so that the resulting strategy suffers a cumulative loss that is almost as good as that of the best expert in the pool. We apply the AA to the case where the experts are all the linear predictors that can change with time. KAARCh is the kernel version of the resulting algorithm. In the kernel case, the experts are all the decision rules in some reproducing kernel Hilbert space that can change over time. We show that KAARCh suffers a cumulative square loss that is almost as good as that of any expert that does not change very rapidly.

1 Introduction

We consider the online protocol where on each trial $t = 1, 2, \ldots$ the learner observes a signal $\mathbf{x}_t$ and attempts to predict the outcome y_t, which is shown to the learner later. The performance of the learner is measured by means of the cumulative square loss. The Aggregating Algorithm (AA), introduced by Vovk in [1] and [2], allows us to merge experts from large pools to obtain optimal strategies. Such an optimal strategy performs nearly as good as the best expert from the class in terms of the cumulative loss.

In [3] the AA is applied to merge all constant linear regressors, i.e., experts θ predicting $\theta' \mathbf{x}_t$ (it is assumed that $\mathbf{x}_t$ and θ are drawn from $\mathbb{R}^n$). The resulting Aggregating Algorithm for Regression (AAR) (also known as the Vovk-Azoury-Warmuth forecaster, see [4, Sect. 11.8]) performs almost as well as the best regressor θ. In [5] the kernel version of AAR, known as the Kernel AAR (KAAR), is introduced and a bound on its performance is derived (see also [6, Sect. 8]). From a computational point of view the algorithm is similar to Ridge Regression. We summarise the results concerning AAR and KAAR in Sect. 2.3.

In this paper, AA is applied to merge a wider class of predictors. We let θ vary between trials. Consider a sequence $\theta_1, \theta_2, \ldots$; let it make the prediction $(\theta_1 + \theta_2 + \ldots + \theta_t)' \mathbf{x}_t$ on trial t. We merge all predictors of this type and obtain an algorithm which is again computationally similar to Ridge Regression. We

M. Hutter, R.A. Servedio, and E. Takimoto (Eds.): ALT 2007, LNAI 4754, pp. 181–195, 2007.
© Springer-Verlag Berlin Heidelberg 2007

call the new algorithm the Aggregating Algorithm for Regression with Changing dependencies (AARCh) and its kernelised version KAARCh. Clearly, our class of experts is very large and we cannot compete in a reasonable sense with every expert from this class. However in Sects. 4 and 5 we show that KAARCh can perform almost as well as any regressor if the latter is not changing very rapidly, i.e., if each $\|\theta_t\|$ is small or only a few are nonzero.

A similar problem is considered in [7], [8], and [9] for classification and regression. In these publications, this problem is referred to as the non-stationary or shifting target problem and the corresponding bounds are called shifting bounds. The work by Herbster and Warmuth in [7] is closest to ours. However, their methods are based on Gradient Descent and therefore their bounds are of a different type. For instance, since our approach is based on the Aggregating Algorithm we get a coeffcient for the term representing the cumulative loss of the experts equal to 1 (see Theorems 3 and 4), whereas those in the bounds of [7, Theorems 14–16] are greater than 1.

In practice, KAARCh can be used to predict parameters that change slowly with time. KAARCh is more computationally expensive than the techniques described in [7], with time and space complexities that grow with time. This is not desirable in an algorithm designed for online learning; however, a practical implementation is described in [10]. Essentially, KAARCh is made to 'forget' older examples that do not affect the prediction too much. In [10] empirical experiments are carried out on an artificial dataset and on the real world problem of predicting the implied volatility of options (the name KAARCh was inspired by the popular GARCH model for predicting volatility in finance).

2   Background

In this section we introduce some preliminaries and related material required for our main results. As usual, all vectors are identified with one-column matrices and $\mathbf{B}'$ stands for the transpose of matrix $\mathbf{B}$. We will not be specifying the size of simple matrices like the identity matrix $\mathbf{I}$ when this is clear from the context.

2.1   Protocol and Loss

We can define online regression by the following protocol. At every moment in time $t = 1, 2, \ldots$, the value of a signal $\mathbf{x}_t \in X$ arrives. Statistician (or Learner) S observes $\mathbf{x}_t$ and then outputs a prediction $\gamma_t \in \mathbb{R}$. Finally, the outcome $y_t \in \mathbb{R}$ arrives. This can be summarised by the following scheme:

> **for** $t = 1, 2, \ldots$ **do**
> S observes $\mathbf{x}_t \in X$
> S outputs $\gamma_t \in \mathbb{R}$
> S observes $y_t \in \mathbb{R}$
> **end for**

The set X is a signal space which is assumed to be known to Statistician in advance. We will be referring to a signal-outcome pair as an example. The performance of S is measured by the sum of squared discrepancies between the predictions and the outcomes (known as square loss). Therefore on trial t Statistician S suffers loss $(y_t - \gamma_t)^2$. Thus after T trials, the total loss of S is

$$L_T(S) = \sum_{t=1}^{T} (y_t - \gamma_t)^2.$$

Clearly, a smaller value of $L_T(S)$ means a better predictive performance.

2.2   Linear and Kernel Predictors

If $X \subseteq \mathbb{R}^n$ we can consider simple linear regressors of the form $\theta \in \mathbb{R}^n$. Given a signal $\mathbf{x} \in X$, such a regressor makes a prediction $\theta'\mathbf{x}$. Linear methods are easy to manipulate mathematically but their use in the real world is limited since they can only model simple dependencies. One solution to this could be to map the data to some high dimensional feature space and then find a simple solution there. This however, can lead to what is known as the curse of dimensionality where both the computational and generalisation performance degrades as the number of features grow [11, Sect. 3.1].

The kernel trick (first used in this context in [12]) is now a widely used technique which can make a linear algorithm operate in feature space without the inherent complexities. Informally, a kernel is a dot product in feature space. Typically, to transform a linear method into a nonlinear one, the linear algorithm is first formulated in such a way that all signals appear only in dot products (known as the dual form). Then these dot-products are replaced by kernels.

For a function $k : X \times X \to \mathbb{R}$ to be a kernel it has to be symmetric, and for all ℓ and all $\mathbf{x}_1, \ldots, \mathbf{x}_\ell \in X$, the kernel matrix $\mathbf{K} = (k(\mathbf{x}_i, \mathbf{x}_j))_{i,j}$, $i, j = 1, \ldots, \ell$ must be positive semi-definite (have nonnegative eigenvalues). For every kernel there exists a unique reproducing kernel Hilbert space (RKHS) F such that k is the reproducing kernel of F. In fact, there is a mapping $\phi : X \to F$ such that kernels can be defined as

$$k(\mathbf{x}, \mathbf{z}) = \langle \phi(\mathbf{x}), \phi(\mathbf{z}) \rangle.$$

A RKHS on a set X is a (separable and complete) Hilbert space of real valued functions on X comprised of linear combinations of k of the form

$$f(\mathbf{x}) = \sum_{i=1}^{l} c_i k(\mathbf{v}_i, \mathbf{x}),$$

where l is a positive integer, $c_i \in \mathbb{R}$ and $\mathbf{v}_i, \mathbf{x} \in X$, and their limits. We will be referring to any function in the RKHS F as D. Intuitively $D(\mathbf{x})$ is a decision rule in F that produces a prediction for the object $\mathbf{x}$. We will be measuring the complexity of D by its norm $\|D\|$ in F. For more information on kernels and RKHS see, for example, [13] and [14].

2.3   The Aggregating Algorithm (AA)

We now give an overview of the Aggregating Algorithm (AA) mostly following [3, Sects. 1 and 2]. Let Ω be an outcome space, Γ be a prediction space and Θ be a (possibly infinite) pool of experts. We consider the following game between Statistician (or Learner) S, Nature, and Θ:

> **for** $t = 1, 2, \ldots$ **do**
>> Every expert $\theta \in \Theta$ makes a prediction $\gamma_t^{(\theta)} \in \Gamma$
>> Statistician S observes all $\gamma_t^{(\theta)}$
>> Statistician S outputs a prediction $\gamma_t \in \Gamma$
>> Nature outputs $\omega_t \in \Omega$
> **end for**

Given a fixed loss function $\lambda : \Omega \times \Gamma \to [0, \infty]$, Statistician aims to suffer a cumulative loss

$$\mathrm{L}_T(S) = \sum_{t=1}^{T} \lambda(\omega_t, \gamma_t)$$

that is not much larger than the loss

$$\mathrm{L}_T(\theta) = \sum_{t=1}^{T} \lambda\left(\omega_t, \gamma_t^{(\theta)}\right)$$

of the best expert $\theta \in \Theta$. The AA takes two parameters, a prior probability distribution P_0 in the pool of experts Θ and a learning rate $\eta > 0$. Let $\beta = e^{-\eta}$.

We will first describe the Aggregating Pseudo Algorithm (APA) that does not output actual predictions but generalised predictions. A generalised prediction $g : \Omega \to \mathbb{R}$ is a mapping giving a value of loss for each possible outcome. At every step t, the APA updates the experts' weights so that those that suffered large loss during the previous step have their weights reduced:

$$P_t(d\theta) = \beta^{\lambda\left(\omega_t, \gamma_t^{(\theta)}\right)} P_{t-1}(d\theta) , \quad \theta \in \Theta.$$

At time t, the APA chooses a generalised prediction by

$$g_t(\omega) = \log_\beta \int_\Theta \beta^{\lambda\left(\omega, \gamma_t^{(\theta)}\right)} P_{t-1}^*(d\theta),$$

where $P_{t-1}^*(d\theta)$ are the normalised weights $P_{t-1}^*(d\theta) = P_{t-1}(d\theta)/P_{t-1}(\Theta)$. This guarantees that for any learning rate $\eta > 0$, prior P_0, and $T = 1, 2, \ldots$ (see [3, Lemma 1])

$$\mathrm{L}_T(\mathrm{APA}) = \log_\beta \int_\Theta \beta^{\mathrm{L}_T(\theta)} P_0(d\theta). \tag{1}$$

To get a prediction from the generalised prediction $g_t(\omega)$ (note that we use ω since we do not yet know the real outcome of step t, ω_t) the AA uses a substitution function Σ mapping generalised predictions into Γ. A substitution function may introduce extra loss; however, in many cases perfect substitution is possible.

We say that the loss function λ is η-mixable if there is a substitution function Σ such that

$$\lambda\left(\omega_t, \Sigma(g_t(\omega))\right) \le g_t(\omega_t) \qquad (2)$$

on every step t, all experts' predictions and all outcomes. The loss function λ is mixable if it is η-mixable for some $\eta > 0$.

Suppose that our loss function is η-mixable. Using (1) and (2) we can obtain the following upper bound on the cumulative loss of the AA:

$$L_T(\mathrm{AA}) \le \log_\beta \int_\Theta \beta^{L_T(\theta)} P_0(d\theta).$$

In particular, when the pool of experts is finite and all experts are assigned equal prior weights, we get, for any $\theta \in \Theta$

$$L_T(\mathrm{AA}) \le L_T(\theta) + \frac{\ln m}{\eta},$$

where m is the size of the pool of experts. This bound can be shown to be optimal in a very strong sense for all algorithms attempting to merge experts' predictions (see [2]).

The Square Loss Game. In this paper we are concerned with the (bounded) square loss game (see [3, Sect. 2.4]), where $\Omega = [-Y, Y]$, $Y \in \mathbb{R}$, $\Gamma = \mathbb{R}$, and $\lambda(\omega, \gamma) = (\omega - \gamma)^2$. The square loss game is η-mixable if and only if $\eta \le 1/(2Y^2)$. A perfect substitution function for this game is

$$\gamma = \frac{g(-Y) - g(Y)}{4Y}. \qquad (3)$$

The Aggregating Algorithm for Regression (AAR). The AA was applied to the problem of linear regression resulting in the Aggregating Algorithm for Regression (AAR). AAR merges all the linear predictors that map signals to outcomes [3, Sect. 3] (a Gaussian prior is assumed on the pool of experts). AAR makes a prediction at time T by

$$\gamma_{\mathrm{AAR}} = \widetilde{\mathbf{y}}' \widetilde{\mathbf{X}} (\widetilde{\mathbf{X}}' \widetilde{\mathbf{X}} + a\mathbf{I})^{-1} \mathbf{x}_T,$$

where a is a positive scalar, $\widetilde{\mathbf{X}} = (\mathbf{x}_1, \mathbf{x}_2, \ldots, \mathbf{x}_T)'$ and $\widetilde{\mathbf{y}} = (y_1, y_2, \ldots, y_{T-1}, 0)'$.

The main property of AAR is that it is optimal in the sense that the total loss it suffers is only a little worse than that of any linear predictor. By the latter we mean a strategy that predicts $\theta' \mathbf{x}_t$ on every trial t, where $\theta \in \mathbb{R}^n$ is some fixed vector. The set of all linear predictors may be identified with $\mathbb{R}^n$.

Theorem 1 ([3, Theorem 1]). *For any $a > 0$ and any point in time T,*

$$L_T(\mathrm{AAR}) \le \inf_\theta (L_T(\theta) + a\|\theta\|^2) + Y^2 \ln \det \left(\frac{1}{a} \sum_{t=1}^T \mathbf{x}_t \mathbf{x}_t' + \mathbf{I}\right).$$

The Kernel Aggregating Algorithm for Regression (KAAR). KAAR, the kernel version of AAR introduced in [5], makes a prediction for the signal $\mathbf{x}_T$ by

$$\gamma_{\text{KAAR}} = \widetilde{\mathbf{y}}'(\widetilde{\mathbf{K}} + a\mathbf{I})^{-1}\widetilde{\mathbf{k}},$$

where

$$\widetilde{\mathbf{K}} = \begin{bmatrix} k(\mathbf{x}_1, \mathbf{x}_1) & \cdots & k(\mathbf{x}_1, \mathbf{x}_T) \\ \vdots & \ddots & \vdots \\ k(\mathbf{x}_T, \mathbf{x}_1) & \cdots & k(\mathbf{x}_T, \mathbf{x}_T) \end{bmatrix}, \quad \text{and} \quad \widetilde{\mathbf{k}} = \begin{bmatrix} k(\mathbf{x}_1, \mathbf{x}_T) \\ \vdots \\ k(\mathbf{x}_T, \mathbf{x}_T) \end{bmatrix}.$$

Like AAR, KAAR has an optimality property. KAAR performs little worse than any decision rule D in the RKHS induced by a kernel function k.

Theorem 2 ([5, Theorem 1] and [6, Sect. 8]). *Let k be a kernel on a space X and D be any decision rule in the RKHS induced by k. Then for every $a > 0$ and any point in time T,*

$$L_T(\text{KAAR}) \leq L_T(D) + a\|D\|^2 + Y^2 \ln \det\left(\frac{1}{a}\widetilde{\mathbf{K}} + \mathbf{I}\right).$$

Corollary 1 ([6, Sect. 8]). *Under the same conditions of Theorem 2 let $c = \sup_{\mathbf{x} \in X} \sqrt{k(\mathbf{x}, \mathbf{x})}$. Then for every $a > 0$, every $d > 0$, every decision rule D such that $\|D\| \leq d$ and any point in time T, we get*

$$L_T(\text{KAAR}) \leq L_T(D) + ad^2 + \frac{Y^2 c^2 T}{a}.$$

If, moreover, T is known in advance, one can choose $a = (Yc/d)\sqrt{T}$ and get

$$L_T(\text{KAAR}) \leq L_T(D) + 2Ycd\sqrt{T}.$$

3 Algorithm

For our new method, we apply the Aggregating Algorithm (AA) to the regression problem where the experts can change with time. We call this method the Aggregating Algorithm for Regression with Changing dependencies (AARCh). Subsequently, we will kernelise this method to get Kernel AARCh (KAARCh). Throughout this section we will be using the lemmas given in the appendix.

3.1 AARCh: Primal Form

The main idea behind AARCh is to apply the Aggregating Algorithm to the case where the pool of experts is made up of all linear predictors that can change independently with time. We assume that outcomes are bounded by Y, i.e., for any t, $y_t \in [-Y, Y]$ (we do not require our algorithm to know Y). We are interested in the square loss, therefore we will be using optimal $\eta = 1/(2Y^2)$ and substitution function (3).

An expert is a sequence $\theta_1, \theta_2, \ldots$, that at time T predicts

$$\mathbf{x}_T'(\theta_1 + \theta_2 + \ldots + \theta_T),$$

where for any t, $\theta_t \in \mathbb{R}^n$ and $\mathbf{x}_T \in \mathbb{R}^n$. To apply the AA to this problem we need to define a lower triangular block matrix $\mathbf{L}$, and θ which is a concatenation of all the θ_t for $t = 1 \ldots T$, such that[1]

$$\mathbf{L}\theta = \begin{bmatrix} \mathbf{I} & \mathbf{0} & \cdots & \cdots & \mathbf{0} \\ \mathbf{I} & \mathbf{I} & \ddots & & \vdots \\ \vdots & \vdots & \ddots & \ddots & \vdots \\ \mathbf{I} & \mathbf{I} & \cdots & \mathbf{I} & \mathbf{0} \\ \mathbf{I} & \mathbf{I} & \cdots & \mathbf{I} & \mathbf{I} \end{bmatrix} \begin{bmatrix} \theta_1 \\ \theta_2 \\ \vdots \\ \theta_{T-1} \\ \theta_T \end{bmatrix} = \begin{bmatrix} \theta_1 \\ \theta_1 + \theta_2 \\ \vdots \\ \theta_1 + \theta_2 + \cdots + \theta_{T-1} \\ \theta_1 + \theta_2 + \cdots + \theta_{T-1} + \theta_T \end{bmatrix}.$$

The matrices $\mathbf{I}$ and $\mathbf{0}$ in $\mathbf{L}$ are the $n \times n$ identity and all-zero matrices respectively. We also need to define $\mathbf{z}_t$ which is $\mathbf{x}_t$ padded with zeros in the following way

$$\mathbf{z}_t = \begin{bmatrix} \underbrace{0 \cdots 0}_{n(t-1)} & \mathbf{x}_t' & \underbrace{0 \cdots 0}_{n(T-t)} \end{bmatrix}',$$

so that

$$\mathbf{z}_t' \mathbf{L}\theta = \mathbf{x}_t'(\theta_1 + \theta_2 + \ldots + \theta_t).$$

Let $a_t > 0$, $t = 1, \ldots, T$, be arbitrary constants. Consider the prior distribution P_0 in the set $\mathbb{R}^{nT}$ of possible weights θ with the Gaussian density

$$P_0(d\theta) = \left(\prod_{t=1}^{T} a_t \right)^{n/2} \left(\frac{\eta}{\pi} \right)^{nT/2} e^{-\eta \sum_{t=1}^{T} a_t \|\theta_t\|^2} d\theta_1 \ldots d\theta_T$$

$$= \left(\left(\frac{\eta}{\pi} \right)^T \prod_{t=1}^{T} a_t \right)^{n/2} e^{-\eta \theta' \mathbf{A}\theta} d\theta,$$

where, letting $\mathbf{I}$ and $\mathbf{0}$ be as above, we have

$$\mathbf{A} = \begin{bmatrix} a_1\mathbf{I} & \mathbf{0} & \cdots & \mathbf{0} \\ \mathbf{0} & a_2\mathbf{I} & \ddots & \vdots \\ \vdots & \ddots & \ddots & \mathbf{0} \\ \mathbf{0} & \cdots & \mathbf{0} & a_T\mathbf{I} \end{bmatrix}.$$

The loss of θ over the first T trials is

$$L_T(\theta) = \sum_{t=1}^{T} (y_t - \mathbf{z}_t'\mathbf{L}\theta)^2$$

$$= \theta'\mathbf{L}' \left(\sum_{t=1}^{T} \mathbf{z}_t\mathbf{z}_t' \right) \mathbf{L}\theta - 2 \left(\sum_{t=1}^{T} y_t\mathbf{z}_t' \right) \mathbf{L}\theta + \sum_{t=1}^{T} y_t^2.$$

---

[1] The sum $\theta_1 + \ldots + \theta_t$ corresponds to the predictor $\mathbf{u}_t$ in [7].

Therefore, the loss of the APA is (recall that $\beta = e^{-\eta}$)

$$L_T(\text{APA}) = \log_\beta \int_{\mathbb{R}^{nT}} \beta^{L_T(\theta)} P_0(d\theta)$$

$$= \log_\beta \int_{\mathbb{R}^{nT}} \left(\left(\frac{\eta}{\pi}\right)^T \prod_{t=1}^{T} a_t \right)^{n/2}$$
$$\times e^{-\eta(\theta' \mathbf{L}'(\sum_{t=1}^{T} \mathbf{z}_t \mathbf{z}_t')\mathbf{L}\theta - 2(\sum_{t=1}^{T} y_t \mathbf{z}_t')\mathbf{L}\theta + \sum_{t=1}^{T} y_t^2 + \theta' \mathbf{A}\theta)} d\theta$$

$$= \log_\beta \int_{\mathbb{R}^{nT}} \left(\left(\frac{\eta}{\pi}\right)^T \prod_{t=1}^{T} a_t \right)^{n/2}$$
$$\times e^{-\eta\theta'(\mathbf{L}'\sum_{t=1}^{T} \mathbf{z}_t \mathbf{z}_t'\mathbf{L} + \mathbf{A})\theta + 2\eta(\sum_{t=1}^{T} y_t \mathbf{z}_t')\mathbf{L}\theta - \eta \sum_{t=1}^{T} y_t^2} d\theta \ .$$

Given the generalised prediction $g_T(\omega)$ which is the APA's loss with variable $\omega \in \mathbb{R}$ replacing y_T and using substitution function (3), the AA's prediction is

$$\gamma_T = \frac{1}{4Y} \log_\beta \frac{\beta^{g_T(-Y)}}{\beta^{g_T(Y)}}$$

$$= \frac{1}{4Y} \log_\beta \frac{\int_{\mathbb{R}^{nT}} e^{-\eta\theta'(\mathbf{L}'\sum_{t=1}^{T} \mathbf{z}_t \mathbf{z}_t'\mathbf{L} + \mathbf{A})\theta + 2\eta(\sum_{t=1}^{T-1} y_t \mathbf{z}_t'\mathbf{L} - Y \mathbf{z}_T'\mathbf{L})\theta} d\theta}{\int_{\mathbb{R}^{nT}} e^{-\eta\theta'(\mathbf{L}'\sum_{t=1}^{T} \mathbf{z}_t \mathbf{z}_t'\mathbf{L} + \mathbf{A})\theta + 2\eta(\sum_{t=1}^{T-1} y_t \mathbf{z}_t'\mathbf{L} + Y \mathbf{z}_T'\mathbf{L})\theta} d\theta} \ .$$

Let

$$Q_1(\theta) = \theta' \left(\mathbf{L}' \sum_{t=1}^{T} \mathbf{z}_t \mathbf{z}_t'\mathbf{L} + \mathbf{A} \right) \theta - 2 \left(\sum_{t=1}^{T-1} y_t \mathbf{z}_t'\mathbf{L} - Y \mathbf{z}_T'\mathbf{L} \right) \theta \ , \text{ and}$$

$$Q_2(\theta) = \theta' \left(\mathbf{L}' \sum_{t=1}^{T} \mathbf{z}_t \mathbf{z}_t'\mathbf{L} + \mathbf{A} \right) \theta - 2 \left(\sum_{t=1}^{T-1} y_t \mathbf{z}_t'\mathbf{L} + Y \mathbf{z}_T'\mathbf{L} \right) \theta \ .$$

By Lemma 1

$$\gamma_T = \frac{1}{4Y} \log_\beta \frac{e^{-\eta \min_{\theta \in \mathbb{R}^{nT}} Q_1(\theta)}}{e^{-\eta \min_{\theta \in \mathbb{R}^{nT}} Q_2(\theta)}}$$

$$= \frac{1}{4Y} \left(\min_{\theta \in \mathbb{R}^{nT}} Q_1(\theta) - \min_{\theta \in \mathbb{R}^{nT}} Q_2(\theta) \right) \ .$$

Finally, by using Lemma 2 we get

$$\gamma_T = \frac{1}{4Y} F \left(\mathbf{L}' \sum_{t=1}^{T} \mathbf{z}_t \mathbf{z}_t'\mathbf{L} + \mathbf{A}, \ -2 \sum_{t=1}^{T-1} y_t \mathbf{z}_t'\mathbf{L}, \ 2Y \mathbf{z}_T'\mathbf{L} \right)$$

$$= \left(\sum_{t=1}^{T-1} y_t \mathbf{z}_t' \right) \mathbf{L} \left(\mathbf{L}' \sum_{t=1}^{T} \mathbf{z}_t \mathbf{z}_t'\mathbf{L} + \mathbf{A} \right)^{-1} \mathbf{L}' \mathbf{z}_T \ . \tag{4}$$

3.2   AARCh: Dual Form

Let us define

$$
\widetilde{\mathbf{Z}} = \begin{bmatrix} \mathbf{z}_1' \\ \mathbf{z}_2' \\ \vdots \\ \mathbf{z}_T' \end{bmatrix}, \quad \sqrt{\mathbf{A}} = \begin{bmatrix} \sqrt{a_1}\mathbf{I} & \mathbf{0} & \cdots & \mathbf{0} \\ \mathbf{0} & \sqrt{a_2}\mathbf{I} & \ddots & \vdots \\ \vdots & \ddots & \ddots & \mathbf{0} \\ \mathbf{0} & \cdots & \mathbf{0} & \sqrt{a_T}\mathbf{I} \end{bmatrix}, \quad \text{and } \widetilde{\mathbf{y}} = \begin{bmatrix} y_1 \\ \vdots \\ y_{T-1} \\ 0 \end{bmatrix}.
$$

We can rewrite (4) in matrix notation to get

$$
\gamma_T = \widetilde{\mathbf{y}}'\widetilde{\mathbf{Z}}\mathbf{L}\left(\mathbf{L}'\widetilde{\mathbf{Z}}'\widetilde{\mathbf{Z}}\mathbf{L} + \mathbf{A}\right)^{-1}\mathbf{L}'\mathbf{z}_T
$$

$$
= \widetilde{\mathbf{y}}'\widetilde{\mathbf{Z}}\mathbf{L}\left(\sqrt{\mathbf{A}}\left(\sqrt{\mathbf{A}}^{-1}\mathbf{L}'\widetilde{\mathbf{Z}}'\widetilde{\mathbf{Z}}\mathbf{L}\sqrt{\mathbf{A}}^{-1} + \mathbf{I}\right)\sqrt{\mathbf{A}}\right)^{-1}\mathbf{L}'\mathbf{z}_T
$$

$$
= \widetilde{\mathbf{y}}'\widetilde{\mathbf{Z}}\mathbf{L}\sqrt{\mathbf{A}}^{-1}\left(\sqrt{\mathbf{A}}^{-1}\mathbf{L}'\widetilde{\mathbf{Z}}'\widetilde{\mathbf{Z}}\mathbf{L}\sqrt{\mathbf{A}}^{-1} + \mathbf{I}\right)^{-1}\sqrt{\mathbf{A}}^{-1}\mathbf{L}'\mathbf{z}_T.
$$

We can now get a dual formulation of this by using Lemma 3:

$$
\gamma_T = \widetilde{\mathbf{y}}'\left(\widetilde{\mathbf{Z}}\mathbf{L}\mathbf{A}^{-1}\mathbf{L}'\widetilde{\mathbf{Z}}' + \mathbf{I}\right)^{-1}\widetilde{\mathbf{Z}}\mathbf{L}\mathbf{A}^{-1}\mathbf{L}'\mathbf{z}_T. \tag{5}
$$

3.3   KAARCh

Since in (5) signals appear only in dot products, we can use the kernel trick to introduce nonlinearity. In this case we get Kernel AARCh (KAARCh) that at time T makes a prediction

$$
\gamma_T = \widetilde{\mathbf{y}}'\left(\bar{\mathbf{K}} + \mathbf{I}\right)^{-1}\bar{\mathbf{k}},
$$

where $\bar{\mathbf{K}} = \left(\left(\sum_{t=1}^{\min(i,j)}\frac{1}{a_t}\right)k(\mathbf{x}_i,\mathbf{x}_j)\right)_{i,j}$, for $i,j = 1,\ldots,T$, i.e.

$$
\bar{\mathbf{K}} = \begin{bmatrix} \frac{1}{a_1}k(\mathbf{x}_1,\mathbf{x}_1) & \frac{1}{a_1}k(\mathbf{x}_1,\mathbf{x}_2) & \cdots & \frac{1}{a_1}k(\mathbf{x}_1,\mathbf{x}_T) \\ \frac{1}{a_1}k(\mathbf{x}_2,\mathbf{x}_1) & \left(\frac{1}{a_1}+\frac{1}{a_2}\right)k(\mathbf{x}_2,\mathbf{x}_2) & \cdots & \left(\frac{1}{a_1}+\frac{1}{a_2}\right)k(\mathbf{x}_2,\mathbf{x}_T) \\ \vdots & \vdots & \ddots & \vdots \\ \frac{1}{a_1}k(\mathbf{x}_T,\mathbf{x}_1) & \left(\frac{1}{a_1}+\frac{1}{a_2}\right)k(\mathbf{x}_T,\mathbf{x}_2) & \cdots & \left(\frac{1}{a_1}+\ldots+\frac{1}{a_T}\right)k(\mathbf{x}_T,\mathbf{x}_T) \end{bmatrix},
$$

and $\bar{\mathbf{k}} = \left(\left(\sum_{t=1}^{i}\frac{1}{a_t}\right)k(\mathbf{x}_i,\mathbf{x}_T)\right)_i$, for $i = 1,\ldots,T$, i.e.

$$
\bar{\mathbf{k}} = \begin{bmatrix} \frac{1}{a_1}k(\mathbf{x}_1,\mathbf{x}_T) \\ \left(\frac{1}{a_1}+\frac{1}{a_2}\right)k(\mathbf{x}_2,\mathbf{x}_T) \\ \vdots \\ \left(\frac{1}{a_1}+\ldots+\frac{1}{a_T}\right)k(\mathbf{x}_T,\mathbf{x}_T) \end{bmatrix}.
$$

4 Upper Bounds

In this section we use the Aggregating Algorithm's properties to derive upper bounds on the cumulative square loss suffered by AARCh and KAARCh, compared to that of any expert in the pool.

4.1 AARCh Loss Upper Bound

Theorem 3. *For any point in time T and any $a_t > 0$, $t = 1, \ldots, T$,*

$$L_T(\text{AARCh}) \leq \inf_\theta \left(L_T(\theta) + \sum_{t=1}^{T} a_t \|\theta_t\|^2 \right)$$

$$+ Y^2 \ln \det \left(\sqrt{A}^{-1} L' \sum_{t=1}^{T} z_t z_t' L \sqrt{A}^{-1} + I \right). \quad (6)$$

Proof. Given the Aggregating Algorithm's properties, we know that

$$L_T(\text{AARCh}) \leq \log_\beta \int_{\mathbb{R}^{nT}} \beta^{L_T(\theta)} P_0(d\theta)$$

$$= \log_\beta \left(\left(\frac{\eta}{\pi} \right)^T \prod_{t=1}^{T} a_t \right)^{n/2}$$

$$\times \int_{\mathbb{R}^{nT}} e^{-\eta \left(\theta' \left(L' \sum_{t=1}^{T} z_t z_t' L + A \right) \theta - 2 \left(\sum_{t=1}^{T} y_t z_t \right) L \theta + \sum_{t=1}^{T} y_t^2 \right)} d\theta.$$

By Lemma 1 this is equal to

$$\inf_\theta \left(L_T(\theta) + \theta' A \theta \right) + \log_\beta \left(\left(\left(\frac{\eta}{\pi} \right)^T \prod_{t=1}^{T} a_t \right)^{n/2} \frac{\pi^{nT/2}}{\sqrt{\det \left(\eta L' \sum_{t=1}^{T} z_t z_t' L + \eta A \right)}} \right)$$

$$= \inf_\theta \left(L_T(\theta) + \sum_{t=1}^{T} a_t \|\theta_t\|^2 \right) + \log_\beta \sqrt{\frac{\left(\eta^T \prod_{t=1}^{T} a_t \right)^n}{\det \left(\eta L' \sum_{t=1}^{T} z_t z_t' L + \eta A \right)}}$$

$$= \inf_\theta \left(L_T(\theta) + \sum_{t=1}^{T} a_t \|\theta_t\|^2 \right) + \frac{1}{2} \log_\beta \left(\frac{\prod_{t=1}^{T} a_t^n}{\det \left(L' \sum_{t=1}^{T} z_t z_t' L + A \right)} \right)$$

$$= \inf_\theta \left(L_T(\theta) + \sum_{t=1}^{T} a_t \|\theta_t\|^2 \right)$$

$$- \frac{1}{2} \log_\beta \left(\frac{\det \left(\sqrt{A} \left(\sqrt{A}^{-1} L' \sum_{t=1}^{T} z_t z_t' L \sqrt{A}^{-1} + I \right) \sqrt{A} \right)}{\prod_{t=1}^{T} a_t^n} \right)$$

$$= \inf_\theta \left(L_T(\theta) + \sum_{t=1}^{T} a_t \|\theta_t\|^2 \right) + Y^2 \ln \det \left(\sqrt{A}^{-1} L' \sum_{t=1}^{T} z_t z_t' L \sqrt{A}^{-1} + I \right).$$

4.2   KAARCh Loss Upper Bound

The following generalises Theorem 3. Note that we cannot repeat the proof for the linear case directly since it involves the evaluation of an integral over the space $\mathbb{R}^{nT}$.

Theorem 4. *Let k be a kernel on a space X, let D_t, $t = 1 \dots T$, be any decision rules in the RKHS F induced by k and let $D = (D_1, D_2, \dots, D_T)'$. Then, for any point in time T and every $a_t > 0$, $t = 1, \dots, T$,*

$$L_T(\text{KAARCh}) \leq L_T(D) + \sum_{t=1}^{T} a_t \|D_t\|^2 + Y^2 \ln \det \left(\bar{\mathbf{K}} + \mathbf{I} \right). \tag{7}$$

Proof. It will be sufficient to prove this for D_t of the form

$$f_t(\mathbf{x}) = \sum_{i=1}^{l^{(t)}} c_i^{(t)} k(\mathbf{v}_i^{(t)}, \mathbf{x}),$$

where $l^{(t)}$ are positive integers, $c_i^{(t)} \in \mathbb{R}$, and $\mathbf{v}_i^{(t)}, \mathbf{x} \in X$ (we use $^{(t)}$ to show that these parameters can be different for each f_t). This is because such finite sums are dense in the RKHS F. If we take $f = (f_1, f_2, \dots, f_T)'$, (7) becomes

$$L_T(\text{KAARCh}) \leq L_T(f) + \sum_{t=1}^{T} a_t \sum_{i,j=1}^{l^{(t)}} c_i^{(t)} c_j^{(t)} k(\mathbf{v}_i^{(t)}, \mathbf{v}_j^{(t)}) + Y^2 \ln \det \left(\bar{\mathbf{K}} + \mathbf{I} \right), \tag{8}$$

where

$$L_T(f) = \sum_{t=1}^{T} \left(y_t - \sum_{i=1}^{l^{(t)}} c_i^{(t)} k(\mathbf{v}_i^{(t)}, \mathbf{x}_t) \right)^2.$$

In the special case when $X = \mathbb{R}^n$ and $k(\mathbf{v}_i, \mathbf{v}_j) = \mathbf{v}_i' \mathbf{v}_j$ for every $\mathbf{v}_i, \mathbf{v}_j \in X$, (8) follows directly from (6). Indeed, a kernel predictor f_t reduces to the linear predictor $\theta_t = \sum_{i=1}^{l^{(t)}} c_i^{(t)} \mathbf{v}_i^{(t)}$ and the term $\sum_{i,j=1}^{l^{(t)}} c_i^{(t)} c_j^{(t)} k \left(\mathbf{v}_i^{(t)}, \mathbf{v}_j^{(t)} \right)$ equals the squared quadratic norm of θ_t. Finally, by Sylvester's determinant identity (see also Lemma 4 for an independent proof of this) we know that

$$\det \left(\bar{\mathbf{K}} + \mathbf{I} \right) = \det \left(\tilde{\mathbf{Z}} \mathbf{L} \mathbf{A}^{-1} \mathbf{L}' \tilde{\mathbf{Z}}' + \mathbf{I} \right)$$

$$= \det \left(\sqrt{\mathbf{A}}^{-1} \mathbf{L}' \tilde{\mathbf{Z}}' \tilde{\mathbf{Z}} \mathbf{L} \sqrt{\mathbf{A}}^{-1} + \mathbf{I} \right).$$

The general case can be obtained by using finite dimensional approximations. Recall that inherent in every kernel is a function ϕ that maps objects to the RKHS F, which is isomorphic to $l_2 = \{\alpha = (\alpha_1, \alpha_2, \dots) | \sum_{i=1}^{\infty} \alpha_i^2 \text{ converges}\}$. Let us consider the sequence on subspaces $R_1 \subseteq R_2 \subseteq \dots \subseteq F$. The set $R_s = \{(\alpha_1, \alpha_2, \dots, \alpha_s, 0, 0, \dots)\}$ may be identified with $\mathbb{R}^s$. Let $p_s : F \to R_s$ be the projection operator $p_s(\boldsymbol{\alpha}) = (\alpha_1, \alpha_2, \ldots, \alpha_s, 0, 0, \ldots)$, $\phi_s : X \to R_s$ be $\phi_s = p_s(\phi)$, and k_s be given by $k_s(\mathbf{v}_1, \mathbf{v}_2) = \langle \phi_s(\mathbf{v}_1), \phi_s(\mathbf{v}_2) \rangle$, where $\mathbf{v}_1, \mathbf{v}_2 \in X$.

Inequality (8) holds for k_s since R_s has a finite dimension. If (8) is violated, then its counterpart with some large s is violated too and this observation completes the proof.

5 Discussion

In this section we shall analyse upper bound (7) in order to obtain an equivalent of Corollary 1. Our goal is to show that KAARCh's cumulative loss is less or equal to that of a wide class of experts plus a term of the order $o(T)$.

Estimating the determinant of a positive definite matrix by the product of its diagonal elements (see [15, Sect. 2.10, Theorem 7]) and using the inequality $\ln(1 + x) \le x$ (in our case x is small, and therefore the resulting bound is tight), we get

$$Y^2 \ln \det \left(\bar{\mathbf{K}} + \mathbf{I}\right) \le Y^2 \sum_{t=1}^{T} \ln \left(1 + c^2 \sum_{i=1}^{t} \frac{1}{a_i}\right)$$

$$\le Y^2 c^2 \sum_{t=1}^{T} \sum_{i=1}^{t} \frac{1}{a_i}$$

$$= Y^2 c^2 \sum_{t=1}^{T} \frac{T - t + 1}{a_t},$$

where $c = \sup_{\mathbf{x} \in X} \sqrt{k(\mathbf{x}, \mathbf{x})}$.

It is natural to single out the first decision rule D_1 and the corresponding coefficient a_1 from the rest. We may think of it as corresponding to the choice of the 'principal' dependency; let the rest of D_t ($t = 2, \ldots, T$) be small correction terms. Let us take equal $a_2 = \ldots = a_t = a$. We get

$$L_T(\text{KAARCh}) \le L_T(D) + \left(a_1 \|D_1\|^2 + \frac{Y^2 c^2 T}{a_1}\right)$$

$$+ \left(a \sum_{t=2}^{T} \|D_t\|^2 + \frac{Y^2 c^2 T(T-1)}{2a}\right). \quad (9)$$

If we bound the norm of D_1 by d_1 and assume that T is known in advance, a_1 may be chosen as in Corollary 1. The second term in the right hand side of (9) can thus be bounded by $O\left(\sqrt{T}\right)$. If we assume that $\sum_{t=2}^{T} \|D_t\|^2 \le s(T)$, then the estimate is minimised by $a = \sqrt{Y^2 c^2 T(T-1)/(2s(T))}$ and the third term in the right hand side of (9) can be bounded by $O\left(T\sqrt{s(T)}\right)$. We therefore get the following corollary:

Corollary 2. *Under the conditions of Theorem 4, let T be known in advance and $c = \sup_{\mathbf{x} \in X} \sqrt{k(\mathbf{x}, \mathbf{x})}$. For every every $d_1 > 0$ and every function $s(T)$, if $\|D_1\| \leq d_1$ and $\sum_{t=2}^{T} \|D_t\|^2 \leq s(T)$, then a_t, for $t = 1, \ldots, T$, can be chosen so that*

$$L_T(\text{KAARCh}) \leq L_T(D) + 2Y c d_1 \sqrt{T} + 2Y c \sqrt{s(T)T(T-1)/2}.$$

If $s(T) = o(1)$, then $L_T(\text{KAARCh}) \leq L_T(D) + o(T)$.

The estimate $s(T) = o(1)$ can be achieved in two natural ways. First, one can assume that each $\|D_t\|$, for $t = 2, \ldots, T$, is small.

Corollary 3. *Under the conditions of Theorem 4, let T be known in advance. For every positive d, d_1, and ε, if $\|D_1\| \leq d_1$ and, for $t = 2, \ldots, T$,*

$$\|D_t\| \leq \frac{d}{T^{0.5+\varepsilon}},$$

then

$$L_T(\text{KAARCh}) \leq L_T(D) + O\left(T^{\max(0.5, (1-\varepsilon))}\right)$$
$$= L_T(D) + o(T).$$

Secondly, one may assume that there are only a few nonzero D_t, for $t = 2, \ldots, T$. In this case, the nonzero D_t can have greater flexibility.

Acknowledgements. We thank Volodya Vovk and Alex Gammerman for valuable discussions. We are grateful to Michael Vyugin for suggesting the problem of predicting implied volatility which inspired this work.

References

1. Vovk, V.: Aggregating strategies. In: Fulk, M., Case, J. (eds.) Proceedings of the 3rd Annual Workshop on Computational Learning Theory, pp. 371–383. Morgan Kaufmann, San Francisco (1990)
2. Vovk, V.: A game of prediction with expert advice. Journal of Computer and System Sciences 56, 153–173 (1998)
3. Vovk, V.: Competitive on-line statistics. International Statistical Review 69(2), 213–248 (2001)
4. Cesa-Bianchi, N., Lugosi, G.: Prediction, Learning, and Games. Cambridge University Press, Cambridge (2006)
5. Gammerman, A., Kalnishkan, Y., Vovk, V.: On-line prediction with kernels and the complexity approximation principle. In: Proceedings of the 20th Conference on Uncertainty in Artificial Intelligence, pp. 170–176. AUAI Press (2004)
6. Vovk, V.: On-line regression competitive with reproducing kernel Hilbert spaces. Technical Report arXiv:cs.LG/0511058 (version 2), arXiv.org (2006)
7. Herbster, M., Warmuth, M.K.: Tracking the best linear predictor. Journal of Machine Learning Research 1, 281–309 (2001)
8. Kivinen, J., Smola, A.J., Williamson, R.C.: Online learning with kernels. IEEE Transactions on Signal Processing 52(8), 2165–2176 (2004)

9. Cavallanti, G., Cesa-Bianchi, N., Gentile, C.: Tracking the best hyperplane with a simple budget perceptron. Machine Learning (to appear)

10. Busuttil, S., Kalnishkan, Y.: Weighted kernel regression for predicting changing dependencies. In: Proceedings of the 18th European Conference on Machine Learning (ECML 2007) (to appear, 2007)

11. Cristianini, N., Shawe-Taylor, J.: An Introduction to Support Vector Machines (and Other Kernel-Based Learning Methods). Cambridge University Press, UK (2000)

12. Aizerman, M., Braverman, E., Rozonoer, L.: Theoretical foundations of the potential function method in pattern recognition learning. Automation and Remote Control 25, 821–837 (1964)

13. Aronszajn, N.: Theory of reproducing kernels. Transactions of the American Mathematical Society 68, 337–404 (1950)

14. Schölkopf, B., Smola, A.J.: Learning with Kernels — Support Vector Machines, Regularization, Optimization and Beyond. MIT Press, Cambridge (2002)

15. Beckenbach, E.F., Bellman, R.: Inequalities. Springer, Heidelberg (1961)

Appendix

Lemma 1. *Let* $Q(\theta) = \theta'\mathbf{A}\theta + \mathbf{b}'\theta + c$, *where* $\theta, \mathbf{b} \in \mathbb{R}^n$, c *is a scalar and* $\mathbf{A}$ *is a symmetric positive definite* $n \times n$ *matrix. Then*

$$\int_{\mathbb{R}^n} e^{-Q(\theta)} d\theta = e^{-Q_0} \frac{\pi^{n/2}}{\sqrt{\det \mathbf{A}}} \ ,$$

where $Q_0 = \min_{\theta \in \mathbb{R}^n} Q(\theta)$.

Proof. Let $\theta_0 \in \arg\min Q$. Take $\xi = \theta - \theta_0$ and $\widetilde{Q}(\xi) = Q(\xi + \theta_0)$. It is easy to see that the quadratic part of $\widetilde{Q}$ is $\xi'\mathbf{A}\xi$. Since $0 \in \arg\min \widetilde{Q}$, the form has no linear term. Indeed, in the vicinity of 0 the linear term dominates over the quadratic term; if $\widetilde{Q}$ has a non-zero linear term, it cannot have a minimum at 0. Since $Q_0 = \min_{\xi \in \mathbb{R}^n} \widetilde{Q}(\xi)$, we can conclude that the constant term in $\widetilde{Q}$ is Q_0. Thus $\widetilde{Q}(\xi) = \xi'\mathbf{A}\xi + Q_0$.

It remains to show that $\int_{\mathbb{R}^n} e^{-\xi'\mathbf{A}\xi} d\xi = \pi^{n/2}/\sqrt{\det \mathbf{A}}$. This can be proved by considering a basis where $\mathbf{A}$ diagonalises (or see [15, Sect. 2.7, Theorem 3]).

Lemma 2. *Let*

$$F(\mathbf{A}, \mathbf{b}, \mathbf{x}) = \min_{\theta \in \mathbb{R}^n} (\theta'\mathbf{A}\theta + \mathbf{b}'\theta + \mathbf{x}'\theta) - \min_{\theta \in \mathbb{R}^n} (\theta'\mathbf{A}\theta + \mathbf{b}'\theta - \mathbf{x}'\theta) \ ,$$

where $\mathbf{b}, \mathbf{x} \in \mathbb{R}^n$ *and* $\mathbf{A}$ *is a symmetric positive definite* $n \times n$ *matrix. Then* $F(\mathbf{A}, \mathbf{b}, \mathbf{x}) = -\mathbf{b}'\mathbf{A}^{-1}\mathbf{x}$.

Proof. It can be shown by differentiation that the first minimum is achieved at $\theta_1 = -\frac{1}{2}\mathbf{A}^{-1}(\mathbf{b} + \mathbf{x})$ and the second minimum at $\theta_2 = -\frac{1}{2}\mathbf{A}^{-1}(\mathbf{b} - \mathbf{x})$. The substitution proves the lemma.

Lemma 3. *Given a matrix* $\mathbf{A}$, *a scalar* a *and* $\mathbf{I}$ *identity matrices of the appropriate size,*

$$(\mathbf{A}\mathbf{A}' + a\mathbf{I})^{-1}\mathbf{A} = \mathbf{A}(\mathbf{A}'\mathbf{A} + a\mathbf{I})^{-1} .$$

Proof.

$$
\begin{aligned}
(\mathbf{A}\mathbf{A}' + a\mathbf{I})^{-1}\mathbf{A} &= (\mathbf{A}\mathbf{A}' + a\mathbf{I})^{-1}\mathbf{A}(\mathbf{A}'\mathbf{A} + a\mathbf{I})(\mathbf{A}'\mathbf{A} + a\mathbf{I})^{-1} \\
&= (\mathbf{A}\mathbf{A}' + a\mathbf{I})^{-1}(\mathbf{A}\mathbf{A}'\mathbf{A} + a\mathbf{A})(\mathbf{A}'\mathbf{A} + a\mathbf{I})^{-1} \\
&= (\mathbf{A}\mathbf{A}' + a\mathbf{I})^{-1}(\mathbf{A}\mathbf{A}' + a\mathbf{I})\mathbf{A}(\mathbf{A}'\mathbf{A} + a\mathbf{I})^{-1} \\
&= \mathbf{A}(\mathbf{A}'\mathbf{A} + a\mathbf{I})^{-1}
\end{aligned}
$$

Lemma 4. *For every matrix* $\mathbf{M}$ *the equality* $\det(\mathbf{I} + \mathbf{M}'\mathbf{M}) = \det(\mathbf{I} + \mathbf{M}\mathbf{M}')$ *holds (where* $\mathbf{I}$ *are identity matrices of the correct size).*

Proof. Suppose that $\mathbf{M}$ is an $n \times m$ matrix. Thus $(\mathbf{I} + \mathbf{M}\mathbf{M}')$ and $(\mathbf{I} + \mathbf{M}'\mathbf{M})$ are $n \times n$ and $m \times m$ matrices respectively. Without loss of generality, we may assume that $n \geq m$ (otherwise we swap $\mathbf{M}$ and $\mathbf{M}'$). Let the columns of $\mathbf{M}$ be m vectors $\mathbf{x}_1, \ldots, \mathbf{x}_m \in \mathbb{R}^n$.

We have $\mathbf{M}\mathbf{M}' = \sum_{i=1}^{n} \mathbf{x}_i \mathbf{x}_i'$. Let us see how the operator $\mathbf{M}\mathbf{M}'$ acts on a vector $\mathbf{x} \in \mathbb{R}^n$. By associativity, $\mathbf{x}_i \mathbf{x}_i' \mathbf{x} = (\mathbf{x}_i' \mathbf{x})\mathbf{x}_i$, where $\mathbf{x}_i' \mathbf{x}$ is a scalar. Therefore, if U is the span of $\mathbf{x}_1, \mathbf{x}_2, \ldots, \mathbf{x}_m$, then $\mathbf{M}\mathbf{M}'(\mathbb{R}^n) \subseteq U$. In a similar way, it follows that $(\mathbf{I} + \mathbf{M}\mathbf{M}')(U) \subseteq U$. On the other hand, if $\mathbf{x}$ is orthogonal to $\mathbf{x}_i$, then $\mathbf{x}_i \mathbf{x}_i' \mathbf{x} = (\mathbf{x}_i' \mathbf{x})\mathbf{x}_i = 0$. Hence $\mathbf{M}\mathbf{M}'(U^{\perp}) = 0$, where $U^{\perp}$ is the orthogonal complement to U with respect to $\mathbb{R}^n$. Consequently, $(\mathbf{I} + \mathbf{M}\mathbf{M}')|_{U^{\perp}} = \mathbf{I}$ (by $\mathbf{B}|_V$ we denote the restriction of an operator $\mathbf{B}$ to a subspace V). Therefore $(\mathbf{I} + \mathbf{M}\mathbf{M}')(U^{\perp}) \subseteq U^{\perp}$.

One can see that both U and $U^{\perp}$ are invariant subspaces of $(\mathbf{I} + \mathbf{M}\mathbf{M}')$. If we choose bases in U and in $U^{\perp}$ and then concatenate them, we get a basis of $\mathbb{R}^n$. In this basis the matrix of $(\mathbf{I} + \mathbf{M}\mathbf{M}')$ has the form

$$\begin{bmatrix} \mathbf{A} & \mathbf{0} \\ \mathbf{0} & \mathbf{I} \end{bmatrix} ,$$

where $\mathbf{A}$ is the matrix of $(\mathbf{I} + \mathbf{M}\mathbf{M}')|_U$. It remains to evaluate $\det(\mathbf{A})$.

First let us consider the case of linearly independent $\mathbf{x}_1, \mathbf{x}_2, \ldots, \mathbf{x}_m$. They form a basis of U and we may use it to calculate the determinant of the operator $(\mathbf{I} + \mathbf{M}\mathbf{M}')|_U$. However,

$$(\mathbf{I} + \mathbf{M}\mathbf{M}')\mathbf{x}_i = \mathbf{x}_i + \sum_{j=1}^{m}(\mathbf{x}_j' \mathbf{x}_i)\mathbf{x}_j$$

and thus the matrix of the operator $(\mathbf{I} + \mathbf{M}\mathbf{M}')|_U$ in the basis $\mathbf{x}_1, \mathbf{x}_2, \ldots, \mathbf{x}_m$ is $(\mathbf{I} + \mathbf{M}'\mathbf{M})$.

The case of linearly dependent $\mathbf{x}_1, \mathbf{x}_2, \ldots, \mathbf{x}_m$ follows by continuity. Indeed, m vectors in an n-dimensional space with $n \geq m$ may be approximated by m independent vectors to any degree of precision and the determinant is a continuous function of the elements of a matrix.

Cluster Identification in
Nearest-Neighbor Graphs

Markus Maier, Matthias Hein, and Ulrike von Luxburg

Max Planck Institute for Biological Cybernetics, Tübingen, Germany
first.last@tuebingen.mpg.de

Abstract. Assume we are given a sample of points from some underlying distribution which contains several distinct clusters. Our goal is
to construct a neighborhood graph on the sample points such that clusters are "identified": that is, the subgraph induced by points from the
same cluster is connected, while subgraphs corresponding to different
clusters are not connected to each other. We derive bounds on the probability that cluster identification is successful, and use them to predict
"optimal" values of k for the mutual and symmetric k-nearest-neighbor
graphs. We point out different properties of the mutual and symmetric
nearest-neighbor graphs related to the cluster identification problem.

1   Introduction

In many areas of machine learning, neighborhood graphs are used to model local relationships between data points. Applications include spectral clustering,
dimensionality reduction, semi-supervised learning, data denoising, and many
others. However, the most basic question about such graph based learning algorithms is still largely unsolved: which neighborhood graph to use for which
application and how to choose its parameters. In this article, we want to make
a first step towards such results in a simple setting we call "cluster identification". Consider a probability distribution whose support consists of several high
density regions (clusters) which are separated by a positive distance from each
other. Given a finite sample, our goal is to construct a neighborhood graph on
the sample such that each cluster is "identified", that is each high density region
is represented by a unique connected component in the graph. In this paper we
mainly study and compare mutual and symmetric k-nearest-neighbor graphs.
For different choices of k we prove bounds on the probability that clusters can
be identified. In toy experiments, the behavior of the bounds as a function of
k corresponds roughly to the empirical frequencies. Moreover, we compare the
different properties of the mutual and the symmetric nearest-neighbor graphs.
Both graphs have advantages in different situations: if one is only interested in
identifying the "most significant" cluster (while some clusters might still not be
correctly identified), then the mutual kNN graph should be chosen. However, if
one wants to identify many clusters simultaneously the bounds show no difference between the two graphs. Empirical evaluations show that in this case the
symmetric kNN graph is to be preferred due to its better connectivity properties.

M. Hutter, R.A. Servedio, and E. Takimoto (Eds.): ALT 2007, LNAI 4754, pp. 196–210, 2007.

There is a huge amount of literature with very interesting results on connectivity properties of random graphs, both for Erdős-Rényi random graphs (Bollobas, 2001) and for geometric random graphs (Penrose, 2003). Applications include percolation theory (Bollobas and Riordan, 2006), modeling ad-hoc networks (e.g., Santi and Blough, 2003, Bettstetter, 2002, Kunniyur and Venkatesh, 2006), and clustering (e.g., Brito et al., 1997 and Biau et al., 2007). In all those cases the literature mainly deals with different kinds of asymptotic results in the limit for $n \to \infty$. However, what we would need in machine learning are finite sample results on geometric random graphs which can take into account the properties of the underlying data distribution, and which ideally show the right behavior even for small sample sizes and high dimensions. In this paper we merely scratch the surface of this long-term goal.

Let us briefly introduce some basic definitions and notation for the rest of the paper. We always assume that we are given n data points $X_1, ..., X_n$ which have been drawn i.i.d. from some underlying density on $\mathbb{R}^d$. Those data points are used as vertices in a graph. By $\text{kNN}(X_j)$ we denote the set of the k nearest neighbors of X_j. The different neighborhood graphs, which are examples of geometric random graphs, are defined as

- the ε-neighborhood graph $G_{\text{eps}}(n, \varepsilon)$: X_i and X_j connected if $\|X_i - X_j\| \leq \varepsilon$,
- the symmetric k-nearest-neighbor graph $G_{\text{sym}}(n, k)$:
  X_i and X_j connected if $X_i \in \text{kNN}(X_j)$ or $X_j \in \text{kNN}(X_i)$,
- the mutual k-nearest-neighbor graph $G_{\text{mut}}(n, k)$:
  X_i and X_j connected if $X_i \in \text{kNN}(X_j)$ and $X_j \in \text{kNN}(X_i)$.

2   Between- and Within-Cluster Connectivity of Mutual and Symmetric kNN-Graphs

This section deals with the connectivity properties of kNN graphs. The proof ideas are basically the same as in Brito et al. (1997). However, since we are more interested in the finite sample case we have tried to make the bounds as tight as possible. We also make all constants explicit, which sometimes results in long expressions, but allows to study the influence of all parameters. In Brito et al. (1997) the main emphasis was put on a rate of k which is sufficient for connectedness of the mutual kNN graphs, resulting in a choice of k that is proportional to $\log(n)$. However, if one is interested in identifying the clusters as the connected components of the mutual kNN graph one should optimize the trade-off between having high probability of being connected within clusters and high probability of having no edges between the different clusters. Most importantly, integrating the properties of the mutual and symmetric kNN graph we derive bounds which work for each cluster individually. This allows us later on to compare both graphs for different scenarios: identification of all clusters vs. the "most significant" one.

We assume that our clusters $C^{(1)}, \ldots, C^{(m)}$ are m disjoint, compact and connected subsets of $\mathbb{R}^d$. The distance of $C^{(i)}$ to its closest neighboring cluster $C^{(j)}$ is denoted by $u^{(i)}$, where the distance between sets S_1, S_2 is measured by

$d(S_1, S_2) = \inf\{\|x - y\| \mid x \in S_1, y \in S_2\}$. Let $p^{(i)}$ be a probability density with respect to the Lebesgue measure in $\mathbb{R}^d$ whose support is $C^{(i)}$. The sample points $\{X_i\}_{i=1}^n$ are drawn i.i.d. from the probability density $p(x) = \sum_{j=1}^m \beta_{(j)} p^{(j)}(x)$, where $\beta_{(j)} > 0$ for all j and $\sum_{j=1}^m \beta_{(j)} = 1$. We denote by $n^{(i)}$ the number of points in cluster $C^{(i)}$ $(i = 1, \ldots, m)$. The kNN radius of a point X_i is the maximum distance to a point in $\text{kNN}(X_i)$. $R_{\min}^{(i)}$ and $R_{\max}^{(i)}$ denote the minimal and the maximal kNN radius of the sample points in cluster $C^{(i)}$. $\text{Bin}(n, p)$ denotes the binomial distribution with parameters n and p. Since we often need tail bounds for the binomial distribution, we set $D(k; n, p) = \text{P}(U \leq k)$ and $E(k; n, p) = \text{P}(U \geq k)$ for a $\text{Bin}(n, p)$-distributed random variable U. Finally, we denote the ball of radius r around x by $B(x, r)$, and the volume of the d-dimensional unit ball by η_d.

In the following we will need upper and lower bounds for the probability mass of balls around points in clusters. These are given by continuous and increasing functions $g_{\min}^{(i)}, \tilde{g}_{\min}^{(i)}, g_{\max}^{(i)} : [0, \infty) \to \mathbb{R}$ with $g_{\min}^{(i)}(t) \leq \inf_{x \in C^{(i)}} \text{P}(B(x, t))$, $\tilde{g}_{\min}^{(i)}(t) \leq \inf_{B(x,t) \subseteq C^{(i)}} \text{P}(B(x, t))$ and $g_{\max}^{(i)}(t) \geq \sup_{x \in C^{(i)}} \text{P}(B(x, t))$.

2.1   Within-Cluster Connectivity of Mutual kNN Graphs

The analysis of connectedness is based on the following observation: If for an arbitrary $z > 0$ the minimal kNN radius is larger than z, then all points in a distance of z or less are connected in the kNN graph. If we can now find a covering of a cluster by balls of radius $z/4$ and every ball contains a sample point, then the distance between sample points in neighboring balls is less than z. Thus the kNN graph is connected. The following proposition uses this observation to derive a bound for the probability that a cluster is disconnected under some technical conditions on the boundary of the cluster. These technical conditions ensure that we do not have to cover the whole cluster but we can ignore a boundary strip (the collar set in the proposition). This helps in finding a better bound for the probability mass of balls of the covering.

Proposition 1 (Within-cluster connectedness of $G_{\text{mut}}(n, k)$). *Assume that the boundary $\partial C^{(i)}$ of cluster $C^{(i)}$ is a smooth $(d - 1)$-dimensional submanifold in $\mathbb{R}^d$ with maximal curvature radius $\kappa^{(i)} > 0$. For $\varepsilon \leq \kappa^{(i)}$, we define the collar set $C^{(i)}(\varepsilon) = \{x \in C^{(i)} \mid d(x, \partial C^{(i)}) \leq \varepsilon\}$ and the maximal covering radius $\varepsilon_{\max}^{(i)} = \max_{\varepsilon \leq \kappa^{(i)}} \{\varepsilon \mid C^{(i)} \backslash C^{(i)}(\varepsilon)$ connected $\}$. Let $z \in (0, 4\varepsilon_{\max}^{(i)})$. Given a covering of $C^{(i)} \backslash C^{(i)}(\frac{z}{4})$ with balls of radius $z/4$, let $\mathcal{F}_z^{(i)}$ denote the event that there exists a ball in the covering that does not contain a sample point. Then*

$$\text{P}\big(\text{Cluster } C^{(i)} \text{ disconnected in } G_{mut}(n, k)\big) \leq \text{P}\big(R_{\min}^{(i)} \leq z\big) + \text{P}\big(\mathcal{F}_z^{(i)}\big). \quad (1)$$

Proof. The proof is based on the fact that the event $\{R_{\min}^{(i)} > z\} \cap \mathcal{F}_z^{(i)}$ implies connectedness of $C^{(i)}$. Namely, sample points lying in neighboring sets of the covering of $C^{(i)} \backslash C^{(i)}(\frac{z}{4})$ have distance less than z. Therefore they are connected by an edge in the mutual kNN graph. Moreover, all sample points lying in the collar set $C^{(i)}(\frac{z}{4})$ are connected to some sample point in $C^{(i)}\setminus C^{(i)}(\frac{z}{4})$.   □

The proof concept of Propositions 1 and 3 does not require smoothness of the boundary of the cluster. However, the more general case requires a different construction of the covering which leads to even more technical assumptions and worse constants.

Proposition 2 (Minimal kNN radius). *For all $z > 0$*

$$\mathrm{P}\big(R_{\min}^{(i)} \leq z\big) \;\leq\; n\,\beta_{(i)}\, E\big(k; n-1, g_{\max}^{(i)}(z)\big).$$

Proof. Assume without loss of generality that $X_1 \in C^{(i)}$ (after a suitable permutation). Define $M_s = |\{j \neq s \mid X_j \in B(X_s, z)\}|$ for $1 \leq s \leq n$. Then

$$\mathrm{P}\big(R_{\min}^{(i)} \leq z \mid n^{(i)} = l\big) \;\leq\; l\,\mathrm{P}\big(M_1 \geq k\big).$$

Since $n^{(i)} \sim \mathrm{Bin}\big(n, \beta_{(i)}\big)$, we have

$$\mathrm{P}\big(R_{\min}^{(i)} \leq z\big) \leq \sum_{l=0}^{n} l\,\mathrm{P}\big(M_1 \geq k\big)\mathrm{P}(n^{(i)} = l) = n\beta_{(i)}\mathrm{P}\big(M_1 \geq k\big).$$

Since $M_1 \sim \mathrm{Bin}\big(n-1, \mathrm{P}\big(B(X_1, z)\big)\big)$, with $\mathrm{P}\big(B(X_1, z)\big) \leq g_{\max}^{(i)}(z)$ we obtain $\mathrm{P}\big(M_1 \geq k\big) \leq E\big(k; n-1, g_{\max}^{(i)}(z)\big)$.   □

Proposition 3 (Covering with balls). *Under the conditions of Proposition 1 there exists a covering of $C^{(i)}\setminus C^{(i)}(\frac{z}{4})$ with N balls of radius $z/4$, such that $N \leq \big(8^d \, \mathrm{vol}\big(C^{(i)}\big)\big) / \big(z^d \eta_d\big)$ and*

$$\mathrm{P}\big(\mathcal{F}_z^{(i)}\big) \;\leq\; N\left(1 - \tilde{g}_{\min}^{(i)}\left(\tfrac{z}{4}\right)\right)^n.$$

Proof. A standard construction using a $z/4$-packing provides us with the covering. Due to the conditions of Proposition 1 we know that balls of radius $z/8$ around the packing centers are disjoint and subsets of $C^{(i)}$. Thus the sum of the volumes of these balls is bounded by the volume of the cluster and we obtain $N(z/8)^d \eta_d \leq \mathrm{vol}\big(C^{(i)}\big)$. Using a union bound over the covering with a probability of $\big(1 - \tilde{g}_{\min}^{(i)}\big(\tfrac{z}{4}\big)\big)^n$ for one ball being empty we obtain the bound.   □

The following proposition gives an easy extension of the result of Proposition 1 to the symmetric k-nearest-neighbor graph:

Proposition 4 (Within-cluster connectedness of $G_{\mathbf{sym}}(n, k)$). *We have*

$$\mathrm{P}\Big(\text{Cluster } C^{(i)} \text{ conn. in } G_{sym}(n, k)\Big) \geq \mathrm{P}\Big(\text{Cluster } C^{(i)} \text{ conn. in } G_{mut}(n, k)\Big)$$

Proof. The edge set of $G_{\mathrm{mut}}(n, k)$ is a subset of the edges of $G_{\mathrm{sym}}(n, k)$. Hence connectedness of $G_{\mathrm{mut}}(n, k)$ implies connectedness of $G_{\mathrm{sym}}(n, k)$.   □

Note that this bound does not take into account the better connectivity properties of the symmetric kNN graph. Therefore one can expect that this bound is quite loose. We think that proving tight bounds for the within-cluster connectivity of the symmetric kNN graph requires a completely new proof concept. See Section 3 for more discussion of this point.

2.2 Between-Cluster Connectivity of kNN Graphs

In this section we state bounds for the probability of edges *between* different clusters. The existence of edges between clusters is closely related to the event that the maximal k-nearest-neighbor radius is greater than the distance to the next cluster. Therefore we first give a bound for the probability of this event in Proposition 5. Then we apply this result to the mutual k-nearest-neighbor graph (in Proposition 6) and to the symmetric k-nearest-neighbor graph (in Proposition 7). It will be evident that the main difference between mutual kNN graphs and symmetric kNN graphs lies in the between-cluster connectivity.

Proposition 5 (Maximal nearest-neighbor radius). *We have*

$$P\big(R_{\mathrm{max}}^{(i)} \geq u^{(i)}\big) \leq n\beta_{(i)} D\Big(k - 1; n - 1, g_{\mathrm{min}}^{(i)}\big(u^{(i)}\big)\Big).$$

The proof is omitted here because it is very similar to the proof of Proposition 2. It can be found in Maier et al. (2007). The previous proposition can be used to compare $G_{\mathrm{mut}}(n, k)$ and $G_{\mathrm{sym}}(n, k)$ with respect to cluster isolation. We say that a cluster $C^{(i)}$ *is isolated in the graph* if there are no edges between sample points lying in $C^{(i)}$ and any other cluster. In $G_{\mathrm{mut}}(n, k)$ isolation of a cluster only depends on the properties of the cluster itself:

Proposition 6 (Cluster isolation in $G_{\mathrm{mut}}(n, k)$). *We have*

$$P\Big(\text{Cluster } C^{(i)} \text{ isolated in } G_{mut}(n, k)\Big) \geq 1 - P\big(R_{\mathrm{max}}^{(i)} \geq u^{(i)}\big)$$

$$\geq 1 - n\,\beta_{(i)} D\Big(k - 1; n - 1, g_{\mathrm{min}}^{(i)}\big(u^{(i)}\big)\Big).$$

Proof. Since the neighborhood has to be mutual, we have no connections between $C^{(i)}$ and another cluster if the maximal kNN radius fulfills $R_{\mathrm{max}}^{(i)} < u^{(i)}$.   □

The next theorem shows that the probability for connections between clusters is significantly higher in the symmetric kNN graph.

Proposition 7 (Cluster isolation in $G_{\mathrm{sym}}(n, k)$). *We have*

$$P\Big(C^{(i)} \text{ isolated in } G_{sym}(n, k)\Big) \geq 1 - \sum_{j=1}^{m} P\big(R_{\mathrm{max}}^{(j)} \geq u^{(j)}\big)$$

$$\geq 1 - n\sum_{j=1}^{m} \beta_{(j)} D\Big(k - 1; n - 1, g_{\mathrm{min}}^{(j)}\big(u^{(j)}\big)\Big).$$

Proof. Let u^{ij} be the distance of $C^{(i)}$ and $C^{(j)}$. The event that $C^{(i)}$ is connected to any other cluster in $G_{\text{sym}}(n, k)$ is contained in the union $\{R_{\text{max}}^{(i)} \geq u^{(i)}\} \cup \{\cup_{j \neq i} \{R_{\text{max}}^{(j)} \geq u^{ij}\}\}$. Using a union bound we have

$$P\left(C^{(i)} \text{ not isolated in } G_{\text{sym}}(n, k)\right) \leq P\left(R_{\text{max}}^{(i)} \geq u^{(i)}\right) + \sum_{j \neq i} P\left(R_{\text{max}}^{(j)} \geq u^{ij}\right).$$

Using first $u^{(j)} \leq u^{ij}$ and then Proposition 6 we obtain the two inequalities. $\square$

Note that the upper bound on the probability that $C^{(i)}$ is isolated is the same for all clusters in the symmetric kNN graph. The upper bound is loose in the sense that it does not respect specific geometric configurations of the clusters where the bound could be smaller. However, it is tight in the sense that the probability that cluster $C^{(i)}$ is isolated in $G_{\text{sym}}(n, k)$ always depends on the *worst* cluster. This is the main difference to the mutual kNN graph, where the properties of cluster $C^{(i)}$ are independent of the other clusters.

3   The Isolated Point Heuristic

In the last sections we proved bounds for the probabilities that individual clusters in the neighborhood graph are connected, and different clusters in the neighborhood graph are disconnected. The bound on the disconnectedness of different clusters is rather tight, while the bound for the within-cluster connectedness of a cluster is tight if n is large, but has room for improvement if n is small. The reason is that the techniques we used to prove the connectedness bound are not well-adapted to a small sample size: we cover each cluster by small balls and require that each ball contains at least one sample point (event $\mathcal{F}_z^{(i)}$ in Section 2). Connectedness of G_{mut} then follows by construction. However, for small n this is suboptimal, because the neighborhood graph can be connected even though it does not yet "cover" the whole data space. Here it would be of advantage to look at connectivity properties more directly. However, this is not a simple task.

The heuristic we propose makes use of the following fact from the theory of random graph models: in both Erdős-Rényi random graphs and random geometric graphs, for large n the parameter for which the graph stops having isolated vertices coincides with the parameter for which the graph is connected (e.g., Bollobas, 2001, p. 161 and Theorem 7.3; Penrose, 2003, p.281 and Theorem 13.17). The isolated point heuristic now consists in replacing the loose bound on the within-cluster connectivity from Section 2 by a bound on the probability of the existence of isolated vertices in the graph, that is we use the heuristic

$$P(C^{(i)} \text{ connected}) \approx P(\text{no isolated points in } C^{(i)}).$$

This procedure is consistent for $n \to \infty$ as proved by the theorems cited above, but, of course, it is only a heuristic for small n.

Proposition 8 (Probability of isolated points in G_{eps}). *We have*

$$P\left(\text{ex. isol. points from } C^{(i)} \text{ in } G_{eps}(n, \varepsilon)\right) \leq \beta_{(i)} \, n \, (1 - g_{\text{min}}^{(i)}(\varepsilon))^{n-1}.$$

Proof. Suppose there are l points in cluster $C^{(i)}$. Then a point X_i from $C^{(i)}$ is isolated if $\min_{1 \leq j \leq n, j \neq i} \|X_i - X_j\| > \varepsilon$. This event has probability less than $(1 - g_{\min}^{(i)}(\varepsilon))^{n-1}$. Thus a union bound yields

$$P\big(\text{ex. isol. points from } C^{(i)} \text{ in } G_{\text{eps}}(n, \varepsilon) \,|\, n^{(i)} = l\big) \leq l \, (1 - g_{\min}^{(i)}(\varepsilon))^{n-1},$$

and we sum over the $\text{Bin}(n, \beta_{(i)})$-distributed random variable $n^{(i)}$. □

For the mutual nearest-neighbor graph, bounding the probability of the existence of isolated points is more demanding than for the ε-graph, as the existence of an edge between two points depends not only on the distance of the points, but also on the location of all other points. We circumvent this problem by transferring the question of the existence of isolated points in $G_{\text{mut}}(n, k)$ to the problem of the existence of isolated vertices of a particular ε-graph. Namely

$$\{\text{ex. isolated points in } G_{\text{mut}}(n, k)\} \implies \{\text{ex. isolated points in } G_{\text{eps}}(n, R_{\min})\}.$$

Proposition 9 (Probability of isolated points in G_{mut}). *Let* $v = \sup\big\{d(x, y) \,|\, x, y \in \cup_{i=1}^m C^{(i)}\big\}$ *and* $b : [0, v] \to \mathbb{R}$ *be a continuous function such that* $P\big(R_{\min}^{(i)} \leq t\big) \leq b(t)$. *Then,*

$$P\big(\text{ex. isol. points from } C^{(i)} \text{ in } G_{mut}(n, k)\big) \leq \beta_{(i)} n \int_0^v \left(1 - g_{\min}^{(i)}(t)\right)^{n-1} db(t).$$

Proof. Let $A_{\text{mut}} = \{\text{ex. isolated points from } C^{(i)} \text{ in } G_{\text{mut}}(n, k)\}$ and $A_{\text{eps}}(t) = \{\text{ex. isolated points from } C^{(i)} \text{ in } G_{\text{eps}}(n, t)\}$. Proposition 8 implies $P\big(A_{\text{mut}} \,|\, R_{\min}^{(i)} = t\big) \leq P\big(A_{\text{eps}}(t)\big) \leq \beta_{(i)} n \,(1 - g_{\min}^{(i)}(t))^{n-1}$. $c(t) = \beta_{(i)} n \,(1 - g_{\min}^{(i)}(t))^{n-1}$ is a decreasing function that bounds $P\big(A_{\text{mut}} \,|\, R_{\min}^{(i)} = t\big)$. Straightforward calculations and standard facts about the Riemann-Stieltjes integral conclude the proof. For details see Maier et al. (2007). □

Note that in the symmetric nearest-neighbor graph isolated points do not exist by definition. Hence, the isolated points heuristic cannot be applied in that case.

4   Asymptotic Analysis

In this section we study the asymptotic behavior of our bounds under some additional assumptions on the probability densities and geometry of the clusters. Throughout this section we assume that the assumptions of Proposition 1 hold and that the densities $p^{(i)}$ satisfy $0 < p_{\min}^{(i)} \leq p^{(i)}(x) \leq p_{\max}^{(i)}$ for all $x \in C^{(i)}$. We define the *overlap function* $O^{(i)}(r)$ by $O^{(i)}(r) = \inf_{x \in C^{(i)}} \big(\text{vol}\,(B(x, r) \cap C^{(i)}) / \text{vol}\,(B(x, r))\big)$. With these assumptions we can establish a relation between the volume of a ball and its probability mass,

$$g_{\min}^{(i)}(t) = \beta_{(i)} O^{(i)}(t) p_{\min}^{(i)} \eta_d t^d \quad \text{and} \quad \tilde{g}_{\min}^{(i)}(t) = \beta_{(i)} p_{\min}^{(i)} t^d \eta_d,$$

$$g_{\max}^{(i)}(t) = \begin{cases} t^d \eta_d \beta_{(i)} p_{\max}^{(i)} & \text{if } t \le u^{(i)} \\ \left(u^{(i)}\right)^d \eta_d \left(\beta_{(i)} p_{\max}^{(i)} - p_{\max}\right) + t^d \eta_d p_{\max} & \text{if } t > u^{(i)}, \end{cases}$$

where $p_{\max} = \max_{1 \le i \le m} \beta_{(i)} p_{\max}^{(i)}$.

In Proposition 1 we have given a bound on the probability of disconnectedness of a cluster which has two free parameters, k and the radius z. Clearly the optimal value of z depends on k. In the following proposition we plug in the expressions for $\tilde{g}_{\min}^{(i)}(t)$ and $g_{\max}^{(i)}(t)$ above and choose a reasonable value of z for every k, in order to find a range of k for which the probability of disconnectedness of the cluster asymptotically approaches zero exponentially fast.

Proposition 10 (Choice of k for asymptotic connectivity). *Define* $1/D^{(i)} = 1 + 4^d(e^2 - 1)p_{\max}^{(i)}/p_{\min}^{(i)}$ *and*

$$k' = \frac{1}{D^{(i)}}(n-1)\beta_{(i)}p_{\min}^{(i)}\eta_d \min\left\{\left(\varepsilon_{\max}^{(i)}\right)^d, \left(\frac{u^{(i)}}{4}\right)^d\right\}.$$

Then if $n \ge e/\left(2^d \beta_{(i)} \operatorname{vol}(C^{(i)})p_{\min}^{(i)}\right)$ there exists $0 < \gamma < 1$ such that

$$P\left(C^{(i)} \text{ conn. in } G_{sym}(n,k)\right) \ge P\left(C^{(i)} \text{ conn. in } G_{mut}(n,k)\right) \ge 1 - 2\,e^{-\gamma\,D^{(i)}\,k},$$

for all $k \in \{1, \dots, n-1\}$ with

$$k' \ge k \ge \frac{1}{D^{(i)}} \frac{\log(2^d \operatorname{vol}(C^{(i)})p_{\min}^{(i)}\beta_{(i)}\,n(1-\gamma))}{(1-\gamma)}. \tag{2}$$

Proof. We give an outline of the proof for the mutual k-nearest-neighbor graph. For details see Maier et al. (2007). The statement for the symmetric k-nearest-neighbor graph then follows with Proposition 4. In the following we set $z^d = 8^d \operatorname{vol}\left(C^{(i)}\right)\alpha\theta/\left(\beta_{(i)}\eta_d\right)$ for a $\theta \in (0, \theta_{\max})$ with $\theta_{\max} = \left(8^d \operatorname{vol}\left(C^{(i)}\right)p_{\max}^{(i)}\right)^{-1}$ and $\alpha = k/(n-1)$. For θ in this interval we can apply a tail bound for the binomial distribution from Hoeffding (1963). Let z denote the radius that corresponds to θ and k. With the tail bound for the binomial and standard inequalities for the logarithm we can show that $\log\left(P\left(R_{\min}^{(i)} \le z\right)\right) \le g(\theta)$, where

$$g(\theta) = \log\left(\frac{\beta_{(i)}}{\theta\alpha}\right) + n\alpha\left(2 + \log 8^d \operatorname{vol}\left(C^{(i)}\right)p_{\max}^{(i)}\theta - 8^d \operatorname{vol}\left(C^{(i)}\right)p_{\max}^{(i)}\theta\right)$$

and $\log(P(\mathcal{F})) \le h(\theta)$, where

$$h(\theta) = \log\left(\frac{\beta_{(i)}}{\theta\alpha}\right) - 2^d n\alpha p_{\min}^{(i)} \operatorname{vol}(C^{(i)})\theta.$$

With straightforward calculations and standard inequalities for the exponential function one can show that for $\theta^* = D^{(i)}/\left(2^d \operatorname{vol}(C^{(i)})p_{\min}^{(i)}\right)$ we have $g(\theta^*) \le$

$h(\theta^*)$. Straightforward calculations show that under the conditions $\gamma \in (0,1)$, $n \geq e/\left(2^d \operatorname{vol}(C^{(i)})(1-\gamma)\beta_{(i)}p_{\min}^{(i)}\right)$ and that k is bounded from below as in Equation (2), we have $h(\theta^*) \leq -\gamma k D^{(i)}$. For all $n \geq e/\left(2^d \beta_{(i)} \operatorname{vol}(C^{(i)})p_{\min}^{(i)}\right)$ we can find $\gamma \in (0,1)$ such that $n \geq e/\left(2^d \operatorname{vol}(C^{(i)})(1-\gamma)\beta_{(i)}p_{\min}^{(i)}\right)$. Using $g(\theta^*) \leq h(\theta^*) \leq -\gamma k D^{(i)}$ we have shown that $P\left(R_{\min}^{(i)} \leq z\right) \leq \exp\left(-\gamma k D^{(i)}\right)$ and $P(\mathcal{F}) \leq \exp\left(-\gamma k D^{(i)}\right)$. Reformulating the conditions $z/4 \leq \varepsilon_{\max}^{(i)}$ and $z \leq u^{(i)}$ in terms of θ^* gives the condition $k \leq k'$. □

The result of the proposition is basically that if we choose $k \geq c_1 + c_2 \log(n)$ with two constants c_1, c_2 that depend on the geometry of the cluster and the respective density, then the probability that the cluster is disconnected approaches zero exponentially in k.

Note that, due to the constraints on the covering radius, we have to introduce an upper bound k' on k, which depends linearly on n. However, the probability of connectedness is monotonically increasing in k, since the k-nearest-neighbor graph contains all the edges of the $(k-1)$-nearest-neighbor graph. Thus the value of the within-connectedness bound for $k = k'$ is a lower bound for all $k > k'$ as well. Since the lower bound on k grows with $\log(n)$ and the upper bound grows with n, there exists a feasible region for k if n is large enough.

Proposition 11 (Maximal kNN radius asymptotically). *Let* $p_2^{(i)} = \beta_{(i)}O^{(i)}\left(u^{(i)}\right)p_{\min}^{(i)}\eta_d\left(u^{(i)}\right)^d$ *and* $k \leq (n-1)p_2^{(i)} + 1$. *Then*

$$P\left(R_{\max}^{(i)} \geq u^{(i)}\right) \leq n\beta_{(i)}e^{-(n-1)\left(\left(p_2^{(i)}\right)^2 e^{-p_2^{(i)}} + p_2^{(i)} - \frac{k-1}{n-1}\right)}.$$

Proof. Using a standard tail bound for the binomial distribution (see Hoeffding, 1963) we obtain from Proposition 5 for $(k-1) \leq (n-1)p_2^{(i)}$

$$P\left(R_{\max}^{(i)} \geq u^{(i)}\right) \leq n\beta_{(i)}e^{-(n-1)\left(\frac{k-1}{n-1}\log\frac{(k-1)}{(n-1)p_2^{(i)}} + (1-\frac{k-1}{n-1})\log\frac{1-(k-1)/(n-1)}{1-p_2^{(i)}}\right)}.$$

Using $\log(1+x) \geq x/(1+x)$ and that $-w^2 e^{-w}$ is the minimum of $x \log(x/w)$ (attained at $x = we^{-w}$) we obtain the result by straightforward calculations. □

4.1 Identification of Clusters as the Connected Component of a Mutual and Symmetric kNN Graph

We say that a cluster $C^{(i)}$ is identified if it is an isolated connected component in the kNN graph. This requires the cluster $C^{(i)}$ to be connected, which intuitively happens for large k. Within-cluster connectedness was considered in Proposition 10. The second condition for the identification of a cluster $C^{(i)}$ is that there are no edges between $C^{(i)}$ and other clusters. This event was considered in Proposition 6 and is true if k is small enough. The following theorems consider the tradeoff for the choice of k for the identification of one and of all clusters in both kNN graph types and derive the optimal choice for k. We say that k is *tradeoff-optimal* if our bounds for within-cluster connectedness and between-cluster disconnectedness are equal.

Theorem 12 (Choice of k for the identification of one cluster in $G_{\text{mut}}(n,k)$). *Define $p_2^{(i)}$ as in Proposition 11 and let n and γ be as in Proposition 10. The tradeoff-optimal choice of k in $G_{\text{mut}}(n,k)$ is given by*

$$k - 1 = (n-1)\,p_2^{(i)}\frac{1 - p_2^{(i)}e^{-p_2^{(i)}}}{1 + \gamma D^{(i)}} - \frac{\log\left(\frac{1}{2}n\beta_{(i)}\right)}{1 + \gamma D^{(i)}},$$

if this choice of k fulfills the conditions in Proposition 10 and $k < (n-1)p_2^{(i)} + 1$. For this choice of k we have

$$\mathrm{P}\left(C^{(i)} \text{ ident. in } G_{mut}(n,k)\right) \geq 1 - 4e^{-(n-1)\frac{\gamma D^{(i)}}{1+\gamma D^{(i)}}\left[p_2^{(i)}(1-p_2^{(i)}e^{-p_2^{(i)}}) - \frac{\log(\frac{1}{2}n\beta_{(i)})}{(n-1)}\right]}$$

Proof. We equate the bounds for within-cluster connectedness of Proposition 10 and the bound for between-cluster edges of Proposition 6 and solve for k. □

The result of the previous theorem is that the tradeoff-optimal choice of k has the form $k = c_3 n - c_4 \log(n) + c_5$ with constants $c_3, c_4 \geq 0$ and $c_5 \in \mathbb{R}$, which depend on the geometry of the cluster and the respective density. Evidently, if n becomes large enough, then k chosen according to this rule fulfills all the requirements in Proposition 10 and Theorem 12.

Theorem 12 allows us to define the "most significant" cluster. Intuitively a cluster is more significant the higher its density and the larger its distance to other clusters. Formally the "most significant" cluster is the one with the best rate for identification, that is the maximal rate of the bound:

$$\max_{1 \leq i \leq m} \frac{\gamma D^{(i)}}{1 + \gamma D^{(i)}}\left[p_2^{(i)}(1 - p_2^{(i)}e^{-p_2^{(i)}}) - \frac{\log(\frac{1}{2}n\beta_{(i)})}{n}\right]$$

The term in front of the bracket is increasing in $D^{(i)}$ and thus is larger, the closer $p_{\max}^{(i)}$ and $p_{\min}^{(i)}$ are, that is for a fairly homogeneous density. The second term in the brackets approaches zero rather quickly in n. It is straightforward to show that the first term in the bracket is increasing in $p_2^{(i)}$. Thus a cluster becomes more significant, the higher the probability mass in balls of radius $u^{(i)}$, that is, the higher $\beta_{(i)}, p_{\min}^{(i)}, u^{(i)}$ and the higher the value of the overlap function $O^{(i)}(u^{(i)})$.

We would like to emphasize that it is a unique feature of the mutual kNN graph that one can minimize the bound independently of the other clusters. This is not the case for the symmetric kNN graph. In particular, in the case of many clusters, a few of which have high probability, the differences in the rates can be huge. If the goal is to identify not all clusters but only the most important ones, that means the ones which can be detected most easily, then the mutual kNN graph has much better convergence properties than the symmetric kNN graph. We illustrate this with the following theorem for the symmetric kNN graph.

Theorem 13 (Choice of k for the identification of one cluster in $G_{\mathsf{sym}}(n,k)$). *Define $\rho_2 = \min_{1 \leq i \leq m} p_2^{(i)}$ and let n and γ be as in Proposition 10. The tradeoff-optimal choice of k in $G_{sym}(n,k)$ is given by*

$$k - 1 = (n-1)\rho_2 \frac{1 - \rho_2 e^{-\rho_2}}{1 + \gamma D^{(i)}} - \frac{\log\left(\frac{n}{2}\right)}{1 + \gamma D^{(i)}},$$

if this choice of k fulfills the conditions in Proposition 10 and $k < (n-1)\rho_2 + 1$. For this choice of k

$$\mathrm{P}\Big(C^{(i)} \text{ identified in } G_{sym}(n,k)\Big) \geq 1 - 4e^{-(n-1)\frac{\gamma D^{(i)}}{1+\gamma D^{(i)}}\left[\rho_2(1-\rho_2 e^{-\rho_2}) - \frac{\log(\frac{n}{2})}{(n-1)}\right]}.$$

Proof. Combining Proposition 7 and Proposition 11 we obtain

$$\mathrm{P}\Big(\text{Cluster } C^{(i)} \text{ not isolated in } G_{\mathsf{sym}}(n,k)\Big) \leq n \sum_{i=1}^{m} \beta_{(i)} e^{-(n-1)\left(\rho_2^2 e^{-\rho_2} + \rho_2 - \frac{k-1}{n-1}\right)}.$$

Equating this bound with the within-connectedness bound in Proposition 10 we obtain the result. □

A comparison with the rate of Theorem 12 for the mutual kNN graph shows that the rate of the symmetric kNN graph depends on the "worst" cluster. This property would still hold if one found a tighter bound for the connectivity of the symmetric kNN graph.

Corollary 14 (Choice of k for the identification of all clusters in $G_{\mathsf{mut}}(n,k)$). *Define $p_{\mathrm{ratio}} = \max_{1 \leq i \leq m} \big(p_{\max}^{(i)}/p_{\min}^{(i)}\big)$ and $\rho_2 = \min_{1 \leq i \leq m} p_2^{(i)}$. Let $1/D = 1 + 4^d(e^2 - 1)p_{\mathrm{ratio}}$ and n,γ be as in Proposition 10. The tradeoff-optimal k for the identification of all clusters in $G_{mut}(n,k)$ is given by*

$$k - 1 = (n-1)\rho_2 \frac{1 - \rho_2 e^{-\rho_2}}{1 + \gamma D} - \frac{\log\left(\frac{n}{2m}\right)}{1 + \gamma D},$$

if this choice of k fulfills the conditions in Proposition 10 for all clusters $C^{(i)}$, $i = 1, \ldots, m$ and $k < (n-1)\rho_2 + 1$. For this choice of k we have

$$\mathrm{P}\Big(\text{All clusters ident. in } G_{mut}(n,k)\Big) \geq 1 - 4m\, e^{-(n-1)\frac{\gamma D}{1+\gamma D}\left[\rho_2(1-\rho_2 e^{-\rho_2}) - \frac{\log(\frac{n}{2m})}{(n-1)}\right]}.$$

Proof. Using a union bound, we obtain from Proposition 10 and Proposition 11

$$\mathrm{P}\Big(\bigcup_{i=1}^{m} C^{(i)} \text{ not isolated}\Big) \leq n \sum_{i=1}^{m} \beta_{(i)} e^{-(n-1)\left(\rho_2^2 e^{-\rho_2} + \rho_2 - \frac{k-1}{n-1}\right)}$$

$$\mathrm{P}\Big(\bigcup_{i=1}^{m} \text{Cluster } C^{(i)} \text{ disconnected in } G_{mut}(n,k)\Big) \leq 2m\, e^{-\gamma k D}$$

We obtain the result by equating these two bounds.   □

The result for the identification of *all* clusters in the mutual kNN graph is not much different from the result for the symmetric kNN graph. Therefore the difference in the behavior of the two graph types is greatest if one is interested in identifying the most important clusters only.

5 Simulations

The long-term goal of our work is to find rules which can be used to choose the parameters k or ε for neighborhood graphs. In this section we want to test whether the bounds we derived above can be used for this purpose, at least in principle. We consider a simple setting with a density of the form $f(x) = \beta \tilde{f}(x) + (1 - \beta)\tilde{f}(x - (u + 2)e_1)$, where $\beta \in (0, 1)$ is the weight of the individual clusters, $\tilde{f}$ is the uniform density on the unit ball in $\mathbb{R}^d$, $e_1 = (1, 0, \ldots, 0)'$, and u is the distance between the clusters.

First we compare the qualitative behavior of the different bounds to the corresponding empirical frequencies. For the empirical setting, we randomly draw n points from the mixture density above, with different choices of the parameters. For all values of k we then evaluate the empirical frequencies P_{emp} for within-cluster connectedness, between-cluster disconnectedness, and the existence of isolated points by repeating the experiment 100 times. As theoretical counterpart we use the bounds obtained above, which are denoted by P_{bound}. To evaluate those bounds, we use the true parameters n, d, β, u, p_{min}, p_{max}. Figure 1 shows the results for $n = 5000$ points from two unit balls in $\mathbb{R}^2$ with a distance of $u = 0.5$ and $\beta = 0.5$. We can see that the bound for within-cluster disconnectedness is loose, but still gets into a non-trivial regime (that is, smaller than 1) for a reasonable k. On the other hand the bound for the existence of isolated points indeed upper bounds the within-cluster disconnectedness and is quite close to the true probability. Hence the isolated point heuristic works well in this example. Moreover, there is a range of values of k where both the empirical frequencies and the bounds for the probabilities become close to zero. This is the region of k we are interested in for choosing optimal values of k in order to identify the clusters correctly. To evaluate whether our bounds can be used for this purpose we sample points from the density above and build the kNN graph for these points. For each graph we determine the range of $k_{min} \leq k \leq k_{max}$ for which both within-cluster connectedness and between-cluster disconnectedness are satisfied, and compute $\hat{k}_{min}$ and $\hat{k}_{max}$ as the mean values over 100 repetitions. To determine "optimal" values for k we use two rules:

$$k_{bound} := \operatorname*{argmin}_{k} \left(P_{bound} \left(\text{connected within} \right) + P_{bound} \left(\text{disconnected between} \right) \right)$$

$$k_{iso} := \operatorname*{argmin}_{k} \left(P_{bound} \left(\text{no isolated points} \right) + P_{bound} \left(\text{disconnected between} \right) \right).$$

The following table shows the results for $G_{\text{mut}}(n, k)$.

n	k_{iso}	k_{bound}	$\hat{k}_{\min}$	$\hat{k}_{\max}$
500	17	25	7.2 ± 1.2	41.0 ± 6.5
1000	29	46	7.3 ± 1.2	71.7 ± 9.4
5000	97	213	8.5 ± 1.2	309.3 ± 16.9
10000	101	425	8.8 ± 1.1	596.6 ± 21.1

We can see that for all values of n in the experiment, both k_{iso} and k_{bound} lie well within the interval of the empirical values $\hat{k}_{\min}$ and $\hat{k}_{\max}$. So in both cases, choosing k by the bound or the heuristic leads to a correct value of k in the sense that for this choice, the clusters are perfectly identified in the corresponding mutual kNN graph.

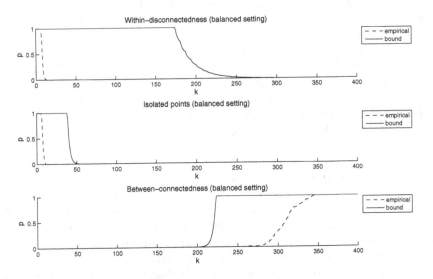

Fig. 1. Bounds and empirical frequencies for $G_{\text{mut}}(n, k)$ for two clusters with $\beta = 0.5$ and $u = 0.5$ (for plotting, we set the bound to 1 if it is larger than 1)

Finally we would like to investigate the difference between G_{mut} and G_{sym}. While the within-cluster connectivity properties are comparable in both graphs, the main difference lies in the between-cluster connectivity, in particular, if we only want to identify the densest cluster in an unbalanced setting where clusters have very different weights. We thus choose the mixture density with weight parameter $\beta = 0.9$, that is we have one very dense and one very sparse cluster. We now investigate the identification problem for the densest cluster. The results are shown in Figure 2. We can see that $G_{\text{sym}}(n, k)$ introduces between-cluster edges for a much lower k than it is the case for $G_{\text{mut}}(n, k)$, which is a large disadvantage of G_{sym} in the identification problem. As a consequence, there is only a very small range of values of k for which the big cluster can be identified.

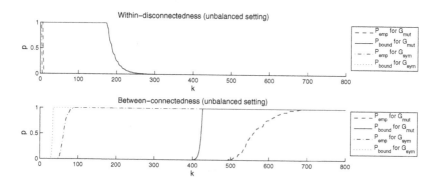

Fig. 2. Within- and between-cluster connectivity for $G_{\mathrm{mut}}(n,k)$ and $G_{\mathrm{sym}}(n,k)$ for two unbalanced clusters with $\beta = 0.9$ and $u = 0.5$. Note that the curves of P_{bound} for $G_{\mathrm{mut}}(n,k)$ and $G_{\mathrm{sym}}(n,k)$ lie on top of each other in the top plot. The scale of the horizontal axis is different from Figure 1.

For G_{mut}, on the other hand, one can see immediately that there is a huge range of k for which the cluster is identified with very high probability. This behavior is predicted correctly by the bounds given above.

6   Conclusions and Further Work

We studied both G_{sym} and G_{mut} in terms of within-cluster and between-cluster connectivity. While the within-cluster connectivity properties are quite similar in the two graphs, the behavior of the between-cluster connectivity is very different. In the mutual kNN graph the event that a cluster is isolated is independent of all the other clusters. This is not so important if one aims to identify *all* clusters, as then also in the mutual graph the worst case applies and one gets results similar to the symmetric graph. However, if the goal is to identify the most significant clusters only, then this can be achieved much easier with the mutual graph, in particular if the clusters have very different densities and different weights.

It is well known that the lowest rate to asymptotically achieve within-cluster connectivity is to choose $k \sim \log(n)$ (e.g., Brito et al., 1997). However, we have seen that the optimal growth rate of k to achieve cluster identification is not linear in $\log(n)$ but rather of the form $k = c_3 n - c_4 \log(n) + c_5$ with constants $c_3, c_4 \geq 0$ and $c_5 \in \mathbb{R}$. This difference comes from the fact that we are not interested in the lowest possible rate for asymptotic connectivity, but in the rate for which the probability for cluster identification is maximized. To this end we can "afford" to choose k higher than absolutely necessary and thus improve the "probability" of within-connectedness. However, as we still have to keep in mind the between-cluster disconnectedness we cannot choose k "as high as we want". The rate now tells us that we can choose k "quite high", almost linear in n.

There are several aspects about this work which are suboptimal and could be improved further:

Firstly, the result on the tradeoff-optimal choice of k relies on the assumption that the density is zero between the clusters We cannot make this assumption if the points are disturbed by noise and therefore the optimal choice of k might be different in that case.

Secondly, the main quantities that enter our bounds are the probability mass in balls of different radii around points in the cluster and the distance between clusters. However, it turns out that these quantities are not sufficient to describe the geometry of the problem: The bounds for the mutual graph do not distinguish between a disc with a neighboring disc in distance u and a disc that is surrounded by a ring in distance u. Obviously, the optimal values of k will differ. It would be possible to include further geometric quantities in the existing proofs, but we have not succeeded in finding simple descriptive expressions. Furthermore, it is unclear if one should make too many assumptions on the geometry in a clustering setting.

Finally, we have indicated in Section 5 how our bounds can be used to choose the optimal value for k from a sample. However, in our experiments we simply took most of the parameters like u or β as given. For a real world application, those parameters would have to be estimated from the sample. Another idea is to turn the tables: instead of estimating, say, the distance u between the clusters for the sample and then predicting an optimal value of k one could decide to go for the most significant clusters and only look for clusters having cluster distance u and cluster weight β bounded from below. Then we can use the bounds not for parameter selection, but to construct a test whether the clusters identified for some value of k are "significant".

References

Bettstetter, C.: On the minimum node degree and connectivity of a wireless multihop network. In: Proceedings of MobiHoc, pp. 80–91 (2002)

Biau, G., Cadre, B., Pelletier, B.: A graph-based estimator of the number of clusters. ESIAM: Prob. and Stat. 11, 272–280 (2007)

Bollobas, B.: Random Graphs. Cambridge University Press, Cambridge (2001)

Bollobas, B., Riordan, O.: Percolation. Cambridge Universiy Press, Cambridge (2006)

Brito, M., Chavez, E., Quiroz, A., Yukich, J.: Connectivity of the mutual k-nearest-neighbor graph in clustering and outlier detection. Stat. Probabil. Lett. 35, 33–42 (1997)

Hoeffding, W.: Probability inequalities for sums of bounded random variables. J. Amer. Statist. Assoc. 58, 13–30 (1963)

Kunniyur, S.S., Venkatesh, S.S.: Threshold functions, node isolation, and emergent lacunae in sensor networks. IEEE Trans. Inf. Th. 52(12), 5352–5372 (2006)

Maier, M., Hein, M., von Luxburg, U.: Cluster identification in nearest-neighbor graphs. Technical Report 163, MPI for Biological Cybernetics, Tübingen (2007)

Penrose, M.: Random Geometric Graphs. Oxford University Press, Oxford (2003)

Santi, P., Blough, D.: The critical transmitting range for connectivity in sparse wireless ad hoc networks. IEEE Trans.

Multiple Pass Streaming Algorithms for Learning Mixtures of Distributions in $\mathbb{R}^d$

Kevin L. Chang

Max Planck Institute for Computer Science
Saarbrücken, Germany
kchang@mpi-sb.mpg.de

Abstract. We present a multiple pass streaming algorithm for learning the density function of a mixture of k uniform distributions over rectangles (cells) in $\mathbb{R}^d$, for any $d > 0$. Our learning model is: samples drawn according to the mixture are placed in *arbitrary order* in a data stream that may only be accessed sequentially by an algorithm with a very limited random access memory space. Our algorithm makes $2\ell + 1$ passes, for any $\ell > 0$, and requires memory at most $\tilde{O}(\epsilon^{-2/\ell} k^2 d^4 + (2k)^d)$. This exhibits a strong memory-space tradeoff: a few more passes significantly lowers its memory requirements, thus trading one of the two most important resources in streaming computation for the other. Chang and Kannan [1] first considered this problem for $d = 1, 2$.

Our learning algorithm is especially appropriate for situations where massive data sets of samples are available, but practical computation with such large inputs requires very restricted models of computation.

1 Introduction

The rise of machine learning as an invaluable data analysis paradigm has co-incided with the proliferation of massive data sets that stress computer systems in ways that render traditional models of computation inadequate. These two important considerations necessitate the theoretical study of algorithms for machine learning and statistical analysis that respect the resource constraints imposed by massive data set computation.

Of paramount importance is the observation that a large data set will not fit into the main memory of a computer system, but rather must be stored on disk or optical drives. For such data, well-designed memory access patterns are crucial, since access to data requires physical movement within storage devices. An algorithm will thus incur large time penalties for each random access; for large data sets, frequent random access is highly undesirable. Random access can be eliminated and I/O optimized by instead reading the data in a sequential fashion. The **multiple pass streaming model** addresses these concerns and is popular in the theoretical computer science literature. The first few problems examined in the streaming model include sorting and selection [2] and approximating frequency moments [3]. In this model, data in storage is modeled as a read-only array that can only be accessed sequentially in passes over the entire

M. Hutter, R.A. Servedio, and E. Takimoto (Eds.): ALT 2007, LNAI 4754, pp. 211–226, 2007.
© Springer-Verlag Berlin Heidelberg 2007

array. The algorithm may make a few passes over the array and use a small random-access memory space (usually sublinear in size, since one cannot hope to store the entire data set in memory) and may take constant time to process each element of the array. The important resources are therefore passes, space, and per-element update time.

An important class of data mining and learning problems arises from generative clustering models. In these models, k clusters are defined by k probability distributions, $F_1, \ldots, F_k$, over some universe Ω, each of which is given a weight $w_i \geq 0$ such that $\sum_1^k w_i = 1$. If the F_is are density functions, then the mixture of these k distributions is defined by the density function $F = \sum_1^k w_i F_i$. The natural interpretation of a point drawn according to the mixture is that distribution F_i is picked with probability w_i, and then a point is drawn according to F_i. We consider the problem of estimating the probability density function of the mixture F given samples drawn according to the mixture.

In this paper, we will study the problem of learning **mixtures of k uniform distributions over axis-aligned rectangles in $\mathbb{R}^d$**, for any $d > 0$. In this case, each F_i is a uniform distribution over some cell in d dimensions $R_i = \{x \in \mathbb{R}^d | a_1 \leq x_1 \leq b_1, \ldots, a_d \leq x_d \leq b_d\}$ for scalars $a_1, b_1, \ldots, a_d, b_d$. The R_is may intersect in arbitrary ways. Since the R_is are arbitrary, learning the R_is and w_is from a set of samples from the mixture is an ill-defined problem, since different sets of rectangles and weights, when "mixed", can form exactly the same distribution. Therefore, we will learn the density function, rather than the components, of the mixture. The output of the algorithm will be a function G that is an estimate of F. G will not be the density function of a mixture of k uniform distributions, but will nonetheless be an approximation to F.

The motivation behind learning mixtures of uniform distributions over rectangles is that these are among the simplest mixtures, and therefore any theory for learning mixture models in massive data set paradigms should start with this. Furthermore, these mixtures are building blocks for more complicated functions; continuous distribution in $\mathbb{R}^d$ can be (heuristically) approximated as a mixture of sufficiently many uniform distributions over rectangles in $\mathbb{R}^d$.

Our learning and computational model is that samples drawn according to the mixture F are placed in a datastream X, in **arbitrary order**.[1] Learning algorithms are required to be multiple-pass streaming algorithms, as described above. The output of the algorithm will be a function G that is an estimate of F, with error measured by L^1 distance: $\int_{\mathbb{R}^d} |F - G|$. An input parameter to the algorithm will be its probability of failure, $\delta > 0$. The approximation G will in general be more complex than simply a mixture of uniform distributions.

Chang and Kannan [1] designed pass-efficient algorithms for learning a mixture of k uniform distributions over intervals in $\mathbb{R}$ and axis-aligned rectangles in $\mathbb{R}^2$. In this paper, we use a similar high level approach, but develop new tools in order to generalize the algorithm to solve problems in arbitrary dimension.

---

[1] Assuming that the data are randomly ordered is not always realistic; for instance if the data were collected from the census, then perhaps it would be ordered by address or some other attribute.

1.1   Our Results

Our main result is a multiple-pass algorithm for learning a mixture of k uniform distributions in $\mathbb{R}^d$ with flexible resource requirements. The number of passes the algorithm may make is a function of an input parameter $\ell > 0$ that is independent of all other variables. The algorithm exhibits the power of multiple passes in the streaming learning model: the error drops exponentially with the number of passes, while the memory required grows very slowly. Viewed another way, the algorithm exhibits a pass-space tradeoff: if the algorithm is alloted just a few more passes and its error is held constant, then its memory requirements drop significantly as a function of ϵ.

We state our results below. We assume that the algorithm knows a number $w > 0$ such that $F(x) \le w$ for all $x \in \mathbb{R}^d$ and that all the probability mass of the mixture is contained in $[0, 1]^d$.

1. We present a $2\ell + 1$ pass algorithm that, with probability at least $1 - \delta$, will learn the mixture's density function to within L^1 distance ϵ and that uses memory at most $\tilde{O}(\epsilon^{-2/\ell}k^2d^4 \log(1/\delta) + (2k)^d)^2$ and per-element update time $O(d^2 \log(kw/(\epsilon\delta)))$. The algorithm requires the data stream to satisfy:
$$|X| = \tilde{\Omega}\left(\left(\frac{10}{8}\right)^{d\ell} \frac{w^{2d} \cdot k^{8d+1} \cdot d^{6d+1}(2\ell+d)}{\epsilon^{4d+1}} \log \frac{1}{\delta}\right).$$

2. These guarantees can be transformed to yield a $2\ell + 1$ algorithm that will learn the mixture with error at most ϵ^ℓ, using space at most:
$$\tilde{O}\left(\frac{k^2d^4}{\epsilon^2} \log(1/\delta) + \frac{\ell kd^2}{\epsilon} \log(1/\delta) + (2k)^d\right).$$

We note that the sample complexity of the algorithm is high: exponential in the dimension d. We justify this constraint to some extent with our massive data set paradigm: we have the luxury of many samples precisely because the data set is very large! Furthermore, if a researcher has a concrete problem to solve (and thus d would be fixed), the algorithm would offer him a tradeoff between the number of passes and the amount of memory required (as a function of ϵ). Our researcher can then tune ℓ to adapt the algorithm's resource requirements to satisfy the constraints imposed by the chosen system.

We note that in [1], the authors proved that any ℓ pass algorithm for a slightly more general version of the learning mixtures of uniform distributions in $\mathbb{R}$ requires $\Omega(1/\epsilon^{1/2\ell})$ bits of memory. This result is thus some indication that the tradeoff that we achieve in this paper has the best order of magnitude (but not a proof of this statement, since the hardness result is for a stronger problem).

1.2   Overview of Methods

The main action of the algorithm is to learn the locations of the boundaries of the constituent mixture rectangles in 2ℓ passes. With this knowledge, **Learn(d, k)** can partition the domain into cells such that $F(x)$ is close to constant on each cell; in one more pass over the data stream, it can easily estimate these constants by counting the number of samples that fall in each of these cells.

---

[2] $\tilde{O}(\cdot), \tilde{\Omega}(\cdot), \tilde{\Theta}(\cdot)$ denote asymptotic notation with polylogarithmic factors omitted.

Learn(d, k) requires Subroutine **FindBoundary(d, k, m)**, which is a 2ℓ pass algorithm for finding the boundary edges of mixture rectangles that lie in a hyperplane that is perpendicular to the mth dimension; in one pass, this algorithm draws a sample from the data stream and uses the sample to partition the domain into a set of roughly $1/\epsilon$ cells that have probability mass on the order of ϵ. It then utilizes one pass subroutine **Invariant(d, k, m)** to test each of these cells for boundaries. Suppose **Invariant(d, k, m)** indicates that a boundary cell lies somewhere in partition cell P. In order to further localize this boundary cell (since it could lie anywhere in the relatively big P), we recurse and partition P and then test the new subcells, which have probability mass $\approx \epsilon^2$. Each time we recurse, we are in essence "zooming in" on cells that we know contain the boundary in order to get a more accurate estimate of where it lies.

The engine of our algorithm is **Invariant(d, k, m)**, which determines if a cell C contains a boundary cell contained in a hyperplane that is perpendicular to the mth dimension. We formulate a statistic that can test this condition; since $|X|$ is large, the test will be very accurate and will tell us if F is within L^1 distance ϵ^ℓ of what we would expect if C did not contain such a boundary cell. Algorithmically computing the statistic in one pass using a small amount of space presents a challenge, and requires the use of a streaming algorithm by Indyk [4] for the small-space approximation of the L_1 lengths of vectors.

The high level overview of the algorithm is similar to the algorithm from [1] for the $d = 1$ case. However, generalizing to arbitrary dimensions requires new ideas and more technical arguments. Our algorithm for the $d = 2$ case is very different from [1], and improves upon the old result.

1.3   Related Work

Many algorithms for the unsupervised learning of mixtures of distributions have appeared in the learning and algorithms theory literature. Algorithms for learning mixtures of Gaussian distributions in $\mathbb{R}^d$ [5,6,7,8] generally estimate the means and covariance matrices of the constituent distributions from samples drawn according to the mixture. Algorithms for classification of sample points to their distribution of origin have been considered [9] for more general distributions. These algorithms are not suitable for massive data sets.

Many one and multiple pass algorithms for database and data mining-inspired problems appear in the theoretical computer science literature. Among the most relevant to this study are the algorithms for histogram maintenance [10]. In the histogram maintenance problem, the algorithm is presented with a datastream of update pairs of the form "add 2 to a_j", where $j \in [n]$.[3] During the pass, the algorithm must maintain a piecewise constant function $F(i)$, with k pieces, that minimizes $\sum_i |F(i) - a_i|$. Gilbert *et al.* gave a one pass algorithm for this problem with approximation ratio $1 + \epsilon$. This work gives the best piecewise constant approximation to arbitrary data (rather than assuming a generative data model) and is thus similar to our problem of learning the density function

---

[3] $[n]$ denotes the set $\{1, \ldots, n\}$.

of a mixture of uniform distributions over intervals in $\mathbb{R}$. A d dimensional variant of the problem has been studied by Thaper *et al.* [11] (their running time is also exponential in d).

Other streaming studies of problems with a statistical flavor include the work of Guha *et al.* [12], who consider one pass algorithms for estimating the entropy of a distribution from samples in a stream.

1.4   Problem Setup

Our algorithm will learn mixtures of distributions over axis-aligned rectangles in $\mathbb{R}^d$. For completeness, we define rectangles:

Definition 1 (d-cell). *For any positive integer $d > 0$, we define a d-**cell** to be a set $K \subset \mathbb{R}^d$ that satisfies $K = \{x \in \mathbb{R}^d | a_1 \le x_1 \le b_1, \ldots, a_d \le x_d \le b_d\}$ for some choice of scalars $a_1, b_1, \ldots, a_d, b_d$. We will sometimes write K as a cross product of d intervals in $\mathbb{R}$: $K = (a_1, b_1) \times \ldots \times (a_d, b_d)$.*

*The **volume** of the d-cell K is given by $vol(K) = |b_1 - a_1| \cdot |b_2 - a_2| \cdot \ldots \cdot |b_d - a_d|$.*

Throughout this paper, the input will be the data stream X of length $|X| = N$, with samples drawn according to a mixture of k uniform distributions in $\mathbb{R}^d$, where the mixture rectangles may intersect arbitrarily. The density function of the mixture will be denoted by F. We assume that the algorithm knows a number $w > 0$ such that $F(x) \le w$ for all $x \in \mathbb{R}$ and that all mixture cells are contained in the cell $[0,1]^d$. We will call the smallest enclosing cell of the mixture the **bounding box**, $R \subseteq [0,1]^d$.

2   Main Algorithm: Learn(d, k)

We will develop the version of our $2\ell + 1$ pass algorithm with error ϵ^ℓ and then show how to transform the parameters to get other guarantees. Before our exposition of the main algorithm, we first introduce the concept of mth component invariance and an algorithm for testing this condition.

Definition 2 (mth component invariance). *A function f is mth component invariant in K if it satisfies the following condition: if x and $y \in K$ satisfy $x_i = y_i$ for all $i \neq m$, then $f(x) = f(y)$. In words, mth component invariance is the condition where $f(x)$ is constant if all components are fixed except the mth component.*

Intuitively, if cell K is invariant in the mth component, then a learning algorithm does not need to consider the mth component when learning the function F in K. A key observation is that if F is invariant in all d components in cell K, then F is, in fact, constant in K.

The learning algorithm relies on the subroutine **FindBoundary(d, k, m)** that will learn a decomposition of the bounding box R into a set of cells such that F is invariant in the mth component of most of the cells. This subroutine is the engine for **Learn(d, k)**; its proof of correctness will be presented in Section 3. To ease the proliferation of complicated expressions, define

$$SC(d, k, \epsilon, \delta, \ell, w) = \left(\frac{10}{8}\right)^{d\ell} \frac{w^{2d} \cdot k^{8d+1} \cdot d^{6d+1}(\ell + d)}{\epsilon^{4\ell d+\ell}} \log 1/\delta,$$

which we will prove is the sample complexity of **FindBoundary(d, k, m)**. Note that $SC(d, k, \epsilon, \delta, \ell, w)$ is exponential in the dimension d. It also contains a term with ϵ^ℓ, but this is necessary, since the error of the algorithm is ϵ^ℓ.

Theorem 1. *FindBoundary(d, k, m) is a 2ℓ pass algorithm that requires at most $\tilde{O}(\frac{d^3k^2}{\epsilon^2} \log(1/\delta) + \frac{\ell k d^2}{\epsilon} \log(1/\delta))$ bits of memory and $O(d \log(kw/\epsilon^\ell \delta)$ per-element update time. If $|X| = \tilde{\Omega}(SC(d, k, \epsilon, \delta, \ell, w))$, then with probability at least $1 - \delta$, it will find a set of cells, $\mathcal{V}$, such that*

1. *For $V_1, V_2 \in \mathcal{V}$ such that $V_1 \neq V_2$, $V_1 \cap V_2 = \emptyset$ and $|\mathcal{V}| \leq 2k$.*
2. *There exists a function F_m such that F_m is invariant in the mth component in each $V \in \mathcal{V}$ and such that $\int_R |F - F_m| \leq \epsilon^\ell/(6d)$.*
3. *$\int_{R \setminus (\cup_{V \in \mathcal{V}} V)} F \leq \epsilon^\ell/(3d)$,*

where R is the bounding box.

The algorithm thus finds a set of disjoint cells, such that for all $V \in \mathcal{V}$, F in V is very close to invariant in the mth component (note that the algorithm guarantees the existence of F_m but only finds $\mathcal{V}$, not F_m). The last condition implies that the cells in $\mathcal{V}$ contain nearly all the weight of F in R.

Our learning algorithm **Learn(d, k)** will run **FindBoundary(d, k, m)** for all $m = 1, \ldots, d$. The output of each call will consist of a set of cells $\mathcal{V}_m$ that are invariant in the mth component. Let $V_m \in \mathcal{V}_m$ be a cell that is invariant in the mth component. For any two indices $1 \leq m_1 < m_2 \leq d$, consider the d-cell $C = V_{m_1} \cap V_{m_2}$. F, restricted to C, is close to a function that is invariant in the m_1th component, and also close to another function that is invariant in the m_2th component. Intuitively, such an F should be close to a function that is invariant in the m_1th *and* m_2th components simultaneously in C.

Extending the reasoning further, let $C = V_1 \cap \ldots \cap V_d$. Then F restricted to C is close to a function that is invariant in the mth component, for all m. In the full version, we prove that this will imply that F is close to *constant* in C.

2.1   The Algorithm

We present **Learn(d, k)** in Figure 1. We will call **FindBoundary(d, k, m)** for each dimension $m = 1, \ldots, d$; from the resulting sets $\tilde{\mathcal{V}}_m$, we will decompose R into cells $\tilde{R}_i$ such that F is close to invariant in $\tilde{R}_i$ in all d components. We will then treat F as if it were constant on $\tilde{R}_i$, and estimate the density in $\tilde{R}_i$ by simply counting the number of sample points that lie in $\tilde{R}_i$.

Input: Datastream X.
1. Run in parallel **FindBoundary(d, k, m)** for all $m = 1, \ldots, d$. Let the output consist of sets $\mathcal{V}_1, \mathcal{V}_2, \ldots, \mathcal{V}_d$.
2. Compute the set of cells consisting of all cells of the form $V_1 \cap V_2 \cap \ldots, V_d$ for all choices of $V_1 \in \mathcal{V}_1, \ldots, V_d \in \mathcal{V}_d$. Call this set of cells $\left\{ \tilde{R}_i \right\}$.
3. In a single pass, count $\left| X \cap \tilde{R}_i \right|$ for all i.
4. (a) On the rectangle $\tilde{R}_i$, estimate the density as $\left| X \cap \tilde{R}_i \right| / \left(|X| \mathrm{vol}\left(\tilde{R}_i \right) \right)$.
   (b) On the set $R \setminus (\cup_i \tilde{R}_i)$, estimate the density as 0.

Fig. 1. Algorithm **Learn(d, k)**

Theorem 2. *If* $|X| = \tilde{\Omega}\left(k^{2d} SC(d, k, \epsilon, \delta, \ell, w) \right)$, *then with probability at least* $1 - \delta$, ***Learn(d, k)*** *will compute an estimate* G *such that* $\int |F - G| \leq \epsilon^\ell$ *in* $2\ell + 1$ *passes, using* $\tilde{O}\left((2k)^d + \frac{k^2 d^4}{\epsilon^2} \log(1/\delta) + \frac{\ell k d^2}{\epsilon} \log(1/\delta) \right)$ *space and per-element update time* $O(d^2 \log(kw/\epsilon^\ell \delta))$.

Proof. From Theorem 1, we know that each call to **FindBoundary(d, k, m)** will output a set of d-cells $\mathcal{V}_m$ such that there exists a function F_m that is invariant in the mth component on each $V \in \mathcal{V}_m$ and that satisfies $\int_R |F_m - F| \leq \epsilon^\ell/6d$. For each rectangle $\tilde{R}_l$ found in Step 2, this is true for all m simultaneously. Note that since $|\mathcal{V}_m| \leq 2k$, $\left| \left\{ \tilde{R}_i \right\} \right| \leq (2k)^d$.

Fix such a rectangle $\tilde{R}_l$. The following property is a precise statement of the intuitive idea that F should be close to constant in $\tilde{R}_l$. The proof has been omitted because of space constraints.

Property 1. There exists a constant $c_{\tilde{R}_l}$ such that

$$\int_{\tilde{R}_l} \left| F - c_{\tilde{R}_l} \right| \leq 2 \sum_{i=1}^d \int_{\tilde{R}_l} |F_i - F|. \tag{1}$$

The VC dimension of the set of all d-cells (intersections of $2d$ *axis-aligned* half spaces in $\mathbb{R}^d$) is $2d$. Since X is drawn according to F, we have chosen our sample complexity so that the VC bound implies that

$$\Pr\left[\sup_i \left| \frac{\left| X \cap \tilde{R}_i \right|}{N} - \int_{\tilde{R}_i} F \right| \leq \frac{\epsilon^\ell}{3(2k)^d} \right] \geq 1 - \delta,$$

where $N = |X|$.

Let $\alpha_i = \left| X \cap \tilde{R}_i \right| / N \mathrm{vol}\left(\tilde{R}_i \right)$ be our algorithm's estimate of F in $\tilde{R}_i$. We now sum our bound on the error of our estimate in each rectangle $\tilde{R}_i$.

$$\sum_i \int_{\tilde{R}_i} |F - \alpha_i| \le \sum_i \int_{\tilde{R}_i} \left(\left| \alpha_i - \frac{\int_{\tilde{R}_i} F}{\mathrm{vol}\left(\tilde{R}_i\right)} \right| + \left| F - \frac{\int_{\tilde{R}_i} F}{\mathrm{vol}\left(\tilde{R}_i\right)} \right| \right)$$

$$\le (2k)^d \frac{\epsilon^\ell}{3(2k^d)} + \sum_i \int_{\tilde{R}_i} |c_{\tilde{R}_i} - F|$$

$$\le \frac{\epsilon^\ell}{3} + 2 \sum_{m=1}^{d} \int_R |F_m - F| \le \frac{2\epsilon^\ell}{3}$$

Lastly, we bound the error induced by estimating F as 0 on the set $R \setminus (\cup \tilde{R}_i)$. Let $\bar{V}_m = R \setminus \{\cup_{V_i \in \mathcal{V}_m} V_i\}$ be the set for which F_m is not invariant in the mth component, so that $R \setminus (\cup \tilde{R}_i) = \cup_m \bar{V}_m$. By Theorem 1, we know that $\int_{\bar{V}_m} F dx \le \epsilon^\ell/(d \cdot 3)$. Thus,

$$\int_{R \setminus (\cup \tilde{R}_i)} F \le \sum_m \int_{\bar{V}_m} F \le \frac{\epsilon^\ell}{3}.$$

The total error of the algorithm is therefore ϵ^ℓ.

Corollary 1. *There exists a $2\ell + 1$ pass algorithm that, with probability at least $1 - \delta$, will learn a mixture of uniform distributions in $\mathbb{R}^d$ with error at most ϵ, using space at most $\tilde{O}(\frac{k^2 d^4}{\epsilon^{2/\ell}} \log(1/\delta) + (2k)^d)$.*

Proof. If we transform the parameter ϵ to $\epsilon^{1/\ell}$, then we may assume that $\ell = O(\log 1/\epsilon)$ (more passes would not decrease the memory requirement).

3   An Algorithm for Learning the Location of Boundary Edges

Algorithm **FindBoundary(d, k, m)** computes a decomposition of R into a set of cells $\mathcal{V}$ that are invariant in the mth component. Roughly stated, it does so by ensuring that each cell $V \in \mathcal{V}$ does not contain an mth-component **boundary** cell of a mixture cell, (which is just the natural geometric concept of the boundary of a rectangle).

Definition 3 (boundary). *The **boundary** of a d-cell $K = (a_1, b_1) \times \ldots \times (a_d, b_d)$ consists of the 2d d-cells defined by $(a_1, b_1) \times \ldots \times (a_i, a_i) \times \ldots \times (a_d, b_d)$ and $(a_1, b_1) \times \ldots \times (b_i, b_i) \times \ldots \times (a_d, b_d)$ for $i = 1, \ldots, d$. The two cells for which $i = m$ are called mth **component boundary** cells.*

Informally, an mth component boundary cell is a $d - 1$ cell embedded in $\mathbb{R}^d$, such that the mth component of the boundary cell is the same for all points in the cell. Our interest in mth component boundary cells is summarized in the following, very intuitive, property. The proof is straightforward and has been omitted from this preliminary version.

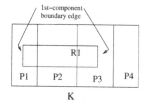

Fig. 2. K and R_1 are 2-cells. The boundaries of K and R_1 are illustrated by their respective rectangular outlines. P_1, P_2, P_3, P_4 comprise a partition of the 1st component of K. Note that the 1st-component boundary edges of R_1 are completely contained in single partition cells.

Lemma 1. *If a d-cell K does not contain any mth component boundary cells of mixture rectangles then F, restricted to K, is invariant in the mth component.*

Definition 4. *A **partition of the mth component** of d-cell $R = (a_1, b_1) \times \ldots \times (a_d, b_d)$ is a set of cells of the form $P_i = \{x \in R | u_i \le x_m < u_{i+1}\}$ (i.e. $P_i = (a_1, b_1) \times \ldots \times (a_{m-1}, b_{m-1}) \times (u_i, u_{i+1}) \times (a_{m+1}, b_{m+1}) \times \ldots \times (a_d, b_d))$, where $a_m = u_1 \le u_2 \le \ldots \le u_{|\mathcal{P}_m|} = b_m$, where $\mathcal{P}_m$ is the set of partition cells $\mathcal{P}_m = \{P_i\}$.*

Note that an mth component boundary cell of a mixture rectangle is completely contained in a single partition rectangle. See Figure 2 for an illustration.

The algorithm **FindBoundary(d, k, m)** requires a subroutine **Invariant(d, k, m)** that will check if F is (approximately) mth component invariant when restricted to the d-cell K. We will defer the proof of the following theorem to Section 4.

Theorem 3. *Let X be a datastream that contains samples from a mixture of k uniform distributions in $\mathbb{R}^d$ with density function H and bounding box R and let $\beta > 0$ be some error parameter. If $|X| = \Omega \left(\frac{kd(kwd)^{2d}}{\beta^{2d}} \log(kdw/\beta\delta) \right)$ then with probability at least $1 - \delta$, algorithm **Invariant(d, k, m)** will **accept** if H is invariant in the mth component and **reject** if there does **not** exists a function $\hat{H}$ that is invariant in the mth component such that $\int_R \left| H(x) - \hat{H}(x) \right| \le \beta$. **Invariant(d, k, m)** requires $\tilde{O}(d \log^2(kdw/\beta) \log(1/\delta))$ bits of memory and $O(d \log(kw/(\beta\delta)))$ per-element update time.*

3.1 The Algorithm

We first give an overview of algorithm **FindBoundary(d, k, m)**. It is organized into ℓ pairs of passes. In the first pass of each pair, it takes a small sample from the datastream and uses it to find a partition $\mathcal{P}_m$, such that all cells have roughly equal probability mass. In a second pass, it tests each partition cell for invariance in the mth component. **Invariant(d, k, m)** uses the large amount of data in the

---

Input: Datastream X, $|X| = \tilde{\Omega}(SC(d, k, \epsilon, \delta, \ell, w))$ **Initialize:** $p \leftarrow 1$.
1. (a) If $p = 1$, set $M = \tilde{\Theta}\left(d^2 k^2 / \epsilon^2\right)$. If $p > 1$, set $M = \tilde{\Theta}\left(1/\epsilon^2\right)$.
   (b) In one pass, draw a sample S of size M from the data stream, uniformly at random.
2. (a) If $p = 1$, set $\alpha \leftarrow c_0 \epsilon / (kd)$. If $p > 1$, set $\alpha \leftarrow c_1 \epsilon$ for constants c_0, c_1 (determined by Lemma 3).
   (b) Compute a partition of the mth component such that for all $P_i \in \mathcal{P}_m$, $|P_i \cap X| = M \cdot \alpha$.
3. Set $C_p \leftarrow \{P_i \cup P_{i+1} | P_i \in \mathcal{P}_m\}$.
4. In a second pass, run algorithm **Invariant(d, k, m)** with error parameter $\left(\frac{8}{10}\right)^\ell \frac{\epsilon^{2\ell}}{150 k^3 d^2}$ on X_{C_i} for all $C_i \in \mathcal{C}_m$ in parallel. Mark C_i that are rejected.
5. Compute $D_p \leftarrow \cup_{C_i \text{marked}} (C_i \cup C_{i-1} \cup C_{i+1})$ to be the union of all marked C_i, as well as C_{i-1} and C_{i+1}.
6. (a) If $p \leq \ell$, set $p \leftarrow p + 1$. In parallel, call **FindBoundary(d, k, m)** on each cell of D_p.
   (b) If $p = \ell$, then output the cells of $R \setminus D_\ell$ as the set $\mathcal{V}$ of cells that are approximately invariant the mth component.

Fig. 3. Algorithm **FindBoundary(d, k, m)**

data stream to perform this test with *very high accuracy*. A cell is rejected only if it contains an mth component boundary cell (the converse is not necessarily true; see Lemma 4). For cells that are rejected, we iterate; the key is that these cells contain much less probability weight than the original bounding box and therefore will be sampled much more densely (but with the same overall sample size) and we will get better estimates for these interesting cells.

In the final iteration, the rejected partition cells C that contain mth component boundary cells have only a very small aggregate weight (roughly ϵ^ℓ / d). Thus, $R \setminus \cup_{\text{rejected}} C$ consists of $2k$ cells that are all close to invariant in the mth component and contain most of the weight of F.

For a cell C, we define the datastream X_C to be the datastream consisting of samples in $X \cap C$. A pass over X can simulate a pass over X_C. We describe the algorithm in Figure 3.

Remark 1. Since **Invariant(d, k, m)** will only reject a cell that contains an mth component boundary cell of a mixture rectangle, each cell $C \subset D_p$ must contain such a boundary cell, of which there are at most $2k$. This implies that $R \setminus D_\ell$ is a union of at most $2k + 1$ cells. Therefore the set of output cells $\mathcal{V}$, which corresponds to cells that are approximately invariant in the mth component, has at most $2k + 1$ cells. (In fact, it can be shown to have only $2k - 1$ cells.)

We call the value of p the **level of the call**.

Lemma 2. *The number of recursive calls at any level is at most $2k$.*

Lemma 3. *With probability at least $1 - \delta/(10d)$, for any cell $C \in \mathcal{C}_p$ created by the algorithm in Step 3 of iteration p:*

$$\left(\frac{8}{10}\right)^p \frac{\epsilon^p}{12 \cdot k \cdot d} \leq \int_C F \leq \frac{\epsilon^p}{12 \cdot k \cdot d}.$$

Proof sketch. The lemma can be proved by an induction argument on the level, coupled with straightforward applications of the VC bound and the fact that the VC dimension of the family of d-cells in $\mathbb{R}^d$ is $2d$.

Lemma 4. *For each $V_i \in \mathcal{V}$, there exists a function $\hat{F}_{V_i}$ that is invariant in the mth component such that $\sum_i \int_{V_i} \left| \hat{F}_{V_i} - F \right| \leq \frac{\epsilon^\ell}{6d}$.*

Due to the structure of the algorithm, each $V \in \mathcal{V}$ can be written as a union of cells (possibly overlapping), $V = \cup_j C_j$, such that each cell C_j is accepted in Step 4 of some iteration of the algorithm. Note that a cell C_j that contains an mth component boundary cell of a mixture rectangle is not always rejected by **Invariant(d, k, m)** in the case when the error induced by estimating F as invariant in the mth component in C_j is very small. Therefore, restricted to *each* C_j, F is close to some function that is invariant in the mth component, but the technical difficulty lies in finding a *single* mth component invariant function $\hat{F}_V$ that is a good approximation to F *for all of* $V = \cup_j C_j$.

Proof sketch. Assume without loss of generality that $m \leftarrow d$.

Recall that all points in a dth component boundary of a mixture rectangle have the same value in the dth component. Since there are at most $2k$ of these boundary cells, there are at most $2k$ such values. Let q_i, $i = 1, \ldots, 2k$ be these values in sorted order. Note that q_1 and q_{2k} correspond to values on the bounding box. Consider the partition of R of the dth component given by $\mathcal{Q} = \{Q_i\}$, where $Q_i = \{x \in R | q_i \leq x_d \leq q_{i+1}\}$ for $i = 1, \ldots, 2k$. Restricted to each of these Q_i, F is invariant in the dth component.

Define the extension of F in Q_i to be the function $\hat{F}_{Q_i} : R \to \mathbb{R}$ given by:

$$\hat{F}_{Q_i}(x) = F(x_1, \ldots, x_{d-1}, y)$$

for any $y \in (q_i, q_{i+1})$. Since $\hat{F}_{Q_i}$ is invariant in the d-th component and $\hat{F}_{Q_i}(x) = F(x)$ for $x \in Q_i$, $\hat{F}_{Q_i}$ is the natural notion of extending F in Q_i to the rest of the bounding box R.

Fix a $V \in \mathcal{V}$. We have designed Step 3 of the algorithm in order to guarantee that for Q_i such that $Q_i \cap V \neq \emptyset$ and $Q_{i+1} \cap V \neq \emptyset$, there exists a cell $C^* \subset V$ such that C^* was accepted by **Invariant(d, k)** and that for some constant c,

$$\int_{C^* \cap Q_i} F \geq c \frac{\epsilon^\ell}{k^2 d}, \int_{C^* \cap Q_{i+1}} F \geq c \frac{\epsilon^\ell}{k^2 d}.$$

Due to the fact that C^* was accepted, it can be shown that $\hat{F}_{Q_i}$ is a good approximation to F restricted to C^*. Since $C^* \cap Q_i$ and $C^* \cap Q_{i+1}$ both carry a large amount of probability mass, it can be shown that $\hat{F}_{Q_i}$ is a good approximation to F in Q_{i+1}, since F is invariant in the dth component in Q_{i+1}. In this manner, an inductive proof will show that $\hat{F}_{Q_i}$ is a good approximation to F in all Q_j such that $Q_j \cap V \neq \emptyset$.

Proof (of Theorem 1). We omit the calculation of memory requirements, but it is straightforward. Theorem 1 follows from combining the previous lemma with the fact that the set $R \setminus (\cup_{V_i \in V} V_i)$ consists of the at most $2k$ cells that were rejected in an ℓth level call to **FindBoundary(d, k, m)**. By Lemma 3, the probability weight of these cells is only $2k \cdot \epsilon^\ell / 6kd$.

4   Checking for mth Component Invariance

We now present the algorithm **Invariant(d, k, m)**. In order to ease our notation, we will assume without loss of generality that $m \leftarrow d$ and that the bounding box of H is given by $R = (0, b_1) \times (0, b_2) \times \ldots \times (0, b_d)$, where $b_i \leq 1$.

Our exposition of **Invariant(d, k)** will compose of three parts: First, we define a sufficient condition for establishing that H is close to invariant in the dth component. We will then propose an estimator $\gamma_{\vec{j}, i}$ and prove that $\gamma_{\vec{j}, i}$ can be used to test the condition. Lastly, we give an algorithm for actually performing the test in a single pass with a small amount of memory. As in [1], the main algorithmic tool that we use is Indyk's one pass algorithm for computing the ℓ_1 length of a vector given as a stream of dynamic updates.

Definition 5. *We define a **regular partition parallel to the dth component** $\mathcal{P}_\eta$ to be the partition of a d-cell $R = (0, b_1) \times \ldots \times (0, b_d)$ into $(1/\eta)^{d-1}$ d-cells $\left\{ P_{\vec{j}} \right\}$: for all $\vec{j} \in [1/\eta]^{d-1}$,*

$$P_{\vec{j}} = \{x \in R | ((j_1 - 1)\eta b_1, j_1 \eta b_1) \times \ldots \times ((j_{d-1} - 1)\eta b_{d-1}, j_{d-1} \eta b_{d-1}) \times (0, b_d)\}.$$

*We will refer to the d-cells $P_{\vec{j}} \in \mathcal{P}_\eta$ as **partition cells**.*

The partition is thus a partition of R into a set of d-cells with the same dimensions, such that each component except the dth is partitioned into $1/\eta$ intervals. Note that $\mathrm{vol}\left(P_{\vec{j}}\right) = \eta^{d-1} \mathrm{vol}(R)$ and that $|\mathcal{P}_\eta| = \frac{1}{\eta^{d-1}}$.

Informally, each partition cell is a long, thin strip, with its long side along the dth component. The main idea is that if F is close to constant in most of these partition cells, then F should be close to invariant in the dth component.

More precisely, **Invariant(d, k)** will check if there exists a constant $c_{P_{\vec{j}}}$ such that $\alpha_{\vec{j}}$, the error of estimating $H(x)$ as the constant $c_{P_{\vec{j}}}$ on $P_{\vec{j}}$, is small, where $\alpha_{\vec{j}} = \int_{P_{\vec{j}}} |H - c_{P_{\vec{j}}}|$. For technical reasons, we will need to classify cells as good or bad and will only be concerned with the good partition cells; this is possible, because the aggregate volume of bad partition cells is small (Lemma 5). Partition cell P is **bad** if it contains an mth component boundary cell of a mixture rectangle, for $m < d$. Otherwise, P is **good**. Note that a good P may contain dth component boundary cells.

Let $\mathcal{G} \subset \mathcal{P}_\eta$ be the set of good partition cells and $\mathcal{B} = \mathcal{P}_\eta \setminus \mathcal{G}$ be the set of bad partition cells.

Lemma 5. $|\mathcal{B}| \leq \frac{2kd}{\eta^{d-2}}$

Proof. Any boundary cell of a mixture rectangle, except dth component boundary cells, can intersect at most $1/\eta^{d-2}$ partition cells. There are at most k mixture cells, each of which has $2d$ boundary cells.

Lemma 6. *Suppose that $\eta \leq \frac{\beta}{2 \cdot k \cdot d \cdot w}$. If there exist constants $c_{\vec{j}}$ for all $P_{\vec{j}} \in \mathcal{G}$ such that*

$$\sum_{P_{\vec{j}} \in \mathcal{G}} \alpha_{\vec{j}} = \sum_{P_{\vec{j}} \in \mathcal{G}} \int_{P_{\vec{j}}} \left| H(x) - c_{\vec{j}} \right| \leq \frac{\beta}{4},$$

then there exists a function $\tilde{H}$ that is invariant in the dth component such that

$$\int_R \left| H(x) - \tilde{H}(x) \right| \leq \beta.$$

Proof sketch. We can prove the lemma by making the obvious choice of dth component invariant function $H(x) = c_{\vec{j}}$ for $x \in P_{\vec{j}}$ such that $P_{\vec{j}} \in \mathcal{G}$, and $H(x) = 0$ otherwise. Bad partition cells do not contribute very much to the error, since they have little aggregate weight from the previous lemma.

4.1   Estimating $\alpha_{\vec{j}}$ from the Datastream

We now describe an estimator $\gamma_{\vec{j},i}$ for $\alpha_{\vec{j}}$ in a good partition d-cell $P_{\vec{j}} \in \mathcal{P}_\eta$, and prove properties of $\gamma_{\vec{j},i}$. Note that we do not provide an algorithm until Section 4.2.

Recall that $P_{\vec{j}} = ((j_1-1)\eta b_1, j_1\eta b_1) \times \ldots \times ((j_{d-1}-1)\eta b_{d-1}, j_{d-1}\eta b_{d-1}) \times (0, b_d)$ for the vector $\vec{j} \in [1/\eta]^{d-1}$. Let $\zeta > 0$ (assume that $1/\zeta$ is an integer). We further partition $P_{\vec{j}} \in \mathcal{P}_\eta$ into $1/\zeta$ d-cells of equal volume.

Definition 6. *For each integer $i \in [1/\zeta]$, define the **sub-partition** d-cell $P_{\vec{j},i}$ by $P_{\vec{j},i} = \left\{ x \in P_{\vec{j}} | (i-1)\zeta b_d \leq x_d \leq i\zeta b_d) \right\}$.*

We define the following random variables:

1. $N_{\vec{j},i} = \left| X \cap P_{\vec{j},i} \right|$

2. $\gamma_{\vec{j},i} = \frac{\left| N_{\vec{j},i} - \zeta \sum_l N_{\vec{j},l} \right|}{N}$

$N_{\vec{j},i}$ is the number of samples that lie in $P_{\vec{j},i}$; since $\zeta \sum_l N_{\vec{j},l}$ is the average number of points in each of the sub-partition cells of $P_{\vec{j}}$, $\gamma_{\vec{j},i}$ is the difference between $N_{\vec{j},i}$ and what we would expect if H were actually constant in $P_{\vec{j}}$. Therefore, if $\sum_i \gamma_{\vec{j},i}$ is small, then $\alpha_{\vec{j}}$ should be close to constant:

Lemma 7. *Let $\zeta \leq \frac{\beta}{64k \cdot w}$ and fix $P_{\vec{j}} \in \mathcal{G}$. If $X = \Omega\left(\frac{d^3 k}{\beta^2 \zeta^2 \eta^{2d-2}} \log(1/\zeta\beta\delta\eta) \right)$, then with probability at least $1 - \delta/2$, $\sum_i \gamma_{\vec{j},i} \geq \alpha_{\vec{j}} + \frac{\beta\eta^{d-1}}{8}$.*

The proof of the Lemma has been omitted due to space constraints. The basic idea of the proof is to consider two cases: $P_{\vec{j},i}$ in which H is constant and $P_{\vec{j},i}$ in which H contains a boundary cell. For the former case, it can be shown that $\gamma_{\vec{j},i}$ is a good estimate of the quantity of interest. For the latter case, such cells can be proved to have a very small aggregate weight and can therefore be disregarded without incurring a large error.

Corollary 2. *With probability at least $1 - \delta/2$, if $\sum_{\vec{j}} \sum_i \gamma_{\vec{j},i} \leq \beta/8$, then there exists a function $\tilde{H}$ that is invariant in the dth component such that* $y \int_R \left| H(x) - \tilde{H}(x) \right| \leq \beta.$

Lemma 8. *With probability at least $1 - \delta/4$, if H is invariant in the dth component, then $\sum_{\vec{j}} \sum_i \gamma_{\vec{j},i} \leq \frac{\beta}{16}$.*

4.2 A One Pass, Small Space Algorithm

Corollary 2 and Lemma 8 prove that an algorithm that **accepts** if $\sum_{\vec{j}} \sum_i \gamma_{\vec{j}} \leq \beta/12$ and **rejects** otherwise will accept if H is invariant in the dth component, and will reject if H is not within β of an invariant function.

A naive one pass algorithm to compute the estimator $\sum_{\vec{j},i} \gamma_{\vec{j},i}$ would explicitly keep one counter for each of the $1/(\eta \cdot \zeta) \, N_{\vec{j},i}$s, which requires too much memory. Indyk [4] designed a one-pass algorithm for approximating the ℓ_1 length of a vector given as a stream of dynamic updates (very similar to the histogram problem mentioned in the related works section). The input is a stream of update pairs $\langle a, i \rangle$, where $a \in [-M, M]$ and $i \in [n]$, that represent the semantics: add a to the ith component of vector $\vec{v} \in \mathbb{R}^n$. The problem is then to approximate $||\vec{v}||_1 = \sum_{i=1}^n |v_i|$ after processing all of the input pairs. The following theorem is an adaptation of a more general result:

Theorem 4 (Indyk[4]). *There exists a one pass algorithm that, with probability at least $1 - \delta$, will find an approximation ι such that $\frac{2}{3}||\vec{v}||_1 \leq \iota \leq \frac{4}{3}||\vec{v}||_1$ using at most $O(\log M \log(n/\delta))$ bits of memory and $O(\log(n/\delta))$ per-element update time.*

The high level idea of this algorithm is that instead of storing all n components of $\vec{v}$, it stores the components of a random projection of $\vec{v}$ to a low dimensional subspace. The random matrix that defines the projection is compressed by only storing the seed of a pseudorandom number generator; the entries of the matrix are generated on the fly, as needed.

We present in Figure 4 the details of our algorithm **Invariant(d, k)**.

Proof (of Theorem 3). Due to the guarantees of Indyk's algorithm, $\frac{\iota}{N}$ will satisfy:

$$\frac{2}{3} \frac{||v||_1}{N} \leq \frac{\iota}{N} \leq \frac{4}{3} \frac{||v||_1}{N}.$$

Input: Datastream X.

1. Set $\eta \leftarrow \frac{\beta}{8 \cdot k \cdot w \cdot d}$ and $\zeta \leftarrow \frac{\beta}{64 \cdot k \cdot w}$
2. Fix any bijection $\phi : [1/\eta]^{d-1} \times [1/\zeta] \rightarrow [1/(\eta^{d-1}\zeta)]$.
3. Initialize Indyk's algorithm to update the vector $\vec{v} \leftarrow \vec{0}$.
4. While data stream is not empty:
   (a) Read sample x from data stream.
   (b) Calculate $\vec{j}$ and i such that $x \in P_{\vec{j},i}$.
   (c) Process the pair $\left\langle 1, \phi(\vec{j}, i) \right\rangle$ with Indyk's algorithm.
   (d) For $l = 1, \ldots, 1/\zeta$, process the pairs: $\left\langle -\zeta, \phi(\vec{j}, l) \right\rangle$ with Indyk's algorithm.
5. Let ι be the output of Indyk's algorithm. **accept** if $\frac{\iota}{N} \leq \beta/12$ and **reject** otherwise.

Fig. 4. Algorithm **Invariant(d, k)**

Note that the $\phi(\vec{j}, i)$th component of v is exactly: $v_{\phi(\vec{j},i)} = N_{\vec{j},i} - \zeta \sum_l N_{\vec{j},l}$. Thus, $\frac{\|v\|_1}{N} = \sum_{\vec{j}} \sum_i \gamma_{\vec{j},i}$. Therefore, the algorithm will accept if $\sum_{\vec{j}} \sum_i \gamma_{\vec{j},i} \leq \beta/16$ and will reject if $\sum_{\vec{j}} \sum_i \gamma_{\vec{j},i} > \beta/8$. The theorem follows from Corollary 2 and Lemma 8.

References

1. Chang, K., Kannan, R.: The space complexity of pass-efficient algorithms for clustering. In: Proceedings of the Seventeenth Annual ACM-SIAM Symposium on Discrete Algorithms, pp. 1157–1166 (2006)
2. Munro, J.I., Paterson, M.: Selection and sorting with limited storage. Theoretical Computer Science 12, 315–323 (1980)
3. Alon, N., Matias, Y., Szegedy, M.: The space complexity of approximating the frequency moments. Journal of Computer and System Sciences 58, 137–147 (1999)
4. Indyk, P.: Stable distributions, pseudorandom generators, embeddings, and data stream computation. Journal of the Association for Computing Machinery 53, 307–323 (2006)
5. Arora, S., Kannan, R.: Learning mixtures of separated nonsphereical Gaussians. Annals of Applied Probability 15, 69–92 (2005)
6. Dasgupta, S.: Learning mixtures of Gaussians. In: Proceedings of the 40th IEEE Symposium on Foundations of Computer Science, pp. 634-644. IEEE Computer Society Press, Los Alamitos (1999)
7. Kannan, R., Salmasian, H., Vempala, S.: The spectral method for general mixture models. In: Auer, P., Meir, R. (eds.) COLT 2005. LNCS (LNAI), vol. 3559, pp. 444-457. Springer, Heidelberg (2005)
8. Vempala, S., Wang, G.: A spectral algorithm for learning mixtures of distributions. Journal of Computer and System Sciences 68, 841–860 (2004)
9. Dasgupta, A., Hopcroft, J.E., Kleinberg, J.M., Sandler, M.: On learning mixtures of heavy-tailed distributions. In: Proceedings of the 46th IEEE Symposium on Foundations of Computer Science, pp. 491–500. IEEE Computer Society Press, Los Alamitos (2005)

10. Gilbert, A.C., Guha, S., Indyk, P., Kotidis, Y., Muthukrishnan, S., Strauss, M.: Fast, small-space algorithms for approximate histogram maintenance. In: Proceedings of the 34th Annual ACM Symposium on the Theory of Computing, pp. 389–398. ACM Press, New York (2002)
11. Thaper, N., Guha, S., Indyk, P., Koudas, N.: Dynamic multidimensional histograms. In: Proceedings of the 2002 ACM SIGMOD international conference on Management of data, pp. 428–439. ACM Press, New York, NY, USA (2002)
12. Guha, S., McGregor, A., Venkatasubramanian, S.: Streaming and sublinear approximation of entropy and information distances. In: Proceedings of the Seventeenth Annual ACM-SIAM Symposium on Discrete Algorithms, pp. 733–742 (2006)

Learning Efficiency of Very Simple Grammars from Positive Data

Ryo Yoshinaka

INRIA-Lorraine
Ryo.Yoshinaka@loria.fr

Abstract. The class of very simple grammars is known to be polynomial-time identifiable in the limit from positive data. This paper gives even more general discussion on the efficiency of identification of very simple grammars from positive data, which includes both positive and negative results. In particular, we present an alternative efficient inconsistent learning algorithm for very simple grammars.

1 Introduction

While efficient identification in the limit from positive data of nonregular languages is a topical issue of grammatical inference, there is no consensus on the definition of efficient identification in the limit. Every definition of efficient learning proposed so far requires an algorithm to update its conjecture in polynomial time at least. Throughout this paper, when we simply say an algorithm *runs in polynomial time*, it means that the update time is bounded by a polynomial in the total size of the given data. However, as Pitt [1] discussed, every successful learning algorithm can be modified into polynomial-time one in the above sense, though it could be intuitively inefficient. His trick lets a learning algorithm compute its conjecture from some small prefix of the input while the rest of the input is used only as an excuse for its long running time. We need to impose some other condition(s) to prevent Pitt's trick.

Most of learning algorithms proposed so far are consistent and conservative. An algorithm is *consistent* if the conjecture of the algorithm is always consistent with the given data. An algorithm is *conservative* if the algorithm changes its conjecture only when the previous conjecture contradicts the newly given example. In fact, the combination of those conditions is restrictive enough to stop Pitt's trick from working. Those conditions are, however, not mandatory. Lange and Wiehagen's [2] algorithm for learning pattern languages from positive data is not consistent but it is *iterative*, i.e., it computes a new conjecture only from the newly given example and the current conjecture. Iterativeness is also strong enough to prevent Pitt's trick.

Moreover, not only each update time, how quickly the learning algorithm converges to the target is also an important issue. One cannot give any upper bound on the gross amount of time that the learner has spent until it converges, because the learner has no control over the given data. Pitt [1] proposed to measure the

M. Hutter, R.A. Servedio, and E. Takimoto (Eds.): ALT 2007, LNAI 4754, pp. 227–241, 2007.

efficiency of learning algorithms by counting the number of *implicit errors of prediction*, where an algorithm makes an implicit error of prediction if the newly given example contradicts the current conjecture. An efficient learning algorithm should make implicit errors of prediction at most polynomial times in the size of the target representation before it converges. De la Higuera's [3] definition requires efficiently learnable languages to admit a *characteristic set* of polynomial size in the size of a smallest representation of the language such that whenever the given data includes the characteristic set, the learner successfully converges to a representation of the target language. Those two definitions originally target learning from complete (i.e., both positive and negative) data and are restrictive enough to exclude Pitt's trick.

It is a natural idea to apply those proposals to learning from positive data only. However, we should be conscious of the difference between learning from complete data and from positive data. For instance, concerning Pitt's definition, while it is not a trivial task to find a representation that are expected to be consistent with the forthcoming positive and negative examples, a representation of the language Σ^* is trivially consistent with any positive examples where Σ is the alphabet of the target language. As we will see later, this entails that the naive straightforward application of Pitt's definition to learning from positive data almost spoils the restriction of polynomial-time updating. On the other hand, although Pitt argued that counting the number of mind changes, i.e., times the learner changes its conjecture, does not make sense when learning regular languages from complete data, his discussion does not work when learning from positive data.

Whatever we adopt as the definition of efficient algorithms, currently not many nonregular subclasses of context-free languages are known that deserve to be called "efficiently" learnable from positive data in nontrivial sense, except for subclasses of even linear grammars whose learning problems are reducible to that of efficiently learnable subclasses of regular languages [4]. Clark and Eyraud [5] presented a polynomial-time learning algorithm for substitutable context-free languages. Their algorithm is consistent, conservative and admits a characteristic set (in the sense of de la Higuera) of polynomial cardinality. Yokomori [6] proposed a polynomial-time learning algorithm for very simple grammars, which is consistent and conservative. After his work, Wakatsuki et al. [7] and Yoshinaka [8] showed that Yokomori's technique is applicable to some related classes of languages.

This paper investigates the learning efficiency of very simple grammars from positive data in various senses. Section 2 discusses possible definitions of efficient identification in the limit from positive data, particularly in comparison with learning from complete data. In Section 3, we define and show some basic properties of very simple grammars and *one-counter simple grammars*, which are very simple grammars that have exactly one nonterminal symbol. Section 4 presents a polynomial-time learning algorithm for one-counter simple grammars that is consistent and conservative and satisfies Pitt's and de la Higuera's definitions if we regard $|\Sigma|$ as a constant. Moreover, one-counter simple grammars admit even more efficient iterative algorithm if we abandon the consistency, as Lange and Wiehagen's algorithm for pattern languages. It runs in polynomial time regardless of whether Σ is fixed or not. At the same time, we give some negative results about the efficiency of learning one-counter simple grammars from positive data, which are applicable to very simple grammars. Section 5 investigates the mathematical properties of Yokomori's [6,9] learning algorithm for very simple grammars and gives some positive and negative results. While his algorithm does not satisfy Pitt's definition of polynomial-time identifiability, we present an alternative polynomial-time algorithm learning very simple grammars from positive data that changes its conjecture at most linear times in the cardinality of the alphabet.

2   Efficient Learning from Positive Data

Preliminaries. $\mathbb{N}$ and $\mathbb{Z}$ denote the sets of positive integers and integers, respectively. $\varnothing$ is the empty set. An *alphabet* Σ is a finite nonempty set of symbols. A string w over Σ is a sequence of symbols in Σ, and Σ^* denotes the set of all strings over Σ. A language over Σ is any subset of Σ^*. For a set Σ, $|\Sigma|$ denotes the cardinality of Σ and for a sequence $w \in \Sigma^*$, $|w|$ denotes the length of w. The size of a finite language K is given as $\|K\| = \sum_{w \in K} |w|$. A *hypothesis space* is a pair $(\mathcal{G}, L)$ where $\mathcal{G}$ is a set of finite descriptions called *representations* and L is a function mapping elements of $\mathcal{G}$ to languages. When no confusion occurs, we denote the hypothesis space simply by $\mathcal{G}$. Throughout this paper, we only consider hypothesis spaces $\mathcal{G}$ whose universal membership problems are decidable, i.e., the language consisting of pairs (G, w) with $G \in \mathcal{G}$ and $w \in L(G)$ is recursive.

A *positive presentation* of a language L_* is a surjection R from $\mathbb{N}$ to L_*. As usual, a positive presentation R is described as the infinite sequence $(R(1), R(2), \ldots)$. Each $R(i)$ is called a *positive example* of L_*. A *learning algorithm* $\mathcal{A}$ on a hypothesis space $\mathcal{G}$ is an algorithm which takes a positive presentation R as input, and outputs some infinite sequence $G_1, G_2, \ldots$ of representations in $\mathcal{G}$, i.e., $\mathcal{A}$ infinitely repeats the cycle where $\mathcal{A}$ receives $R(n)$ and outputs a representation G_n in $\mathcal{G}$ for $n = 1, 2, \ldots$. We denote the nth output of $\mathcal{A}$ on a positive presentation R by $\mathcal{A}(R, n)$. A learning algorithm $\mathcal{A}$ *converges to G on a presentation R* if for all but finitely many n, $\mathcal{A}(R, n) = G$. $\mathcal{A}$ *identifies a class $\mathcal{L}$ of languages in the limit from positive data* if for every positive presentation R of every $L_* \in \mathcal{L}$, there is $G \in \mathcal{G}$ such that $L(G) = L_*$ and $\mathcal{A}$ converges to G on R. We also say $\mathcal{A}$ *identifies a class $\mathcal{G}$ of representations in the limit* if $\mathcal{A}$ identifies the class of the languages represented by the elements of $\mathcal{G}$ in the limit.

We say that $\mathcal{A}$ *updates its conjecture in polynomial time* (or simply $\mathcal{A}$ *is a polynomial-time algorithm*) iff it computes $\mathcal{A}(R, n)$ in polynomial time in $|R(1)| + \cdots + |R(n)|$ for any R and n. For a language K, a representation G is said to be *consistent with K* iff $K \subseteq L(G)$. $\mathcal{A}$ is *consistent* if $\mathcal{A}(R, n)$ is consistent with $\{R(1), \ldots, R(n)\}$ for any R and n. $\mathcal{A}$ is *conservative* if $\mathcal{A}(R, n+1) = \mathcal{A}(R, n)$ whenever $\mathcal{A}(R, n)$ is consistent with $\{R(n+1)\}$ for any R and n.

A *complete presentation* of a language L_* over Σ is a function R from $\mathbb{N}$ to $\{0,1\} \times \Sigma^*$ such that $w \in L_*$ iff $R(i) = (1,w)$ for some i and $w \in \Sigma^* - L_*$ iff $R(i) = (0,w)$ for some i. For each $R(i) = (j,w)$, w is called a *positive example* of L_* if $j = 1$ and w is called a *negative example* of L_* if $j = 0$. A representation G is said to be *consistent* with $R_k = \{R(1), \ldots, R(k)\}$ iff every positive example appearing in the set R_k is in $L(G)$ and no negative example appearing in the set R_k is in $L(G)$. The notions of *identification in the limit from complete data* etc. are defined similarly to identification in the limit from positive data etc.

Efficient Learning from Positive Data. This section discusses possible definitions of "efficient" identification in the limit from positive data. There are several different definitions that formalize the notion of efficient identification. To update its conjecture in polynomial time in the size of the input is a common property among them, however Pitt [1] showed that every successful learning algorithm can be modified so that it runs in polynomial time. Roughly speaking, his trick lets the algorithm compute the conjecture from appropriately small prefix of the input so that the computation is done in polynomial time in the size of the whole input. Hereafter we refer to this trick as *Pitt's trick*.

Against this problem, Pitt has introduced the notion of *implicit errors of prediction*. Note that the original definition concerns learning from complete data. An algorithm makes an *implicit error of prediction at step n* iff $\mathcal{A}(R,n)$ is not consistent with $\{R(n+1)\}$. A class $\mathcal{G}$ of representations is *polynomial-time identifiable in the limit in Pitt's sense* iff $\mathcal{G}$ admits a polynomial-time learning algorithm $\mathcal{A}$ such that

– for any presentation of $L(G)$ for $G \in \mathcal{G}$, $\mathcal{A}$ makes implicit errors of prediction at most polynomial times in $\|G\|$, where $\|G\|$ is the description size of G.

Yokomori [10] proposed to relax Pitt's definition. A class $\mathcal{G}$ of representations is *polynomial-time identifiable in the limit in Yokomori's sense* iff $\mathcal{G}$ admits a polynomial-time learning algorithm $\mathcal{A}$ such that

– for any presentation R of $L(G)$ for $G \in \mathcal{G}$ and $n \in \mathbb{N}$, the number of implicit errors of prediction made by $\mathcal{A}$ on the first n examples is bounded by a polynomial in $\|G\|l$ where $l = \max\{|R(1)|, \ldots, |R(n)|\}$.

On the other hand, de la Higuera's [3] definition demands the existence of characteristic sets of polynomial size. As Pitt's definition, the original definition concerns learning from complete data. A class $\mathcal{G}$ of representations is *identifiable in the limit from polynomial time and data* iff $\mathcal{G}$ admits a polynomial-time learning algorithm $\mathcal{A}$ such that

– $\mathcal{A}$ is consistent,
– for each $G \in \mathcal{G}$, there is a finite set K (called a *characteristic set*) of examples such that
  • $\|K\|$ is bounded by a polynomial in $\|G\|$,
  • G is consistent with K,

- for any presentation R of $L(G)$ and $n \in \mathbb{N}$, whenever $K \subseteq \{R(1), \ldots, R(n)\}$, $L(\mathcal{A}(R, n)) = L(G)$ and $\mathcal{A}(R, n) = \mathcal{A}(R, n + 1)$.

One might expect that those definitions work well also when learning from positive data only. We however should be conscious of the difference between learning from complete data and learning from positive data.

We say that an algorithm *changes its mind* if the new conjecture is different from the previous one. First we propose to count the number of mind changes rather than implicit errors of prediction when learning from positive data. Pitt [1] refuted the idea that considers the number of mind changes as a measure of the efficiency of an algorithm, when he proposed to count the number of implicit errors of prediction. His discussion shows that when learning from complete data, if the equivalence problem for the target class is decidable, then it admits a polynomial-time learning algorithm that changes its mind at most linear times in the size of the target representation. However, his technique does not work when learning from positive data. Giving an upper bound on the number of mind changes of a learning algorithm is not a trivial issue when learning from positive data.

Besides, counting the total number of implicit errors of prediction is not more restrictive than counting the number of mind changes when learning from positive data. For every learning algorithm $\mathcal{A}_1$ we get another algorithm $\mathcal{A}_2$ that makes implicit errors of prediction at most $|\Sigma|$ more times than $\mathcal{A}_1$ changes its conjecture on the same positive presentation. We design $\mathcal{A}_2$ so that it outputs the same conjecture as $\mathcal{A}_1$ whenever the output of $\mathcal{A}_1$ is consistent with the input, and otherwise outputs G_Σ with $L(G_\Sigma) = \Sigma^*$ where Σ is the set of letters appearing in the input.[1] $\mathcal{A}_2$ is consistent. When $\mathcal{A}_2$ makes an implicit error of prediction, there are two cases: (i) the current conjecture is G_Σ and the newly given example includes a letter not in Σ, or (ii) the current conjecture G is from $\mathcal{A}_1$. The former case (i) occurs at most $|\Sigma|$ times and the latter case (ii) occurs not more than $\mathcal{A}_1$ changes its conjecture, because $\mathcal{A}_2$ will never output the same conjecture G again. Note that if the universal membership problem for the hypothesis space of $\mathcal{A}_1$ is solved in polynomial time and $\mathcal{A}_1$ updates its conjecture in polynomial time, then $\mathcal{A}_2$ also runs in polynomial time. Therefore, we propose to count the number of mind changes rather than implicit errors of prediction when learning from positive data due to the simplicity of the definition.

Here we see the difference of the roles of consistency between learning from complete data and learning from positive data. When learning from complete data, it is not a trivial task to find a representation that is expected to be consistent with the forthcoming positive and negative examples. On the other hand, when learning from positive data, the language Σ^* trivially contains any

---

[1] Note that we allow an algorithm to output a conjecture outside the target class, so any algorithm on a standard representation formalism is allowed to conjecture the language Σ^*. Prohibition of outputting representation outside the target class does not seem to be an essential restriction, because in fact many efficiently learnable classes include the language Σ^*, e.g., zero-reversible languages [11], pattern languages [12, 2], substitutable context-free languages [5], etc.

positive examples including ones that will be given in the future as long as the alphabet is not expanded. When learning from complete data, counting number of implicit errors of prediction is powerful enough to prevent Pitt's trick. When learning from positive data, counting mind changes is at least as restrictive as counting implicit errors of prediction, but, is *not* powerful enough to prevent Pitt's trick. We need to find some further conditions, though giving an upper bound on the number of mind changes of a learning algorithm is not a trivial issue when learning from positive data.

One persuasive solution for this problem would be to impose both consistency and conservativeness to polynomial-time learning algorithms. Consistency only is not restrictive enough, because one can modify any successful algorithm learning from positive data so that it is consistent and runs in polynomial time by combining Pitt's trick and the technique converting $\mathcal{A}_1$ into $\mathcal{A}_2$ if the universal membership problem for the hypothesis space of the original algorithm is solved in polynomial time. We also note that conservativeness only is neither restrictive enough. One can modify any successful conservative algorithm learning from positive data so that it runs in polynomial time with preserving the conservativeness. The technique is almost the same as the above construction of polynomial-time consistent algorithms. Here instead of outputting G_Σ, the modified algorithm should output $G_\varnothing$ with $L(G_\varnothing) = \varnothing$. The combination of conservativeness and consistency forces the algorithm to output a consistent conjecture whose language does not properly contains any other consistent languages in the target class. This rejects Pitt's trick and tricky algorithms such as $\mathcal{A}_2$. In fact consistency and conservativeness are very common properties among various learning algorithms proposed so far. Under this condition, the numbers of mind changes and implicit errors of prediction coincide.

On the other hand, consistency is, however, not mandatory. Lange and Wiehagen's [2] polynomial-time algorithm for learning pattern languages from positive data is not consistent but it is worth calling efficient. We say an algorithm is *iterative* iff it computes a new conjecture only from the newly given example and the current conjecture. Iterativeness is also strong enough to prevent tricky polynomial-time algorithms. Their algorithm is iterative and moreover admits a characteristic set of polynomial size.

In fact, the second condition (characteristic set) of de la Higuera's definition itself is restrictive enough to prohibit Pitt's trick. Although it is easy to satisfy the consistency by the technique constructing $\mathcal{A}_2$, de la Higuera's definition itself works well when learning from positive data.

Summarizing the above discussion, we have the following possible properties of learning algorithms from positive data.

Definition 1. Suppose that a learning algorithm $\mathcal{A}$ identifies $\mathcal{G}$ in the limit from positive data. $\mathcal{A}$ satisfies the properties $\mathcal{P}_{\mathrm{CC}}$, $\mathcal{P}_{\mathrm{UP}}$, $\mathcal{P}_{\mathrm{MC+}}$, $\mathcal{P}_{\mathrm{MC-}}$, $\mathcal{P}_{\mathrm{CS}}$, $\mathcal{P}_{\mathrm{IT}}$, respectively, iff

- ($\mathcal{P}_{\mathrm{UP}}$): it updates its conjecture in polynomial time in the size of the input,
- ($\mathcal{P}_{\mathrm{CC}}$): it is consistent and conservative,

- ($\mathcal{P}_{\text{MC+}}$): for any $G \in \mathcal{G}$, any positive presentation R of $L(G)$, $|\{ i \mid \mathcal{A}(R, i) \neq \mathcal{A}(R, i+1), i \in \mathbb{N} \}|$ is bounded by a polynomial in $\|G\|$,
- ($\mathcal{P}_{\text{MC-}}$): for any $G \in \mathcal{G}$, any positive presentation R of $L(G)$, any $n \in \mathbb{N}$, $|\{ i \mid \mathcal{A}(R, i) \neq \mathcal{A}(R, i+1), 1 \leq i < n \}|$ is bounded by a polynomial in $\|G\|l$ where $l = \max\{|R(1)|, \dots, |R(n)|\}$,
- ($\mathcal{P}_{\text{CS}}$): for any $G \in \mathcal{G}$, there is a finite subset K of $L(G)$ such that
  - $\|K\|$ is bounded by a polynomial in $\|G\|$,
  - for any presentation R of $L(G)$ and $n \in \mathbb{N}$, whenever $K \subseteq \{R(1), \dots, R(n)\}$, we have $L(\mathcal{A}(R, n)) = L(G)$ and $\mathcal{A}(R, n) = \mathcal{A}(R, n+1)$,
- ($\mathcal{P}_{\text{IT}}$): it computes new conjecture only from the newly given example and the current conjecture.

If an algorithm satisfies $\mathcal{P}_{\text{UP}}$ and $\mathcal{P}_{\text{IT}}$, the size of the input is the sum of the sizes of the newly given example and of the current conjecture.

In the remainder of this paper, we discuss the learning efficiency of very simple grammars in terms of the above properties.

3   Very Simple Grammars

A *context-free grammar (*CFG*)* is a quadruple $G = (N, \Sigma, P, S)$, where N is a finite set of *nonterminal symbols*, Σ a finite set of *terminal symbols*, $P \subseteq N \times (N \cup \Sigma)^*$ a finite set of *production rules*, and $S \in N$ the *start symbol*. $\Rightarrow_G$ denotes the one step derivation and $\Rightarrow_G^*$ the reflexive and transitive closure of $\Rightarrow_G$. The *language generated by* G is the set $L(G) = \{ w \in \Sigma^* \mid S \Rightarrow_G^* w \}$. A CFG G is *reduced* iff for every $A \in N \cup \Sigma$, there are $x, y, z \in \Sigma^*$ such that $S \Rightarrow_G^* xAz \Rightarrow_G^* xyz$. The description size of a CFG G is defined as $\|G\| = \sum_{A \to \zeta \in P} |A\zeta|$.

We use early lower Italic letters for terminal symbols, late lower Italic letters for sequences of terminal symbols, early upper Italic letters for nonterminal symbols, and early lower Greek letters for sequences of nonterminal symbols.

Definition 2. A CFG $G = (N, \Sigma, P, S)$ in Greibach normal form is a *very simple grammar (*VSG*)* iff

$$A \to a\alpha, \; B \to a\beta \in P \text{ implies } A = B \text{ and } \alpha = \beta.$$

Moreover a VSG G is a *one-counter simple grammar (*OCSG*)* iff $N = \{S\}$.

For every reduced VSG, $|N| \leq |P| = |\Sigma|$ holds.

Yokomori's [6, 9] learning algorithm for VSGs considers a condition on the length of each rule of a grammar consistent with a set of positive examples. He represents the lengths of rules with a vector of $|\Sigma|$ dimensions. For notational convenience, we express the same idea with a homomorphism, rather than a vector.

A homomorphism $\sharp$ mapping from Σ^* to $\mathbb{Z}$ ($\sharp(xy) = \sharp(x) + \sharp(y)$ for all $x, y \in \Sigma^*$) is called a *shape* iff $\sharp(a) \geq -1$ for all $a \in \Sigma$.

For a VSG $G = (N, \Sigma, P, S)$, *the shape of* G denoted by $\sharp_G$ is defined as

$$\sharp_G(a) = |\alpha| - 1 \quad \text{if } A \to a\alpha \in P \text{ for some } A \in N.$$

We say that a shape $\sharp$ is *compatible with a language* $L \subseteq \Sigma^*$ iff for all $w \in L$

- $\sharp(w) = -1$,
- $\sharp(w') \geq 0$ for any proper prefix w' of w.

The notion of compatible shapes is the same as that of solution vectors by Yokomori [6,9]. The following lemma establishes a close relationship between finding a consistent VSG with a finite language and finding a compatible shape with that language.

Lemma 1. *Let L be a language over Σ and $\sharp$ a shape on Σ. There is a VSG (OCSG) G such that $\sharp_G = \sharp$ and $L \subseteq L(G)$ iff $\sharp$ is compatible with L.*

Lemma 2. *Suppose that $\sharp$ is compatible with a language L. For every $a \in \Sigma$, $\sharp(a) < \min\{ |y| \mid xay \in L \}$.*

It is a trivial task to decide whether a shape is compatible with a finite language. Together with $\sharp(a) \geq -1$, Lemma 2 ensures that one can enumerate all the shapes compatible with the given finite set of positive examples.

4   Learning One-Counter Simple Grammars

There is a simple strategy for identification in the limit from positive data of OCSGs. For a fixed shape, we have a unique (up to the renaming of the start symbol) OCSG of that shape. By Lemma 1, finding a consistent grammar and finding a compatible shape are exactly the same task.

Theorem 3. *The class of OCSGs admits a learning algorithm with the properties $\mathcal{P}_{UP}$, $\mathcal{P}_{CC}$, $\mathcal{P}_{MC+}$, $\mathcal{P}_{CS}$ when we regard $|\Sigma|$ as a constant.*

Proof. We define the *size* of a shape $\sharp$ by $\|\sharp\| = \sum_{a \in \Sigma}(\sharp(a) + 2)$, which coincides with the size of the OCSG of that shape. We give a specification of our algorithm $\mathcal{A}_{MSG}$ here. If the current conjecture is consistent with the newly given example, $\mathcal{A}_{MSG}$ keeps that conjecture. Otherwise, $\mathcal{A}_{MSG}$ first finds a compatible shape $\sharp$ consistent with the input that is minimum with respect to the size $\|\sharp\|$. Second it outputs a grammar with rules $S \to aS^{1+\sharp(a)}$ for $a \in \Sigma$.

The fact that $\mathcal{A}_{OCSG}$ identifies the class of OCSGs follows from $\mathcal{P}_{CS}$.

($\mathcal{P}_{CC}$): This property is obvious by definition.

($\mathcal{P}_{UP}$): Let K be the set of given positive examples. By Lemma 2, every compatible shape $\sharp$ satisfies $-1 \leq \sharp(a) \leq l - 2$ where $l = \max\{ |w| \mid w \in K \}$. Thus the size of the search space of compatible shapes is bounded by $l^{|\Sigma|-1}$ (because there is at least one terminal b with $xb \in K$, to which every compatible shape assigns -1). Checking compatibility of a shape $\sharp$ with K is done in linear time in $\|K\|$. Therefore, the update time is bounded by $O(\|K\|^{|\Sigma|})$.

($\mathcal{P}_{MC+}$): Each shape $\sharp$ has at most $\|\sharp\|^{|\Sigma|}$ shapes $\sharp'$ such that $\|\sharp'\| \leq \|\sharp\|$. If the target grammar G has shape $\sharp$, $\mathcal{A}_{OCSG}$ never outputs a grammar of a shape of larger size than $\sharp$. Therefore, by $\|\sharp\| = \|G\|$ and $\mathcal{P}_{CC}$, it changes its mind at most $\|G\|^{|\Sigma|}$ times.

$(\mathcal{P}_{\text{CS}})$: For an OCSG G, let $b \in \Sigma$ be such that $\sharp_G(b) = -1$ and K_G be

$$K_G = \{ ab^{1+\sharp_G(a)} \mid a \in \Sigma \} \subseteq L(G).$$

We have $\|K_G\| \leq \|G\|$. Clearly K_G admits exactly one compatible shape, which is $\sharp_G$. Whenever a superset of K_G is given to the algorithm, it outputs G. $\square$

Theorem 3 holds if we assume $|\Sigma|$ as a constant. It is natural to ask if we can lift this assumption. Here we give two negative results. One is concerning the update time and the other is about the number of mind changes.

Proposition 4. *The following problem is NP-complete:*

- *Instance: a finite alphabet Σ and a finite language K over Σ,*
- *Question: does K admit a compatible shape?*

Proof. The problem is clearly in NP. We show the NP-hardness by reduction from a well known NP-complete problem, *the satisfiability problem*. Let X be a finite set of *Boolean variables* and $\bar{X} = \{ \bar{p} \mid p \in X \}$. A *clause* C is a non-empty subset of $X \cup \bar{X}$ and a *formula* F is a finite collection of clauses. A *valuation* on X is a function ϕ from $X \cup \bar{X}$ to $\{0, 1\}$ such that $\{\phi(p), \phi(\bar{p})\} = \{0, 1\}$ for all $p \in X$. An instance of the satisfiability problem is a pair of a set X of Boolean variables and a formula F over X. The satisfiability problem is the decision problem of determining whether there is a valuation ϕ on X such that for each clause $C \in F$, there is $q \in C$ such that $\phi(q) = 1$.

For a given formula $F = \{ C_1, \dots, C_k \}$ over $X = \{ p_1, \dots, p_m \}$, let

$$\Sigma_X = \{ a_i, \bar{a}_i \mid 1 \leq i \leq m \} \cup \{b\},$$
$$K_F = \{ a_i \bar{a}_i bb, \bar{a}_i a_i bb \mid 1 \leq i \leq m \}$$
$$\cup \{ \sigma(q_1) \dots \sigma(q_l) b \tau(q_1) \dots \tau(q_l) b^l \mid C_j = \{ q_1, \dots, q_l \}, 1 \leq j \leq k \},$$

where $\sigma(p_i) = a_i$, $\tau(p_i) = \bar{a}_i$ for $p_i \in X$ and $\sigma(\bar{p}_i) = \bar{a}_i$, $\tau(\bar{p}_i) = a_i$ for $\bar{p}_i \in \bar{X}$. This reduction can be done in polynomial time. It is not difficult to see that there is a valuation satisfying F iff K_F has a compatible shape. $\square$

Corollary 5. *There is no polynomial-time algorithm that takes Σ and a finite language K over Σ and outputs a consistent OCSG (if any) unless $P = NP$.*

Proposition 6. *Unless we regard $|\Sigma|$ as a constant, there is no successful learning algorithm for OCSGs satisfying $\mathcal{P}_{\text{MC}-}$ such that it always outputs a consistent OCSG.*

Proof. Let $\mathcal{A}$ be a learning algorithm such that it always outputs a consistent OCSG (if any). Let $\Sigma_m = \{ a_i, \bar{a}_i \mid 1 \leq i \leq m \} \cup \{b\}$. We show that for any m, there is an OCSG on Σ_m of description size $O(m)$ and a sequence of positive examples of length $O(m)$ on which $\mathcal{A}$ changes its mind at least $2^m - 1$ times. Let $K_0 = \{ a_i \bar{a}_i bb, \bar{a}_i a_i bb \mid 1 \leq i \leq m \}$ be the set of the first $2m$ positive examples. The set of shapes compatible with K_0 is given by

$$\Gamma = \{ \sharp \mid \{ \sharp(a_i), \sharp(\bar{a}_i) \} = \{0, 1\} \text{ for } 1 \leq i \leq m \text{ and } \sharp(b) = -1 \}.$$

For $\natural \in \Gamma$, let $w_\natural = x_1 \ldots x_m b y_1 \ldots y_m b^m$ where $x_i = a_i$ and $y_i = \bar{a}_i$ if $\natural(a_i) = 0$, and $x_i = \bar{a}_i$ and $y_i = a_i$ if $\natural(a_i) = 1$. Then, for any $\natural_0, \natural_1, \ldots, \natural_k \in \Gamma$, $\natural_0$ is a shape compatible with $K_0 \cup \{w_{\natural_1}, \ldots, w_{\natural_k}\}$ iff $\natural_0 \notin \{\natural_1, \ldots, \natural_k\}$. After $\mathcal{A}$ outputs a grammar of shape $\natural \in \Gamma$, $w_\natural$ may be given as the succeeding example. At this step, $\mathcal{A}$ must abandon this conjecture and change its mind. Indeed this is possible for $|\Gamma| - 1 = 2^m - 1$ times. The target grammar can be the one constructed on the last shape in Γ such that $w_\natural$ has not appeared in the presentation yet. □

Despite the above two propositions, abandon of the consistency enables us to construct an efficient iterative learning algorithm.

Theorem 7. *The class of* OCSGs *admits a learning algorithm* $\mathcal{B}_{\text{OCSG}}$ *with the properties* $\mathcal{P}_{\text{UP}}$, $\mathcal{P}_{\text{MC+}}$, $\mathcal{P}_{\text{CS}}$, $\mathcal{P}_{\text{IT}}$ *even when we regard* $|\Sigma|$ *as a variable.*

Proof. We give the specification of the algorithm $\mathcal{B}_{\text{OCSG}}$. Let $G = (\{S\}, \Sigma_G, P, S)$ be the current conjecture and w the newly given example.

1. If $b \notin \Sigma_G$ for $w = xb$, then add b to Σ_G and $S \to b$ to P.
2. If there is a $a \notin \Sigma_G$ such that $w \in (\Sigma_G \cup \{a\})^* - \Sigma_G^*$, then add a to Σ_G and $S \to aS^m$ to P where m is uniquely determined by the equation $\natural_G(w) = -1$.
3. Output the updated grammar G.

The properties $\mathcal{P}_{\text{UP}}$ and $\mathcal{P}_{\text{IT}}$ are obvious. Whenever $\mathcal{B}_{\text{OCSG}}$ adds a new rule to the conjecture, it is indeed a rule of the target grammar modulo renaming the nonterminal symbol. Thus $\mathcal{B}_{\text{OCSG}}$ changes its mind at most $|\Sigma|$ times ($\mathcal{P}_{\text{MC+}}$). K_G in the proof of Theorem 3 is also a characteristic set for $\mathcal{B}_{\text{OCSG}}$ ($\mathcal{P}_{\text{CS}}$). □

The algorithm $\mathcal{B}_{\text{OCSG}}$ has similar properties to the learning algorithm for pattern languages by Lange and Wiehagen [2]. Although the membership problem for a pattern language is NP-complete [12], they overcome the intractability of pattern languages by giving up the consistency. Indeed their algorithm satisfies $\mathcal{P}_{\text{UP}}$, $\mathcal{P}_{\text{CS}}$, $\mathcal{P}_{\text{IT}}$ when we regard $|\Sigma|$ as a variable. One different property between Lange and Wiehagen's and our algorithms is that the output of their algorithm on a sequence of positive examples is the same as the one on a permutation of the sequence, while $\mathcal{B}_{\text{OCSG}}$ does not have this property. For instance, the output by $\mathcal{B}_{\text{OCSG}}$ for the sequence abc, bc is different from the one for bc, abc. The lack of this property is not necessarily a defect of our algorithm $\mathcal{B}_{\text{OCSG}}$. To smoothly learn a complex concept, to begin with a simpler portion of it is a better strategy than the opposite in general.

5   Learning Very Simple Grammars

Yokomori's Algorithm. Yokomori [6, 9] has shown that the class of VSGs admits a learning algorithm with the properties $\mathcal{P}_{\text{UP}}$ and $\mathcal{P}_{\text{CC}}$ if we regard $|\Sigma|$ as a constant. We begins this section with roughly describing his algorithm and show that his algorithm also satisfies $\mathcal{P}_{\text{MC-}}$. To make the further discussion concise, we give slightly different notation and specification for describing his algorithm.

Though the simplified algorithm given below would be less efficient than the original from the practical point of view, we still give the same theoretical upper bound on the running time as the original.

Yokomori introduced "simulation process" for constructing the least consistent VSG for each compatible shape. Let K be a finite language and $\sharp$ be a shape compatible with K. The least VSG G such that $\sharp_G = \sharp$ and $K \subseteq L(G)$ is uniquely determined (modulo renaming of nonterminals) by *simulating* derivations of elements of K. Before starting simulation, assume that $G = (N_\sharp, \Sigma, P_\sharp, S)$ where

$$N_\sharp = \{ X_a \mid a \in \Sigma \} \cup \{ Z_{a,j} \mid a \in \Sigma,\, 0 \le j \le \sharp(a) \} \cup \{ S \},$$
$$P_\sharp = \{ X_a \to a Z_{a,0} \dots Z_{a,\sharp(a)} \mid a \in \Sigma \}.$$

Then, we merge nonterminals in $N_\sharp$ so that G can derive all the elements of K.

Let $G = (N, \Sigma, P, S)$ and $G' = (N', \Sigma, P', S')$ be two VSGs. We write $G \preccurlyeq G'$ iff for any $a, b \in \Sigma$ and $A \in N$, whenever $A \to a\alpha$, $A \to b\beta \in P$ for some $\alpha, \beta \in N^*$, there is $A' \in N'$ such that $A' \to a\alpha'$, $A' \to b\beta' \in P'$ for some $\alpha', \beta' \in N'^*$. If $G \preccurlyeq G'$ and $G' \not\preccurlyeq G$, then we write $G \prec G'$. It is easy to see that if $G_1, \dots, G_m$ are VSGs on Σ such that $G_1 \prec G_2 \prec \cdots \prec G_m$, then $m \le |\Sigma|$.

Lemma 8 (Yokomori [6]). *For any reduced VSGs G and G', $L(G) \subseteq L(G')$ implies $G \preccurlyeq G'$ and moreover $L(G) \subsetneq L(G')$ implies $G \prec G'$.*

To choose a minimal (with respect to the language) VSG among many candidates, Yokomori's original algorithm chooses a VSG that is minimal with respect to the relation $\preccurlyeq$. Here we propose an even simpler criterion.

Corollary 9. *Let G and G' be reduced VSGs on the same alphabet. If $L(G) \subsetneq L(G')$, then $|N| > |N'|$.*

The learning algorithm $\mathcal{A}_{\mathrm{VSG}}$, which is a slight simplification of Yokomori's original algorithm, for VSGs runs as follows. Let G_0 be the previous conjecture (for the first conjecture, assume $G_0 = (\{S\}, \varnothing, \varnothing, S)$) and K be the given set of positive examples.

1. If $K \subseteq L(G_0)$, then output G_0.
2. Otherwise, enumerate all the shapes $\sharp_1, \dots, \sharp_k$ compatible with K.
3. Construct consistent VSGs $G_1, \dots, G_k$ on $\sharp_1, \dots, \sharp_k$, respectively, by simulating derivation of the elements of K.
4. Among $G_1, \dots, G_k$, output any VSG that has the largest number of nonterminals.

Theorem 10. *The algorithm $\mathcal{A}_{\mathrm{VSG}}$ identifies VSGs in the limit from positive data. Moreover, $\mathcal{A}_{\mathrm{VSG}}$ satisfies $\mathcal{P}_{\mathrm{CC}}$, $\mathcal{P}_{\mathrm{UP}}$, $\mathcal{P}_{\mathrm{MC-}}$, if we regard $|\Sigma|$ as a constant.*

Proof. Yokomori has given a proof for this theorem except for the property $\mathcal{P}_{\mathrm{MC-}}$. Note that $\mathcal{A}_{\mathrm{VSG}}$ updates its conjecture in $O(\|K\|^{|\Sigma|})$ steps for the input K. We prove the property $\mathcal{P}_{\mathrm{MC-}}$. Let $G_1, \dots, G_m$ be the sequence of pairwise distinct VSGs output by $\mathcal{A}_{\mathrm{VSG}}$ in this order for the sequence of positive examples $w_1, \dots, w_n$. Let us write $G_i \sim G_j$ if $\sharp_{G_i} = \sharp_{G_j}$. Since each $\sharp_{G_i}$ is compatible with some non-empty subset of $\{w_1, \ldots, w_n\}$, by Lemma 2, we have $|\mathcal{G}/\sim| \leq l^{|\Sigma|}$ where $\mathcal{G} = \{G_1, \ldots, G_m\}$ and $l = \max\{|w_1|, \ldots, |w_n|\}$. Moreover, if $G_i \sim G_j$ and $i < j$, by the procedure of simulation, we have $G_i \prec G_j$. Therefore, each element of $\mathcal{G}/\sim$ contains at most $|\Sigma|$ grammars. We get $|\mathcal{G}| \leq |\Sigma| l^{|\Sigma|}$ and thus the algorithm changes its conjecture at most $|\Sigma| l^{|\Sigma|}$ times.   □

$\mathcal{A}_{\mathrm{VSG}}$ does not satisfy $\mathcal{P}_{\mathrm{MC}+}$. It is easy to see that the language $L_* = a\{b, c\}^* d$ is very simple. We show that for every natural number m, there is a positive presentation of L_* on which $\mathcal{A}_{\mathrm{VSG}}$ can change its conjecture more than m times until it converges. Let the positive examples $w_0, \ldots, w_m, w_{m+1} \in L_*$ be given to $\mathcal{A}_{\mathrm{VSG}}$ in this order where $w_i = ac^i bc^{m-i} d$ for $i \leq m$ and $w_{m+1} = ad$. It is easy to check that for each $K_k = \{w_0, \ldots, w_k\}$ with $1 \leq k \leq m$, every VSG constructed by the simulation procedure of $\mathcal{A}_{\mathrm{VSG}}$ has at most 3 nonterminals. Therefore, $\mathcal{A}_{\mathrm{VSG}}$ may output the VSG G_k with the following rules

$$S \to aX_b^{k+1}X_d, \ X_b \to bX_b^{m-k}, \ X_b \to c, \ X_d \to d,$$

which indeed has 3 nonterminals. But $w_{k+1} \notin L(G_k)$ and thus $\mathcal{A}_{\mathrm{VSG}}$ has to change its conjecture. Therefore, the number of mind changes of $\mathcal{A}_{\mathrm{VSG}}$ can be more than m until it converges. Moreover, one can give another very simple language by modifying the above example so that $\mathcal{A}_{\mathrm{VSG}}$ *must* change the conjecture more than m times for any m whichever $\mathcal{A}_{\mathrm{VSG}}$ chooses among VSGs with the largest number of nonterminals.

Alternative Algorithm. In the rest of this section, we propose an alternative learning algorithm $\mathcal{B}_{\mathrm{VSG}}$ for VSGs that changes its conjecture at most $|\Sigma|$ times ($\mathcal{P}_{\mathrm{MC}+}$). As discussed in Section 2, the property $\mathcal{P}_{\mathrm{MC}+}$ is not powerful enough for making the property $\mathcal{P}_{\mathrm{UP}}$ meaningful. Here we introduce another property.

Let $\mathcal{L}$ be a class of languages. We say that a finite subset K of a language $L \in \mathcal{L}$ is a *universal characteristic set*[2] of L with respect to $\mathcal{L}$ iff for any language $L' \in \mathcal{L}$, $K \subseteq L'$ implies $L \subseteq L'$. A learning algorithm satisfies the property $\mathcal{P}_{\mathrm{UCS}}$ iff whenever the input includes a universal characteristic set, the algorithm converges to a grammar representing the target language. Note that the notion of universal characteristic sets does not depend on any particular algorithm different from characteristic sets (K in the definition of $\mathcal{P}_{\mathrm{CS}}$, Definition 1). Although the existence of universal characteristic set is not a necessary condition for identifiability from positive data, the property $\mathcal{P}_{\mathrm{UCS}}$ is strong enough to prevent Pitt's trick if the target class admits a universal characteristic set for each language. Here we show that every VSG G indeed admits a computable universal characteristic set. Let K_0 be any finite subset of $L(G)$ in which every $a \in \Sigma$ appears. By Lemma 2 we have a finite number of shapes compatible with K_0. Moreover each shape has a finite number of VSGs of that shape modulo renaming nonterminals. Thus we can enumerate all the VSGs G' on Σ such that $K_0 \subseteq L(G')$. Then we

---

[2] A universal characteristic set is exactly what is called a *characteristic sample* in Angluin [11]. To avoid confusion with the notion of *characteristic sets* by de la Higuera, we refrain from using the term "characteristic sample" here.

define K_G by adding some $w_{G'} \in L(G) - L(G')$ to K_0 for each G' such that $K_0 \subseteq L(G')$ and $L(G) \not\subseteq L(G')$. Here the inclusion problem for VSGs is decidable [13]. Then, the resultant set K_G is a universal characteristic set of $L(G)$. We note that if a learning algorithm with the property $\mathcal{P}_{CC}$ always outputs a conjecture in the target class (such as $\mathcal{A}_{VSG}$), it also satisfies $\mathcal{P}_{UCS}$.

Now we define the learning algorithm $\mathcal{B}_{VSG}$ for VSGs. It is obtained by replacing the procedure (4) of $\mathcal{A}_{VSG}$ with

4. If there is $j \in \{1, \ldots, k\}$ such that $L(G_j) \subseteq L(G_i)$ for all $i \in \{1, \ldots, k\}$, then output G_j, else output G_0,

where $G_1, \ldots, G_k$ are VSGs constructed by the previous step and G_0 is the previous conjecture.

Theorem 11. *The algorithm $\mathcal{B}_{VSG}$ identifies VSGs in the limit from positive data with the properties $\mathcal{P}_{UP}$, $\mathcal{P}_{MC+}$, $\mathcal{P}_{UCS}$, where $\mathcal{P}_{UP}$ holds if we regard $|\Sigma|$ as a constant and $\mathcal{P}_{MC+}$ holds if we regard $|\Sigma|$ as a variable.*

Proof. Since each VSG admits a universal characteristic set, the algorithm identifies VSGs in the limit from positive data.

($\mathcal{P}_{UP}$): The only difference from $\mathcal{A}_{VSG}$ is the 4th step. It is known that the inclusion of two VSGs constructed on the set K of positive examples is decidable in $O(\|K\|^p)$ for a natural number p [13,8]. It is enough to execute this sub-algorithm at most $2k \leq 2\|K\|^{|\Sigma|-1}$ times to decide whether a least VSG with respect to the language exists. Therefore, the algorithm $\mathcal{B}_{VSG}$ runs in $O(\|K\|^{|\Sigma|+p})$ steps.

($\mathcal{P}_{MC+}$): Let $G_1, \ldots, G_m$ be the pairwise distinct VSGs output by the algorithm in this order. Let K_i be the set of positive examples on which G_i is first output. Since G_i is the least VSG such that $K_i \subseteq L(G_i)$, for any $j \in \{i, \ldots, m\}$, $K_i \subseteq L(G_j)$ implies $L(G_i) \subseteq L(G_j)$. By the conservativeness of the algorithm, we have $L(G_i) \subsetneq L(G_{i+1})$. Therefore, $L(G_1) \subsetneq L(G_2) \subsetneq \cdots \subsetneq L(G_m)$ and thus $G_1 \prec \cdots \prec G_m$ by Lemma 8. We have $m \leq |\Sigma|$.

($\mathcal{P}_{UCS}$): It is clear by definition. $\square$

As discussed in Section 2, one can modify the algorithm $\mathcal{B}_{VSG}$ so that it makes implicit errors of prediction at most $O(|\Sigma|)$ times, and in that case we lose the conservativeness in exchange for the consistency.

Although the upper bound on the number of mind changes by $\mathcal{B}_{VSG}$ is much smaller than the one by $\mathcal{A}_{VSG}$, it is disputable if $\mathcal{B}_{VSG}$ is more efficient than $\mathcal{A}_{VSG}$. In fact, whenever $\mathcal{B}_{VSG}$ converges to a VSG on a set of positive examples, $\mathcal{A}_{VSG}$ also converges on the same examples, though the converse does not hold.

Nevertheless, the algorithm $\mathcal{B}_{VSG}$ is still worth considering. First, the convergence of the output of $\mathcal{B}_{VSG}$ is actually sufficiently quick by the property $\mathcal{P}_{UCS}$. $\mathcal{B}_{VSG}$ outputs an inconsistent grammar only when it has insufficient positive examples to uniquely determine a conjecture. In that case, there is no reason why the algorithm should output a certain grammar among other equally plausible grammars that generate incomparable languages.

Second, the property $\mathcal{P}_{MC+}$ is a non-trivial property of identification of VSGs in the limit from positive data. Moreover whenever the conjecture is updated, the new one generates a larger language than the old one. By this incrementality, one can modify $\mathcal{B}_{VSG}$ so that it ignores all the positive examples that can be generated by the current conjecture and constructs the new conjecture from the current conjecture and examples with which the current conjecture is inconsistent. This allows one to save the memory space. $\mathcal{A}_{VSG}$ does not admit this modification.

Third, $\mathcal{B}_{VSG}$ is very cautious against over-generalization. While the output of $\mathcal{B}_{VSG}$ may be inconsistent with the given examples, its language is always a subset of the target language, hence the output grammar generates only "safe" strings, which are in the target. On the other hand, the output of $\mathcal{A}_{VSG}$ may generate a language incomparable with the target language.

Fourth, one can know from the behavior of $\mathcal{B}_{VSG}$ whether or not the presented positive examples form a universal characteristic set of a very simple language, because $\mathcal{B}_{VSG}$ outputs a consistent VSG iff the input is a universal characteristic set of the conjectured language. Here even an inconsistent output is informative.

6   Conclusions and Discussion

We have discussed possible definitions of polynomial-time learning algorithms from positive data through presenting two kinds of learning algorithms for each of very simple grammars and one-counter simple grammars. One is consistent and conservative, and one is not consistent. Our discussion of the former type of algorithms augments Yokomori's original work [6, 9] by revealing a property of his algorithm for VSGs, while an upper bound on the efficiency of consistent and conservative learning algorithms for OCSGs is given.

On the other hand, the algorithm $\mathcal{B}_{OCSG}$ shows a similarity to the learning algorithm for pattern languages by Lange and Wiehagen [2]. For those algorithms, there are two kinds of examples, "good" ones and "bad" ones, depending on the current conjecture. When the new example is bad, the algorithms simply ignore the example, and they update their conjectures only when the example is good. Ignoring bad examples is the key for overcoming the intractability of the target class. Concerning this issue, Wiehagen and Zeugmann [14] discuss the importance of inconsistent strategy for learning from complete data.

The algorithm $\mathcal{B}_{VSG}$ also behaves inconsistently. $\mathcal{B}_{VSG}$ does not run faster than $\mathcal{A}_{VSG}$ and it requires positive examples at least as much as $\mathcal{A}_{VSG}$ for convergence. However, $\mathcal{B}_{VSG}$ shows several virtues as discussed in the previous section. We would like to emphasize that an inconsistent output can be more reliable than a consistent output. For some set of positive examples, every consistent grammar possibly generates a language incomparable with the target language. In that case, an inconsistent conjecture that surely generates a subset of the target language might be better in some application.

We also note that the modification from $\mathcal{A}_{VSG}$ to $\mathcal{B}_{VSG}$ is applicable to the related algorithms by Wakatsuki et al. [7] and Yoshinaka [8], because the target classes of them admit an efficient algorithm for the inclusion problem.

Concerning the property $\mathcal{P}_{cs}$, it is known that the class of VSGs cannot satisfy that property unless $|\Sigma|$ is regarded as a constant. It remains open how small a characteristic set of a VSG can be for a learning algorithm.

Acknowledgement

The author is deeply grateful to Takashi Yokomori for helpful discussion on the topic of this paper and Takeshi Shibata for careful reading and helpful comments on a draft of this paper. He also would like to thank the anonymous reviewers for their valuable advice.

References

1. Pitt, L.: Inductive inference, DFAs, and computational complexity. In: Jantke, K.P. (ed.) AII 1989. LNCS, vol. 397, pp. 18–44. Springer, Heidelberg (1989)
2. Lange, S., Wiehagen, R.: Polynomial-time inference of arbitrary pattern languages. New Generation Computing 8, 361–370 (1991)
3. de la Higuera, C.: Characteristic sets for polynomial grammatical inference. Machine Learning 27, 125–138 (1997)
4. Takada, Y.: Grammatical inference for even linear languages based on control sets. Information Processing Letters 28, 193–199 (1988)
5. Clark, A., Eyraud, R.: Identification in the limit of substitutable context-free languages. In: Jain, S., Simon, H.U., Tomita, E. (eds.) ALT 2005. LNCS (LNAI), vol. 3734, pp. 283–296. Springer, Heidelberg (2005)
6. Yokomori, T.: Polynomial-time identification of very simple grammars from positive data. Theoretical Computer Science 298, 179–206 (2003)
7. Wakatsuki, M., Teraguchi, K., Tomita, E.: Polynomial time identification of strict deterministic restricted one-counter automata in some class from positive data. In: Paliouras, G., Sakakibara, Y. (eds.) ICGI 2004. LNCS (LNAI), vol. 3264, pp. 260–272. Springer, Heidelberg (2004)
8. Yoshinaka, R.: Polynomial-time identification of an extension of very simple grammars from positive data. In: Sakakibara, Y., Kobayashi, S., Sato, K., Nishino, T., Tomita, E. (eds.) ICGI 2006. LNCS (LNAI), vol. 4201, pp. 45–58. Springer, Heidelberg (2006)
9. Yokomori, T.: Erratum to Polynomial-time identification of very simple grammars from positive data [Theoret. Comput. Sci. 298 (2003) 179–206]. Theoretical Computer Science 377(1-3), 282–283 (2007)
10. Yokomori, T.: On polynomial-time learnability in the limit of strictly deterministic automata. Machine Learning 19, 153–179 (1995)
11. Angluin, D.: Inference of reversible languages. Journal of the Association for Computing Machinery 29(3), 741–765 (1982)
12. Angluin, D.: Finding patterns common to a set of strings. Journal of Computer and System Sciences 21(1), 46–62 (1980)
13. Wakatsuki, M., Tomita, E.: A fast algorithm for checking the inclusion for very simple deterministic pushdown automata. IEICE transactions on information and systems E76-D(10), 1224–1233 (1993)
14. Wiehagen, R., Zeugmann, T.: Ignoring data may be the only way to learn efficiently. Journal of Experimental and Theoretical Artificial Intelligence 6, 131–144 (1994)

Learning Rational Stochastic Tree Languages

François Denis and Amaury Habrard

Laboratoire d'Informatique Fondamentale de Marseille
CNRS, Aix-Marseille Université
{francois.denis,amaury.habrard}@lif.univ-mrs.fr

Abstract. We consider the problem of learning stochastic tree languages, i.e. probability distributions over a set of trees $T(\mathcal{F})$, from a sample of trees independently drawn according to an unknown target P. We consider the case where the target is a rational stochastic tree language, i.e. it can be computed by a rational tree series or, equivalently, by a multiplicity tree automaton. In this paper, we provide two contributions. First, we show that rational tree series admit a canonical representation with parameters that can be efficiently estimated from samples. Then, we give an inference algorithm that identifies the class of rational stochastic tree languages in the limit with probability one.

1 Introduction

In this paper, we stand in the field of probabilistic grammatical inference and we focus on the learning of stochastic tree languages. A *stochastic tree language* is a probability distribution over the set of trees $T(\mathcal{F})$ built on a ranked finite alphabet $\mathcal{F}$. Given a sample of trees independently drawn according to an unknown stochastic language P, we aim at finding an estimate of P in a given class of models such as *probabilistic tree automata*. Carrasco *et al.* have proposed to learn *deterministic* stochastic tree automata [1]. Specific work for *probabilistic k-testable tree languages* was presented in [2] and for learning stochastic grammars in [3]. However, to our knowledge, no efficient inference algorithm capable of identifying the whole class of probabilistic tree automata is known.

Here, we can make a parallel with results on stochastic languages on strings. Indeed, there exists no efficient algorithm capable of identifying the whole class of probabilistic automata on strings either and the main reason is that we cannot define a canonical structure for these models. Most former results deal with specific subclasses of the class of probabilistic automata. Recently, it has been proposed to consider a larger class of models: the class $S_{\mathbb{R}}^{rat}$ of rational stochastic languages [4]. In the field of strings, a *rational stochastic language* is a stochastic language that can be computed by a *multiplicity automaton*, whose parameters may be positive or negative. Rational stochastic languages have a minimal canonical representation while such canonical representations do not exist for probabilistic automata. And it has been shown that the class of rational stochastic languages can be inferred in the limit with probability 1 [5,6]. The aim of this paper is to study an extension of these results to the case of trees.

M. Hutter, R.A. Servedio, and E. Takimoto (Eds.): ALT 2007, LNAI 4754, pp. 242–256, 2007.

A tree series is a mapping from $T(\mathcal{F})$ to $\mathbb{R}$. Rational tree series have been studied in [7,8]. As far as we know, very few approaches have focused on the learning of tree series but we can mention two papers that stand in a variant of the MAT learning model of Angluin: [9] in a general case and [10] in a deterministic case. But, to the best of our knowledge, this is the first attempt for learning rational stochastic tree languages. Note that the adaptation to trees is not trivial. Prefixes and suffixes of a string are also strings. The equivalent notions for trees are *subtrees* and *contexts* (a context c is a tree one leaf of which acts as a variable, so that substituting a tree t to the variable yields a new tree $c[t]$), which are not similar objects. In the case of words, it can be shown that any rational series r has a canonical representation that can be built on derived rational series of the form $\dot{u}r$ such that $\dot{u}r(v) = r(uv)$ for any string v. The corresponding notion for trees could be rational series of the form $\dot{c}r$ where c is a context, which associates $r(c[t])$ with each tree t. However, it seems impossible to build a canonical representation on them and we need to consider much more sophisticated objects. Let $\mathbb{R}\langle\langle T(\mathcal{F})\rangle\rangle$ be the vector space composed of all series defined on $T(\mathcal{F})$, let $r \in T(\mathcal{F})$ be a tree rational series, let W be the subspace of $\mathbb{R}\langle\langle T(\mathcal{F})\rangle\rangle$ spanned by all the series of the form $\dot{c}r$.

The first result of this paper shows that a canonical representation of r can be defined on the dual vector space W^* composed of all the linear forms defined on W. We show that given an order on $T(\mathcal{F})$, a canonical basis $\{\overline{t_1}, \ldots, \overline{t_n}\}$ - whose elements naturally correspond to trees - can be defined for W^*. This point is important from a machine learning perspective. We show that such a basis can be extracted from any sufficiently large sample of trees drawn according to the target. This leads us to the inference part of our paper.

Our second contribution consists in proposing an inference algorithm which identifies in the limit any rational stochastic tree language with probability one. We show that there exists a sample size above which, the structure of the canonical representation is identified with probability one. Moreover, we show that the parameters output by the algorithm converge to the true parameters at a convergence rate equal to $O(|S|^{\gamma})$ for any $\gamma \in]-1/2, 0[$.

The paper is organized as follows. In Section 2, we introduce some preliminaries on tree series. The canonical linear representation for rational tree series is presented in Section 3. We propose our inference algorithm in Section 4. We conclude by a discussion and a description of future work in Section 5.

2 Preliminaries

2.1 Formal Power Series on Trees

See [11] for references on trees. Let $\mathcal{F} = \mathcal{F}_0 \cup \mathcal{F}_1 \cup \cdots \cup \mathcal{F}_p$ be a ranked alphabet where the elements in $\mathcal{F}_m$ are the function symbols of *arity* m. Let $T(\mathcal{F})$ be the set of all the *trees* that can be constructed from $\mathcal{F}$. Let us define the *height* of a tree t by: $height(t) = 0$ if $t \in \mathcal{F}_0$ and $height(t) = 1 + Max\{height(t_i) | i = 1..m\}$ if $t = f(t_1, \ldots, t_m)$. For any integer n, let us define $T^n(\mathcal{F})$ (resp. $T^{\leq n}(\mathcal{F})$) the set of trees whose height is equal to n (resp. $\leq n$).

Let \$ be a zero arity function symbol not in $\mathcal{F}_0$. A *context* is an element of $T(\mathcal{F} \cup \{\$\})$ such that the symbol \$ appears exactly once. We denote by $C(\mathcal{F})$ the set of all the contexts that can be defined over $\mathcal{F}$. Let t be a tree and let c be a context, $c[t]$ denotes the tree obtained by substituting the symbol \$ in the context c by the tree t. A subset A of $T(\mathcal{F})$ is *prefixial* if for any $c \in C(\mathcal{F})$ and any $t \in T(\mathcal{F})$, $c[t] \in A \Rightarrow t \in A$.

A *formal power tree series* on $T(\mathcal{F})$ is a mapping $r : T(\mathcal{F}) \to \mathbb{R}$. The set of all formal power series on $T(\mathcal{F})$ is denoted by $\mathbb{R}\langle\langle T(\mathcal{F})\rangle\rangle$. It is a vector space, when provided with addition and multiplication by a scalar.

Let V be a finite dimensional vector space over $\mathbb{R}$. We denote by $\mathcal{L}(V^p; V)$ the set of p-linear mappings from V^p to V. Let $\mathcal{L} = \cup_{p \geq 0} \mathcal{L}(V^p; V)$. We denote by V^* the dual space of V, i.e. the vector space composed of all the linear forms defined on V.

Definition 1. *A linear representation of $T(\mathcal{F})$ is a couple (V, μ), where V is a finite dimensional vector space over $\mathbb{R}$, and where $\mu : \mathcal{F} \to \mathcal{L}$ maps $\mathcal{F}_p$ into $\mathcal{L}(V^p; V)$ for each $p \geq 0$.*

Thus for each $f \in \mathcal{F}_p$, $\mu(f) : V^p \to V$ is p-linear. It can easily be shown that μ extends uniquely to a morphism $\mu : T(\mathcal{F}) \to V$ by the formula

$$\mu(f(t_1, \ldots, t_p)) = \mu(f)(\mu(t_1), \ldots, \mu(t_p)). \tag{1}$$

The μ function can be extended to work over contexts. Let $\overline{\mu} : C(\mathcal{F}) \to \mathcal{L}(V; V)$ be inductively by $\overline{\mu}(\$)(v) = v$ and $\overline{\mu}(f(t_1, \ldots, t_{i-1}, c, t_{i+1}, \ldots, t_n))(v) = \mu(f)(\mu(t_1), \ldots, \mu(t_{i-1}), \overline{\mu}(c)(v), \mu(t_{i+1}), \ldots, \mu(t_n))$.

It can be shown that for any context c and any term t, $\overline{\mu}(c)(\mu(t)) = \mu(c[t])$.

Let (V, μ) be a linear representation of $T(\mathcal{F})$ and let $V_{T(\mathcal{F})}$ be the vector subspace of V spanned by $\mu(T(\mathcal{F}))$. It can be shown that $(V_{T(\mathcal{F})}, \mu)$ is also a linear representation of $T(\mathcal{F})$. Let A be a prefixial subset of $T(\mathcal{F})$ and let V_A be the subspace of V spanned by $\mu(A)$. Suppose that for any integer m, any $f \in \mathcal{F}_m$ and any $t_1, \ldots, t_m \in A$, $\mu(f(t_1, \ldots, t_m)) \in V_A$. Then, $V_A = V_{T(\mathcal{F})}$. As a consequence, a basis of $V_{T(\mathcal{F})}$ can be extracted from $\mu(A)$. Therefore, given a linear representation (V, μ) of $T(\mathcal{F})$, a basis of $V_{T(\mathcal{F})}$ can be computed within polynomial time.

Definition 2. *Let r be a formal series over $T(\mathcal{F})$, r is a recognizable tree series if there exists a triple (V, μ, λ), where (V, μ) is a linear representation of $T(\mathcal{F})$, and $\lambda : V \to \mathbb{R}$ is a linear form, such that $r(t) = \lambda(\mu(t))$ for all t in $T(\mathcal{F})$.*

We say that (V, μ, λ) is *trimmed* if (i) $V = V_{T(\mathcal{F})}$ and (ii) $\forall v \in V \setminus \{0\}, \exists c \in C(\mathcal{F}), \lambda\overline{\mu}(c)(v) \neq 0$.

Rational tree series have been studied in [7]. It has been shown that the notions of recognizable tree series and rational tree series coincide. From now on, we shall refer to them by using the term of *rational* tree series. Note also that rational series on strings can be seen as particular cases of rational series on trees and hence, counterexamples designed in the first field can be directly exported in the second one.

Example 1. Let $\mathcal{F} = \{a, b, g(\cdot), f(\cdot, \cdot)\}$, let $V = \mathbb{R}^2$ and let (e_1, e_2) be a basis of V. We define μ, λ and r by:

$$\mu(a) = 2e_1/3, \mu(b) = e_2/2, \mu(g)(e_1) = e_2/2, \mu(g)(e_2) = 0,$$

$$\mu(f)(e_i, e_j) = \begin{cases} e_1/3 \text{ if } i = 1 \text{ and } j = 2 \\ 0 \text{ otherwise} \end{cases}$$

and

$$\lambda(e_1) = 1, \lambda(e_2) = 0 \text{ and } r(t) = \lambda\mu(t) \text{ for any term } t.$$

We have $\mu(f(a, b)) = \mu(f)(\mu(a), \mu(b)) = e_1/9$ and $\mu(f(a, g(a))) = \mu(f)(\mu(a),$ $\mu(g)(\mu(a))) = 2e_1/27$.

Hence, $r(a) = 2/3$, $r(b) = 0$, $r(f(a, b)) = 1/9$, $r(f(a, g(a))) = 2/27$.

Definition 3. *A* multiplicity tree automaton *(MA) over $\mathcal{F}$ is a tuple $\mathcal{A} = (Q, \mathcal{F}, \tau, \delta)$ where Q is a set of states, τ is a mapping from Q to $\mathbb{R}$ and δ is a mapping from $\cup_{m \geq 0} \mathcal{F}_m \times Q^m \times Q$ to $\mathbb{R}$.*

A multiplicity automaton is a device that can be used to compute tree series. They can be interpreted in a bottom-up or a top-down way, since $\delta(f, q_1, \ldots, q_m, q) = w$ can be rewritten as a bottom-up rule or a top-down rule:

$$f(q_1, \ldots, q_m) \xrightarrow{w} q \text{ or } q \xrightarrow{w} f(q_1, \ldots, q_m).$$

A *probabilistic tree automaton* (PA) is an MA $\mathcal{A} = (Q, \mathcal{F}, \tau, \delta)$ which satisfies the following conditions:

- δ and τ take their values in $[0, 1]$,
- $\sum_{q \in Q} \tau(q) = 1$,
- for any $q \in Q$, $\sum_{f(q_1, \ldots, q_m) \xrightarrow{w} q} w = 1$.

Multiplicity automata and linear representations are two equivalent ways to represent rational series. For example, let (V, μ, λ) be a linear representation of the formal series r defined on $T(\mathcal{F})$ and let $B = (e_1, \ldots, e_n)$ be a basis of V. A multiplicity automaton $\mathcal{A} = (Q, \mathcal{F}, \lambda, \delta)$ can be associated with (V, μ, λ, B) as follows:

- $Q = \{e_1, \ldots, e_n\}$,
- $\delta(f, e_{i_1}, \ldots, e_{i_m}, e_j) = w_j$ for any $f \in \mathcal{F}_m$ where $\mu(f)(e_{i_1}, \ldots, e_{i_m}) = \sum_j w_j e_j$.

Conversely, an equivalent linear representation can be associated with any multiplicity automaton.

Example 2. It can easily be shown that the linear representation described in Example 1 is equivalent to the probabilistic automaton defined by: $Q = \{e_1, e_2\}$, $\tau(e_1) = 1$, $\tau(e_2) = 0$ and

$$\delta = \{e_1 \xrightarrow{2/3} a, e_1 \xrightarrow{1/3} f(e_1, e_2), e_2 \xrightarrow{1/2} b, e_2 \xrightarrow{1/2} g(e_1)\}.$$

2.2   Rational Stochastic Tree Languages

Definition 4. *A stochastic tree language* over $T(\mathcal{F})$ *is a tree series* $r \in K\langle\langle T(\mathcal{F})\rangle\rangle$ *such that for any* $t \in T(\mathcal{F})$, $0 \leq r(t) \leq 1$ *and* $\sum_{t \in T(\mathcal{F})} r(t) = 1$.

Therefore, a *rational stochastic tree language* is a stochastic tree language which admits a linear representation. Stochastic languages that can be computed by a probabilistic automaton are rational. However, the converse is false: there exists a rational stochastic tree language that cannot be computed by a probabilistic automaton [4]. Moreover, it can be shown that the rational series computed by a PA is not always a stochastic language. For example, it can easily be shown that the PA defined by $Q = \{q\}$, $\tau(q) = 1$, $\delta = \{q \xrightarrow{\alpha} a, q \xrightarrow{1-\alpha} f(q,q)\}$ defines a stochastic language iff $\alpha \geq 1/2$. When $\alpha < 1/2$, $\sum_{t \in T(\mathcal{F})} r(t) < 1$ [12].

Let P be a stochastic tree language over $T(\mathcal{F})$. We consider infinite samples S composed of trees independently drawn according to P. For any integer m, let S_m be the sample composed of the m first elements of S. We denote by P_{S_m} the empirical distribution on $T(\mathcal{F})$ associated with S_m. Let $\mathcal{A} = (A_i)_{i \in I}$ be a family of subsets of $T(\mathcal{F})$. It can be shown [13,14] that for any confidence parameter δ and any integer m, with a probability greater than $1 - \delta$, for any $i \in I$,

$$|P_{S_m}(A_i) - P(A_i)| \leq C\sqrt{\frac{d - \log\frac{\delta}{4}}{m}}. \tag{2}$$

where d is the Vapnik-Chervonenkis dimension of $\mathcal{A}$ and C is a universal constant. In particular, with a probability greater than $1 - \delta$, for any $t \in T(\mathcal{F})$,

$$|P_{S_m}(t) - P(t)| \leq C\sqrt{\frac{1 - \log\frac{\delta}{4}}{m}}. \tag{3}$$

Let $\Psi(d, \epsilon, \delta) = \frac{C^2}{\epsilon^2}(d - \log\frac{\delta}{4})$. One can easily verify that if $m \geq \Psi(d, \epsilon, \delta)$, with a probability greater than $1 - \delta$, $|P_{S_m}(A_i) - P(A_i)| \leq \epsilon$ for any index i.

Borel-Cantelli Lemma states that if $(A_k)_{k \in \mathbb{N}}$ is a family of events such that $\sum_k P(A_k) < \infty$, the probability that a finite number of events A_k occur is equal to 1.

Check that for any α such that $-1/2 < \alpha < 0$ and any $\beta < -1$, if we define $\epsilon_k = k^\alpha$ and $\delta_k = k^\beta$, then there exists K such that for all $k \geq K$, we have $k \geq \Psi(1, \epsilon_k, \delta_k)$. For such choices of α and β, we have $\lim_{k \to \infty} \epsilon_k = 0$ and $\sum_{k \geq 1} \delta_k < \infty$. Therefore, from Borel-Cantelli Lemma, it can easily be shown that with probability 1, there exists K such that for any $k \geq K$, for any $t \in T(\mathcal{F})$,

$$|P_{S_k}(t) - P(t)| \leq \epsilon_k. \tag{4}$$

3   A Canonical Linear Representation for Rational Tree Series

The main goal of the paper is to show that any rational stochastic tree language P can be inferred in the limit from an infinite sample drawn according to P with probability 1. The first step is to define the *canonical linear representation* of a rational tree series r, whose components only depend on r.

3.1   Defining the Canonical Representation

Let $c \in C(\mathcal{F})$. We define the (linear) mapping $\dot{c} : \mathbb{R}\langle\langle T(\mathcal{F})\rangle\rangle \to \mathbb{R}\langle\langle T(\mathcal{F})\rangle\rangle$ by:

$$\dot{c}(r)(t) = r(c[t]).$$

Lemma 1. *Let (V, μ, λ) be a linear representation of the rational series r. For any context c, $\dot{c}r$ is rational and $(V, \overline{\mu}(c) \circ \mu, \lambda)$ is a linear representation of $\dot{c}r$.*

Proof. Indeed, for any term t, $\dot{c}r(t) = r(c[t]) = \lambda\mu(c[t]) = \lambda(\overline{\mu}(c) \circ \mu)(t)$. $\qquad\square$

Let r be a formal power series on $T(\mathcal{F})$. Let us denote by W_r the vector subspace of $\mathbb{R}\langle\langle T(\mathcal{F})\rangle\rangle$ spanned by $\{\dot{c}r | c \in C(\mathcal{F})\}$.

Lemma 2. *If r is rational, then the dimension of W_r is finite.*

Proof. Let (V, μ, λ) be a linear representation of r. For any context c, $\lambda\overline{\mu}(c) \in V^*$. Since the dimension of V^* is finite, there exist $c_1, \ldots c_n$ s. t. for any $c \in C(\mathcal{F})$, there exists $\alpha_1, \ldots, \alpha_n$ s.t. $\lambda\overline{\mu}(c) = \sum_i \alpha_i \lambda\overline{\mu}(c_i)$. Check that $\{\dot{c_1}r, \ldots, \dot{c_n}r\}$ spans W_r. $\qquad\square$

Let W_r^* be the dual space of W_r, i.e. the set of all linear forms defined over W_r. For any $t \in T(\mathcal{F})$, let $\overline{t} \in W_r^*$ be defined by: $\forall s \in W_r, \overline{t}(s) = s(t)$.

Lemma 3. *Let $f(u_1, \ldots, u_i, \ldots, u_p), t_1, \ldots, t_n \in T(\mathcal{F})$ and suppose that $\overline{u_i} = \sum_{j=1}^n \alpha_j \overline{t_j}$ for some index i. Then, $\overline{f(u_1, \ldots, u_i, \ldots, u_p)} = \sum_{j=1}^n \alpha_j \overline{f(u_1, \ldots, t_j, \ldots, u_p)}$.*

Proof. Let c_i be the context $f(u_1, \ldots, \$, \ldots, u_n)$ where $\$$ is at the i-th position. For any $s \in W_r$,

$$\overline{f(u_1, \ldots, u_i, \ldots, u_p)}(s) = \overline{u_i}(\dot{c_i}s) = \sum_{j=1}^n \alpha_j \overline{t_j}(\dot{c_i}s) = \sum_{j=1}^n \alpha_j \overline{f(u_1, \ldots, t_j, \ldots, u_p)}(s). \square$$

Suppose that the dimension of W_r is finite and let $\{c_1^{-1}r, \ldots, c_n^{-1}r\}$ be a basis of W_r. One can show that there exists n terms $t_1, \ldots, t_n$ such that the rank of the matrix $(c_i^{-1}r(t_j))_{1 \leq i,j \leq n}$ is n. Therefore, $(\overline{t_1}, \ldots, \overline{t_n})$ is a basis of W_r^*.

Let r be a rational series. We know that the dimension of W_r is finite. Let $t_1, \ldots, t_n$ be n terms such that $(\overline{t_1}, \ldots, \overline{t_n})$ is a basis of W_r^*. We define a linear representation (W_r^*, ν, τ) of r as follows:

- for any $f \in \mathcal{F}_p$, define $\nu(f) \in \mathcal{L}((W_r^*)^p; W_r^*)$ by $\nu(f)(\overline{t_{i_1}}, \ldots, \overline{t_{i_p}}) = \overline{f(t_{i_1}, \ldots, t_{i_p})}$.
- $\tau \in (W_r^*)^* = W_r$ by $\tau(\overline{t}) = r(t)$.

Lemma 4. *For any term $t \in T(\mathcal{F})$, $\nu(t) = \overline{t}$.*

Proof. Let $t = f(s_1, \ldots, s_p) \in T(\mathcal{F})$ and let $\overline{s_i} = \sum_{j=1}^n \alpha_i^j \overline{t_i}$. Using the previous lemma, we have

$$\nu(f)(\overline{s_1}, \ldots, \overline{s_p}) = \sum_{j_1, \ldots, j_p} \alpha_1^{j_1} \ldots \alpha_p^{j_p} \overline{f(t_{j_1}, \ldots, t_{j_p})} = \overline{f(s_1, \ldots, s_p)}$$

Remark that ν and τ do not depend on any basis chosen for W_r^*.

Theorem 1. (W_r^*, ν, τ) *is a trimmed linear representation of* r *which is called the* canonical linear representation *of* r.

Proof. For any term t, $\tau(\nu(t)) = \tau(\bar{t}) = r(t)$. Therefore, (W_r^*, ν, τ) is a linear representation of r. By construction, $\nu(T(\mathcal{F}))$ spans W_r^*. Now, let $w \in W_r^* \setminus \{0\}$ and let $\{\bar{t_1}, \dots, \bar{t_n}\}$ be a basis of W_r^*. There exist $\alpha_1, \dots, \alpha_n$ not all zero s.t. $w = \sum \alpha_i \bar{t_i}$. Since $\{\bar{t_1}, \dots, \bar{t_n}\}$ is linearly independent, there exists a context c such that $\sum \alpha_i \bar{t_i}(c) = \tau \bar{\nu}(c)(w) \neq 0$. Therefore, (W_r^*, ν, τ) is trimmed. $\square$

Given a total order $\leq$ on $T(\mathcal{F})$, there exists a unique subset B of $\nu(T(\mathcal{F}))$ which is a basis of W_r^* and such that for any $s \in T(\mathcal{F})$, $\bar{s} \in B$ or $\{\bar{s}\} \cup \{\bar{t} \in B | t \leq s\}$ is linearly dependent. We say that B is the canonical basis of W_r^* (wrt $\leq$).

3.2   Building the Canonical Representation

Given an n-dimensional trimmed linear representation (V, μ, λ) for r, it is possible to build the canonical representation of r in time polynomial in n^p where p is the maximal arity of symbols in $\mathcal{F}$. The proof of this result relies on the following lemma:

Lemma 5. *Let* (V, μ, λ) *be an* n-*dimensional trimmed linear representation* (V, μ, λ) *for* r *and let* $t_1, \dots, t_m \in T(\mathcal{F})$. *Then,* $\{\bar{t_1}, \dots, \bar{t_m}\}$ *is linearly independent iff* $\{\mu(t_1), \dots, \mu(t_m)\}$ *is linearly independent.*

Proof. Suppose that $\{\mu(t_1), \dots, \mu(t_m)\}$ is linearly independent in V and let $\alpha_1, \dots, \alpha_m$ be such that $\sum \alpha_{i=1}^m \bar{t_i} = 0$. For any context c, $\sum_i \alpha_i \bar{t_i}(c) = \sum_i r(c[t_i])$ $= \lambda \bar{\mu}(c)(\sum_i \alpha_i \mu(t_i)) = 0$. Therefore, $\sum_i \alpha_i \mu(t_i) = 0$ since (V, μ, λ) is trimmed and $\alpha_i = 0$ for $i = 1, \dots, n$ since $\{\mu(t_1), \dots, \mu(t_m)\}$ is linearly independent: hence, $\{\bar{t_1}, \dots, \bar{t_m}\}$ is linearly independent.

Suppose that $\{\mu(t_1), \dots, \mu(t_m)\}$ is linearly dependent and let $\sum_i \alpha_{i=1}^m \mu(t_i) = 0$ where the α_i are not all zero. For any context c,

$$\sum_i \alpha_i \bar{t_i}(c) = \sum_i \alpha_i r(c[t_i]) = \sum_i \alpha_i \lambda \mu(c[t_i]) = \lambda \bar{\mu}(c)(\sum_i \alpha_i \mu(t_i)) = 0.$$

Therefore, $\sum_{i=1}^m \alpha_i \bar{t_i} = 0$. $\square$

Proposition 1. *Given an* n-*dimensional trimmed linear representation* (V, μ, λ) *for the rational series* r, *a basis for* W_r^* *can be computed in time polynomial in* n^p.

Proof. One can verify that Algorithm 1 computes a basis of W_r^*. $\square$

One can remark that the linear representation is only used to check whether $B \cup \overline{f(t_1, \dots, t_p)}$ is linearly independent. Therefore, the linear representation can be replaced by an oracle that says whether $B \cup \overline{f(t_1, \dots, t_p)}$ is linearly independent.

Data   : A trimmed linear representation (V, μ, λ) for r
Result   : A basis B of W_r^*
begin
    $B \leftarrow \emptyset$;   is_a_basis $\leftarrow$ False;
    while *not is_a_basis* **do**
        is_a_basis $\leftarrow$ True;
        for *every* $f \in \mathcal{F}$ **do**
            let $p = arity(f)$;
            for $t_1, \ldots, t_p \in B$ **do**
                if $B \cup \overline{f(t_1, \ldots, t_p)}$ *is linearly independent* **then**
                    $B = B \cup \overline{f(t_1, \ldots, t_p)}$;   is_a_basis $\leftarrow$ False;

end

Algorithm 1. Building a canonical linear representation of r

Such an oracle could be achieved, in a variant of the MAT learning model of Angluin, by using a *membership oracle* which would compute $r(t)$ for any tree t and an *equivalence oracle* which would say whether the current representation computes r, and would provide a counterexample $(t, r(t))$ otherwise. See [9,10] for related work.

Example 3. Let us consider the previous example.

- $\overline{a} \neq 0$ since $\overline{a}(\$) = 2/3$.
- $\{\overline{a}, \overline{b}\}$ is linearly independent since $\overline{a}(f(a, \$)) = 0$ and $\overline{b}(f(a, \$)) = 1/9$.
- We have $\overline{f(a, a)} = \overline{g(b)} = \overline{f(b, a)} = \overline{f(b, b)} = 0$.
- We have also $\overline{g(a)} = 2\overline{b}/3$ and $\overline{f(a, b)} = \overline{a}/6$.

Therefore, $\{\overline{a}, \overline{b}\}$ is a basis of the canonical linear representation of r.

4   Inference of Rational Tree Series in the Limit

In this section, we show how to identify in the limit a canonical linear representation of a rational stochastic tree language P from an infinite sample S of trees independently drawn according to P.

Let (W^*, ν, τ) be the canonical linear representation of the target. Given a total order $\leq$ on $T(\mathcal{F})$ satisfying $height(t) < height(t') \Rightarrow t \leq t'$, the aim of the algorithm is to identify the canonical basis $B = \{\overline{t_1}, \ldots, \overline{t_n}\}$ of W^* associated with $\leq$. Let t_{max} be the maximal element of $\{t_1, \ldots, t_n\}$. Let S be an infinite sample independently drawn according to P and let S_m be the sample composed of the m first elements of S. We have to show that with probability one, there exists an integer N such that for any $m \geq N$, the following properties can be identified from S_m:

- $B = \{\overline{t_1}, \ldots, \overline{t_n}\}$ is linearly independent,
- for any $t \leq t_{max}$, $B \cup \{\overline{t}\}$ is linearly dependent,

– for any $f \in \mathcal{F}$ and any $1 \leq i_1, \ldots, i_p \leq n$, $B \cup \{\overline{f(t_{i_1}, \ldots, t_{i_p})}\}$ is linearly dependent, where p is the arity of f.

Given these relations, a linear representation (W^*, ν_m, τ_m) can be computed. Then, we have to show that the (multi-) linear mappings $\nu_m(f)$ for any $f \in \mathcal{F}$ and τ_m converge to the correct ones.

Since we are working on finite samples S_m, we cannot consider exact linear dependencies. Let T be a finite subset of $T(\mathcal{F})$, let S_m be a finite sample composed of m trees independently drawn from the target, let $t \in T(\mathcal{F})$, let $\{x_s | s \in T\}$ be a set of variables and let $\epsilon > 0$. We denote by $I(T, t, S_m, \epsilon)$ the following set of inequalities:

$$I(T, t, S_m, \epsilon) = \{|\overline{t}(\dot{c}P_S) - \sum_{s \in T} x_s \overline{s}(\dot{c}P_S)| \leq \epsilon | c \in C(S_m)\}$$

where P_S is the empirical distribution on S_m and where $C(S) = \{c \in C(\mathcal{F}) | \exists t \in T(\mathcal{F}) \text{ s.t. } c[t] \in S_m\}$.

Let S be an infinite sample of the target P. Suppose that $\{\overline{t}\} \cup \{\overline{s} | s \in T\}$ is linearly independent. We show that, with probability 1, there exists $\epsilon > 0$ and a sample size from which $I(T, t, S_m, \epsilon)$ has no solution.

Lemma 6. *Let P be a stochastic language and let $\{t_0, t_1, \ldots, t_n\}$ be a set of trees such that $\{\overline{t_0}, \overline{t_1}, \ldots, \overline{t_n}\}$ is linearly independent. Then, with probability one, for any infinite sample S of P, there exists a positive number ϵ and an integer M such that for every $m \geq M$, $I(\{t_1, \ldots, t_n\}, t_0, S_m, \epsilon)$ has no solution.*

Proof. Let S be an infinite sample of P. Suppose that for every $\epsilon > 0$ and every integer M, there exists $m \geq M$ such that $I(\{t_1, \ldots, t_n\}, t_0, S_m, \epsilon)$ has a solution. Then, for any integer k, there exists $m_k \geq k$ such that $I(\{t_1, \ldots, t_n\}, t_0, S_{m_k}, 1/k)$ has a solution $(\alpha_{1,k}, \ldots, \alpha_{n,k})$.

Let $\rho_k = Max\{1, |\alpha_{1,k}|, \ldots, |\alpha_{n,k}|\}$, $\gamma_{0,k} = 1/\rho_k$ and $\gamma_{i,k} = -\alpha_{i,k}/\rho_k$ for $1 \leq i \leq n$. For every k, $Max\{|\gamma_{i,k}| : 0 \leq i \leq n\} = 1$. Check that for any context $c: \forall k \geq 0, \left|\sum_{i=0}^{n} \gamma_{i,k} \overline{t_i}(\dot{c}P_{S_{m_k}})\right| \leq \frac{1}{\rho_k k} \leq \frac{1}{k}$.

There exists a subsequence $(\alpha_{1,\phi(k)}, \ldots, \alpha_{n,\phi(k)})$ of $(\alpha_{1,k}, \ldots, \alpha_{n,k})$ such that $(\gamma_{0,\phi(k)}, \ldots, \gamma_{n,\phi(k)})$ converges to $(\gamma_0, \ldots, \gamma_n)$. We show below that we should have $\sum_{i=0}^{n} \gamma_i \overline{t_i}(\dot{c}P) = 0$ for every context c, which is contradictory with the independence assumption since $Max\{\gamma_i : 0 \leq i \leq n\} = 1$ and hence, some γ_i is not zero.

Let $c \in C(\mathcal{F})$. With probability 1, there exists an integer k_0 such that $c \in C(S_{m_k})$ for any $k \geq k_0$. For such a k, we can write

$$\gamma_i \overline{t_i}(\dot{c}P) = (\gamma_i \overline{t_i}(\dot{c}P) - \gamma_i \overline{t_i}(\dot{c}P_{S_{m_k}})) + (\gamma_i - \gamma_{i,\phi(k)}) \overline{t_i}(\dot{c}P_{S_{m_k}}) + \gamma_{i,\phi(k)} \overline{t_i}(\dot{c}P_{S_{m_k}})$$

and therefore $\left|\sum_{i=0}^{n} \gamma_i \overline{t_i}(\dot{c}P)\right| \leq \sum_{i=0}^{n} |\overline{t_i}(\dot{c}P - \dot{c}P_{S_{m_k}})| + \sum_{i=0}^{n} |\gamma_i - \gamma_{i,\phi(k)}| + \frac{1}{k}$ which converges to 0 when k tends to infinity. □

Let S be an infinite sample of the target P. Suppose that $\overline{t} = \sum_{s \in T} \alpha_s \overline{s}$. We show that, with probability 1, for any $\gamma \in]-1/2, 0[$, there exists a sample size M from which, $I(T, t, S_m, m^\gamma)$ has a solution for any $m \geq M$.

Lemma 7. *Let P be a stochastic language and let $t_0, t_1, \ldots, t_n$ be a set of trees such that there exist $\alpha_1, \ldots, \alpha_n \in \mathbb{R}$ such that $\overline{t_0} = \sum_{i=1}^{n} \alpha_i \overline{t_i}$. Then, for any $\gamma \in]-1/2, 0[$, with probability one, for any infinite sample S of P, there exists K s.t. $I(\{t_1, \ldots, t_n\}, t_0, S_k, k^\gamma)$ has a solution for every $k \geq K$.*

Proof. Let S an infinite sample of P. Let $\alpha_0 = 1$ and let $R = Max\{|\alpha_i| : 0 \leq i \leq n\}$. With probability one, there exists K_1 s.t. $\forall k \geq K_1$, $k \geq \Psi(1, [k^\gamma(n + 1)R]^{-1}, [(n + 1)k^2]^{-1})$ (see definition of Ψ in Section 2). Let $k \geq K_1$, for any $c \in C(\mathcal{F})$,

$$|\overline{t_0}(\dot{c}P_{S_k}) - \sum_{i=1}^{n} \alpha_i \overline{t_i}(\dot{c}P_{S_k})| \leq |\overline{t_0}(\dot{c}P_{S_k}) - \overline{t_0}(\dot{c}P)| + \sum_{i=1}^{n} |\alpha_i||\overline{t_i}(\dot{c}P_{S_k}) - \overline{t_i}(\dot{c}P)|.$$

From the definition of Ψ, with probability greater than $1 - \frac{1}{k^2}$, for any $i = 0, \ldots, n$ and any context c, $|\overline{t_i}(\dot{c}P_{S_k}) - \overline{t_i}(\dot{c}P)| \leq [k^{-\gamma}(n + 1)R]^{-1}$ and therefore $|\overline{t_0}(\dot{c}P_{S_k}) - \sum_{i=1}^{n} \alpha_i \overline{t_i}(\dot{c}P_{S_k})| \leq k^\gamma$. For any integer $k \geq K_1$, let E_k be the event: $|\overline{t_0}(\dot{c}P_{S_k}) - \sum_{i=1}^{n} \alpha_i \overline{t_i}(\dot{c}P_{S_k})| > k^\gamma$. Since $Pr(E_k) < 1/k^2$, from the Borel-Cantelli Lemma, the probability that a finite number of E_k occurs is 1.

Therefore, with probability 1, there exists an integer K such that for any $k \geq K$, $I(\{t_1, \ldots, t_n\}, t_0, S_k, k^\gamma)$ has a solution. $\square$

In the next lemma, we focus on the convergence of the parameters found when resolving an inequation system.

Lemma 8. *Let $P \in \mathcal{S}(T(\mathcal{F}))$, let $t_0, t_1, \ldots, t_n$ such that $\{\overline{t_1}, \ldots, \overline{t_n}\}$ is linearly independent and let $\alpha_1, \ldots, \alpha_n \in \mathbb{R}$ be such that $\overline{t_0} = \sum_{i=1}^{n} \alpha_i \overline{t_i}$. Then, for any $\gamma \in]-1/2, 0[$, with probability one, for any infinite sample S of P, there exists an integer K such that $\forall k \geq K$, any solution $\widehat{\alpha_1}, \ldots, \widehat{\alpha_n}$ of $I(\{t_1, \ldots, t_n\}, t_0, S_k, k^\gamma)$ satisfies $|\widehat{\alpha_i} - \alpha_i| < O(k^\gamma)$ for $1 \leq i \leq n$.*

Proof. Let $c_1, \ldots, c_n \in C(\mathcal{F})$ be such that the square matrix M defined by $M[i, j] = \overline{t_j}(\dot{c_i}P)$ for $1 \leq i, j \leq n$ is invertible. Let $A = (\alpha_1, \ldots, \alpha_n)^t$, $U = (\overline{t_0}(\dot{c_1}P), \ldots, \overline{t_0}(\dot{c_n}P))^t$. We have $M \times A = U$. Let S be an infinite sample of P, let $k \in \mathbb{N}$ and let $\widehat{\alpha_1}, \ldots, \widehat{\alpha_n}$ be a solution of $I(\{t_1, \ldots, t_n\}, t_0, S_k, k^\gamma)$. Let M_k be the square matrix defined by $M_k[i, j] = \overline{t_j}(\dot{c_i}P_{S_k})$ for $1 \leq i, j \leq n$, let $A_k = (\widehat{\alpha_1}, \ldots, \widehat{\alpha_n})^t$ and $U_k = (\overline{t_0}(\dot{c_1}P_{S_k}), \ldots, \overline{t_0}(\dot{c_n}P_{S_k}))^t$. We have

$$\|M_k A_k - U_k\|^2 = \sum_{i=1}^{n} [\overline{t_0}(\dot{c_i}P_{S_k}) - \sum_{j=1}^{n} \widehat{\alpha_j}\overline{t_j}(\dot{c_i}P_{S_k})]^2 \leq nk^{2\gamma}.$$

Check that $A - A_k = M^{-1}(MA - U + U - U_k + U_k - M_k A_k + M_k A_k - MA_k)$ and therefore, for any $1 \leq i \leq n$

$$|\alpha_i - \widehat{\alpha_i}| \leq \|A - A_k\| \leq \|M^{-1}\|(\|U_0 - U_k\| + n^{1/2}k^\gamma + \|M_k - M\|\|A_k\|).$$

Now, by using Equation 4 and Borel-Cantelli Lemma as in the proof of Lemma 7, with probability 1, there exists K such that for all $k \geq K$, $\|U_0 - U_k\| < O(k^\gamma)$ and $\|M_k - M\| < O(k^\gamma)$. Therefore, for all $k \geq K$, any solution $\widehat{\alpha_1}, \ldots, \widehat{\alpha_n}$ of $I(\{t_1, \ldots, t_n\}, t_0, S_k, k^\gamma)$ satisfies $|\widehat{\alpha_i} - \alpha_i| < O(k^\gamma)$ for $1 \leq i \leq n$. $\square$

Data   : S a finite sample of k trees, $\gamma \in]-1/2, 0[$
Result   : a linear representation (V, λ, μ)
begin

    $a_0 \leftarrow min(\mathcal{F}_0 \cap Subtrees(S));$
    $B \leftarrow \{\overline{a_0}\}; \quad \mu(a_0) \leftarrow \overline{a_0}; \quad \lambda_{\overline{a_0}} \leftarrow P_s(a_0);$
    $FS \leftarrow \bigcup_{f \in \mathcal{F}_p, p \geq 0}\{f(t_{j_1}, \ldots, t_{j_p}) | \overline{t_{i_j}} \in B\}; \quad FS \leftarrow FS \backslash \{a_0\};$
    while $FS \neq \emptyset$ **do**

        $t \leftarrow min(FS); \quad FS \leftarrow FS \backslash \{t\};$
        if $I(B, t, S, k^\gamma)$ *has no solution* **then**
            $B \leftarrow B \cup \{\overline{t}\}; \quad \mu(t) \leftarrow \overline{t}; \quad \lambda_{\overline{t}} \leftarrow P_S(t);$
            $FS \leftarrow FS \bigcup_{f \in \mathcal{F}_p, p \geq 1}\{f(t_{j_1}, \ldots, t_{j_p}) | \overline{t_{j_i}} \in B\};$

        else
            Let $(\alpha_{t_i})_{t_i \in B}$ a solution of I; $\quad \mu(t) \leftarrow \sum_{t_i \in B} \alpha_{t_i} \overline{t_i};$

end

Algorithm 2. Learning algorithm Algo(S, γ)

The learning algorithm is presented in Algorithm 2 and works as follows. We suppose that a total order is defined over $T(\mathcal{F})$ such that $height(t) < height(t') \Rightarrow t \leq t'$. To begin with, we extract the first constant symbol a_0 of the learning sample and we put it in the basis set B. We define the frontier set (FS) to be composed of all the trees of the form $f(a_0, \ldots, a_0)$. Note that FS contains all the constant symbols different from a_0. Then, the algorithm processes the frontier set while it is not empty. For each tree t in this set, we check if it can approximately be expressed according to a linear combination of the elements of the current basis. If the answer is no, we add t to the basis and we enlarge the frontier set by adding all the trees of the form $f(t_1, \ldots, t_m)$ where every $t_i \in B$. Otherwise, we use the linear relation obtained from the inequation system to complete the definition of μ.

We can now present the theorem of convergence in the limit.

Theorem 2. *Let P be a rational stochastic tree language defined on $T(\mathcal{F})$, let (V, μ, λ) be the canonical linear representation of P, let $B = \{\overline{t_1}, \ldots, \overline{t_n}\}$ the canonical basis of V (associated with some known total order on $T(\mathcal{F})$) and let $\gamma \in]-1/2, 0[$. Then, with probability one, for any infinite sample S of P, there exists an integer K such that for any $k \geq K$, Algo(S_k, γ) identifies B. Moreover, let (V, μ_k, λ_k) be the linear representation output by the algorithm. There exists a constant C such that $|\mu_k(f)(t_{i_1}, \ldots, t_{i_n}) - \mu(f)(t_{i_1}, \ldots, t_{i_n})| \leq Ck^\gamma$ and $|\lambda_k(t_i) - \lambda(t_i)| \leq Ck^\gamma$ for any $f \in \mathcal{F}$ and any elements t_i, t_{i_j} of B.*

Proof. Lemmas 6 and 7 prove that the basis B will be identified from some step with probability one. Lemma 8 can then be used to prove the last part of the theorem. $\square$

When P is a rational stochastic tree language which takes its values in the set of rational numbers $\mathbb{Q}$, the algorithm can be completed to exactly identify it. The proof is based on the representation of real numbers by continuous fractions. See [15] for a survey on continuous fractions and [16] for a similar application.

Let (ϵ_n) be a sequence of non negative real numbers which converges to 0, let $x \in \mathbb{Q}$, let (y_n) be a sequence of elements of $\mathbb{Q}$ such that $|x - y_n| \leq \epsilon_n$ for all but finitely many n. It can be shown that there exists an integer N such that, for any $n \geq N$, x is the unique rational number $\frac{p}{q}$ which satisfies $\left|y_n - \frac{p}{q}\right| \leq \epsilon_n \leq \frac{1}{q^2}$. Moreover, the unique solution of these inequalities can be computed from y_n.

Let P be a rational stochastic tree language which takes its values in $\mathbb{Q}$, let $\gamma \in]-1/2, 0[$, let S be an infinite sample of P and let (V, μ_k, λ_k) the linear representation output by the algorithm on input (S_k, γ). Let (V, μ'_k, λ'_k) be the representation derived from (V, μ_k, λ_k) by replacing every parameter $\alpha_k = \mu_k(\overline{f(t_{i_1}, \ldots, t_{i_n})})$ or $\alpha_k = \lambda_k(t_i)$ with a solution $\frac{p}{q}$ of $\left|\alpha_k - \frac{p}{q}\right| \leq k^{\gamma/2} \leq \frac{1}{q^2}$ and let $Algo'$ be the corresponding algorithm.

Theorem 3. *Let P be a rational stochastic tree language which takes its values in $\mathbb{Q}$, let $\gamma \in]-1/2, 0[$, and let (V, μ, λ) be its canonical linear representation. Then, with probability one, for any infinite sample S of P, there exists an integer K such that $\forall k \geq K$, $Algo'(S_k, \gamma)$ returns (V, μ, λ).*

Proof. From the previous theorem, for every parameter α of (V, μ, λ), the corresponding parameter α_k in (V, μ_k, λ_k) satisfies $|\alpha - \alpha_k| \leq Ck^\gamma$ for some constant C, from some step k, with probability one. Therefore, if k is sufficiently large, we have $|\alpha - \alpha_k| \leq k^{\gamma/2}$ and there exists an integer K such that $\alpha = p/q$ is the unique solution of $\left|\alpha - \frac{p}{q}\right| \leq k^{\gamma/2} \leq \frac{1}{q^2}$. Therefore, the parameter corresponding to α in the linear representation output by $Algo'(S_k, \gamma)$ is α itself. $\square$

Example 4. To illustrate the principle of our algorithm. Consider the following learning sample made up of 20 trees (the number of occurrences of each term is indicated inside brackets):

$$\{a[13], f(a, b)[4], f(a, g(a))[1], f(a, g(f(a, g(a))))[1], f(f(f(a, g(a)), b), b)[1]\}.$$

In a first step the algorithm puts $\overline{a}$ in the basis and sets $\mu(a) = \overline{a}$.

Next, the algorithm considers the constant symbol b. To check if $\overline{b}$ should belong to the basis, the algorithm constructs a set of inequations with the contexts definable in the learning set. For sake of simplicity, we will not consider all the contexts, but only 3 of them $c_0 = \$$, $c_1 = f(\$, b)$, $c_2 = f(a, \$)$. We obtain the following inequation system:

$$\left|\overline{b}(\dot{c}_0 p_S) - X_{\overline{a}}\overline{a}(\dot{c}_0 p_S)\right| = \left|p_S(c_0[b]) - X_{\overline{a}}p_S(c_0[a])\right| = \left|0 - X_{\overline{a}}\tfrac{13}{20}\right| \leq \epsilon$$
$$\left|\overline{b}(\dot{c}_1 p_S) - X_{\overline{a}}\overline{a}(\dot{c}_1 p_S)\right| = \left|p_S(c_1[b]) - X_{\overline{a}}p_S(c_1[a])\right| = \left|\tfrac{4}{20} - X_{\overline{a}}0\right| \leq \epsilon$$
$$\left|\overline{b}(\dot{c}_2 p_S) - X_{\overline{a}}\overline{a}(\dot{c}_2 p_S)\right| = \left|p_S(c_2[b]) - X_{\overline{a}}p_S(c_2[a])\right| = \left|0 - X_{\overline{a}}\tfrac{4}{20}\right| \leq \epsilon$$

If we set ϵ to 0.1, the systems admits no solution and then $\overline{b}$ is added to the basis with $\lambda_{\overline{b}} = 0$.

The algorithm examines the terms $f(a,a)$, $g(a)$, $f(a,b)$, $f(b,a)$, $f(b,b)$, $g(b)$. Since, the values of p_S according to the 3 contexts is null for $f(a,a)$ $f(b,a)$, $f(b,b)$ and $g(b)$ we do not show the inequation systems.

For $g(a)$ the system obtained is:

$$|\overline{g(a)}(\dot{c}_0 p_S) - X_{\overline{a}}\overline{a}(\dot{c}_0 p_S) - X_{\overline{b}}\overline{b}(\dot{c}_0 p_S)| = |0 - X_{\overline{a}}\tfrac{13}{20} - X_{\overline{a}}0| \le \epsilon$$
$$|\overline{g(a)}(\dot{c}_1 p_S) - X_{\overline{a}}\overline{a}(\dot{c}_1 p_S) - X_{\overline{b}}\overline{b}(\dot{c}_1 p_S)| = |0 - X_{\overline{a}}\tfrac{4}{20} - X_{\overline{b}}| \le \epsilon$$
$$|\overline{g(a)}(\dot{c}_2 p_S) - X_{\overline{a}}\overline{a}(\dot{c}_2 p_S) - X_{\overline{b}}\overline{b}(\dot{c}_2 p_S)| = |\tfrac{1}{20} - X_{\overline{a}}0 - X_{\overline{b}}\tfrac{4}{20}| \le \epsilon$$

$X_{\overline{a}} = 0$ and $X_{\overline{b}} = \tfrac{1}{4}$ is a solution of the system, then the algorithm sets $\mu(g)(\overline{a}) = \tfrac{1}{4}\overline{b}$.

For $f(a,b)$, the inequation system is:

$$|\overline{f(a,b)}(\dot{c}_0 p_S) - X_{\overline{a}}\overline{a}(\dot{c}_0 p_S) - X_{\overline{b}}\overline{b}(\dot{c}_0 p_S)| = |\tfrac{4}{20} - X_{\overline{a}}\tfrac{13}{20} - X_{\overline{a}}0| \le \epsilon$$
$$|\overline{f(a,b)}(\dot{c}_1 p_S) - X_{\overline{a}}\overline{a}(\dot{c}_1 p_S) - X_{\overline{b}}\overline{b}(\dot{c}_1 p_S)| = |0 - X_{\overline{a}}\tfrac{4}{20} - X_{\overline{b}}0| \le \epsilon$$
$$|\overline{f(a,b)}(\dot{c}_2 p_S) - X_{\overline{a}}\overline{a}(\dot{c}_2 p_S) - X_{\overline{b}}\overline{b}(\dot{c}_2 p_S)| = |0 - X_{\overline{a}}0 - X_{\overline{b}}\tfrac{4}{20}| \le \epsilon$$

$X_{\overline{a}} = \tfrac{4}{13}$ and $X_{\overline{b}} = 0$ is a solution of the system, then the algorithm sets $\mu(f)(\overline{a},\overline{b}) = \tfrac{4}{13}\overline{a}$. The representation obtained is finally:

$$\mu(a) = \overline{a}, \quad \mu(b) = \overline{b}, \quad \mu(g)(\overline{a}) = \tfrac{1}{4}\overline{b}, \quad \mu(f)(\overline{a},\overline{b}) = \tfrac{4}{13}\overline{a}, \quad \lambda_{\overline{a}} = \tfrac{13}{20}, \quad \lambda_{\overline{b}} = 0.$$

5   Discussion, Future Work and Conclusion

We have proved a theoretical result: rational stochastic tree languages are identifiable in the limit with probability one. The inference algorithm we use runs within polynomial time and approximates the parameters of the model with usual statistical rates of convergence. How can it be used in practical cases? Can it be improved?

First of all, the algorithm highly relies on an inequation system which aims at detecting linear combinations

$$I(T, t, S_n, \epsilon) = \{|\overline{t}(\dot{c}P_{S_n}) - \sum_{s \in T} x_s \overline{s}(\dot{c}P_{S_n})| \le \epsilon | c \in C(S_n)\}.$$

However, this system uses contexts which can be poorly represented in current samples. We can overcome this drawback by using *generalized contexts*, i.e. contexts containing several variables.

Let $\$_0, \$_1, \ldots, \$_k$ be zero arity function symbols not in $\mathcal{F}_0$. A generalized context is an element of $T(\mathcal{F} \cup \{\$_0, \$_1, \ldots, \$_k\})$ such that $\$_0$ appears exactly once and each other new symbol appears at most once. Now, for any stochastic languages P and any generalized context c, we define

$$\overline{t}(\dot{c}P) = \dot{c}P(t) = \sum_{t_1, \ldots, t_k \in T(\mathcal{F})} P(c[\$_0 \leftarrow t, \$_1 \leftarrow t_1, \ldots, \$_k \leftarrow t_k]).$$

We can then replace the inequation system $I(T, t, S_n, \epsilon)$ with

$$I(T, t, S_n, \epsilon) = \{|\overline{t}(\dot{c}P_{S_n}) - \sum_{s \in T} x_s \overline{s}(\dot{c}P_{S_n})| \le \epsilon | c \in C_k^g(S_n)\}$$

where $C_k^g(S_n)$ is the set of generalized context with k variables occurring in S_n.

If the number of new variables in not bounded, the VC-dimension of the set of generalized contexts is unbounded. However, it can easily be shown that the VC-dimension of the set of generalized contexts with k variables is bounded by $2k+1$. Therefore, we can adjust the number of variables to the size of the current learning sample in the inference algorithm in order to avoid overfitting.

Next, the rational series r output by the inference algorithm is not a stochastic language. Moreover, it may happen that the sum $\sum_{t \in T(\mathcal{F})} r(t)$ diverges. We conjecture that as soon as the size of the learning sample is large enough, with a high probability, the sum $\sum_{t \in T(\mathcal{F})} r(t)$ is absolutely convergent, i.e. $\sum_{t \in T(\mathcal{F})} |r(t)|$ converges. Moreover, let (V, μ, λ) be the canonical linear representation of a rational tree series r and let $B = \{\overline{t_1}, \ldots, \overline{t_n}\}$ be a basis of V. For any tree t and any index i, let α_i^t be such that $\overline{t} = \sum_{i=1}^n \alpha_i^t \overline{t_i}$. We have $r(t) = \sum_{i=1}^n \alpha_i^t r(t_i)$. We also conjecture that $\sum_{t \in T(\mathcal{F})} \alpha_i^t$ is absolutely convergent for any index i so that, $s_i = \sum_{t \in T(\mathcal{F})} \alpha_i^t$ is defined without ambiguity. One can show that s_i can be efficiently estimated.

Given these properties, it is possible to normalize the linear representation output by the algorithm in such a way that it computes a series $\overline{r}$ satisfying $\sum_{t \in T(\mathcal{F})} |\overline{r}(t)| < \infty$ and $\sum_{t \in T(\mathcal{F})} \overline{r}(t) = 1$. Let (V, μ_N, λ_N) be defined by

- $\forall f \in \mathcal{F}$, $[\mu_N(f)(\overline{t_{j_1}}, \ldots, \overline{t_{j_p}})]_i = [\mu(f)(\overline{t_{j_1}}, \ldots, \overline{t_{j_p}})]_i \cdot \pi_{k=1}^p s_{j_k}/s_i$.
- $\lambda_N(\overline{t_i}) = \lambda(\overline{t_i}) \times s_i$ for any element of λ_N.

It can easily be shown that (V, μ_N, λ_N) computes r and that $\sum_{\overline{t_{j_1}}, \ldots, \overline{t_{j_p}} \in B} [\mu_N(f(\overline{t_{j_1}}, \ldots, \overline{t_{j_p}}))]_i = 1$.

We can then adjust the linear form λ by multiplying each of its coordinates by a constant in order to get a series $\overline{r}$ which sums to 1.

However, it may happen that the series $\overline{r}$ takes negative values. We call such a series, a *pseudo-stochastic language*. From these languages, we can extract a probability distribution $P_{\overline{r}}$ such that $P_{\overline{r}}(t) = 0$ if $\overline{r}(t) < 0$ and otherwise $P_{\overline{r}}(t) = b_t \overline{r}(t)$ with a normalization that compensates the loss of the negative values. We may compute this distribution iteratively when developing a tree. Suppose that at a given step, we are building a tree with some leaves labeled by states. We choose to develop a new branch from any of these states. We consider all the transitions leaving from the considered state grouped by symbols. If all the possible expansions with a given symbol lead to a negative value, then we omit this symbol and we renormalized the probabilities of the other expansions. Note that when r defines a stochastic language, $P_{\overline{r}} = r$ since there will be no negative values. See [6] for a more detailed description of this point, in the case of pseudo-stochastic languages defined on strings.

To conclude, we have studied in this paper the inference of a stochastic tree language P from a sample of trees independently drawn according to P. We have proposed to work in the class of rational stochastic tree languages that are stochastic languages computed by rational tree series. We have presented two contributions. First, we have shown that rational tree series admit a canonical linear representation. Then, we have proposed an inference algorithm which identifies in the limit the class of rational stochastic tree languages. Our future work will concern improvements of our approach in practical cases as evoked in the previous discussion.

References

1. Carrasco, R., Oncina, J., Calera-Rubio, J.: Stochastic inference of regular tree languages. Machine Learning 44(1/2), 185–197 (2001)
2. Rico-Juan, J., Calera, J., Carrasco, R.: Probabilistic k-testable tree-languages. In: Oliveira, A.L. (ed.) ICGI 2000. LNCS (LNAI), vol. 1891, pp. 221–228. Springer, Heidelberg (2000)
3. Abe, N., Mamitsuka, H.: Predicting protein secondary structure using stochastic tree grammars. Machine Learning Journal 29(2-3), 275–301 (1997)
4. Denis, F., Esposito, Y.: Rational stochastic languages. Technical report, LIF - Université de Provence (2006), http://hal.ccsd.cnrs.fr/ccsd-00019728
5. Denis, F., Esposito, Y., Habrard, A.: Learning rational stochastic languages. In: Lugosi, G., Simon, H.U. (eds.) COLT 2006. LNCS (LNAI), vol. 4005, Springer, Heidelberg (2006)
6. Habrard, A., Denis, F., Esposito, Y.: Using pseudo-stochastic rational languages in probabilistic grammatical inference. In: Sakakibara, Y., Kobayashi, S., Sato, K., Nishino, T., Tomita, E. (eds.) ICGI 2006. LNCS (LNAI), vol. 4201, pp. 112–124. Springer, Heidelberg (2006)
7. Berstel, J., Reutenauer, C.: Recognizable formal power series on trees. Theoretical Computer Science 18, 115–148 (1982)
8. Ésik, Z., Kuich, W.: Formal tree series. Journal of Automata, Languages and Combinatorics 8(2), 219–285 (2003)
9. Habrard, A., Oncina, J.: Learning multiplicity tree automata. In: Sakakibara, Y., Kobayashi, S., Sato, K., Nishino, T., Tomita, E. (eds.) ICGI 2006. LNCS (LNAI), vol. 4201, Springer, Heidelberg (2006)
10. Drewes, F., Vogler, H.: Learning deterministically recognizable tree serie. Journal of Automata, Languages and Combinatorics (2007)
11. Comon, H., Dauchet, M., Gilleron, R., Jacquemard, F., Lugiez, D., Tison, S., Tommasi, M.: Tree Automata Techniques and Applications (1997), Available from: http://www.grappa.univ-lille3.fr/tata
12. Wetherell, C.S.: Probabilistic languages: A review and some open questions. ACM Comput. Surv. 12(4), 361–379 (1980)
13. Vapnik, V.: Statistical Learning Theory. John Wiley, New York, NY (1998)
14. Lugosi, G.: Pattern classification and learning theory. In: Principles of Nonparametric Learning, Springer, Heidelberg (2002)
15. Hardy, G., Wright, M.: An introduction to the theory of numbers. Oxford University Press, Oxford (1979)
16. Denis, F., Esposito, Y.: Learning classes of probabilistic automata. In: Shawe-Taylor, J., Singer, Y. (eds.) COLT 2004. LNCS (LNAI), vol. 3120, pp. 124–139. Springer, Heidelberg (2004)

One-Shot Learners Using Negative Counterexamples and Nearest Positive Examples

Sanjay Jain[1,*] and Efim Kinber[2]

[1] School of Computing, National University of Singapore, Singapore 117590
sanjay@comp.nus.edu.sg

[2] Department of Computer Science, Sacred Heart University, Fairfield, CT
06432-1000, U.S.A.
kinbere@sacredheart.edu

Abstract. As some cognitive research suggests, in the process of learning languages, in addition to *overt* explicit negative evidence, a child often receives *covert* explicit evidence in form of corrected or rephrased sentences. In this paper, we suggest one approach to formalization of overt and covert evidence within the framework of *one-shot* learners via subset and membership queries to a teacher (oracle). We compare and explore general capabilities of our models, as well as complexity advantages of learnability models of one type over models of other types, where complexity is measured in terms of number of queries. In particular, we establish that "correcting" positive examples give sometimes more power to a learner than just negative (counter)examples and access to full positive data.

1   Introduction

There are two major formal models that have been used over the years to address various aspects of human learning: Gold's model [Gol67] of *identification in the limit*, that treats learning as a limiting process of creating and modifying hypotheses about the target concept, and Angluin's model [Ang88] of learning via queries that views learning as a finite (rather than an infinite limiting) process, however, allowing interaction between a learner and a teacher (formally, an oracle) in form of questions and answers. Unlike in Gold's model, the learner in the latter model cannot change its mind: it can ask a finite number of questions, but, ultimately, it must produce a sole right conjecture (note that a different approach to learning via oracles was suggested in [GS91]). Such learners have been named *one-shot* in [LZ04]. There has been a good deal of research on one-shot learners using primarily *superset* queries (when a learner asks if a particular language is a superset of the target concept) and *disjointness* queries (when a learner asks if a particular language is disjoint with the target concept) ([LZ05, JLZ05]). In this paper, we study and compare one-shot learners that receive different types of answers to *subset* and *membership* queries. Learners making subset queries

---

* Supported in part by NUS grant number R252-000-212-112 and 251RES070107.

M. Hutter, R.A. Servedio, and E. Takimoto (Eds.): ALT 2007, LNAI 4754, pp. 257–271, 2007.

(testing if a particular language is a subset of the target concept) are concerned with a possibility of *overgeneralizing* — that is, including into conjecture data not belonging to the target concept. Membership queries test if a particular datum belongs to the target concept — it is, perhaps, the most natural type of a question to the teacher. We refer to these models as **SubQ** and, respectively, **MemQ**.

While for a membership query, a natural answer would be just *yes* or *no*, a learner making a subset query can also receive a *negative counterexample* showing where a learner errs. In her original model [Ang88], D. Angluin suggested that a learner could receive an *arbitrary* negative counterexample for a subset query, when several negative counterexamples were possible. In addition to this, traditional, type of answers to subset queries (considered in several variants of models using subset queries, e.g., in [JK07a]), we also consider the following types of answers:

— a learner receives the *least* negative counterexample (this type of counterexamples was considered, in particular, in [JK07a]); we refer to this model as **LSubQ**.
— in addition to a negative counterexample, a learner receives a "correction", the positive example *nearest* to the negative one; this approach stems from the following observation discussed, in particular, in [RP99]: while learning a language, in addition to *overt* explicit negative evidence (when a parent points out that a certain statement by a child is grammatically incorrect), a child often receives also *covert* explicit evidence in form of corrected or rephrased utterances. As languages in our learning models are represented by subsets of the set of all positive integers, our concept of the nearest positive example seems to be appropriate in the given context. We apply the same idea to membership queries: we consider a model where a learner receives the nearest positive example if it gets the answer 'no' to a membership query. We refer to these two models as **NPSubQ** and, respectively, **NPMemQ**. A similar approach to "correction" queries was suggested in [BBBD05, BBDT06]: a learner, in response to a membership query, receives the least (in the lexicographic order) *extension* of the queried datum belonging to the target language. One can argue, however, that a (rephrased) correcting sentence, while obviously being close to the queried wrong one, is not necessarily an extension of it. Thus, in our model, we require the "correction" to be just close to a wrong datum.
— in the above approach, a teacher may have difficulties providing the nearest (correcting) positive example, as it can still be too complex — far larger than the negative example. Therefore, we consider also a variant of learning via queries, where the nearest positive example not exceeding the size of the negative example is provided (if any). We refer to the variants of this model for subset and, respectively, membership queries as **BNPSubQ** and **BNPMemQ** (**B** here stands for *bounded*).

In our most general models, we assume that, in addition to subset and/or membership queries, a one-shot learner also has access to potentially all positive examples in the target concept. It can be easily seen that, when a learner can make an indefinite number of subset or membership queries, this positive data presented to a one-shot learner becomes essentially irrelevant. However, we will also study the cases when the number of queries will be uniformly bounded, and in this context access to additional positive data may be important.

We concentrate primarily on indexable classes of recursive languages [Ang80, ZL95]; an example of an indexable class is the class of all regular languages. In this context, it is natural to require a learner to output a conjecture that is an index of the target concept in the given numbering. It is also natural to require that subset queries are made about languages L_i from the given indexed family $\mathcal{L}$ (as defined in the original Angluin's model) — these languages can be viewed as potential conjectures.

Our primary goal is to compare variants of one-shot learners receiving variants of answers to subset and membership queries discussed above. First, we compare capabilities of these learners, establishing where learners in one model can learn classes of languages not learnable within the framework of another model. Secondly, we study when and how learners of one type can learn same classes of languages more efficiently than the learners of the other type, where efficiency is measured in terms of number of queries made during the learning process. In the latter context, unlike the case when the number of queries is not bounded, if a learner can have access to (potentially) all positive data, this can significantly affect its learning power. Therefore, we also consider positive data (*texts*) as an additional factor for learners with a uniform bound on the number of queries.

The paper is structured as follows. In Sections 2 and 3 we provide necessary notation and define our models of one-shot learners via subset and membership queries. In Section 4 we compare learning powers of different models defined in Section 3. First we show that if the number of queries is not uniformly bounded, then access to a text for the language does not enhance the capabilities of a learner in any of our models. Thus, in cases where we compare learning power of different models, we can assume that the learner does not receive the text of a language. Specifically, we establish that

(a) least counterexamples provided in response to subset queries can help to learn classes of languages that cannot be learned if the teacher, in addition to arbitrary negative counterexample, provides the nearest positive example to the negative counterexample too;

(b) learners receiving arbitrary nearest positive and bounded nearest positive examples for subset or membership queries and corresponding counterexamples or, respectively, answers 'no' are incomparable.

(c) learners using membership queries can sometimes learn more than the ones using subset queries getting least counterexamples and the nearest positive examples; conversely, learners using subset queries and getting the weakest type of feedback can sometimes learn more than the ones using membership queries and getting the strongest type of feedback in the framework of our models.

For all the results above, examples of exhibited classes are such that the learners on the negative sides cannot be helped even if they have access to full positive data representing the target concept.

In Section 5 we primarily study the following problem: when a type of queries QA does or does not give advantage over a type of queries QB when the number of queries of the type QA is uniformly bounded? Our main results can be summarized as follows:

(a) one subset query providing the least counterexample can have more learning power than any number of subset queries returning arbitrary counterexamples, even if the nearest positive examples and access to full positive data are provided;

(b) one membership query and either the nearest positive example (for the answer 'no'), or access to full positive data can have more learning power than any number of subset queries returning least counterexamples and the *bounded* nearest positive examples, or returning arbitrary counterexamples and the nearest positive examples — even in the presence of full positive data; for showing advantage over least counterexamples and the nearest positive examples, we need either one membership query and access to full text of the target language or two membership queries and the nearest positive examples in case of 'no' answer.

(c) on the other hand, a finite number of subset queries returning least counterexamples and the nearest positive examples can be used to learn any class of languages learnable using only *one* membership query and the nearest positive example; if no nearest positive examples to membership queries are provided, then $2^r - 1$ subset queries are enough to learn any class learnable using r membership queries; if the bounded nearest positive examples to membership queries are provided, then a finite number of subset queries is enough to learn any class learnable using a bounded number of membership queries;

(d) still, one membership query returning a *bounded* positive example can have more learning power than a previously fixed bounded number of subset queries returning least negative counterexamples and the nearest positive examples — even in presence of full positive data.

In this section, we also demonstrate that $k + 1$ membership or subset queries can do more than k queries of the same type — even when the strongest additional information (within the framework of our models) is provided. On the other hand, for both membership and subset queries, it is shown that no bounded number of membership or subset queries with the strongest additional information can reach the power of learners using unlimited number of queries of respective types.

We also study the following problem: when a class $\mathcal{L}$ is learnable using query type QB, can one speed up the learning process (in terms of usage of number of queries) by using query type QA? We address questions such as when classes which are learnable using small number of queries of a type QA, require arbitrarily large number of queries of a type QB. For example, we address questions about existence of classes which can be learned using 1 query of a type QA or finite number of queries of a type QB, but cannot be learned using bounded number of queries of the type QB.

Overall, we hope that our results and multitude of different examples of classes witnessing separations will give the reader a good insight on how one-shot learners using membership and subset queries operate. Section 6, Conclusion, is devoted to discussion of our results and possible directions for future research.

We refer the reader to [JK07b] for omitted proofs.

2   Notation and Preliminaries

Any unexplained recursion theoretic notation is from [Rog67]. The symbol N denotes the set of natural numbers, $\{0, 1, 2, 3, \ldots\}$. Cardinality of a set S is denoted by $\mathrm{card}(S)$. By φ we denote an acceptable numbering of all partial recursive functions [Rog67]. φ_i denotes the partial recursive function computed by program i in the φ system. W_i denotes $\mathrm{domain}(\varphi_i)$. W_i is thus the i-th recursively enumerable set, in some acceptable numbering of recursively enumerable (r.e.) sets. Symbol $\mathcal{E}$ will denote the set of all r.e. languages. Symbol L, with or without decorations, ranges over $\mathcal{E}$. Symbol $\mathcal{L}$, with or without decorations, ranges over subsets of $\mathcal{E}$. We let $K = \{i \mid i \in W_i\}$. Note that K is a recursively enumerable but not recursive set [Rog67].

$\mathcal{L}$ is called an indexed family of recursive languages (abbreviated: indexed family) iff there exists an indexing $(L_i)_{i \in N}$ of languages such that: (i) $\{L_i \mid i \in N\} = \mathcal{L}$; (ii) One can effectively determine, in i and x, whether $x \in L_i$.

We now present some concepts from language learning theory. Gold considered the following definition of presentation of data to a learner. A *text* T for a language L is a mapping from N into $(N \cup \{\#\})$ such that L is the set of natural numbers in the range of T. The *content* of a text T, denoted by $\mathrm{content}(T)$, is the set of natural numbers in the range of T; that is, the language which T is a text for.

There are several criteria for learning considered in the literature. We will be mainly concerned with finite learning [Gol67]. In this model, the learner gets a text for the language as input. After reading some initial portion of the text, the learner outputs a conjecture and stops. If this conjecture is correct, then we say that the learner **TxtFIN**-identifies the language from the given text. A learner **TxtFIN**-identifies a language if it **TxtFIN**-identifies the language from all texts for the language. A learner **TxtFIN**-identifies $\mathcal{L}$, if it **TxtFIN**-identifies each $L \in \mathcal{L}$. **TxtFIN** denotes the set of all classes $\mathcal{L}$ such that some learner **TxtFIN**-identifies $\mathcal{L}$.

An issue in the above model is the hypotheses space from where the conjecture of the learner comes from. For this paper, we are mainly concerned about learning indexed families of recursive languages and assume a class preserving hypotheses space. That is, we assume that there exists a hypotheses space $H_0, H_1, \ldots$, representing all the languages in an indexed family $\mathcal{L}$ such that

(i) one can effectively, from i and x, determine whether $x \in H_i$;
(ii) $\mathcal{L} = \{H_i \mid i \in N\}$.

3   Definitions for Query Learning

In addition to access to texts representing full positive data, the learners in our model, following [Ang88], will also use two types of queries to the teacher (formally, oracle): subset queries and membership queries.

We only consider queries in the context of class preserving learning. That is, for learning a class $\mathcal{L}$, all the hypotheses are assumed to be from the above mentioned hypotheses space $H_0, H_1, \ldots$. The subset queries are now restricted to the form "$H_i \subseteq$ input language?". Correspondingly, the membership queries are also assumed to come from the hypotheses space: the learner only asks membership queries of the form "$x \in$ input language?", for some $x \in \bigcup_{i \in N} H_i$. Note that this approach is somewhat different from traditional membership queries, where any member of N may be queried. If a learner uses hypotheses from the indexed family, it is natural to require that it only tests if elements belonging to candidate conjectures belong to the target concept. Moreover, if one allows membership queries for any member of N, then one can obtain all positive and negative data (so-called *informant*) for the input language, and thus the model (including the cases for the (bounded) nearest positive examples) would collapse to learning from informant, except for the case of learning the empty language, $\emptyset$. For these reasons, for learning $\mathcal{L}$, we restrict our study to considering membership queries only for elements of $\bigcup_{L \in \mathcal{L}} L$.

The 'yes'/'no' answer provided to the learner is based on whether the answer to the query is true or false. For subset queries (about H_i), in case of 'no' answer (meaning that H_i is not a subset of the input language), the teacher also provides a negative counterexample, which is a member of H_i, but not a member of the input language. Here, we make distinction between two different cases: when the least counterexample is provided (we refer to such queries as **LSubQ**) and when an arbitrary counterexample is provided (we refer to such queries as **SubQ**).

In addition, for 'no' answers to subset queries, we often also consider providing the learner with the nearest positive example to the negative counterexample. In the context of membership queries, if the answer is 'no', the learner is then provided the positive example nearest to the queried element x. We will denote it by using the prefix **NP** to the query type. We also consider the variant of providing the *bounded* nearest positive example, when the nearest positive example not exceeding the negative counterexample (or negative element, in the context of membership queries) is provided (this is denoted by using the prefix **BNP** to the query type).

In addition text may or may not be provided to the learner: this is denoted by using **Txt** in the criterion name.

The above will provide us with the following criteria for one-shot learnability: **SubQFIN, LSubQFIN, MemQFIN, NPSubQFIN, NPLSubQFIN, NPMemQFIN, BNPSubQFIN, BNPLSubQFIN, BNPMemQFIN** and **SubQTxtFIN, LSubQTxtFIN, MemQTxtFIN, NPSubQTxtFIN, NPLSubQTxtFIN, NPMemQTxtFIN, BNPSubQTxtFIN, BNPLSubQTxtFIN, BNPMemQTxtFIN**. Below we formally give the definition of **NPLSubQFIN**. Other criteria can be defined similarly.

Definition 1. (a) **M NPLSubQFIN**-identifies $\mathcal{L}$, iff for some class preserving hypotheses space $H_0, H_1, \ldots$, for all $L \in \mathcal{L}$, **M** asks a sequence of subset queries, where the answer to each query H_i is as follows:

(i) 'yes', if $H_i \subseteq L$

(ii) 'no', if $H_i \not\subseteq L$. In addition, the learner is provided with $x = \min(H_i - L)$ as a negative counterexample and a y such that y is the earliest element in the sequence $x - 1, x + 1, x - 2, x + 2, \ldots, 0, 2x, 2x + 1, 2x + 2, \ldots$ which belongs to L (if there is no such y, then a special answer 'none' is provided to the learner).

After asking a finite number of queries, the learner outputs an index i such that $H_i = L$.

(b) **NPLSubQFIN** $= \{\mathcal{L} \mid$ some **M NPLSubQFIN**-identifies $\mathcal{L}\}$.

Note that later queries may depend on earlier answers (and, in the case of text being provided, on the elements of the text already read by the learner).

We sometimes consider limiting the number of queries made by the learner. For example, **NPMemQ^{k}TxtFIN** denotes the criterion **NPMemQTxtFIN** where the number of queries made by the learner is limited to k.

4   Relationships Among Various Query Criteria

Following Theorem shows that when there is no bound on the number of queries, texts do not help: providing text does not increase learning power of one-shot learners.

Theorem 1. *Suppose* **Q** $\in$ {**SubQ, LSubQ, MemQ**}. *Then,* **QFIN** $=$ **QTxtFIN**; **NPQFIN** $=$ **NPQTxtFIN**; **BNPQFIN** $=$ **BNPQTxtFIN**.

4.1   Variants of SubQ

In this subsection, we explore relationships between different variants of **SubQ**. First we show that learners getting least counterexamples can sometimes learn more than any learner getting arbitrary counterexamples, as well as the nearest positive examples, bounded or not, and access to full positive data.

Theorem 2. **LSubQFIN** $-$ (**NPSubQTxtFIN** $\cup$ **BNPSubQTxtFIN**) $\neq \emptyset$.

Proof. Let $L = \{100i + 1, 100i + 2, 100i + 3 \mid i \in N\}$.

Let $L_i = L - \{100i + 1, 100i + 3\}$. Let $X_i = L \cup \{100i + 1\}$.

Let $\mathcal{L} = \{L\} \cup \{L_i \mid i \in N\} \cup \{X_i \mid i \in K\}$.

Clearly, $\mathcal{L}$ is an indexed family. To see that $\mathcal{L} \in$ **LSubQFIN**, first query whether L is a subset of the input language. If L is a subset of the input language, then the input language must be L. Otherwise, the least counterexample is either $100i + 1$ or $100i + 3$, for some i. In the former case, the input language must be L_i. In the latter case, the input language must be X_i.

Now suppose by way of contradiction that $\mathcal{L} \in$ **NPSubQTxtFIN** (**BNPSubQTxtFIN**) as witnessed by **M**. Then following algorithm solves K.

On input i: Simulate **M** on a text for L_i, where answers to the queries are as follows — queries which do not contain $100i+3$ are answered 'yes'; queries which contain $100i+3$ are answered 'no', along with the counterexample $100i+3$ and the (bounded) nearest positive example $100i+2$. If, in this simulation, **M** ever queries a language which contains $100i+1$ but not $100i+3$, then output $i \in K$. If the simulation stops with a conjecture, without ever querying a language which contains $100i+1$ but not $100i+3$, then output $i \notin K$.

Now, as **M** **NPSubQTxtFIN**-identifies (**BNPSubQTxtFIN**-identifies) $\mathcal{L}$, **M**, on input language L_i, must eventually output a conjecture. During the process, if **M** queries a language containing $100i+1$, but not $100i+3$, then we have $i \in K$, as **M** is allowed only to query languages within the class $\mathcal{L}$. On the other hand, if **M** outputs a conjecture without querying about X_i, then, since the answers given to the queries of **M** are consistent with the input language being L_i or X_i, we must have that $X_i \notin \mathcal{L}$ (otherwise, **M** cannot identify both L_i and X_i). Thus $i \notin K$. ∎

Our next result shows that learners getting arbitrary counterexamples and the unbounded positive examples nearest to them can do sometimes more than the ones getting the bounded nearest positive example, even if the latter ones receive the least counterexamples and have access to full positive data.

Theorem 3. NPSubQFIN – BNPLSubQTxtFIN $\neq \emptyset$.

Proof. Let $L = \{100i + x \mid i \in N, x \in \{1,3\}\}$.
Let $L_i = L - \{100i + 1\}$. Let $X_i = L_i \cup \{100i + 2\}$.
Let $\mathcal{L} = \{L\} \cup \{L_i \mid i \in N\} \cup \{X_i \mid i \in K\}$.

It is easy to verify that $\mathcal{L}$ is an indexed family. Now, we show that $\mathcal{L} \in$ **NPSubQFIN**. First, a learner can query L. If L is contained in the input language, then the input language must be L. If there is a counterexample, it must be $100i+1$, for some i. Now the input language is X_i if the nearest positive example was $100i + 2$. Otherwise, the input language must be L_i.

We can show that $\mathcal{L} \notin$ **BNPLSubQTxtFIN**, as a learner for $\mathcal{L}$ can be used to solve K as follows: on input L_i, the learner either asks a query for X_i (in which case $i \in K$), or outputs a conjecture for L_i without ever asking a query for X_i (in which case X_i must not be in $\mathcal{L}$, as the input and answers are consistent with both L_i and X_i being the input language; thus $i \notin K$). We omit the details. ∎

As our next result shows, in the above result, the learners getting the unbounded nearest positive examples can be replaced by the ones getting the bounded nearest positive examples, and vice versa.

Therefore, the learners via subset queries and getting counterexamples and the nearest positive data of these two types are incomparable.

Theorem 4. BNPSubQFIN– NPLSubQTxtFIN $\neq \emptyset$.

From the above results, we can immediately derive the following corollary.

Corollary 1. SubQFIN $\subset$ NPSubQFIN. LSubQFIN $\subset$ NPLSubQFIN. SubQFIN $\subset$ BNPSubQFIN. LSubQFIN $\subset$ BNPLSubQFIN.

Our next result shows that learners getting the nearest positive examples, in addition to arbitrary counterexamples, can sometimes learn more than the ones getting the least counterexamples and full positive data, but no nearest positive examples.

Theorem 5. $\mathbf{NPSubQFIN} \cap \mathbf{BNPSubQFIN} - \mathbf{LSubQTxtFIN} \neq \emptyset$.

4.2 MemQ vs SubQ

In this subsection, we compare powers of one-shot learners using membership and subset queries. We establish that, within the framework of our models, weakest learners using one type of queries can sometimes do more than the strongest learners using the other type of queries. First, we show that the learners using membership queries can sometimes be stronger than the ones using subset queries.

Theorem 6. $\mathbf{MemQFIN} - (\mathbf{NPLSubQTxtFIN} \cup \mathbf{BNPLSubQTxtFIN}) \neq \emptyset$.

Proof. Let $L = \{0\} \cup \{100i + 2, 100i + 3, 100i + 5 \mid i \in N\}$.
   Let $L_i = \{100i + 1, 100i + 5\}$. Let $X_i = \{100i + 1, 100i + 2, 100i + 5\}$.
   Let $\mathcal{L} = \{L\} \cup \{L_i \mid i \in N\} \cup \{X_i \mid i \in K\}$.
   We first show that $\mathcal{L} \in \mathbf{MemQFIN}$. Learner queries 0. If the answer is 'yes', then input language must be L; Otherwise, learner determines an i such that $100i + 1$ is in the input language. Then querying $100i + 2$ determines whether the input language is L_i or X_i.
   We can show that $\mathcal{L} \notin \mathbf{NPLSubQTxtFIN}$ ($\mathbf{BNPLSubQTxtFIN}$), as a learner for $\mathcal{L}$ can be used to solve K as follows: on input L_i, the learner either asks a query for X_i (in which case $i \in K$), or outputs a conjecture for L_i without ever asking a query for X_i (in which case X_i must not be in $\mathcal{L}$, as the input text and answers are consistent with both L_i and X_i being the input language; thus $i \notin K$). We omit the details. ∎

The following theorem demonstrates the advantage of subset queries over membership queries.

Theorem 7. $\mathbf{SubQFIN} - (\mathbf{NPMemQTxtFIN} \cup \mathbf{BNPMemQTxtFIN}) \neq \emptyset$.

4.3 Different Variants of MemQ

In this subsection, we compare different variants of learners using membership queries. Our first two results show that, as in the case of subset queries, learners getting the unbounded or bounded nearest positive examples (in addition to answers 'no') can learn more than learners getting the nearest positive example of the other type and access to full positive data.

Theorem 8. $\mathbf{NPMemQFIN} - \mathbf{BNPMemQTxtFIN} \neq \emptyset$.

Theorem 9. **BNPMemQFIN** – **NPMemQTxtFIN** $\neq \emptyset$.

Next result shows that the nearest positive examples of either type give sometimes more power to learners over those getting access to full positive data but not getting the nearest positive examples: the nearest positive example, coupled with the answer 'no' to a specific query, and obtained at the "right" time, can be more helpful than access to a text for the target language.

Theorem 10. **NPMemQFIN** $\cap$ **BNPMemQFIN** – **MemQTxtFIN** $\neq \emptyset$.

Results of this section can also be generalized to the case when r.e. classes of r.e. languages are considered. Some of our results for bounded queries below also work for r.e. classes of r.e. languages. However, we do not yet know if all of our results on bounded queries can be obtained when r.e. classes of r.e. languages are considered.

5   Complexity Hierarchy

This section gives the relationship between different criteria of query learning considered in the paper: **MemQ**, **SubQ**, **LSubQ**, with or without the (bounded) nearest positive example, and with or without text, when the number of queries may be bounded. We begin with the following useful proposition.

Proposition 1. *Suppose* $\text{card}(\mathcal{L}) \leq k + 1$. *Then* $\mathcal{L} \in$ **MemQ**^{k}**FIN** *and* $\mathcal{L} \in$ **SubQ**^{k}**FIN**.

If the number of membership queries is uniformly bounded and the nearest positive examples are bounded, then learnable classes are finite.

Proposition 2. *Suppose* $k \in N$.
  (a) If $\mathcal{L} \in$ **MemQ**^{k}**FIN**, *then* $\text{card}(\mathcal{L}) \leq 2^k$.
  (b) If $\mathcal{L} \in$ **BNPMemQ**^{k}**FIN**, *then* $\mathcal{L}$ *must be finite.*

Our next result shows how one-shot learners using uniformly bounded membership queries can simulate the ones using uniformly bounded number of subset queries when the target class is finite.

Theorem 11. *Suppose* $\mathcal{L}$ *is finite and* $\mathcal{L} \in$ **SubQ**^{k}**FIN**. *Then* $\mathcal{L} \in$ **MemQ**$^{2^k-1}$**TxtFIN**.

Now we turn our attention to arbitrary (possibly infinite) target classes. First we note that a learner using just one subset query can sometimes do more than any learner using the strongest type of membership queries within the framework of our models — including access to a text of the input language.

Theorem 12. **SubQ**^{1}**FIN** – (**NPMemQTxtFIN** $\cup$ **BNPMemQTxtFIN**) $\neq \emptyset$.

The next result demonstrates what advantages of a bounded number of membership queries over subset queries are possible. The following picture is quite complex. In particular, we show that r membership queries can be simulated by $2^r - 1$ subset queries (we also show that this estimate is tight). However, if, in addition to membership queries, a learner gets either access to text of the input language, or the nearest positive examples, then, in most cases, just one membership query gives advantage over a learner using subset queries (and getting strongest feedback and having access to text of the input language). On the other hand, a learner using a finite number of subset queries and getting *least* counterexamples and the nearest positive examples can simulate any learner using just one membership query and getting the nearest positive examples; still, such a simulation is not always possible if a learner can use two such membership queries. Also, learners using a uniformly bounded number of membership queries and getting *bounded* positive examples can always be simulated by learners using subset queries if the number of queries of the latter type is not bounded.

Theorem 13. *(a)* $\quad$ **MemQ^{1}TxtFIN** $\quad - \quad$ (**NPLSubQTxtFIN** $\cup$ **BNPLSubQTxtFIN**) $\neq \emptyset$.

(b) **NPMemQ^{1}FIN** $-$ (**NPSubQTxtFIN** $\cup$ **BNPLSubQTxtFIN**) $\neq \emptyset$.

(c) **NPMemQ^{2}FIN** $-$ (**NPLSubQTxtFIN** $\cup$ **BNPLSubQTxtFIN**) $\neq \emptyset$.

(d) **NPMemQ^{1}FIN** $\subseteq \bigcup_{r \in N}$ **NPLSubQ^{r}FIN**.

(e) For all k, **NPMemQ^{1}FIN** $\quad - \quad$ (**NPLSubQ^{k}TxtFIN** $\cup$ **BNPLSubQ^{k}TxtFIN**) $\neq \emptyset$.

(f) For all k, **BNPMemQ^{1}FIN** $\quad - \quad$ (**NPLSubQ^{k}TxtFIN** $\cup$ **BNPLSubQ^{k}TxtFIN**) $\neq \emptyset$.

(g) For all k, **BNPMemQ^{k}FIN** $\subseteq \bigcup_{r \in N}$ **SubQ^{r}FIN**.

(h) For $r \geq 1$, **MemQ^{r}FIN** $-$ (**BNPLSubQ$^{2^r-2}$TxtFIN** $\cup$ **NPLSubQ$^{2^r-2}$TxtFIN**) $\neq \emptyset$.

(i) For $r \geq 1$, **MemQ^{r}FIN** $\subseteq$ **SubQ$^{2^r-1}$FIN**.

Proof. We only show part (d). Suppose the queried element by **NPMemQ^{1}FIN** learner is y. Let A be the language in $\mathcal{L}$ which contains y (there must be such an A as queries are posed only for elements belonging to at least one language in $\mathcal{L}$; furthermore such an A is unique as the learner learns the class using only one membership query). For $i \leq 2y$, let A_i be the language in $\mathcal{L}$ which does not contain y, but contains i as the element nearest to y (here, recall that the nearest elements to y have the order $y - 1$, $y + 1$, $y - 2$, $y + 2$, ...). Now the **NPLSubQFIN** learning algorithm is:

1. Query $A, A_0, A_1, \ldots, A_{2y}$. Find which of these are subsets of the input language. If none, then proceed to the step 2. Otherwise, the subset sequence uniquely determines the input language. (If A is a subset of the input language, then the input language must be A; otherwise it is A_i, where A_i is a subset of the input language and, among all j such that $j \leq 2y$ and $A_j \subseteq$ input language, i is the nearest element to y; note that this A_i would be the conjecture returned by the **NPMemQ^{1}FIN** learner, when the membership query for y is answered 'no', with the nearest positive example being i — no other nearer element to

y exists in the input language, as otherwise the conjecture of $\mathbf{NPMemQ^1FIN}$ learner would not be a subset of the input language).

2. If the algorithm reaches this step, then A is not a subset of the input language. The least negative counterexample to query A has to be an element $\leq y$. Let the corresponding nearest positive example be z. Note that $z > 2y$ (otherwise step 1 would have handled the case). Give 'no' answer to the membership query for y by the $\mathbf{NPMemQ^1FIN}$ learner, with the nearest positive example being z (since z is also the nearest positive example in the input language for 'no' answer to the membership query for y). Output the conjecture outputted by the $\mathbf{NPMemQ^1FIN}$ learner.

Note that the number of queries made above by the $\mathbf{NPLSubQFIN}$ learner is bounded by $2y + 2$.   ∎

Now we will compare learners using queries of the same type. Our next result shows that one subset query providing the least counterexample can do more than subset queries returning arbitrary counterexamples, even if the nearest positive examples are returned, and a text of the input language is accessible.

Theorem 14. $\mathbf{LSubQ^1FIN} - (\mathbf{NPSubQTxtFIN} \cup \mathbf{BNPSubQTxtFIN}) \neq \emptyset$.

The next result establishes hierarchies on the number of queries. We show that, for both types of queries, $k + 1$ queries give more than k (even if a learner using k queries gets additional feedback and has access to full positive data).

Theorem 15. *For all $k \in N$,*
    (a) $\mathbf{SubQ^{k+1}FIN} - (\mathbf{NPLSubQ^kTxtFIN} \cup \mathbf{BNPLSubQ^kTxtFIN}) \neq \emptyset$.
    (b) $\mathbf{MemQ^{k+1}FIN} - (\mathbf{NPMemQ^kTxtFIN} \cup \mathbf{BNPMemQ^kTxtFIN}) \neq \emptyset$.

For both types of queries, uniformly bounded number of queries is not enough to achieve full learning power.

Theorem 16. *(a)* $\mathbf{SubQFIN}$ $-$ $\bigcup_{k \in N} (\mathbf{NPLSubQ^kTxtFIN}$ $\cup$
$\mathbf{BNPLSubQ^kTxtFIN}) \neq \emptyset$.
    (b) $\mathbf{MemQFIN} - \bigcup_{k \in N} (\mathbf{NPMemQ^kTxtFIN} \cup \mathbf{BNPMemQ^kTxtFIN}) \neq \emptyset$.

Next two results compare power of the nearest and the bounded nearest positive examples for learners using the same type of queries. Our next result shows that one subset query returning negative counterexample and the nearest positive example of either type can do more than any learner using subset queries, least counterexamples, nearest positive examples of the other type, and full positive data.

Theorem 17. $\mathbf{NPSubQ^1FIN} - \mathbf{BNPLSubQTxtFIN} \neq \emptyset$.
    $\mathbf{BNPSubQ^1FIN} - \mathbf{NPLSubQTxtFIN} \neq \emptyset$.

For membership queries, the picture is more complex. One *unbounded* nearest positive example can do more than any number of bounded nearest positive examples — even in the presence of full positive data. However, the learners making membership queries and getting the *bounded* nearest positive examples can be simulated by learners using a finite number of just simple membership queries; still, no uniform bound on the number of queries in such a simulation is possible (even if the learner receives also the nearest positive examples and has access to full positive data) — moreover, if the learner using just one bounded positive example has also access to full positive data, then no simulation like the one above is possible.

Theorem 18. *(a)* $\mathbf{NPMemQ^1FIN} - \mathbf{BNPMemQTxtFIN} \neq \emptyset$.
  (b) For $k \in N$, $\mathbf{BNPMemQ^kFIN} \subseteq \bigcup_{r \in N} \mathbf{MemQ^rFIN}$.
  (c) For $k \in N$, $\mathbf{BNPMemQ^1FIN} - \mathbf{NPMemQ^kTxtFIN} \neq \emptyset$.
  (d) $\mathbf{BNPMemQ^1TxtFIN} - \mathbf{NPMemQTxtFIN} \neq \emptyset$.

Theorem 1 showed that $\mathbf{TxtFIN} \subseteq \mathbf{MemQFIN} \cap \mathbf{SubQFIN}$. Now we show that no uniformly bounded number of queries of either type suffices for simulation of full positive data.

Theorem 19. $\mathbf{TxtFIN} - \bigcup_{k \in N}(\mathbf{NPLSubQ^kFIN} \cup \mathbf{BNPLSubQ^kFIN} \cup \mathbf{NPMemQ^kFIN} \cup \mathbf{BNPMemQ^kFIN}) \neq \emptyset$.

We now consider when a class is learnable by using query type QB, but needs a high number of queries, whereas if we had used query type QA, then a small number of queries suffice. Ideally, for diagonalizations (showing that one type of queries may be stronger than the other one) and complexity speedup, we would like theorems of the following types: (A) $QA^1\mathbf{FIN} \cap QB\mathbf{FIN}$ diagonalizes against $\bigcup_{k \in N}(\mathbf{BNP}QB^k\mathbf{TxtFIN} \cup \mathbf{NP}QB^k\mathbf{TxtFIN})$; (B) $QA^1\mathbf{FIN} \cap QB^{k+1}\mathbf{FIN}$ diagonalizes against $\mathbf{BNP}QB^k\mathbf{TxtFIN} \cup \mathbf{NP}QB^k\mathbf{TxtFIN}$. That would give us a perfect set of speedup effects. However, this is not always possible, and we get as close to the above as possible (we do not have the best possible results for QB being $\mathbf{MemQ}$, and QA being $\mathbf{SubQ}$). Note also that the results of type (C): $QA^1\mathbf{FIN}$ diagonalizes against $\mathbf{BNP}QB\mathbf{TxtFIN} \cup \mathbf{NP}QB\mathbf{TxtFIN}$ — have been obtained earlier in this section, where possible: see Theorems 12, 13, and 14. Due to space restrictions, here we only deal with problems of form (A) above.

Theorem 20. $\mathbf{LSubQ^1FIN} \cap \mathbf{SubQFIN} - \bigcup_{k \in N}(\mathbf{NPSubQ^kTxtFIN} \cup \mathbf{BNPSubQ^kTxtFIN}) \neq \emptyset$.

Theorem 21. *(a)* $\mathbf{MemQ^1TxtFIN}$ $\cap$ $\mathbf{SubQFIN}$ $-$ $\bigcup_{k \in N}(\mathbf{NPLSubQ^kTxtFIN} \cup \mathbf{BNPLSubQ^kTxtFIN}) \neq \emptyset$.
  (b) $\mathbf{NPMemQ^1FIN} \cap \mathbf{SubQFIN} - \bigcup_{k \in N}(\mathbf{NPSubQ^kTxtFIN} \cup \mathbf{BNPLSubQ^kTxtFIN}) \neq \emptyset$.
  (c) $\mathbf{NPMemQ^2FIN} \cap \mathbf{SubQFIN} - \bigcup_{k \in N}(\mathbf{NPLSubQ^kTxtFIN} \cup \mathbf{BNPLSubQ^kTxtFIN}) \neq \emptyset$.
  (d) $\mathbf{MemQFIN} \cap \mathbf{SubQFIN} - \bigcup_{k \in N}(\mathbf{NPLSubQ^kTxtFIN} \cup \mathbf{BNPLSubQ^kTxtFIN}) \neq \emptyset$.

The above theorem is optimal, as $\bigcup_{r \in N} \textbf{BNPMemQ}^r\textbf{FIN}$ $\subseteq$ $\bigcup_{k \in N} \textbf{SubQ}^k\textbf{FIN}$, (see Propositions 1 and 2) and $\textbf{NPMemQ}^1\textbf{FIN}$ $\subseteq$ $\bigcup_{k \in N} \textbf{NPLSubQ}^k\textbf{FIN}$ (see Theorem 13).

Theorem 22. $\textbf{SubQ}^1\textbf{FIN} \cap \textbf{MemQFIN} - \bigcup_{k \in N}(\textbf{NPMemQ}^k\textbf{TxtFIN} \cup \textbf{BNPMemQ}^k\textbf{TxtFIN}) \neq \emptyset$.

We also studied problems of the type (B) mentioned above. Questions about what happens when $QA = \textbf{SubQ}$ and $QB = \textbf{MemQ}$ are not fully resolved yet for this situation. Relevant results, discussion, and open problems can be found in [JK07b].

We also have results similar to the ones obtained in this section, when one considers separation of nearest positive examples versus bounded nearest positive examples, and vice versa, rather than based on number of queries. However, we do not have a complete picture there either.

6 Conclusion

In this paper, we extended Angluin's model of learning via subset and membership queries, allowing teachers, in addition to just answers 'no' or arbitrary counterexamples (as suggested by Angluin in her original query model in [Ang88]) to return *least* counterexamples and/or the *nearest* ("correcting") positive examples together with answer 'no' or a counterexample. We explored how different variants of corresponding learning models fair against each other in terms of their general learning capabilities and in terms of their complexity advantages, where number of queries is used as the complexity measure (in the latter case, possible access to a text for the target language becomes a significant factor, contributing an interesting component to the interplay of different learning tools). As we established, "correcting" nearest positive examples can sometimes do more than just negative (counter)examples and access to full positive data. We also established that, though, in most cases, just one query of one type can do more than any number of queries of another type with the strongest possible feedback, typically even coupled with access to text for the target language, the general picture is more complex — for example, sometimes one query is not enough, while two queries suffice — or one query is enough to achieve advantage (general, or complexity) if a learner has also access to full positive data.

In our model, indexed families do not contain languages not belonging to the target class (such families are known in literature as *class preserving*). It would be interesting to explore how our approach works in the context of learning via *class comprising* hypotheses spaces, when an (indexed) hypotheses space can contain languages not belonging to the target class.

Our approach to representation of *covert* feedback from a teacher in form of the *nearest positive* examples is, of course, only one of several possible ways to address this problem (a somewhat different approach was suggested in [BBBD05, BBDT06]). It would be interesting also to define and explore formalizations of one-shot learnability via queries, where positive feedback were semantically close or structurally similar to the negative datum, rather than being close based on coding. Such models may also be interesting in the context of learning some important specific indexed classes, for example, patterns, finite automata, or regular expressions.

Acknowledgements. We thank the anonymous referees for helpful comments.

References

[Ang80]   Angluin, D.: Inductive inference of formal languages from positive data. Information and Control 45, 117–135 (1980)

[Ang88]   Angluin, D.: Queries and concept learning. Machine Learning 2, 319–342 (1988)

[BBBD05] Becerra-Bonache, L., Bibire, C., Dediu, A.H.: Learning dfa from corrections. In: Fernau, H. (ed.) Theoretical Aspects of Grammar Induction (TAGI), pp. 1–12 (2005). WSI–2005–14

[BBDT06] Becerra-Bonache, L., Dediu, A.H., Tîrnăucă, C.: Learning dfa from correction and equivalence queries. In: Sakakibara, Y., Kobayashi, S., Sato, K., Nishino, T., Tomita, E. (eds.) ICGI 2006. LNCS (LNAI), vol. 4201, pp. 281–292. Springer, Heidelberg (2006)

[Gol67]   Gold, E.M.: Language identification in the limit. Information and Control 10, 447–474 (1967)

[GS91]   Gasarch, W., Smith, C.: Learning via queries. Journal of the ACM, 649–674 (1991)

[JK07a]   Jain, S., Kinber, E.: Learning languages from positive data and negative counterexamples. Journal of Computer and System Sciences (to appear, 2007)

[JK07b]   Jain, S., Kinber, E.: One-shot learners using negative counterexamples and nearest positive examples. Technical Report TRA3/07, School of Computing, National University of Singapore (2007)

[JLZ05]   Jain, S., Lange, S., Zilles, S.: Gold-style and query learning under various constraints on the target class. In: Jain, S., Simon, H.U., Tomita, E. (eds.) ALT 2005. LNCS (LNAI), vol. 3734, pp. 226–240. Springer, Heidelberg (2005)

[LZ04]   Lange, S., Zilles, S.: Formal language identification: Query learning vs gold-style learning. Information Processing Letters 91, 285–292 (2004)

[LZ05]   Lange, S., Zilles, S.: Relations between gold-style learning and query learning. Information and Computation 203, 211–237 (2005)

[Rog67]   Rogers, H.: Theory of Recursive Functions and Effective Computability. McGraw-Hill (1967) (reprinted by MIT Press in 1987)

[RP99]   Rohde, D.L.T., Plaut, D.C.: Language acquisition in the absence of explicit negative evidence: how important is starting small? Cognition 72, 67–109 (1999)

[ZL95]   Zeugmann, T., Lange, S.: A guided tour across the boundaries of learning recursive languages. In: Lange, S., Jantke, K.P. (eds.) Algorithmic Learning for Knowledge-Based Systems. LNCS, vol. 961, pp. 190–258. Springer, Heidelberg (1995)

Polynomial Time Algorithms for Learning k-Reversible Languages and Pattern Languages with Correction Queries*

Cristina Tîrnăucă[1] and Timo Knuutila[2]

[1] Research Group on Mathematical Linguistics, Rovira i Virgili University
Pl. Imperial Tàrraco 1, Tarragona 43005, Spain
[2] Department of Information Technology, University of Turku
Joukahaisenkatu 3-5 B, FI-20014 Turku, Finland
cristina.bibire@estudiants.urv.es,timo.knuutila@it.utu.fi

Abstract. We investigate two of the language classes intensively studied by the algorithmic learning theory community in the context of learning with correction queries. More precisely, we show that any pattern language can be inferred in polynomial time in length of the pattern by asking just a linear number of correction queries, and that k-reversible languages are efficiently learnable within this setting. Note that although the class of all pattern languages is learnable with membership queries, this cannot be done in polynomial time. Moreover, the class of k-reversible languages is not learnable at all using membership queries only.

Keywords: Correction queries, k-reversible languages, pattern languages, polynomial algorithms.

1 Introduction

Without any doubt, there is no formal model that can capture all aspects of human learning. Nevertheless, the overall aim of researchers working in algorithmic learning theory has been to gain a better understanding of what learning really is. Actually, the field itself has been introduced as an attempt to construct a precise model for the notion of "being able to speak a language" [9].

Among the most celebrated models (Gold's model of *learning from examples* [9], Angluin's *query learning* model [4], Valiant's *PAC learning* model [18]), the best one for describing the child-adult interaction within the process of child acquiring his native language is the one proposed in [4]. There, *the learner* receives information about a target concept by asking queries of a specific kind (depending on the chosen query model type) which will be truthfully answered

---

* The preparation of this paper was done while the first author was visiting the Department of Mathematics of Turku University, and was supported in part by the European Science Foundation (ESF) for the activity entitled 'Automata: from Mathematics to Applications', and by the FPU Fellowship AP2004-6968 from the Spanish Ministry of Education and Science.

M. Hutter, R.A. Servedio, and E. Takimoto (Eds.): ALT 2007, LNAI 4754, pp. 272–284, 2007.

by *the teacher*. After finitely many queries, the learner is required to return its hypothesis, and this should be the correct one.

The first query learning algorithm, called L^*, was able to identify any minimal complete deterministic finite automaton (DFA) in polynomial time, using membership queries (MQs) and equivalence queries (EQs) [4]. Meanwhile, other types of queries have been introduced: subset, superset, disjointness and exhaustive queries [5], structured MQs [15], *etc.*, and also various target concepts have been investigated: non-deterministic finite automata [19], context-free grammars [15], two-tape automata [20], regular tree languages [8,17], *etc.*

Still, none of the above mentioned queries reflects one important aspect of children language acquisition, namely that although children are not explicitly provided negative information, they are corrected when they make mistakes. Following this idea, L. Becerra-Bonache, A.H. Dediu and C. Tîrnăucă introduced in [7] a new type of query, the so-called *correction query* (CQ), and showed that DFAs are learnable in polynomial time using CQs and EQs. Continuing the investigation on CQs, C. Tîrnăucă and S. Kobayashi found necessary and sufficient conditions for an indexable class of recursive languages to be learnable with CQs only [16]. Also, they showed some relations existing between this model and other well-known (query and Gold-style) learning models.

In contrast with the approach in [16], where the learnability was studied regardless time complexity, we focus in this paper on algorithms for identifying language classes in polynomial time. Thus, the rest of the paper is structured as follows. Preliminary notions and results are presented in Section 2. In Section 3 we give a polynomial time algorithm for learning the class of k-reversible languages with CQs. Section 4 contains an algorithm for learning the class of pattern languages, along with discussions about correctness, termination and time analysis. In Section 5 we present some results on the learnability with MQs of the classes investigated in the previous sections. We conclude with remarks and future work ideas (Section 6).

2 Preliminaries

Familiarity with standard recursion and language theoretic notions is assumed (good introductory books are [10,12], for example).

Let Σ be a finite alphabet of symbols. By Σ^* we denote the set of all finite strings of symbols from Σ. A *language* is any set of strings over Σ. The length of a string w is denoted by $|w|$, and the concatenation of two strings u and v by uv or $u \cdot v$. The empty string (i.e., the unique string of length 0) is denoted by λ. If $w = uv$ for some $u, v \in \Sigma^*$, then u is a *prefix* of w and v is a *suffix* of w.

A set S is said to be *prefix-closed* if for all strings u in S and all prefixes v of u, the string v is also in S. The notion of *suffix-closed* set is defined similarly.

By $\Sigma^{\le k}$ we denote the set $\{w \in \Sigma^* \mid |w| \le k\}$, by $Pref(L)$ the set $\{u \mid \exists v \in \Sigma^*$ such that $uv \in L\}$ of all prefixes of a language $L \subseteq \Sigma^*$, and by $Tail_L(u) = \{v \mid uv \in L\}$ the left-quotient of L and u. Thus, $Tail_L(u) \neq \emptyset$ if and only if $u \in Pref(L)$.

A *deterministic finite automaton* is a 5-tuple $\mathcal{A} = (Q, \Sigma, \delta, q_0, F)$, where Q is a finite set of *states*, Σ is a finite alphabet, $q_0 \in Q$ is the *initial* state, $F \subseteq Q$ is the set of *final states*, and δ is a partial function, called *transition function*, that maps $Q \times \Sigma$ to Q. This function can be extended to strings by writing $\delta(q, \lambda) = q$, and $\delta(q, u \cdot a) = \delta(\delta(q, u), a)$ for all $q \in Q$, $u \in \Sigma^*$ and $a \in \Sigma$. A string $u \in \Sigma^*$ is *accepted* by $\mathcal{A}$ if $\delta(q_0, u) \in F$. The set of strings accepted by $\mathcal{A}$ is denoted by $L(\mathcal{A})$ and called a *regular language*. The number of states of an automaton $\mathcal{A}$ is also called the *size* of $\mathcal{A}$. A DFA $\mathcal{A} = (Q, \Sigma, \delta, q_0, F)$ is *complete* if for all q in Q and a in Σ, $\delta(q, a)$ is defined, i.e., δ is a total function. For any regular language L, there exists a minimum state DFA $\mathcal{A}_L$ such that $L(\mathcal{A}_L) = L$ (see [10], pp. 65-71).

A state q is called *reachable* if there exists $u \in \Sigma^*$ such that $\delta(q_0, u) = q$ and *co-reachable* if there exists $u \in \Sigma^*$ such that $\delta(q, u) \in F$. A reachable state that is not co-reachable is a *sink* state. Note that in a minimum DFA there is at most one sink state, and all states are reachable.

Given a language $L \subseteq \Sigma^*$, one can define the following relation on strings: $u_1 \equiv_L u_2$ if and only if for all u in Σ^*, $u_1 \cdot u \in L \Leftrightarrow u_2 \cdot u \in L$. It is easy to show that $\equiv_L$ is an equivalence relation, and thus it divides the set of all finite strings in Σ^* into one or more equivalence classes. We denote by $[u]_L$ (or simply $[u]$, when there is no confusion) the equivalence class of the string u (i.e., $\{u' \mid u' \equiv_L u\}$), and by $\Sigma^*/_{\equiv_L}$ the set of all equivalence classes induced by $\equiv_L$ on Σ^*.

The Myhill-Nerode Theorem states that the number of equivalence classes of $\equiv_L$ (also called the *index* of L) is equal to the number of states of $\mathcal{A}_L$. As a direct consequence, a language L is regular if and only its index is finite.

Assume that Σ is a totally ordered set, and let $\prec_{lex}$ be the lexicographical order on Σ^*. Then, the *lex-length order* $\prec$ on Σ^* is defined by: $u \prec v$ if either $|u| < |v|$, or else $|u| = |v|$ and $u \prec_{lex} v$. In other words, strings are compared first according to length and then lexicographically.

If $f : A \to B$ is a function, by $f(X)$ we denote the set $\{f(x) \mid x \in X\}$. Moreover, we say that f and g are equal if they have the same domain A, and $f(x) = g(x)$ for all $x \in A$.

2.1   Query Learning

Let $\mathcal{C}$ be a class of recursive languages over Σ^*. We say that $\mathcal{C}$ is an *indexable class* if there is an effective enumeration $(L_i)_{i \geq 1}$ of all and only the languages in $\mathcal{C}$ such that membership is uniformly decidable, i.e., there is a computable function that, for any $w \in \Sigma^*$ and $i \geq 1$, returns 1 if $w \in L_i$, and 0 otherwise. Such an enumeration will subsequently be called an *indexing* of $\mathcal{C}$. In the sequel we might say that $\mathcal{C} = (L_i)_{i \geq 1}$ is an indexable class and understand that $\mathcal{C}$ is an indexable class and $(L_i)_{i \geq 1}$ is an indexing of $\mathcal{C}$.

In the query learning model a learner has access to an oracle that truthfully answers queries of a specified kind. A *query learner* M is an algorithmic device that, depending on the reply on the previous queries, either computes a new query, or returns a hypothesis and halts.

More formally, let $C = (L_i)_{i \geq 1}$ be an indexable class, M a query learner, and let $L \in C$. We say that M learns L using some type of queries if it eventually halts and its only hypothesis, say i, correctly describes L, i.e., $L_i = L$. So, M returns its unique and correct guess i after only finitely many queries. Moreover, M learns C using some type of queries if it learns every $L \in C$ using queries of the specified type. In the sequel we consider:

- *Membership queries.* The input is a string w, and the answer is 'yes' or 'no', depending on whether or not w belongs to the target language L.
- *Correction queries.* The input is a string w, and the answer is the smallest string (in lex-length order) of the set $Tail_L(w)$ if $w \in Pref(L)$, and the special symbol $\theta \notin \Sigma$ otherwise. We denote the correction of a string w with respect to the language L by $C_L(w)$.

The collection of all indexable classes C for which there is a query learner M such that M learns C using MQs (CQs) is denoted by $MemQ$ ($CorQ$, respectively).

3   Learning k-Reversible Languages with CQs

Angluin introduces the class of k-reversible languages (henceforth denoted by k-Rev) in [3], and shows that it is inferable from positive data in the limit. Later on, she proves that there is no polynomial algorithm that exactly identifies DFAs for 0-reversible languages using only equivalence queries [6].

We study the learnability of the class k-Rev in the context of learning with CQs, and show that there is a polynomial time algorithm which identifies any k-reversible language after asking a finite number of CQs.

Although the original definition of k-reversible languages uses the notion of k-reversible automata, we will give here only a purely language-theoretic characterization.

Theorem 1 (Angluin, [3]). *Let L be a regular language. Then L is in k-Rev if and only if whenever $u_1 v w, u_2 v w$ are in L and $|v| = k$, $Tail_L(u_1 v) = Tail_L(u_2 v)$.*

Let Σ be an alphabet, and $L \subseteq \Sigma^*$ be the target k-reversible language. For any string u in Σ^*, we define the function $row_k(u) : \Sigma^{\leq k} \to \Sigma^* \cup \{\theta\}$ by $row_k(u)(v) = C_L(uv)$. We show that each equivalence class in $\Sigma^*/{\equiv_L}$ is uniquely identified by the values of function row_k on $\Sigma^{\leq k}$.

Proposition 1. *Let L be a k-reversible language. Then, for all $u_1, u_2 \in \Sigma^*$, $u_1 \equiv_L u_2$ if and only if $row_k(u_1) = row_k(u_2)$.*

Proof. Let us first notice that for all regular languages L and for any $k \in \mathbb{N}$, $u_1 \equiv_L u_2 \Rightarrow row_k(u_1) = row_k(u_2)$ (by the definition of function row_k), so we just have to show that $row_k(u_1) = row_k(u_2) \Rightarrow u_1 \equiv_L u_2$.

Indeed, suppose there exist $u_1, u_2 \in \Sigma^*$ such that $row_k(u_1) = row_k(u_2)$ and $u_1 \not\equiv_L u_2$. Hence, there must exist w such that either

- $u_1w \in L$ and $u_2w \notin L$, or
- $u_1w \notin L$ and $u_2w \in L$.

Let us assume the former case (the other one is similar).

1) If $|w| \le k$, then $w \in \Sigma^{\le k}$, and since $row_k(u_1) = row_k(u_2)$ we get in particular $row_k(u_1)(w) = row_k(u_2)(w)$, that is $C_L(u_1w) = C_L(u_2w)$. But $u_1w \in L$ implies $C_L(u_1w) = \lambda$, and so $C_L(u_2w) = \lambda$ which is in contradiction with $u_2w \notin L$.

2) If $|w| > k$, then there must exist $v, w' \in \Sigma^*$ such that $w = vw'$ and $|v| = k$. Moreover, by assumption $u_1vw' \in L$ and $u_2vw' \notin L$, so $u_1v \not\equiv_L u_2v$. On the other hand since $row_k(u_1) = row_k(u_2)$ and $v \in \Sigma^{\le k}$, we have $row_k(u_1)(v) = row_k(u_2)(v)$, that is $C_L(u_1v) = C_L(u_2v) = v'$. Because $u_1v \cdot w' \in L$, $Tail_L(u_1v) \ne \emptyset$ and hence $C_L(u_1v) \in \Sigma^*$. Since $L \in k\text{-}Rev$, $u_1vv' \in L, u_2vv' \in L$ and $|v| = k$, we get $Tail_L(u_1v) = Tail_L(u_2v)$ (cf. Theorem 1) which is in contradiction with $u_1v \not\equiv_L u_2v$. $\qquad\square$

This result tells us that if $\mathcal{A}_L = (Q, \Sigma, \delta, q_0, F)$ is the minimal complete automaton for the k-reversible language L, then the values of function $row_k(u)$ on $\Sigma^{\le k}$ uniquely identify the state $\delta(q_0, u)$. We use this property to show that k-reversible languages are learnable in polynomial time with CQs.

3.1   The Algorithm

The algorithm follows the lines of L^*. We have an *observation table* denoted by (S, E, C) in which lines are indexed by the elements of a prefix-closed set S, columns are indexed by the elements of a suffix-closed set E, and the element of the table situated at the intersection of line u with column v is $C_L(uv)$.

We start with $S = \{\lambda\}$ and $E = \Sigma^{\le k}$, and then increase the size of S by adding elements with distinct row values. An important difference between our algorithm and L^* is that in our case the set E is never modified during the run of the algorithm (in L^*, E contains only one element in the beginning, and it is gradually enlarged when needed).

We say that the observation table (S, E, C) is *closed* if for all $u \in S$ and $a \in \Sigma$, there exists $u' \in S$ such that $row_k(u') = row_k(ua)$. Moreover, (S, E, C) is *consistent* if for all $u_1, u_2 \in S$, $row_k(u_1) \ne row_k(u_2)$. It is clear that if the table (S, E, C) is consistent and S has exactly n elements, where n is the index of L, then the strings in S are in bijection with the elements of $\Sigma^*/_{\equiv_L}$.

For any closed and consistent table (S, E, C), we construct the automaton $\mathcal{A}(S, E, C) = (Q, \Sigma, \delta, q_0, F)$ as follows. $Q := \{row_k(u) \mid u \in S\}$, $q_0 := row_k(\lambda)$, $F := \{row_k(u) \mid u \in S \text{ and } C_L(u) = \lambda\}$, and $\delta(row_k(u), a) := row_k(ua)$ for all $u \in S$ and $a \in \Sigma$.

To see that this is a well-defined automaton, note that since S is a non-empty prefix-closed set, it must contain λ, so q_0 is defined. Because S is consistent, there are no two elements u_1, u_2 in S such that $row_k(u_1) = row_k(u_2)$. Thus, F is well defined. Since the observation table (S, E, C) is closed, for each $u \in S$ and $a \in \Sigma$, there exists u' in S such that $row_k(ua) = row_k(u')$, and because it is consistent, this u' is unique. So δ is well defined.

Remark 1. The following statements are true.

1) $row_k(u)$ is a sink state if and only if $C_L(u) = \theta$;
2) $\delta(q_0, u) = row_k(u)$ for all u in $S \cup S\Sigma$.

We present a polynomial time algorithm that learns any k-reversible language L after asking a finite number of CQs.

Algorithm 1. An algorithm for learning the class k-*Rev* with CQs

---

1: $S := \{\lambda\}$, $E := \Sigma^{\leq k}$
2: *closed* := TRUE
3: update the table by asking CQs for all strings in $\{uv \mid u \in S \cup S\Sigma, v \in E\}$
4: **repeat**
5:     **if** $\exists u \in S$ and $a \in \Sigma$ such that $row_k(ua) \notin row_k(S)$ **then**
6:         add ua to S
7:         update the table by asking CQs for all strings in $\{uaa'v \mid a' \in \Sigma, v \in E\}$
8:         *closed* := FALSE
9:     **end if**
10: **until** *closed*
11: output $\mathcal{A}(S, E, C)$ and halt.

Note that since the algorithm adds to S only elements with distinct row values, the table (S, E, C) is always consistent. We will see that as long as $|S| < n$, it is not closed.

Lemma 1. *If $|S| < n$, then (S, E, C) is not closed.*

Proof. Let us assume that there exists $m < n$ such that $|S| = m$ and the table (S, E, C) is closed. Let $\mathcal{A}_L = (Q', \Sigma, \delta', q_0', F')$ be the minimal complete automaton accepting L, and $\mathcal{A}(S, E, C) = (Q, \Sigma, \delta, q_0, F)$.

We define the function $\varphi : Q \to Q'$ by $\varphi(row_k(u)) := \delta'(q_0', u)$. Note that φ is well-defined because there are no two strings u_1, u_2 in S such that $row_k(u_1) = row_k(u_2)$. Moreover, it is injective since $\varphi(row_k(u_1)) = \varphi(row_k(u_2))$ implies $\delta'(q_0', u_1) = \delta'(q_0', u_2)$ which is equivalent to $[u_1] = [u_2]$, and cf. Proposition 1, to $row_k(u_1) = row_k(u_2)$. We show that φ is a morphism of automata from $\mathcal{A}(S, E, C)$ to $\mathcal{A}_L$, that is: $\varphi(q_0) = q_0'$, $\varphi(F) \subseteq F'$, and $\varphi(\delta(row_k(u), a)) = \delta'(\varphi(row_k(u)), a)$ for all $u \in S$ and $a \in \Sigma$.

Clearly, $\varphi(q_0) = \varphi(row_k(\lambda)) = \delta'(q_0', \lambda) = q_0'$. Let us now take $row_k(u)$ in F, that is, $u \in S$ and $C_L(u) = \lambda$. Since $\varphi(row_k(u)) = \delta'(q_0', u)$ and $u \in L$, it follows that $\varphi(row_k(u)) \in F'$. Finally, $\varphi(\delta(row_k(u), a)) = \varphi(row_k(ua)) = \varphi(row_k(v))$ for some v in S such that $row_k(ua) = row_k(v)$ (the table is closed), and $\delta'(\varphi(row_k(u)), a) = \delta'(\delta'(q_0', u), a) = \delta'(q_0', ua)$. It is enough to see that $\varphi(row_k(v)) = \delta'(q_0', v) = \delta'(q_0', ua)$ (because by Proposition 1, $row_k(v) = row_k(ua)$ implies $[v] = [ua]$, and $\mathcal{A}_L$ is the minimal automaton accepting L) to conclude the proof.

We have constructed an injective morphism from $\mathcal{A}(S, E, C)$ to $\mathcal{A}_L$ such that $|Q| = m < n = |Q|$. Since both $\mathcal{A}(S, E, C)$ and $\mathcal{A}_L$ are complete automata, this leads to a contradiction. $\square$

We show that Algorithm 1 cannot be used for the whole class of regular languages.

Lemma 2. *Algorithm 1 does not work in general for arbitrary regular languages.*

Proof. Indeed, let us assume that Algorithm 1 can identify any regular language. Let us fix $k \geq 0$, and consider the language $L_k = \{ab^k a, ab^k b, b^{k+1} a\}$ which is finite, and hence regular. The minimal complete DFA of L_k is represented in Figure 1.

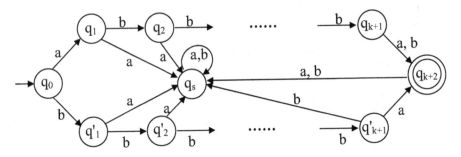

Fig. 1. The automaton $\mathcal{A}_{L_k}$

Since the strings $a \cdot b^k \cdot a$ and $b \cdot b^k \cdot a$ are both in L_k, and $Tail_{L_k}(a \cdot b^k) = \{a, b\} \neq \{a\} = Tail_{L_k}(b \cdot b^k)$, the language L_k is not k-reversible (Theorem 1).

When running the algorithm on L_k, the set S is initialized with the value $\{\lambda\}$. Then, since both $row_k(a)$ and $row_k(b)$ are different from $row_k(\lambda)$, one of the two elements is added to S. Note that for all u in $\Sigma^{\leq k}$, $row_k(a)(u) = row_k(b)(u)$ because:

- if $u = b^i$ with $0 \leq i \leq k$, then $C_{L_k}(au) = b^{k-i}a = C_{L_k}(bu)$, and
- if $u = \Sigma^{\leq k} \backslash \{b^i \mid 0 \leq i \leq k\}$, then $C_{L_k}(au) = \theta = C_{L_k}(bu)$.

Hence, $row_k(a) = row_k(b)$. But this implies that in the automaton output by the algorithm, the strings a and b represent the same state, a contradiction. $\square$

In the following sections we show that the algorithm runs in polynomial time, and terminates with the minimal automaton for the target language as its output.

3.2   Correctness and Termination

We have seen that as long as $|S| < n$, the table is not closed, so there will always be an u in S and a symbol a in Σ such that $row_k(ua) \notin row_k(S)$. Since the cardinality of the set S is initially 1, and increases by 1 with each "repeat-until" loop (lines 4–10), it will eventually be n, and hence the algorithm is guaranteed to terminate.

We claim that when $|S| = n$, the observation table (S, E, C) is closed and consistent, and $\mathcal{A}(S, E, C)$ is isomorphic to $\mathcal{A}_L$. Indeed if $|S| = n$, then the set $\{row_k(u) \mid u \in S\}$ has cardinality n, since the elements of S have distinct row values. Thus for all $u \in S$ and $a \in \Sigma$, $row_k(ua) \in row_k(S)$ (otherwise $[ua]$ would be the $(n+1)^{\text{th}}$ equivalence class of $\Sigma^*/_{\equiv_L}$), and hence the table is closed.

To see that $\mathcal{A}(S, E, C)$ and $\mathcal{A}_L$ are isomorphic, let us take $\mathcal{A}(S, E, C) = (Q, \Sigma, \delta, q_0, F)$, $\mathcal{A}_L = (Q', \Sigma, \delta', q_0', F')$, and the function $\varphi : Q \to Q'$ defined by $\varphi(row_k(u)) := \delta'(q_0', u)$ for all $u \in S$. As in the proof of Lemma 1, it can be shown that φ is a well-defined and injective automata morphism. Since the two automata have the same number of states, φ is also surjective, and hence bijective. Let us now show that $\varphi(F) = F'$. Indeed, take $q \in F'$. Because φ is bijective, there exists u in S such that $\varphi(row_k(u)) = q$. It follows immediately that $\delta'(q_0', u) \in F'$, and hence $u \in L$. Thus, $C_L(u) = \lambda$ and $row_k(u) \in F$. Clearly, $\varphi(row_k(u)) = q \in \varphi(F)$. So, $F' \subseteq \varphi(F)$, and since $\varphi(F) \subseteq F'$, $\varphi(F) = F'$ which concludes the proof.

3.3 Time Analysis and Query Complexity

Let us now discuss the time complexity of the algorithm. While the cardinality of S is smaller than n, the algorithm searches for a string u in S and a symbol a in Σ such that $row_k(ua)$ is distinct from all $row_k(v)$ with $v \in S$. This can be done using at most $|S|^2 \cdot |\Sigma| \cdot |E|$ operations: there are $|S|$ possibilities for choosing u (and the same number for v), $|\Sigma|$ for choosing a, and $|E|$ operations to compare $row_k(ua)$ with $row_k(v)$. If we take $|\Sigma| = l$, the total running time of the "repeat-until" loop can be bounded by $(1^2 + 2^2 + \ldots + (n-1)^2) \cdot l \cdot (1 + l + l^2 + \ldots + l^k)$. Note that by "operations" we mean string comparisons, since they are generally acknowledged as being the most costly tasks.

On the other hand, to construct $\mathcal{A}(S, E, C)$ we need n comparisons for determining the final states, and at most $n^2 \cdot |\Sigma| \cdot |E|$ operations for constructing the transition function. This means that the total running time of the algorithm is bounded by $n + l \cdot \frac{l^{k+1} - 1}{l - 1} \cdot \frac{n(n+1)(2n+1)}{6}$, that is $O(n^3 l^k)$.

As for the number of queries asked by the algorithm, it can be bounded by $|S \cup S\Sigma| \cdot |E|$ (i.e., by the size of the final observation table), so the query complexity of the algorithm is $O(nl^k)$.

4 Pattern Languages

Initially introduced by Angluin [1] to show that there are non-trivial classes of languages learnable from text in the limit, the class of pattern languages has been intensively studied in the context of language learning ever since. Polynomial time algorithms have been given for learning pattern languages using one or more examples and queries [13], or just superset queries [5], or for learning k-variables pattern languages from examples [11], *etc.*

We assume a finite alphabet Σ such that $|\Sigma| \geq 2$, and a countable, infinite set of *variables* $X = \{x, y, z, x_1, y_1, z_1, \ldots, \}$. A *pattern* π is any non-empty string over $\Sigma \cup X$. The *pattern language* $L(\pi)$ consists of all the words obtained by replacing the variables in π with arbitrary strings in Σ^+. Let us denote by $\mathcal{P}$ the set of all pattern languages over a fixed alphabet Σ.

We say that the pattern π is in *normal form* if the variables occurring in π are precisely $x_1, \ldots, x_k$, and for every j with $1 \leq j < k$, the leftmost occurrence of x_j in π is left to the leftmost occurrence of x_{j+1}.

Next we show that there exists an algorithm which learns $\mathcal{P}$ using a finite number of CQs.

4.1 The Algorithm

Suppose that the target language is a pattern language $L(\pi)$, where π is in normal form. Then the following algorithm outputs the pattern π after asking a finite number of CQs.

Algorithm 2. An algorithm for learning the class $\mathcal{P}$ with CQs

---

1: $w := C_L(\lambda), n := |w|, var := 0$
2: **for** $i := 1$ to n **do**
3:    $\pi[i] := null$
4: **end for**
5: **for** $i := 1$ to n **do**
6:    **if** $(\pi[i] = null)$ **then**
7:        choose $a \in \Sigma \backslash \{w[i]\}$ arbitrarily
8:        $v := C_L(w[1 \ldots i - 1]a), m := |v|$
9:        **if** $(|v| = |w[i+1, \ldots, n]|)$ **then**
10:            $var := var + 1, \pi[i] := x_{var}$
11:            **for all** $j \in \{1, \ldots, m\}$ for which $v[j] \neq w[i+j]$ **do**
12:                $\pi[i+j] := x_{var}$
13:            **end for**
14:        **else**
15:            $\pi[i] := w[i]$
16:        **end if**
17:    **end if**
18: **end for**
19: output π

4.2 Correctness and Termination

The correctness of the algorithm is based on the following observation. If w is the smallest string (in lex-length order) in $L(\pi)$ and $n = |w|$, then for all i in $\{1, \ldots, n\}$, we have:

– if $\pi[i]$ is a variable x such that i is the position of the leftmost occurrence of x in π, then $|C_L(w[1, \ldots, i - 1]a)| = |w[i + 1, \ldots, n]|$ for any symbol

$a \in \Sigma$; moreover, we can detect the other occurrences of the variable x in π by just checking the positions where the strings $C_L(w[1, \ldots, i-1]a)$ and $w[i+1, \ldots, n]$ do not coincide, where a is any symbol in $\Sigma \backslash \{w[i]\}$;
- if $\pi[i] = a$ for some a in Σ, then for all $b \in \Sigma \backslash \{a\}$, $C_L(w[1, \ldots, i-1]b)$ is either θ, or longer than $w[i+1, \ldots, n]$.

Obviously, the algorithm terminates in finite steps.

4.3 Time Analysis and Query Complexity

For each symbol in the pattern, the algorithm makes at most $n+1$ comparisons, where n is the length of the pattern. This implies that the total running time of the algorithm is bounded by $n(n+1)$, that is $O(n^2)$.

It is easy to see that the query complexity is linear in the length of the pattern since the algorithm does not ask more than $n+1$ CQs.

5 Learning with CQs Versus Learning with MQs

The notion of CQ appeared as an extension of the well-known and intensively studied MQ. The inspiration for introducing them comes from a real life setting (which is the case for MQs also): when children make mistakes, the adults do not reply by a simple 'yes' or 'no' (the agreement is actually implicit), but they also provide them with a corrected word. Clearly, CQs can be thought as some more informative MQs. So, it is only natural to compare the two learning settings (learning with CQs vs. learning with MQs), and to analyze their expressive power.

The first step in this direction has already been done: C. Tîrnăucă and S. Kobayashi showed in [16] that learning with CQs is strictly more powerful than learning with MQs, when we neglect the time complexity.

In this section we make a step further towards understanding the differences and similarities between these two learning models by taking into consideration the efficiency of the learning algorithms, that is, the time complexity. For this, we need some further terminology.

Let $\mathcal{C} = (L_i)_{i \geq 1}$ be an indexable class. We say that $\mathcal{C}$ is *polynomially learnable with MQs* (or *with CQs*) if there exists a polynomial time algorithm which learns $\mathcal{C}$ using MQs (CQs, respectively). We denote the collection of all indexable classes $\mathcal{C}$ which are polynomially learnable with MQs by *PolMemQ* (*PolCorQ* is defined similarly).

Recall that if the correction for a given string u is λ, then the string is in the language, and the oracle's answer would be 'yes'; in all other cases, the string is not in the language, and the answer would be 'no'. Since the answer to any CQ gives us also the answer to the corresponding MQ, it follows immediately that the class *PolMemQ* is included in *PolCorQ*. We show that the inclusion is strict using pattern languages as the separating case.

Theorem 2. *The class $\mathcal{P}$ is in PolCorQ\PolMemQ.*

Proof. It is clear that $\mathcal{P}$ is in *PolCorQ* since Algorithm 2 is a polynomial time algorithm which identifies any pattern language using COs (see Section 4).

Assume now that $\mathcal{P}$ is in *PolMemQ*, and consider the class of singletons $\mathcal{S}$ of fixed length n over the alphabet Σ. Because every language $L = \{w\}$ in $\mathcal{S}$ can be written as a pattern language ($L = L(w)$, where w is a pattern without any variables), $\mathcal{S}$ is also in *PolMemQ*. But Angluin shows that, if l is the cardinality of the alphabet, then any algorithm which learns $\mathcal{S}$ using MQs needs to ask at least $l^n - 1$ MQs [2], which leads to a contradiction. □

Note that although $\mathcal{P}$ is not polynomially learnable with MQs, it is in *MemQ* (see [14], page 266). However, there are classes of languages in *PolCorQ* which cannot be learned at all (polynomially or not) using MQs, as we will see in the sequel.

Theorem 3. *The class k-Rev is in PolCorQ\MemQ.*

Proof. Since Algorithm 1 learns any k-reversible language using CQs in polynomial time (see Section 3), it follows immediately that *k-Rev* is in *PolCorQ*.

To show that *k-Rev* is not in *MemQ*, we use Mukouchi's characterization of the class *MemQ* in terms of pairs of definite finite tell-tales. A pair $\langle T, F \rangle$ is said to be a *pair of definite finite tell-tales of L_i* if:

(1) T_i is a finite subset of L_i, F_i is a finite subset of $\Sigma^* \backslash L_i$, and
(2) for all $j \geq 1$, if L_j is consistent with the pair $\langle T, F \rangle$ (that is, $T \subseteq L_j$ and $F \subseteq \Sigma^* \backslash L_j$), then $L_j = L_i$.

Mukouchi proves in [14] that an indexable class $\mathcal{C} = (L_i)_{i \geq 1}$ belongs to *MemQ* if and only if a pair of definite finite tell-tales of L_i is uniformly computable for any index i.

So, let us assume that *k-Rev* is in *MemQ*. Consider the alphabet Σ such that $\{a, b\} \subseteq \Sigma$, and the language $L = \{a\}$. Clearly, L is in *k-Rev* for all $k \geq 0$ and hence a pair of definite finite tell-tales $\langle T, F \rangle$ is computable for L. This means that $T \subseteq L$ and F is a finite set included in $\Sigma^* \backslash \{a\}$. Let us take $m = \max\{|w| \mid w \in F\}$ and the language $L' = \{a, ba^m b\}$. It is clear that L' is in *k-Rev* for all $k \geq 0$, and that it is consistent with $\langle T, F \rangle$. Moreover, $L' \neq L$ which leads to a contradiction. □

On the other hand, very simple classes of languages cannot be learned in polynomial time using CQs. For example, if we take $\bar{\mathcal{S}}$ to be $\bar{\mathcal{S}} = (L_w)_{w \in \Sigma^*}$, where $L_w = \Sigma^* \backslash \{w\}$, then any algorithm would require at least $1 + l + l^2 + \ldots + l^n$ CQs in order to learn L_w, where $n = |w|$ and $l = |\Sigma|$.

6   Concluding Remarks

We have investigated the learnability of some well-known language classes in the query learning setting. Figure 2 illustrates a synthesis of the results obtained.

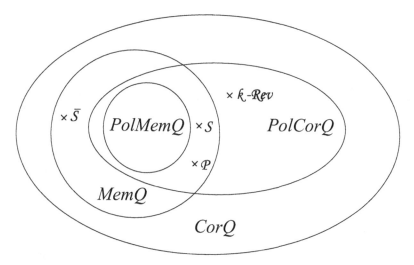

Fig. 2. CQ learning vs MQ learning

The class of pattern languages was known to be learnable with MQs. We gave a polynomial time algorithm for learning $\mathcal{P}$ using CQs, and showed that they cannot be efficiently learned with MQs. Moreover, we proved that k-reversible languages are efficiently learnable with CQs, and not learnable (at all) with MQs.

For the future, we would like to see what happens with the learnability results obtained so far when we change the correcting string. A possible direction could be to choose as correction the closest string in the edit distance.

Acknowledgments

Many thanks to Magnus Steinby for fruitful discussions and valuable advices.

References

1. Angluin, D.: Finding patterns common to a set of strings (extended abstract). In: STOC '79. Proc. 11th Annual ACM Symposium on Theory of Computing, pp. 130–141. ACM Press, New York, NY, USA (1979)
2. Angluin, D.: A note on the number of queries needed to identify regular languages. Information and Control 51(1), 76–87 (1981)
3. Angluin, D.: Inference of reversible languages. Journal of the ACM 29(3), 741–765 (1982)
4. Angluin, D.: Learning regular sets from queries and counterexamples. Information and Computation 75(2), 87–106 (1987)
5. Angluin, D.: Queries and concept learning. Machine Learning 2(4), 319–342 (1988)
6. Angluin, D.: Negative results for equivalence queries. Machine Learning 5(2), 121–150 (1990)

7. Beccera-Bonache, L., Dediu, A.H., Tîrnăucă, C.: Learning DFA from correction and equivalence queries. In: Sakakibara, Y., Kobayashi, S., Sato, K., Nishino, T., Tomita, E. (eds.) ICGI 2006. LNCS (LNAI), vol. 4201, pp. 281–292. Springer, Heidelberg (2006)

8. Drewes, F., Högberg, J.: Query learning of regular tree languages: How to avoid dead states. Theory of Computing Systems 40(2), 163–185 (2007)

9. Gold, E.M.: Language identification in the limit. Information and Control 10(5), 447–474 (1967)

10. Hopcroft, J.E., Ullman, J.D.: Introduction to Automata Theory, Languages, and Computation. Addison-Wesley, Reading, Massachusetts (1979)

11. Kearns, M., Pitt, L.: A polynomial-time algorithm for learning k-variable pattern languages from examples. In: COLT '89. Proc. 2nd Annual Workshop on Computational Learning Theory, pp. 57–71. Morgan Kaufmann Publishers Inc., San Francisco, CA, USA (1989)

12. Martín-Vide, C., Mitrana, V., Păun, G. (eds.): Formal Languages and Applications. Studies in Fuzzyness and Soft Computing, vol. 148, Springer, Heidelberg (2004)

13. Marron, A., Ko, K.I.: Identification of pattern languages from examples and queries. Information and Computation 74(2), 91–112 (1987)

14. Mukouchi, Y.: Characterization of finite identification. In: Jantke, K.P. (ed.) AII 1992. LNCS, vol. 642, pp. 260–267. Springer, Heidelberg (1992)

15. Sakakibara, Y.: Learning context-free grammars from structural data in polynomial time. Theoretical Computer Science 76, 223–242 (1990)

16. Tîrnăucă, C., Kobayashi, S.: A characterization of the language classes learnable with correction queries. In: TAMC '07. LNCS, vol. 4484, pp. 398–407. Springer, Heidelberg (2007)

17. Tîrnăucă, C.I., Tîrnăucă, C.: Learning regular tree languages from correction and equivalence queries. Journal of Automata, Languages and Combinatorics, Special Issue WATA (2006) (to appear, 2007)

18. Valiant, L.G.: A theory of the learnable. Communications of the ACM 27(11), 1134–1142 (1984)

19. Yokomori, T.: Learning non-deterministic finite automata from queries and counterexamples. Machine Intelligence 13, 169–189 (1994)

20. Yokomori, T.: Learning two-tape automata from queries and counterexamples. Mathematical Systems Theory 29(3), 259–270 (1996)

Learning and Verifying Graphs Using Queries with a Focus on Edge Counting

Lev Reyzin* and Nikhil Srivastava**

Department of Computer Science
Yale University, New Haven, CT 06520, USA
{lev.reyzin,nikhil.srivastava}@yale.edu

Abstract. We consider the problem of learning and verifying hidden graphs and their properties given query access to the graphs. We analyze various queries (edge detection, edge counting, shortest path), but we focus mainly on edge counting queries. We give an algorithm for learning graph partitions using $O(n \log n)$ edge counting queries. We introduce a problem that has not been considered: verifying graphs with edge counting queries, and give a randomized algorithm with error ϵ for graph verification using $O(\log(1/\epsilon))$ edge counting queries. We examine the current state of the art and add some original results for edge detection and shortest path queries to give a more complete picture of the relative power of these queries to learn various graph classes. Finally, we relate our work to Freivalds' 'fingerprinting technique' – a probabilistic method for verifying that two matrices are equal by multiplying them by random vectors.

1 Introduction

Graph learning appears in many different contexts. Suppose we are presented with a circuit containing a set of chips on a board. We can test the resistance between two chips with an ammeter. In as few measurements as possible, we want to learn whether the entire circuit is connected, or whether we need to power the components separately. This can be seen as a graph learning problem, in which the chips are vertices of a hidden graph and the ammeter measurements are queries into the graph, which tell whether a pair of vertices is connected by a path. If we are given a strong enough ammeter to tell not only whether two chips are connected, but also how far apart they are in the underlying circuit, we get the stronger 'shortest path' queries.

In a different setting [3], testing which pairs of chemicals react in a solution is modeled by 'edge detection' queries. Here, vertices correspond to chemicals, edges designate chemical reactions, and a set of chemicals 'reacts' iff it induces an edge. Applications of this model extend to bioinformatics, where learning a

---

* Supported by a Yahoo! Research Kern Family Scholarship.
** This material is based upon work supported in part by the National Science Foundation under Grant No. 0707522.

M. Hutter, R.A. Servedio, and E. Takimoto (Eds.): ALT 2007, LNAI 4754, pp. 285–297, 2007.

hidden matching [2] turns out to be useful in DNA sequencing. With each setup we have different tools and target concepts to learn.

Our goal is to explore several graph-learning problems and queries. We consider the following types of queries, defined on graphs $G = (V, E)$:

- **Edge detection query (ED):** Check if there is edge between any two vertices in $S \subseteq V$. *This model has applications in genome sequencing and was studied in [1,2,3,4,10].*
- **Edge counting query (EC):** Return the number of edges in the subgraph induced by $S \subseteq V$. *This has extensive uses in bioinformatics and was studied in [6,11].*
- **Shortest Path query (SP):** Return the length of shortest path in G between two vertices; if no path exists, return ∞. *This is the canonical model in the evolutionary tree literature; see [12,13,14].*

The second kind of task we consider is graph verification. Suppose we are interested in learning the structure of some protein networks, and after months of careful measurement, we complete our learning task. If we then find out there is a small chance we made a mistake in our measurements or if we have reason to believe our equipment may have been broken during experimentation, can we verify the structures we've learned more efficiently than learning them over again? More concretely, we are interested in how efficiently can we decide whether a graph presented to us is indeed the "true graph." This is a natural question to ask, especially since real world data is often noisy, or we sometimes have reason to mistrust results we are given. Every learning problem induces a new verification problem.

We consider different classes of graphs for our learning and verification tasks. The first class is **arbitrary graphs**, where there are no restrictions on the topology of the graph. Any algorithm that learns or verifies an arbitrary graph can also be used for more restricted settings. We also consider learning **trees**, where we know the graph we are trying to learn is a tree, but we are not aware of its topology. This is a natural setting for learning structures that we know do not have underlying cycles, for example evolutionary trees. Finally, we consider the problem of learning the **partition** of a graph into connected components. Here, we do not restrict the underlying class of graphs, but instead relax the learning problem. This is a natural question in settings where different partitions represent qualitative differences, for example in electrical networks, a power generator in one partition cannot power any nodes outside its own partition. Note that this also subsumes the natural question of whether or not a graph is connected.

In this paper we fill in some gaps in the literature on these problems and introduce the verification task for these queries. We also introduce the problem of learning partitions and present results in the **EC** query case. We then show what problems remain open. After presenting a summary of the past work done on these problems, we divide our results into two sections: Graph Learning and Graph Verification.

2   Previous Work

In one of the earliest works in graph discovery, Hein [12] tackles the problem of learning a degree d restricted tree with **SP** queries. He describes an $O(dn \lg n)$ algorithm that builds the tree by inserting one node at a time, in a carefully chosen order under which each insertion takes $O(d \lg n)$ queries. Among other results, King et al. [13] provide a matching lower bound by showing that solving this problem requires solving multiple partition problems whose difficulty they then analyze.

Angluin and Chen [3] show that $O(\lg n)$ adaptive **ED** queries per edge are sufficient to learn an arbitrary hidden graph. Their algorithm repeatedly divides the graph into independent subgraphs (i.e., it colors the graph), so as to eliminate interference to **ED** queries from previously discovered edges, and uses a variant of binary search to find new edges within each subgraph. It is worth noting that this is not far from an information-theoretic lower bound of $\Omega(\epsilon \lg n)$ **ED** queries per edge for the family of graphs with $n^{2-\epsilon}$ edges. A later paper [4] generalizes these results to hypergraphs using different techniques.

The work of Angluin and Chen is preceded by a few papers [1,2,10] that tackle learning restricted families of graphs, such as stars, cliques, and matchings. Alon et al. [2] provide lower and upper bounds of $.32\binom{n}{2}$ and $(1/2 + o(1))\binom{n}{2}$ respectively on learning a matching using nonadaptive **ED** queries, and a tight bound of $\Theta(n \lg n)$ **ED** queries in expectation if randomization is allowed. Alon and Asodi [1] prove similar bounds for the classes of stars and cliques. Grebinski and Kucherov [10] study reconstructing Hamiltonian paths with **ED** queries. It turns out that many of these results are subsumed by those of [3] if we ignore constant factors.

Grebinski and Kucherov [11] also study the problem of learning a graph using **EC** queries and give tight bounds of $\Theta(dn)$ and $\Theta(n^2 / \lg n)$ nonadaptive queries for d-degree-bounded and general graphs respectively. They also prove tight $\Theta(n)$ bounds for learning trees. Their constructions make heavy use of separating matrices. In [6], Grebinski and Kucherov present a survey on learning various restricted cases of graphs, including Hamiltonian cycles, matchings, stars, and k−degenerate graphs, with **ED** and **EC** queries.

In the graph verification setting, Beerliova et al. [5] consider the problem of discovering and verifying networks using distance queries. In this setting that models discovering nodes on the internet, the learner can query a vertex, and the answer to the query is the set of all edges whose endpoints have different graph-theoretic distance from the query vertex. They show there is no $o(\log n)$ competitive algorithm unless $P = NP$.

Both the learning and verification tasks also bear some relation to the field of Property Testing, where the object is to examine small parts of the adjacency matrix of a graph to determine a global property of the graph. For a survey of this area, see [9].

3   Graph Learning

We first note that **EC** queries are at least as strong than **ED** queries and that the problem of learning an arbitrary graph is at least as hard as learning trees or partitions. Hence, in this paper, any lower bounds for stronger queries and easier targets apply to weaker queries and harder target classes. Conversely, any upper bounds we establish for weaker queries and harder problems apply for stronger queries and more restricted classes.

We first establish that $\Theta(n^2)$ **SP** and **ED** queries is essentially tight for learning arbitrary graphs and partitions.

Proposition 1. $\Omega(n^2)$ **SP** queries are needed to learn the **partition** of a hidden graph on n vertices.

Proof. We prove this by an adversarial argument; the adversary simply answers '∞' (i.e., not connected) for all pairs of vertices i, j. If fewer than $\binom{n}{2}$ queries are made, then some pair i, j is not queried, and the algorithm cannot differentiate between the graph with no edges and the graph with a single edge $\{i, j\}$ (for which $\mathbf{SP}(i, j) = 1$). But these graphs have different partitions.   □

If k is the number of components in a graph, there is an obvious algorithm that does better for $k < n$, even without knowledge of k:

Proposition 2. $O(nk)$ **SP** queries are sufficient to determine the **partition** of a hidden graph on n vertices, if k is the number of components in the graph.

Proof. We use a simple iterative algorithm:

- Step 1: Place 1 in its own component.[1]
- Step $i > 1$: Query $\mathbf{SP}(i, w)$ for an item w from each existing component; if $\mathbf{SP}(i, w) \neq \infty$, place i in the corresponding component and move to the next step. Otherwise, create a new component containing i and move to the next step.

Correctness is trivial. For complexity, note that there at most k components at any step (since there are at most k components at phase n and components are never destroyed); hence n vertices take at most nk queries.   □

Proposition 3. $\Omega(n^2)$ **ED** queries are needed to learn the **partition** of a hidden graph on n vertices.

Proof. Consider the class of graphs on n vertices consisting of two copies of $K_{\frac{n}{2}}$, which we will call C_1 and C_2, and one possible edge between C_1 and C_2. If there is an edge, all the vertices are in a single component; otherwise there are two components. Any algorithm that learns the partition must distinguish between the two cases. Observe that an **ED** query on a set S containing more than one vertex from either C_1 or C_2 will not yield any information since an

---

[1] We use numbers $1, 2, \ldots, n$ to represent the vertices of the graph.

edge is guaranteed to be present in S and any such query will be answered with a 'yes'. Hence, all informative queries must contain one vertex from C_1 and one vertex from C_2. An adversary can keep on answering 'no' to all such queries, and unless all possible pairs are checked, an edge may be present between C_1 and C_2. Hence, the algorithm cannot tell whether the graph has one component or two until it asks all $\approx (\frac{n}{2})^2 = \Omega(n^2)$ queries. □

It turns out that **EC** queries are considerably more powerful than **ED** queries for this problem.

Proposition 4. $\Omega(n)$ **EC** *queries are needed to learn the* **partition** *of a hidden graph on n vertices.*

Proof. We use an information-theoretic argument. The number of partitions of an n element set is given by the Bell number B_n; according to de Bruijn [7]:

$$\ln B_n = \Omega(n \ln n)$$

Since each **EC** query gives a $\lg(\binom{n}{2}) = 2\lg n$ bit answer, we need $\Omega(\frac{\lg(B_n)}{2\lg n}) = \Omega(\frac{n \lg n}{\lg n}) = \Omega(n)$ queries. □

Theorem 5. $O(n \lg n)$ **EC** *queries are sufficient to learn the* **partition** *of a hidden graph on n vertices.*

Proof. Consider the following n−phase algorithm, in which the components of $G[1 \ldots i]$ are determined in phase i.

- *Phase 1*: Set $\mathcal{C} = \{c_1\}$ with $c_1 = \{1\}$. $\mathcal{C}$ will keep track of the components $c_1, c_2, \ldots$ known at any phase, and we will let $\mathcal{C} + v$ denote $\{v\} \cup \bigcup_{c_i \in \mathcal{C}} c_i$.
- *Phase $(i+1)$*: Let $v = (i+1)$, and query **EC**$(\mathcal{C} + v)$. If **EC**$(\mathcal{C} + v) = $ **EC**$(\mathcal{C})$ (i.e., there are no edges between v and $\mathcal{C}$), add a new component $c = \{v\}$ to $\mathcal{C}$.

  Otherwise, split $\mathcal{C}$ into roughly equal halves $\mathcal{C}_1$ and $\mathcal{C}_2$ and query **EC**$(\mathcal{C}_1 + v)$, **EC**$(\mathcal{C}_2 + v)$. Pick any half $h \in \{1, 2\}$ for which **EC**$(\mathcal{C}_h + v) > $ **EC**$(\mathcal{C}_h)$ and repeat recursively until **EC**$(\{c_j\} + v) > $ **EC**(c_j) for a single component $c_j \in \mathcal{C}^2$. This implies that there are edges between c_j and v; we will call c_j a *live* component.

  Repeat on $\mathcal{C} \setminus \{c_j\}$ to find another live component $c_{j'}$, if it exists; repeat again on $\mathcal{C} \setminus \{c_j, c_{j'}\}$ and so on until no further live components remain (or equivalently, no new edges are found). Remove all live components from $\mathcal{C}$ and add a new component $\{v\} \cup \bigcup_{\text{live } c_j} c_j$.

Correctness is simple, by induction on the phase: we claim that $\mathcal{C}$ contains the components of $G[1 \ldots i]$ at the end of phase i. This is trivial for $i = 1$. For $i > 1$, suppose $\mathcal{C} = \{c_1, \ldots, c_m\}$ at the beginning of phase i, and by the inductive hypothesis $\mathcal{C}$ contains precisely the components of $G[1 \ldots (i-1)]$. The

---

[2] Notice that this is essentially a binary search.

components that do not have edges to v are unaffected by its introduction in $G[1\ldots i]$, and these are not changed by the algorithm. All other components are connected to v and therefore to each other in $G[1\ldots i]$; but these are marked 'live' and subsequently merged into a single component at the end of the phase. This completes the proof.

To analyze complexity, we use a "potential argument." Let Δ_i denote the increase in the number of components in $\mathcal{C}$ during phase i. There are three cases:

- $\Delta_i = 1$: There are no live components (v has no edges to any component in $\mathcal{C}$), and this is determined with a single $\mathbf{EC}(\mathcal{C} + v)$ query.
- $\Delta_i = 0$: There is exactly 1 live component (v connects to exactly one member of $\mathcal{C}$). Since there are at most n components to search, it takes $O(\lg n)$ queries to find this component.
- $\Delta_i < 0$: There are $k > 1$ live components with edges to v, bringing the number of components down by $k - 1$.[3] Finding each one takes $O(\lg n)$ queries, for a total of $O(k \lg n) = O((-\Delta_i + 1) \lg n)$.

The total number of queries is

$$\sum_{i:\Delta_i=1} 1 + \sum_{i:\Delta_i=0} (\lg n) + \sum_{i:\Delta_i<0} O((-\Delta_i + 1)\lg n)$$

The first two sums are bounded by $O(n \lg n)$ since there are n phases, and the last one becomes

$$O(n\lg n) + O(\lg n)\sum_{\Delta_i<0}(-\Delta_i).$$

But $\sum_{\Delta_i<0}(-\Delta_i)$, the total *decrease* in the number of components, cannot be greater than n since the total *increase* is bounded by n (one new component per phase) and the final number of components is nonnegative. So the total number of queries is $O(n\lg n)$, as desired.

To see that this analysis is tight, consider the case where G has exactly $n/2$ components, with $\Delta_i = 1$ for $i < n/2$, $\Delta_i = 0$ for $i \geq n/2$. The first $n/2$ phases take only $O(n/2)$ queries, but the remaining $n/2$ take $O(\lg(n/2))$ queries each, for a total of $O(n/2 \lg(n/2) + n/2) = O(n\lg n)$ queries.   □

Proposition 6. $O(|E| \lg n)$ **EC** *queries are sufficient to learn a hidden* **graph** *on n vertices.*

Proof. The algorithm of Angluin and Chen ([3]) achieves this since **EC** queries are more powerful than **ED** queries, but we present a simpler method here that exploits the counting ability of **EC**. The key observation is that we can learn the degree of any vertex v in two queries:

$$d(v) = \mathbf{EC}(V) - \mathbf{EC}(V \setminus \{v\})$$

---

[3] The k components previously in $\mathcal{C}$ are replaced by a single component, hence $\Delta_i = -(k-1)$.

We use this to find all of the neighbors of v, using a binary search similar to that in the algorithm of theorem 5. Split $V \setminus \{v\}$ into halves V_1, V_2 and query $\mathbf{EC}(V_1+v), \mathbf{EC}(V_2+v)$. Pick a half such that $\mathbf{EC}(V_i+v) > \mathbf{EC}(V_i)$ and recurse until $\mathbf{EC}(w+v) > 0$ for some vertex w. This implies that w is a neighbor of v. Repeat the procedure on $V \setminus \{w, v\}$ to find more neighbors, and so on, until $d(v)$ neighbors are found.

We can reconstruct the graph by finding the neighbors of each vertex; this uses a total of

$$\sum_v d(v) \lg n = \lg n \sum_v d(v) = 2|E| \lg n = O(|E| \lg n)$$

queries, as desired.   □

It follows from the above proof that the degree sequence of a graph can be computed in $2n$ queries, and consequently any property that is determined by it takes only linear queries.

Proposition 7. $\Omega(n^2)$ **SP** *queries are needed to learn a hidden* **tree**.

Proof. Consider a graph G on $2n+1$ vertices, which are of three kinds: a single center vertex s, n 'inner' vertices $x_1 \ldots x_n$, and n 'outer' vertices $y_1 \ldots y_n$. The center and inner vertices form a star (with edges $\{x_i, s\}$) and the outer vertices are matched with the inner vertices (for each y_i there is a unique x_{j_i} such that $\{x_{j_i}, y_i\}$ is an edge; no x_{j_i} is repeated).

Suppose a learning algorithm knows that G is a quasi-star. There are only three kinds of **SP** queries: $\mathbf{SP}(s, x_i) = 1$, $\mathbf{SP}(s, y_i) = 2$, and

$$\mathbf{SP}(x_i, y_j) = \begin{cases} 1 \text{ if } \{x_i, y_j\} \text{ is an edge} \\ 3 \text{ otherwise} \end{cases}$$

The only query that gives any information is the last kind, and the problem reduces to that of learning a matching using **ED** queries, which we know by [2] takes $\Omega(n^2)$ queries.   □

Table 1. Summary of results. n denotes the number of vertices, $|E|$ the number of edges, d the degree restriction, and k the number of components

Query	partition	graph	tree
ED	$\Theta(n^2)$	$\Theta(|E| \lg n), \Theta(n^2)[3]$	$\Theta(n \lg n)$
EC	$O(n \lg n)$ $\Omega(n)$	$O(|E| \lg n), O(\frac{n^2}{\lg n}), O(dn)[3,11]$ $\Omega(dn), \Omega(\frac{n^2}{\lg n})[11]$	$\Theta(n)$
SP	$\Theta(nk)$	$\Theta(n^2)$	$\Theta(n^2), \Theta(dn \lg n) [12,13]$

Table 1 shows the known bounds for the problems we consider. We can see that tight asymptotic bounds exist for all of these learning problems, except for learning partitions with **EC**.

We note that learning a tree becomes significantly easier when the degrees of its vertices are restricted, and in many cases, knowing a bound on the degree of a graph can help with the learning problem.

4   Graph Verification

In this setting, a verifier is presented a graph $G(V, E)$ and asked to check whether it is the same as a hidden graph $G^*(V, E^*)$, given query access to G^*. In this section, we explore the complexity of graph verification using various queries. Mainly, we show that while verifying unrestricted graphs is hard using **SP** and **ED** queries, there is a fast randomized algorithm that uses **EC** queries.

Proposition 8. *Verifying an arbitrary **graph** takes $\Theta(n^2)$ **SP** queries and $\Theta(n^2)$ **ED** queries.*

Proof. Consider the problem of verifying a clique, when the hidden graph is a clique with some edge (u, v) removed, and the verifier knows this. $\mathbf{SP}(u', v') = 2$ if and only if $u' = u$ and $v' = v$. A simple adversarial argument shows that $\Omega(n^2)$ queries are necessary. Similarly, for **ED** queries, let $S = \{u, v\}$. The answer to query $\mathbf{ED}(U)$, where $|U| \neq 2$ is predetermined. Otherwise, $\mathbf{ED}(U) = 0$ if and only if $U = S$. There are $\binom{n}{2}$ choices for S such that $|S| = 2$; hence $\Omega(n^2)$ are needed. For both **SP** and **ED** queries the $O(n^2)$ algorithm of checking all pairs of vertices is obvious.   □

Given that **SP** queries are most often considered in evolutionary tree learning, we also consider the problem of verifying a tree with **SP** queries. In this setting, the verifier knows the hidden graph is a tree and is presented with a tree to verify.

Proposition 9. *Verifying a **tree** takes $\Theta(n)$ **SP** queries.*

Proof. Consider the problem of verifying a path graph (from the class of path graphs). This reduces to verifying that a given ordering of the vertices is correct. If the answers to each query are consistent with the graph to be verified, each query verifies at most two vertices in the ordering. An adversary can choose whether or not to swap any pair of vertices that have not been queried and either stay consistent with the input path graph or not until at least $n/2$ **SP** queries have been performed. Conversely, we can verify each edge individually in $n - 1$ queries.   □

We now consider the problem of verifying a graph with **EC** queries. Here, we see that **EC** queries are quite powerful for verifying arbitrary graphs.

Theorem 10. *Any **graph** can be verified by a randomized algorithm using 1 **EC** query, with success probability 1/4.*

Proof. We define $\mathbf{EC}(V, G)$ to be the query $\mathbf{EC}(V)$ on graph G. The algorithm is simple. We let Q be a random subset of vertices of V, with each vertex chosen independently with probability $\frac{1}{2}$. We query $\mathbf{EC}(Q, G^*)$ and compute $\mathbf{EC}(Q, G)$. If the two quantities are not equal, we say G and G^* are different. Otherwise we say they are the same. We will show that if $G = G^*$ the algorithm always returns the correct answer, and otherwise gives the correct answer with probability at least $\frac{1}{4}$.

Consider the symmetric difference $S = (V, E \Delta E^*)$. Let $A = \{(u,v) \in E \setminus E^* : u, v \in Q\}$ and $B = \{(u, v) \in E^* \setminus E : u, v \in Q\}$. If $G = G^*$ then $|A| = |B| = 0$ and we are always right in saying the graphs are identical; otherwise $G \neq G^*$ and $E \Delta E^* \neq \varnothing$, so by the following lemma $|E \Delta E^*| = |A| + |B|$ is odd with probability $\frac{1}{4}$. But this immediately implies that $|A| \neq |B|$, as desired.   □

Lemma 11. *Let $G(V, E)$ be a graph with at least one edge. Let $G'(V', E')$ be the subgraph induced by taking each vertex in G independently with probability $\frac{1}{2}$. If G is non-empty, the probability that $|E'|$ is odd is at least $\frac{1}{4}$.*

Proof. Fix an ordering $v_1 \ldots v_n$ so that $(v_{n-1}, v_n) \in E$. Select each of $v_1 \ldots v_{n-2}$ independently with probability $1/2$, and let H' be the subgraph induced by the selected vertices. Suppose the probability that H' contains an odd number of edges (i.e., $\texttt{parity}(H') = 1$) is p.

Let i (resp. j) be the number of edges between v_{n-1} and H' (resp. v_n and H'). Consider two cases:

- $i \equiv j \mod 2$ If both are chosen an odd number of edges is added to H' and $\texttt{parity}(H') = 1 - \texttt{parity}(G')$. This happens with probability $1/4$.
- $i \not\equiv j \mod 2$. Assume w.l.o.g. that i is odd and j is even. Then, if v_{n-1} is chosen and v_n is *not* chosen, an odd number of edges is added to H', and again $\texttt{parity}(H') = 1 - \texttt{parity}(G')$. This happens with probability $1/4$.

On the other hand, if neither v_{n-1} nor v_n is chosen then $\texttt{parity}(G') = \texttt{parity}(H')$, and this happens with probability $1/4$. So upon revealing the last two vertices, the parity of H' is flipped with probability at least $1/4$ and not flipped with probability at least $1/4$, independently of what happens in H'. Let F denote the event that it is flipped (i.e., that $\texttt{parity}(H') \neq \texttt{parity}(G')$). Then,

$$\begin{aligned}
\mathbb{P}[\texttt{parity}(G') = 1] &= \mathbb{P}[\texttt{parity}(G') = 1 | \texttt{parity}(H') = 1]\mathbb{P}[\texttt{parity}(H') = 1] \\
&\quad + \mathbb{P}[\texttt{parity}(G') = 1 | \texttt{parity}(H') = 0]\mathbb{P}[\texttt{parity}(H') = 0] \\
&= \mathbb{P}[\overline{F}|\texttt{parity}(H') = 1]p + \mathbb{P}[F|\texttt{parity}(H') = 0](1 - p) \\
&= \mathbb{P}[\overline{F}]p + \mathbb{P}[F](1 - p) \quad \text{by independence} \\
&\geq 1/4(p + 1 - p) = 1/4
\end{aligned}$$

as desired.   □

This finishes the proof of Theorem 10. Since this result has 1-sided error, we can easily boost the $\frac{1}{4}$ probability to any constant, and Corollary 12 follows immediately.

Corollary 12. *Any graph can be verified by a randomized algorithm with error* ϵ *using* $O(\log(\frac{1}{\epsilon}))$ **EC** *queries.*

4.1   Relation to Fingerprinting

Suppose A and B are $n \times n$ matrices over a field $\mathbb{F}$. It is known that if $A \neq B$, then for a vector $v \in \{0,1\}^n$ chosen uniformly at random we have

$$\mathbb{P}[Av \neq Bv] \geq 1/2.$$

This is Freivalds' fingerprinting technique [8]. It is was originally developed as a technique for verifying matrix multiplications, and can be used for testing for equality of any two matrices.

An easy extension of this method says that for vectors $v, w \in \{0,1\}^n$ chosen independently uniformly at random, if $A \neq B$ we have

$$\begin{aligned}
\mathbb{P}[w^T Av \neq w^T Bv] &= \mathbb{P}[w^T Av \neq w^T Bv | Av = Bv]\mathbb{P}[Av = Bv] \\
&\quad + \mathbb{P}[w^T Av \neq w^T Bv | Av \neq Bv]\mathbb{P}[Av \neq Bv] \\
&\geq 0 \times \mathbb{P}[Av = Bv] + \frac{1}{2} \times \frac{1}{2} \\
&= \frac{1}{4}
\end{aligned}$$

This bears a strong resemblance to graph verification with **EC** queries. Let A and B be the incidence matrices of G and G^*, respectively. Then an **EC** query Q corresponds to multiplication on the left and right by the characteristic vector of Q, and the algorithm becomes: choose $v \in \{0,1\}^n$ uniformly at random and return 'same' iff $v^T Av = v^T Bv$. By Theorem 10 if $A \neq B$ then $Pr[v^T Av \neq v^T Bv] \geq \frac{1}{4}$.

This raises a natural question. For *arbitrary* $n \times n$ matrices A and B over a field, if $A \neq B$, then for a vector $v \in \{0,1\}^n$ chosen uniformly at random, is $\mathbb{P}[v^T Av \neq v^T Bv] \geq 1/4$ (or some other constant > 0)?

This turns out not to be the case. Consider the two matrices

$$A = \begin{pmatrix} 0 & 1 & 0 \\ 0 & 0 & 1 \\ 1 & 0 & 0 \end{pmatrix} \quad B = \begin{pmatrix} 0 & 0 & 1 \\ 1 & 0 & 0 \\ 0 & 1 & 0 \end{pmatrix}$$

$A \neq B$, but it is not hard to check that for any vector $v \in \{0,1\}^n$, $v^T Av = v^T Bv$. In fact, this holds true for adjacency matrices of 'opposite' directed cycles on > 3 vertices. A graph theoretic interpretation of this fact is that if the number of directed edges on any induced subset of the two opposite directed cycles is the same, then an **EC** query will always return the same answer for the two different cycles. Needless to say, this property is not limited to the adjacency matrices of directed cycles: in fact, it holds for any two matrices A and B such that $A + A^T = B + B^T$, since

$$v^T(A + A^T)v = v^T Av + v^T A^T v = v^T Av + (v^T Av)^T = 2v^T Av$$

for all v, so that $v^T Av = v^T Bv$ for all v.

Hence, we know that standard fingerprinting techniques do not imply Theorem 10. Furthermore, the proof to Theorem 10 generalizes easily to weighted graphs and a more general form of **EC** queries, where the answer to the query is the sum of the weights of its induced edges. Since any symmetric matrix can be viewed as an adjacency matrix of an undirected graph, we have the following fingerprinting technique for symmetric matrices.

Theorem 13. *Let A and B be $n \times n$ symmetric matrices over a field such that $A \neq B$,[4] then for v chosen uniformly at random from $v \in \{0,1\}^n$, $Pr[v^T A v \neq v^T B v] \geq \frac{1}{4}$.*

Proof. Let $C = A - B \neq 0$, and note that $v^T A v \neq v^T B v \iff v^T C v \neq 0$. Identify C with the weighted graph $G = (V, E)$, where $V = \{v_1 \ldots v_n\}$ and $E = \{(u, v) : C(u, v) \neq 0\}$, and $\text{wt}(u, v) = C(u, v)$. We proceed as in the proof of Lemma 11. Fix $v_1 \ldots v_n$ so that $\text{wt}(v_{n-1}, v_n) \neq 0$, and let H' be as before. Define:

$$\text{wt}(H) = \sum_{(u,v) \in H} \text{wt}(u, v); \quad \text{wt}(w, H) = \sum_{(w,v) \in G, v \in H} \text{wt}(w, v).$$

The first quantity is a generalization of **parity**, the second of the number of edges from a vertex to a subgraph. Let $T = \text{wt}(v_{n-1}, H') + \text{wt}(v_n, H') + \text{wt}(v_{n-1}, v_n)$, and consider two cases:

- $T = 0$. Since $\text{wt}(v_{n-1}, v_n) \neq 0$, we know that at least one of the other terms must be nonzero. Assume w.l.o.g. that this is $\text{wt}(v_n, H')$. So choosing v_n but not v_{n-1} is will make $\text{wt}(G') \neq \text{wt}(H')$, and this happens with probability $1/4$.
- $T \neq 0$. Choosing both v_n and v_{n-1} sets $\text{wt}(G') = \text{wt}(H') + T \neq \text{wt}(H')$. This happens with probability $1/4$.

Again, we choose *neither* vertex with probability $1/4$, in which case $\text{wt}(G') = \text{wt}(H')$. Finally,

$$
\begin{aligned}
\mathbb{P}[\text{wt}(G') \neq 0] &= \mathbb{P}[\text{wt}(G') \neq 0 | \text{wt}(H') \neq 0]\mathbb{P}[\text{wt}(H') \neq 0] \\
&\quad + \mathbb{P}[\text{wt}(G') \neq 0 | \text{wt}(H') = 0]\mathbb{P}[\text{wt}(H') = 0] \\
&\geq \mathbb{P}[\text{wt}(G') = \text{wt}(H') | \text{wt}(H') \neq 0]\mathbb{P}[\text{wt}(H') \neq 0] \\
&\quad + \mathbb{P}[\text{wt}(G') \neq \text{wt}(H') | \text{wt}(H') = 0]\mathbb{P}[\text{wt}(H') = 0] \\
&= \mathbb{P}[\text{wt}(G') = \text{wt}(H')]\mathbb{P}[\text{wt}(H') \neq 0] \\
&\quad + \mathbb{P}[\text{wt}(G') \neq \text{wt}(H')]\mathbb{P}[\text{wt}(H') = 0] \qquad \text{by independence} \\
&\geq 1/4(\mathbb{P}[\text{wt}(H') = 0] + \mathbb{P}[\text{wt}(H') \neq 0]) = 1/4
\end{aligned}
$$

as desired.   □

---

[4] Or, more generally, any matrices A and B with $A + A^T \neq B + B^T$.

5 Discussion

There is a tantalizing asymptotic gap of $O(\lg n)$ in our bounds for **EC** queries for learning the partition of the graph. It would also be interesting to know under which, if any, query models it is easier to learn the number of components than the partition itself. There is also the open question whether for general graphs, the $O(|E|\lg n)$ bound can be improved to $O(E)$ for **EC** queries. This is the open question asked by Bouvel et. al. [6] on whether a hidden graph of *average* degree d can be learned with $O(dn)$ **EC** queries.[5]

Some other problems left to be considered are learning and verification problems for other restricted classes of graphs. For example, of theoretical interest is the problem of verifying trees with **ED** queries. There is an obvious $O(n)$ brute-force algorithm, but it may be possible to do better. Also, other classes of graphs have been studied in the literature (see the Section 2) including Hamiltonian paths, matchings, stars, and cliques. It may be revealing to see the power of the queries considered herein for learning and verifying these restricted classes of graphs.

It would also be useful to look at this problem from a more economic perspective. Since edge counting queries are strictly more powerful than edge detecting queries, they ought to be more expensive in some natural framework. Taking costs into account and allowing learners to be able to choose queries with the goal of both learning the graph and minimizing cost should be an interesting research direction.

Finally, our work shows that graph verification is possible even for many classes of directed graphs. It would be interesting to redefine these queries for directed graphs and explore their power.

Acknowledgments

We would like to thank Dana Angluin, Pradipta Mitra, and Daniel Spielman for useful discussions and comments. We would also like to thank Dana Angluin and Jiang Chen for suggesting Proposition 7.

References

1. Alon, N., Asodi, V.: Learning a hidden subgraph. SIAM J. Discrete Math. 18(4), 697–712 (2005)
2. Alon, N., Beigel, R., Kasif, S., Rudich, S., Sudakov, B.: Learning a hidden matching. SIAM J. Comput. 33(2), 487–501 (2004)
3. Angluin, D., Chen, J.: Learning a hidden graph using O(log n) queries per edge. In: COLT, pp. 210–223 (2004)
4. Angluin, D., Chen, J.: Learning a hidden hypergraph. Journal of Machine Learning Research 7, 2215–2236 (2006)

[5] [6] restrict themselves to a non-adaptive framework, where all queries must be asked simultaneously.

5. Beerliova, Z., Eberhard, F., Erlebach, T., Hall, A., Hoffmann, M., Mihalák, M., Ram, L.S.: Network discovery and verification. In: Kratsch, D. (ed.) WG 2005. LNCS, vol. 3787, pp. 127–138. Springer, Heidelberg (2005)
6. Bouvel, M., Grebinski, V., Kucherov, G.: Combinatorial search on graphs motivated by bioinformatics applications: A brief survey. In: Kratsch, D. (ed.) WG 2005. LNCS, vol. 3787, pp. 16–27. Springer, Heidelberg (2005)
7. de Bruijn, N.G.: Asymptotic Methods in Analysis. Dover, Mineola, NY (1981)
8. Freivalds, R.: Probabilistic machines can use less running time. In: IFIP Congress, pp. 839–842 (1977)
9. Goldreich, O., Goldwasser, S., Ron, D.: Property testing and its connection to learning and approximation. J. ACM 45(4), 653–750 (1998)
10. Grebinski, V., Kucherov, G.: Reconstructing a hamiltonian cycle by querying the graph: Application to dna physical mapping. Discrete Applied Mathematics 88(1-3), 147–165 (1998)
11. Grebinski, V., Kucherov, G.: Optimal reconstruction of graphs under the additive model. Algorithmica 28(1), 104–124 (2000)
12. Hein, J.J.: An optimal algorithm to reconstruct trees from additive distance data. Bulletin of Mathematical Biology 51(5), 597–603 (1989)
13. King, V., Zhang, L., Zhou, Y.: On the complexity of distance-based evolutionary tree reconstruction. In: SODA '03. Proceedings of the fourteenth annual ACM-SIAM symposium on Discrete algorithms, Philadelphia, PA, USA, Society for Industrial and Applied Mathematics, pp. 444–453 (2003)
14. Reyzin, L., Srivastava, N.: On the longest path algorithm for reconstructing trees from distance matrices. Inf. Process. Lett. 101(3), 98–100 (2007)

Exact Learning of Finite Unions of Graph Patterns from Queries

Rika Okada[1,*], Satoshi Matsumoto[2], Tomoyuki Uchida[3], Yusuke Suzuki[3], and Takayoshi Shoudai[4]

[1] Dept. of Computer and Media Technologies, Hiroshima City University, Japan
licca_okada@toc.cs.hiroshima-cu.ac.jp
[2] Dept. of Mathematical Sciences, Tokai University, Japan
matumoto@ss.u-tokai.ac.jp
[3] Dept. of Intelligent Systems, Hiroshima City University, Japan
{uchida,y-suzuki}@cs.hiroshima-cu.ac.jp
[4] Dept. of Informatics, Kyushu University, Japan
shoudai@i.kyushu-u.ac.jp

Abstract. A linear graph pattern is a labeled graph such that its vertices have constant labels and its edges have either constant or mutually distinct variable labels. An edge having a variable label is called a variable and can be replaced with an arbitrary labeled graph. Let $\mathcal{GP}(\mathcal{C})$ be the set of all linear graph patterns having a structural feature $\mathcal{C}$ like "having a tree structure", "having a two-terminal series parallel graph structure" and so on. The graph language $GL_\mathcal{C}(g)$ of a linear graph pattern g in $\mathcal{GP}(\mathcal{C})$ is the set of all labeled graphs obtained from g by substituting arbitrary labeled graphs having the structural feature $\mathcal{C}$ to all variables in g. In this paper, for any set $\mathcal{T}_*$ of m linear graph patterns in $\mathcal{GP}(\mathcal{C})$, we present a query learning algorithm for finding a set S of linear graph patterns in $\mathcal{GP}(\mathcal{C})$ with $\bigcup_{g \in \mathcal{T}_*} GL_\mathcal{C}(g) = \bigcup_{f \in S} GL_\mathcal{C}(f)$ in polynomial time using at most $m + 1$ equivalence queries and $O(m(n + n^2))$ restricted subset queries, where n is the maximum number of edges of counterexamples, if the number of labels of edges is infinite. Next we show that finite sets of graph languages generated by linear graph patterns having tree structures or two-terminal series parallel graph structures are not learnable in polynomial time using restricted equivalence, membership and subset queries.

1 Introduction

Many electronic data become accessible on Internet. Electronic data such as HTML/XML files, bioinformatics and chemical compounds have graph structures but have no rigid structure. Hence, such data are called *graph structured data*. Especially, graph structured data such as HTML/XML files having tree

---

* Rika Okada is currently working at Sanyo Girls' Junior and Senior High School, Hiroshima, Japan.

M. Hutter, R.A. Servedio, and E. Takimoto (Eds.): ALT 2007, LNAI 4754, pp. 298–312, 2007.
© Springer-Verlag Berlin Heidelberg 2007

structures are called *tree structured data*. In the fields of data mining and knowledge discovery, many researchers have developed techniques based on machine learning for analyzing such graph structured data. If we can construct oracles which answer any query in practical time, we can design efficient and effective data mining tools based on query learning algorithms using such oracles. The purpose of our work is to present fundamental learning algorithms for data mining from graph structured data. In this paper, we consider polynomial time learnabilities of finite unions of graph patterns having structured variables, which are knowledge representations for graph structured data, in exact learning model of Angluin [2].

A linear graph pattern is defined as a labeled graph such that its vertices have constant labels and its edges have either constant or mutually distinct variable labels. An edge (u, v) having a variable label is called a *variable*, denoted by $\langle u, v \rangle$, and can be replaced with an arbitrary labeled graph. For example, in Fig. 1, we give a linear graph pattern g having two variables $\langle u_1, u_2 \rangle$ and $\langle v_1, v_2 \rangle$ with variable labels x and y, respectively. In the figures of this paper, a variable is represented by a box with lines to its elements. The numbers at these lines indicate the order of the vertices of which a variable consists. The symbol inside a box shows the label of the variable. We can obtain a new linear graph pattern from a linear graph pattern g by substituting an arbitrary linear graph pattern to a variable in g. For example, in Fig. 1, the labeled graph G_3 is obtained from the linear graph pattern g by replacing the variables $\langle u_1, u_2 \rangle$ and $\langle v_1, v_2 \rangle$ of g with the labeled graphs G_1 and G_2, respectively.

Web documents like HTML/XML files are expressed by labeled graphs having tree structures. In applications for electrical network and scheduling problems, input data are formalized by labeled graphs having two-terminal series parallel (TTSP for short) graph structures. In order to represent structural features of graph structured data such as "having tree structures" and "having TTSP graph structures", we define *simple* Formal Graph System, which is a restricted class of Formal Graph System (FGS for short) presented by Uchida et al. [15]. FGS is a kind of logic programming systems which directly deals with graph patterns instead of terms in first-order logic. A finite set of clauses on FGS is called an *FGS program*. As examples of simple FGS programs, we give a simple FGS program $\mathcal{OT}$ in Fig. 2 generating all ordered rooted trees and a simple FGS program $\mathcal{TTSP}$ in Fig. 3 generating all TTSP graphs such as F_1, F_2, F_3, F_4, and F_5 in Fig. 3, where TTSP graphs are constructed by recursively applying "series" and "parallel" operations (see [5]). For a simple FGS program Γ, let $\mathcal{GP}(\Gamma)$ be the set of all linear graph patterns obtained from any labeled graph generated by Γ by replacing some edges in it with mutually distinct variables, that is, $\mathcal{GP}(\Gamma)$ contains all linear graph patterns with the graph structural feature "generated by Γ". The graph language $GL_\Gamma(g)$ of a linear graph pattern g in $\mathcal{GP}(\Gamma)$ is the set of all labeled graphs whose graph structures are generated by Γ and which are obtained from g by substituting arbitrary labeled graphs whose graph structures are generated by Γ to all variables in g.

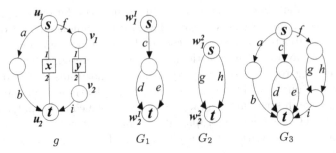

Fig. 1. Linear graph pattern g and labeled graphs G_1, G_2, G_3 over Λ. In the figures in this paper, a variable is represented by a box with lines to its elements. The numbers at these lines indicate the order of the vertices of which a variable consists. The symbol inside a box shows the label of the variable. In this figure, we omit the labels of vertices except two labels s, t.

$$
\mathcal{OT} = \left\{
\begin{array}{l}
q(\overset{0}{\textcircled{s}}\rightarrow \textcircled{t}) \longleftarrow , \\[4pt]
q(\overset{0}{\textcircled{s}}\text{-}^1\boxed{x}^2\text{-}\overset{0}{\bigcirc}\text{-}^1\boxed{y}^2\text{-}\textcircled{t}) \longleftarrow q(\overset{0}{\textcircled{s}}\text{-}^1\boxed{x}^2\text{-}\textcircled{t}),\ q(\overset{0}{\textcircled{s}}\text{-}^1\boxed{y}^2\text{-}\textcircled{t}),\\[4pt]
q(\overset{0}{\textcircled{s}}\text{-}^1\boxed{x}^2\text{-}\overset{0}{\bigcirc}\text{-}^1\boxed{y}^2\text{-}\textcircled{l}) \longleftarrow q(\overset{0}{\textcircled{s}}\text{-}^1\boxed{x}^2\text{-}\textcircled{l}),\ q(\overset{0}{\textcircled{s}}\text{-}^1\boxed{y}^2\text{-}\textcircled{l}),\\[4pt]
q(\overset{0}{\textcircled{s}}\!\!\begin{array}{c}{}^1\boxed{x}^2\,\textcircled{t}\\[-2pt]{}^1\boxed{y}^2\,\textcircled{l}\end{array}) \longleftarrow q(\overset{0}{\textcircled{s}}\text{-}^1\boxed{x}^2\text{-}\textcircled{t}),\ q(\overset{0}{\textcircled{s}}\text{-}^1\boxed{y}^2\text{-}\textcircled{t})
\end{array}
\right\}
$$

Fig. 2. Simple FGS program $\mathcal{OT}$. The symbol o over internal vertices indicates that the vertex has ordered children. The broken arrow shows that the order of the leaf labeled with t is less than that of the leaf labeled with l.

In exact learning model of Angluin [2], a learning algorithm accesses to oracles, which answer specific kinds of queries, and collects information about a target. Let Γ be a simple FGS program and $\mathcal{T}_*$ a subset of $\mathcal{GP}(\Gamma)$. A learning algorithm is said to *exactly identify* the target set $\mathcal{T}_*$ if it outputs a set of linear graph patterns $S \subseteq \mathcal{GP}(\Gamma)$ such that the union of graph languages of all linear graph patterns in S is equal to that in $\mathcal{T}_*$ and halts, after it asks a certain number of queries to oracles. In this paper, for a simple FGS program Γ and any set $\mathcal{T}_*$ of m linear graph patterns in $\mathcal{GP}(\Gamma)$, we present a query learning algorithm which exactly identifies $\mathcal{T}_*$ in polynomial time using at most $m + 1$ equivalence queries and at most $m(n + rn^2)$ restricted subset queries, where n is the maximum number of edges of counterexamples and r is the number of clauses in Γ (i.e., r is a constant), if the number of labels of edges is infinite. Firstly, the algorithm gets a counterexample h_i ($1 \le i \le m$) as an answer of an equivalence query for an empty set, that is, h_i is a labeled graph generated by some linear graph

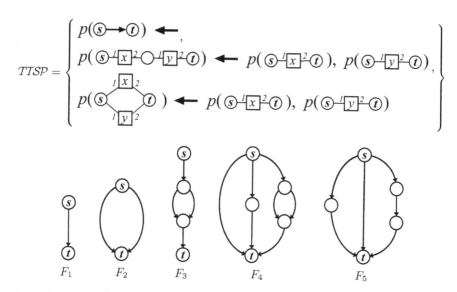

Fig. 3. Simple FGS program *TTSP* and TTSP graphs F_1, F_2, F_3, F_4, F_5. In this figure, we omit the labels of edges and vertices except two labels s and t of vertices.

pattern g_i in $\mathcal{T}_*$. Secondly, the algorithm recursively reconstructs h_i by replacing edges of h_i with variables or subgraphs of h_i generated by Γ with variables and by asking a certain number of restricted subset queries. Next, asking an equivalence query, the algorithm gets a new counterexample h_j ($1 \leq j \neq i \leq m$) if the equivalence oracle does not answer "yes". Finally, the algorithm halts, if the algorithm exactly identifies $\mathcal{T}_*$.

Next, we show that, by asking a certain number of restricted equivalence, membership and subset queries, finite sets of linear graph patterns in $\mathcal{GP}(\mathcal{OT})$ and $\mathcal{GP}(TTSP)$ are not learnable in polynomial time.

In [10], we already showed that any finite set of m linear graph patterns in $\mathcal{GP}(\mathcal{OT})$ is exactly identifiable at most $m + 1$ equivalence queries and using at most $2mn^2$ restricted subset queries, where n is the maximum number of edges in counterexamples, if the number of labels of edges is infinite. The results of this paper are improvements and extensions of the results in [10]. Moreover, in [11], we considered polynomial time learnabilies of finite unions of *non-linear* graph patterns having ordered tree structures, that is, ordered rooted tree patterns in which variables are allowed to have the same variable labels. In [11], we showed that any finite set of m graph patterns having ordered tree structures is exactly identifiable using $O(m^2n^4 + 1)$ superset queries and $O(m + 1)$ restricted equivalence queries, where n is the maximum number of edges in counterexamples, if the number of labels of edges is infinite.

As for related works, the work [9,16] studied the learnabilities of graph structured patterns in the framework of polynomial time inductive inference. Also the work [13,14] showed the classes of linear graph patterns in $\mathcal{GP}(\mathcal{OT})$ and linear graph patterns in $\mathcal{GP}(TTSP)$ are polynomial time inductively inferable from positive data, respectively. As an application, the work [12] proposed a tag tree pattern, which is an extension of a linear graph pattern in $\mathcal{GP}(OT)$, and gave a data mining method from tree structured data. As for other related works, the works [1,3] show the exact learnability of tree structured patterns, which are incomparable to linear graph patterns having tree structures, in the exact learning model.

This paper is organized as follows: In Section 2, we formally define a linear graph pattern as a labeled graph having structural variables, and then define its graph language. Moreover, we briefly introduce a exact learning model treated in this paper. In Section 3, we consider the learnabilities of finite unions of graph languages of linear graph patterns in the framework of exact learning model. In Section 4, we consider the insufficiency of learning of finite unions of some graph languages of linear graph patterns in exact learning model. In Section 5, we conclude this work and give future works.

2   Preliminaries

We introduced *term graphs* and *term graph languages* in [15] in order to develop efficient graph algorithms for grammatically defined graph classes. In this section, based on term graphs and term graph languages, we define labeled graph patterns as graphs having structural variables, and then introduce their graph languages. For a set S, $|S|$ denotes the number of elements of S.

2.1   Linear Graph Patterns

Let Λ and $\mathcal{X}$ be infinite alphabets whose elements are called *constant labels* and *variable labels*, respectively. We assume that $\Lambda \cap \mathcal{X} = \emptyset$. Let $G = (V, E)$ be a directed labeled graph consisting of a set V of vertices and a set E of edges such that G has no loop but multiple edges are allowed. We denote by ψ_G a vertex labeling assigning a constant label in Λ to each vertex in V and by φ_G an edge labeling assigning either a constant label or a variable label in $\Lambda \cup \mathcal{X}$ to each edge in E. A *graph pattern over* $\Lambda \cup \mathcal{X}$ *obtained from* G is defined as a triplet $g = (V_g, E_g, H_g)$ where $V_g = V$, $E_g = \{e \in E \mid \varphi_G(e) \in \Lambda\}$ and $H_g = E - E_g$. An element of H_g is called a *variable*. We note that $\psi_g(u) = \psi_G(u)$ for each vertex $u \in V_g$, $\varphi_g(e) = \varphi_G(e) \in \Lambda$ for each edge $e \in E_g$ and $\varphi_g(h) = \varphi_G(h) \in \mathcal{X}$ for each variable $h \in H_g$. We use notations (u, v) and $\langle s, t \rangle$ to represent an edge in E_g and a variable in H_g consisting of two vertices u, v and s, t in V_g, respectively. Here after, since the background graph G can be easily found from a triplet $g = (V_g, E_g, H_g)$, we omit the description of the background graph G. A graph pattern g over $\Lambda \cup \mathcal{X}$ is said to be *linear* if all variables in g have mutually distinct variable labels in $\mathcal{X}$. In particular, a graph pattern over $\Lambda \cup \mathcal{X}$ with no variable is regarded as a (standard) labeled graph over Λ. We denote the set of all linear graph patterns over $\Lambda \cup \mathcal{X}$ by $\mathcal{GP}_{\Lambda \cup \mathcal{X}}$ and the set of all labeled graphs over Λ by $\mathcal{G}_\Lambda$. In this paper, we deal with only linear graph patterns over

$\Lambda \cup \mathcal{X}$, and then we call a linear graph pattern over $\Lambda \cup \mathcal{X}$ a *graph pattern*, simply. A graph pattern having no edge is said to be *simple*. A graph pattern g is said to be *primitive* if g is a simple graph pattern consisting of two vertices and only one variable, (i.e. $|V_g| = 2$, $|E_g| = 0$ and $|H_g| = 1$, where $g = (V_g, E_g, H_g)$).

Two graph patterns $f = (V_f, E_f, H_f)$ and $g = (V_g, E_g, H_g)$ are said to be *isomorphic*, denoted by $f \equiv g$, if there is a bijection π from V_f to V_g, such that (1) $(u, v) \in E_f$ if and only if $(\pi(u), \pi(v)) \in E_g$, (2) $\psi_f(u) = \psi_g(\pi(u))$ for each vertex $u \in V_f$ and $\varphi_f((u, v)) = \varphi_g((\pi(u), \pi(v)))$ for each edge $(u, v) \in E_f$, and (3) $\langle u, v \rangle \in H_f$ if and only if $\langle \pi(u), \pi(v) \rangle \in H_g$. A bijection π satisfying (1)–(3) is called an *isomorphism* from f to g. Two isomorphic graph patterns are considered to be identical.

Let f and g be graph patterns having at least two vertices. Let $\sigma = [u, v]$ be a pair of distinct vertices in g. The form $x := [g, \sigma]$ is called a *binding* for a variable label x in $\mathcal{X}$. A new graph pattern, denoted by $f\{x := [g, \sigma]\}$, is obtained by applying the binding $x := [g, \sigma]$ to f in the following way: Let $e = \langle s, t \rangle$ be a variable in f with the variable label x, i.e., $\varphi_f(e) = x$. Let g' be a copy of g. And let u' and v' be the vertices of g' corresponding to u and v of g, respectively. For the variable $e = \langle s, t \rangle$, we attach g' to f by removing the variable e from f and identifying the vertices s and t with the vertices u' and v' of g', respectively. For two bindings $x := [g, [u_g, v_g]]$ and $x := [f, [u_f, v_f]]$, we write $(x := [g, [u_g, v_g]]) \equiv (x := [f, [u_f, v_f]])$ if there exists an isomorphism π from g to f such that $\pi(u_g) = u_f$ and $\pi(v_g) = v_f$. A *substitution* θ is a finite set of bindings $\{x_1 := [g_1, \sigma_1], x_2 := [g_2, \sigma_2], \dots, x_n := [g_n, \sigma_n]\}$, where x_i's are mutually distinct variable labels in $\mathcal{X}$. For a graph pattern f and a substitution θ, we denote by $f\theta$ the graph pattern obtained from f and θ by applying all bindings in θ to f simultaneously. For example, for the graph pattern g in Fig. 1 and labeled graphs G_1, G_2, G_3 in Fig. 1, G_3 is isomorphic to the graph pattern $g\theta$ obtained by applying $\theta = \{x := [G_1, [w_1^1, w_2^1]], y := [G_2, [w_1^2, w_2^2]]\}$ to g (i.e., $G_3 \equiv g\theta$).

For graph patterns f and g, we write $f \preceq g$ if there exists a substitution θ such that $f \equiv g\theta$. Especially, we write $f \prec g$ if $f \preceq g$ and $g \not\preceq f$. For example, for the graph patterns G_3 and g given in Fig. 1, we can see that $G_3 \prec g$ because of $G_3 \equiv g\{x := [G_1, [w_1^1, w_2^1]], y := [G_2, [w_1^2, w_2^2]]\}$ and $g \not\preceq G_3$.

2.2 Graph Languages over Λ

The purpose of this subsection is to define graph languages over an alphabet Λ of infinitely many constant labels (i.e., $|\Lambda| = \infty$). First of all, in order to represent structural features of graph structured data like "having tree structures", "having TTSP graph structures" and so on, we introduce *simple* Formal Graph System, which is a restricted class of Formal Graph System (FGS for short) presented by Uchida et al. [15]. FGS is a kind of logic programming systems which directly deals with graph patterns instead of terms in first-order logic.

Let Π be a set of unary predicate symbols and Σ a finite subset of Λ. An *atom* is an expression of the form $p(g)$, where p is a unary predicate symbol in Π and g is a graph pattern over $\Sigma \cup \mathcal{X}$. For two atoms $p(g)$ and $q(f)$, we write

$$UT = \begin{cases} p(\text{⑤}\!\!\rightarrow\!\!\text{①}) \leftarrow, \\ p(\text{⑤}\!-\!\boxed{x}\!-\!\text{○}\!-\!\boxed{y}\!-\!\text{①}) \leftarrow p(\text{⑤}\!-\!\boxed{x}\!-\!\text{①}),\ p(\text{⑤}\!-\!\boxed{y}\!-\!\text{①}), \\ p(\text{⑤}\!-\!\boxed{x}\!-\!\text{○}\!-\!\boxed{y}\!-\!\text{①}) \leftarrow p(\text{⑤}\!-\!\boxed{x}\!-\!\text{①}),\ p(\text{⑤}\!-\!\boxed{y}\!-\!\text{①}), \\ p(\text{⑤}\!\begin{smallmatrix}\boxed{x}\ \,\text{①}\\ \boxed{y}\ \,\text{①}\end{smallmatrix}) \leftarrow p(\text{⑤}\!-\!\boxed{x}\!-\!\text{①}),\ p(\text{⑤}\!-\!\boxed{y}\!-\!\text{①}) \end{cases}$$

Fig. 4. Simple FGS program UT

$p(g) \equiv q(f)$ if $p = q$ and $g \equiv f$ hold. Let $A, B_1, B_2, \ldots, B_n$ be atoms, where $n \geq 0$. Then, a *graph rewriting rule* is a clause of the form $A \leftarrow B_1, B_2, \ldots, B_n$. We call the atom A the *head* and the right part $B_1, B_2, \ldots, B_n$ the *body* of the graph rewriting rule. For a graph pattern $g = (V_g,\ E_g,\ H_g)$ and a variable label $x \in \mathcal{X}$, the number of variables of g labeled with x is denoted by $o(g, x)$ (i.e., $o(g, x) = |\{h \in H_g \mid \varphi_g(h) = x\}|$). Because any graph pattern is assumed to be linear in this paper, we have $o(g, x) = 1$ if x appears in g, otherwise $o(g, x) = 0$. A graph rewriting rule $p(g) \leftarrow q_1(f_1),\ q_2(f_2), \ldots,\ q_n(f_n)$ is said to be *simple* if the following conditions (1)-(3) hold: (1) f_i is primitive for any $i = 1, 2, \ldots, n$, (2) g consists of two vertices and the edge between them if $n = 0$, otherwise g is simple, and (3) for any variable $x \in X$, $o(g, x) = 1$ if and only if $o(f_1, x) + o(f_2, x) + \cdots + o(f_n, x) = 1$. A *FGS program* is a finite set of graph rewriting rules. An FGS program Γ is said to be *simple* if any graph rewriting rule in Γ is simple. For example, we give some simple FGS programs in Figs. 2–5.

We define substitutions for graph rewriting rules in a similar way to those in logic programming [7]. For an atom $p(g)$, a graph rewriting rule $A \leftarrow B_1, \ldots, B_n$ and a substitution θ, we define $p(g)\theta = p(g\theta)$ and $(A \leftarrow B_1, \ldots, B_n)\theta = A\theta \leftarrow B_1\theta, \ldots, B_n\theta$. Let Γ be an FGS program. The relation $\Gamma \vdash C$ for a graph rewriting rule C is inductively defined as follows.

(1) If $C \in \Gamma$, then $\Gamma \vdash C$.
(2) If $\Gamma \vdash C$, then $\Gamma \vdash C\theta$ for any substitution θ.
(3) If $\Gamma \vdash A \leftarrow B_1, \ldots, B_i, \ldots, B_n$ and $\Gamma \vdash B_i \leftarrow C_1, \ldots, C_m$,
    then $\Gamma \vdash A \leftarrow B_1, \ldots, B_{i-1}, C_1, \ldots, C_m, B_{i+1}, \ldots, B_n$.

For an FGS program Γ and its predicate symbol p in Π, $GL(\Gamma, p)$ denotes the subset $\{g \in \mathcal{G}_\Sigma \mid \Gamma \vdash p(g) \leftarrow\}$ of $\mathcal{G}_\Sigma$. We say that a subset $L \subseteq \mathcal{G}_\Sigma$ is an *FGS language* if there exists an FGS program Γ and its predicate symbol p such that $L = GL(\Gamma, p)$. The FGS language $GL(\Gamma, p)$ is simply denoted by $GL(\Gamma)$ if we need not clarify the predicate symbol p. For example, for the simple FGS programs OT in Fig. 2, $TTSP$ in Fig. 3, UT in Fig. 4 and $MT = UT \cup OT \cup \mathcal{R}$ (here, $\mathcal{R}$ in Fig. 5), the FGS languages $GL(OT, q)$, $GL(TTSP, p)$, $GL(UT, p)$ and $GL(MT, r)$ of OT, $TTSP$, UT and MT are the sets of all rooted ordered trees, all TTSP graphs (see [5]), all rooted unordered trees, and all rooted mixed

$$\mathcal{R} = \begin{cases} r(\text{⊙-}\boxed{x}^2\text{-}①) \leftarrow p(\text{⊙-}\boxed{x}^2\text{-}①), \\ r(\text{⊙-}\boxed{x}^2\text{-}①) \leftarrow q(\text{⊙-}\boxed{x}^2\text{-}①), \\ p(\text{⊙-}\boxed{x}^2\overset{o}{\text{⊙-}}\boxed{y}^2\text{-}①) \leftarrow p(\text{⊙-}\boxed{x}^2\text{-}①),\ q(\overset{o}{\text{⊙-}}\boxed{y}^2\text{-}①), \\ p(\text{⊙-}\boxed{x}^2\overset{o}{\text{⊙-}}\boxed{y}^2\text{-}①) \leftarrow p(\text{⊙-}\boxed{x}^2\text{-}①),\ q(\overset{o}{\text{⊙-}}\boxed{y}^2\text{-}①), \\ q(\overset{o}{\text{⊙-}}\boxed{x}^2\text{-⊙-}\boxed{y}^2\text{-}①) \leftarrow q(\overset{o}{\text{⊙-}}\boxed{x}^2\text{-}①),\ p(\text{⊙-}\boxed{y}^2\text{-}①), \\ q(\overset{o}{\text{⊙-}}\boxed{x}^2\text{-⊙-}\boxed{y}^2\text{-}①) \leftarrow q(\overset{o}{\text{⊙-}}\boxed{x}^2\text{-}①),\ p(\text{⊙-}\boxed{y}^2\text{-}①) \end{cases}$$

Fig. 5. Simple FGS program $\mathcal{R}$

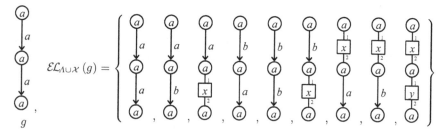

Fig. 6. The set $\mathcal{EL}_{\Lambda\cup\mathcal{X}}(g)$ of all graph patterns over $\Lambda\cup\mathcal{X}$ obtained from the labeled graph g over $\{a\}$ in case of $\Lambda=\{a,b\}$

trees each of whose internal vertices has ordered or unordered children (see [14]), respectively.

Next, we construct graph languages over an infinite alphabet Λ from FGS languages over Σ. For a labeled graph $g \in \mathcal{G}_\Sigma$, $\mathcal{EL}_{\Lambda\cup\mathcal{X}}(g)$ and $\mathcal{EL}_\Lambda(g)$ denote the sets of all graph patterns over $\Lambda \cup \mathcal{X}$ and all labeled graphs over Λ which are obtained from g by ignoring all edge labels of g and relabeling all edges with arbitrary labels in $\Lambda \cup \mathcal{X}$ and Λ, respectively. When $|\Lambda| = \infty$, this indicates that $\mathcal{EL}_{\Lambda\cup\mathcal{X}}(g)$ and $\mathcal{EL}_\Lambda(g)$ are infinite subsets of $\mathcal{GP}_{\Lambda\cup\mathcal{X}}$ and $\mathcal{G}_\Lambda$, respectively. In Fig. 6, as an example in case of $\Lambda = \{a,b\}$, we give the set $\mathcal{EL}_{\Lambda\cup\mathcal{X}}(g)$ of all graph patterns over $\Lambda\cup\mathcal{X}$ obtained from the labeled graph g over $\{a\}$. For a simple FGS program Γ, let $\mathcal{GP}_{\Lambda\cup\mathcal{X}}(\Gamma) = \bigcup_{g\in GL(\Gamma)} \mathcal{EL}_{\Lambda\cup\mathcal{X}}(g)$ and $\mathcal{G}_\Lambda(\Gamma) = \bigcup_{g\in GL(\Gamma)} \mathcal{EL}_\Lambda(g)$. We denote all the finite subsets of $\mathcal{GP}_{\Lambda\cup\mathcal{X}}(\Gamma)$ by $\mathcal{FGP}_{\Lambda\cup\mathcal{X}}(\Gamma)$. For a simple FGS program Γ and a graph pattern $g \in \mathcal{GP}_{\Lambda\cup\mathcal{X}}(\Gamma)$, let $\mathcal{L}_\Lambda(\Gamma,g) = \{f \in \mathcal{G}_\Lambda(\Gamma) \mid f \preceq g\} \subseteq \mathcal{G}_\Lambda$ and we call $\mathcal{L}_\Lambda(\Gamma,g)$ the *graph language of Γ and g*. For a simple FGS program Γ and a finite subset S of $\mathcal{GP}_{\Lambda\cup\mathcal{X}}(\Gamma)$, we define $\mathcal{L}_\Lambda(\Gamma,S)$ as the union of graph languages of Γ and $g \in S$ with respect to S (i.e., $\mathcal{L}_\Lambda(\Gamma,S) = \bigcup_{g\in S} \mathcal{L}_\Lambda(\Gamma,g)$) and we call it the *graph language over Γ and S*. In particular, we assume that $\mathcal{L}_\Lambda(\Gamma,\phi) = \phi$.

Let Γ be a simple FGS program, g a graph pattern in $\mathcal{GP}_{\Lambda \cup \mathcal{X}}(\Gamma)$ and S in $\mathcal{FGP}_{\Lambda \cup \mathcal{X}}(\Gamma)$. Then, we consider the following property: $\mathcal{L}_\Lambda(\Gamma, g) \subseteq \mathcal{L}_\Lambda(\Gamma, f)$ for some $f \in S$ if and only if $\mathcal{L}_\Lambda(\Gamma, g) \subseteq \mathcal{L}_\Lambda(\Gamma, S)$. This property is important in the learning of unions of graph languages and called *compactness*, which was proposed in [4]. The following lemma shows that the graph language over a simple FGS program Γ and $S \in \mathcal{FGP}_{\Lambda \cup \mathcal{X}}(\Gamma)$ has compactness. We can prove the following lemma by slightly modifying the proof of Lemma 1 in [10].

Lemma 1. *Let Γ be a simple FGS program, S in $\mathcal{FGP}_{\Lambda \cup \mathcal{X}}(\Gamma)$ and $|\Lambda|$ infinite. Then, for a graph pattern g in $\mathcal{GP}_{\Lambda \cup \mathcal{X}}(\Gamma)$, $\mathcal{L}_\Lambda(\Gamma, g) \subseteq \mathcal{L}_\Lambda(\Gamma, S)$ if and only if there exists a graph pattern f in S with $g \preceq f$.*

In this paper, we consider polynomial time learnabilities of the class of graph languages for a simple FGS program Γ and a finite subset S of $\mathcal{GP}_{\Lambda \cup \mathcal{X}}(\Gamma)$. We remark that we do not consider the learnabilities of FGS languages.

2.3   Learning Model

Let Γ be a simple FGS program. In what follows, let $\mathcal{T}_* \subseteq \mathcal{GP}_{\Lambda \cup \mathcal{X}}(\Gamma)$ (i.e., $\mathcal{T}_* \in \mathcal{FGP}_{\Lambda \cup \mathcal{X}}(\Gamma)$) denotes a finite set of graph patterns to be identified, and we say that $\mathcal{T}_*$ is a *target*. In the exact learning model via queries due to Angluin [2], learning algorithms can access to *oracles* that will answer queries about the target $\mathcal{T}_*$. In this paper, we consider the following queries.

1. **Membership query:** The input is a labeled graph $g \in \mathcal{G}_\Lambda(\Gamma)$. The output is **yes** if $g \in \mathcal{L}_\Lambda(\Gamma, \mathcal{T}_*)$, otherwise **no**. The oracle which answers the membership query is called a *membership oralce*.
2. **Subset query and Restricted subset query:** The input of both queries is a finite subset S of $\mathcal{GP}_{\Lambda \cup \mathcal{X}}(\Gamma)$. The output of a subset query is **yes** if $\mathcal{L}_\Lambda(\Gamma, S) \subseteq \mathcal{L}_\Lambda(\Gamma, \mathcal{T}_*)$, otherwise a labeled graph, called a *counterexample*, in $(\mathcal{L}_\Lambda(\Gamma, S) - \mathcal{L}_\Lambda(\Gamma, \mathcal{T}_*))$. The oracle which answers the subset query is called a *subset oracle*. The output of a restricted subset query is **yes** if $\mathcal{L}_\Lambda(\Gamma, S) \subseteq \mathcal{L}_\Lambda(\Gamma, \mathcal{T}_*)$, otherwise **no**. The oracle which answers the restricted subset query is called a *restricted subset oracle*.
3. **Equivalence query and Restricted equivalence query:** The input of both queries is a finite subset S of $\mathcal{GP}_{\Lambda \cup \mathcal{X}}(\Gamma)$. The output of a equivalence query is **yes** if $\mathcal{L}_\Lambda(\Gamma, S) = \mathcal{L}_\Lambda(\Gamma, \mathcal{T}_*)$, otherwise a labeled graph, called a *counterexample*, in $(\mathcal{L}_\Lambda(\Gamma, S) \cup \mathcal{L}_\Lambda(\Gamma, \mathcal{T}_*)) - (\mathcal{L}_\Lambda(\Gamma, S) \cap \mathcal{L}_\Lambda(\Gamma, \mathcal{T}_*))$. The oracle which answers the equivalence query is called a *equivalence oracle*. The output of a restricted equivalence query is **yes** if $\mathcal{L}_\Lambda(\Gamma, S) = \mathcal{L}_\Lambda(\Gamma, \mathcal{T}_*)$, otherwise **no**. The oracle which answers the restricted equivalence query is called a *restricted equivalence oracle*.

A learning algorithm $\mathcal{A}$ is said to *exactly identify* a target $\mathcal{T}_*$ *in polynomial time* if $\mathcal{A}$ outputs a set $S \in \mathcal{FGP}_{\Lambda \cup \mathcal{X}}(\Gamma)$ in polynomial time with $\mathcal{L}_\Lambda(\Gamma, S) = \mathcal{L}_\Lambda(\Gamma, \mathcal{T}_*)$.

Algorithm LEARN_UNION

Assumption: A simple FGS program Γ and a target $\mathcal{T}_* \in \mathcal{FGP}_{\Lambda \cup \mathcal{X}}(\Gamma)$.
Given: Oracles for $\text{Equiv}_{\mathcal{T}_*}$ and $\text{rSub}_{\mathcal{T}_*}$ for $\mathcal{T}_*$.
Output: A set $S \in \mathcal{FGP}_{\Lambda \cup \mathcal{X}}(\Gamma)$ with $\mathcal{L}_\Lambda(\Gamma, S) = \mathcal{L}_\Lambda(\Gamma, \mathcal{T}_*)$.

begin
1. $S := \emptyset$;
2. **while** $\text{Equiv}_{\mathcal{T}_*}(S) \neq$ **yes do**
3.   **begin**
4.   Let g be a counterexample;
5.   **foreach** edge e of g **do**
6.   **if** $\text{rSub}_{\mathcal{T}_*}(\{g/\{e\}\}) =$ **yes then** $g := g/\{e\}$;
7.   **repeat**
8.   **foreach** $f \in \{g' \mid g \dashv_\Gamma g'\}$ **do**
9.   **if** $\text{rSub}_{\mathcal{T}_*}(\{f\}) =$ **yes then begin** $g := f$; **break end**
10.   **until** g does not change;
11. $S := S \cup \{g\}$
12.   **end**;
13. **output** S
end.

Fig. 7. Algorithm LEARN_UNION

---

3   Learning Finite Unions of Graph Languages

In this section, for a fixed simple FGS program Γ, we consider the learnabilities of finite unions of graph languages over Γ and $S \in \mathcal{FGP}_{\Lambda \cup \mathcal{X}}(\Gamma)$ in the framework of exact learning model. For a simple FGS program Γ and a target $\mathcal{T}_*$ in $\mathcal{FGP}_{\Lambda \cup \mathcal{X}}(\Gamma)$, we present a polynomial time learning algorithm LEARN_UNION in Fig. 7 which outputs a set S in $\mathcal{FGP}_{\Lambda \cup \mathcal{X}}(\Gamma)$ such that $\mathcal{L}_\Lambda(\Gamma, S) = \mathcal{L}_\Lambda(\Gamma, \mathcal{T}_*)$ holds, by asking several queries to a restricted subset oracle, denoted by $\text{rSub}_{\mathcal{T}_*}$, and an equivalence oracle, denoted by $\text{Equiv}_{\mathcal{T}_*}$. The formal definitions of notations used in LEARN_UNION are stated later. We assume that $|\Lambda|$ is infinite.

First, we consider the internal foreach-loop at lines 5 and 6 in the algorithm LEARN_UNION. For two graph patterns $g = (V_g, E_g, H_g)$ and f in $\mathcal{GP}_{\Lambda \cup \mathcal{X}}$, we write $g \lhd f$ if f is isomorphic to a graph pattern g' obtained from g by replacing an edge $(u, v) \in E_g$ with a new variable $\langle u, v \rangle$ labeled with a new variable label in $\mathcal{X}$ (i.e., $g' = (V_g, E_g - \{(u, v)\}, H_g \cup \{\langle u, v \rangle\})$). That is, f is a generalized graph pattern of g such that $g \preceq f$. In order to show the replaced edge (u, v) explicitly, f is denoted by $g/\{(u, v)\}$. Let $\lhd^*$ be the reflexive and transitive closure of $\lhd$ on $\mathcal{GP}_{\Lambda \cup \mathcal{X}}$. Then, we have the following lemma.

Lemma 2. *For graph patterns g, g_1, $g_2 \in \mathcal{GP}_{\Lambda \cup \mathcal{X}}$, if $g \lhd^* g_1$ and $g \lhd^* g_2$, then there exists a graph pattern $g' \in \mathcal{GP}_{\Lambda \cup \mathcal{X}}$ such that $g_1 \lhd^* g'$ and $g_2 \lhd^* g'$ hold.*

Proof. Since $g \lhd^* g_1$, there exists a subset $I_1 = \{e_{g_1,1}, e_{g_1,2}, \ldots, e_{g_1,k}\}$ of E_g such that $g/\{e_{g_1,1}\}/\{e_{g_1,2}\}/\ldots/\{e_{g_1,k}\} \equiv g_1$ holds, where E_g is the set of edges in

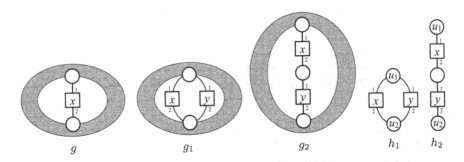

Fig. 8. Graph patterns g, g_1, g_2, h_1, h_2

g. Moreover, there exists also a subset $I_2 = \{e_{g_2,1}, e_{g_2,2}, \ldots, e_{g_2,r}\}$ of E_g such that $g/\{e_{g_2,1}\}/\{e_{g_2,2}\}/\ldots/\{e_{g_2,r}\} \equiv g_2$ holds. For a set $I_1 \cup I_2 = \{e_1, e_2, \ldots, e_s\}$ ($s \leq k + r$) and $g' \equiv g/\{e_1\}/\{e_2\}/\ldots/\{e_s\}$, we have $g_1 \lhd^* g'$ and $g_2 \lhd^* g'$.   □

This lemma shows that the binary relation $\lhd$ over $\mathcal{GP}_{\Lambda \cup \mathcal{X}}$ has the Church-Rosser property. We can easily prove the following lemma.

Lemma 3. *For two graph patterns g, f in $\mathcal{GP}_{\Lambda \cup \mathcal{X}}$, we have $g \preceq f$ if $g \lhd^* f$.*

For a graph pattern g given after executing the internal foreach-loop at lines 5 and 6 in the algorithm LEARN_UNION, from Lemma 2, we can see that $\mathcal{L}_\Lambda(\Gamma, g) \subseteq \mathcal{L}_\Lambda(\Gamma, \mathcal{T}_*)$ and $\mathcal{L}_\Lambda(\Gamma, g/\{e\}) \not\subseteq \mathcal{L}_\Lambda(\Gamma, \mathcal{T}_*)$ for any edge e in g. Then, by modifying the proof of Lemma 3 in [10], we can prove the following lemma.

Lemma 4. *Let Γ be a simple FGS program, $g = (V_g, E_g, H_g)$ a graph pattern in $\mathcal{GP}_{\Lambda \cup \mathcal{X}}(\Gamma)$ and S in $\mathcal{FGP}_{\Lambda \cup \mathcal{X}}(\Gamma)$. If $\mathcal{L}_\Lambda(\Gamma, g) \subseteq \mathcal{L}_\Lambda(\Gamma, S)$ and $\mathcal{L}_\Lambda(\Gamma, g/\{e\}) \not\subseteq \mathcal{L}_\Lambda(\Gamma, S)$ for any edge $e \in E_g$, then there exists a graph pattern $g' = (V_{g'}, E_{g'}, H_{g'})$ in S such that $g \preceq g'$ and $|E_g| = |E_{g'}|$ hold.*

Second, we consider the internal repeat-loop between lines 7 and 10 in the algorithm LEARN_UNION. Let Γ be a simple FGS program. Let g be a graph pattern in $\mathcal{GP}_{\Lambda \cup \mathcal{X}}(\Gamma)$, g' a graph pattern in $\mathcal{GP}_{\Lambda \cup \mathcal{X}}$ and x a variable label appearing in g'. We write $g \dashv_\Gamma g'$ if there exists a graph rewriting rule D in Γ such that $g \equiv g'\{x := [h, \sigma]\}$, that is, if g is a graph pattern obtained from g' by replacing the variable having the variable label x with h, where h is a simple graph pattern appearing in the head of D. For example, for graph patterns g, g_1, g_2 given in Fig. 8, we have $g_1 \dashv_{TTSP} g$ and $g_2 \dashv_{TTSP} g$ (i.e., $g_1 \equiv g\{x := [h_1, (u_1, u_2)]\}$ and $g_2 \equiv g\{x := [h_2, (u_1, u_2)]\}$), from the second and the third graph rewriting rules in $TTSP$, where h_1, h_2 are simple graph patterns given in Fig. 8 and $TTSP$ is the simple FGS program in Fig. 3.

For graph patterns g, $g' \in \mathcal{GP}_{\Lambda \cup \mathcal{X}}(\Gamma)$, if $g \dashv_\Gamma g'$ then g' is a generalized graph pattern of g such that $g \preceq g'$. Therefore we have the following lemma.

Lemma 5. *Let Γ be a simple FGS program. For two graph patterns g, g' in $\mathcal{GP}_{\Lambda \cup \mathcal{X}}(\Gamma)$, if $g \dashv_\Gamma g'$ holds then $g \preceq g'$ holds.*

For a graph pattern g given after executing the internal repeat-loop between lines 7 and 10 in the algorithm LEARN_UNION, we can see that $\mathcal{L}_\Lambda(\Gamma, f) \not\subseteq \mathcal{L}_\Lambda(\Gamma, \mathcal{T}_*)$ for any graph pattern $f \in \mathcal{GP}_{\Lambda \cup \mathcal{X}}(\Gamma)$ with $g \dashv_\Gamma f$.

Lemma 6. *Let Γ be a simple FGS program. Let $g = (V_g, E_g, H_g)$ be a graph pattern in $\mathcal{GP}_{\Lambda \cup \mathcal{X}}(\Gamma)$ and S a set in $\mathcal{FGP}_{\Lambda \cup \mathcal{X}}(\Gamma)$ such that there exists a graph pattern $g' = (V_{g'}, E_{g'}, H_{g'})$ in S with $g \preceq g'$ and $|E_g| = |E_{g'}|$. Then, if $\mathcal{L}_\Lambda(\Gamma, f) \not\subseteq \mathcal{L}_\Lambda(\Gamma, S)$ for any graph pattern $f \in \mathcal{GP}_{\Lambda \cup \mathcal{X}}(\Gamma)$ with $g \dashv_\Gamma f$, then $g \equiv g'$ holds.*

Proof. Since $g \preceq g'$ holds, there exists a substitution $\theta = \{x_1 := [f_1, \sigma_1], x_2 := [f_2, \sigma_2], \ldots, x_n := [f_n, \sigma_n]\}$ such that $g \equiv g'\theta$ holds. Since $|E_g| = |E_{g'}|$, $|H_g| \geq |H_{g'}|$ holds. We assume that $|H_g| > |H_{g'}|$ holds. Then, we can see that there exists a binding $x_\ell := [f_\ell, \sigma_\ell]$ in θ such that $|V_{f_\ell}| \geq 2$, $|E_{f_\ell}| = 0$ and $|H_{f_\ell}| \geq 2$ hold, where $f_\ell = (V_{f_\ell}, E_{f_\ell}, H_{f_\ell})$. Hence, since $g' \in \mathcal{GP}_{\Lambda \cup \mathcal{X}}(\Gamma)$, there exists a substitution θ' such that $g \equiv g'\theta \dashv_\Gamma g'\theta' \in \mathcal{GP}_{\Lambda \cup \mathcal{X}}(\Gamma)$ holds. Since from Lemma 5, $g \preceq g'\theta' \preceq g' \in S$ holds, we have $\mathcal{L}_\Lambda(L, g) \subseteq \mathcal{L}_\Lambda(L, g'\theta') \subseteq \mathcal{L}_\Lambda(L, g')$. This is a contradiction. Thus, we can see that $|H_g| = |H_{g'}|$. We have $g \equiv g'$.   □

Let Γ be a simple FGS program. For two sets $P, Q \in \mathcal{FGP}_{\Lambda \cup \mathcal{X}}(\Gamma)$, if there exists a graph pattern $f \in Q$ such that $f \preceq g$ for any $g \in P$, we write $P \sqsubseteq Q$. If $P \sqsubseteq Q$ and $Q \not\sqsubseteq P$, then we write $P \sqsubset Q$. Then, from the above lemmas, the following theorem holds.

Theorem 1. *Let Γ be a simple FGS program. The algorithm LEARN_UNION in Fig. 7 exactly identifies any set $\mathcal{T}_* \in \mathcal{FGP}_{\Lambda \cup \mathcal{X}}(\Gamma)$ in polynomial time using at most $m+1$ equivalence queries and at most $m(n+rn^2)$ restricted subset queries, where $m = |\mathcal{T}_*|$, n is the maximum number of edges of counterexamples and $r = |\Gamma|$, if the number of labels of edges is infinite.*

Proof. We consider the i-th iteration from the line 2 to the line 12 of the algorithm LEARN_UNION, where $i \geq 1$. Let S_i be a hypothesis given to $\text{Equiv}_{\mathcal{T}_*}$ at the line 2 of LEARN_UNION and $g_i = (V_i, E_i, H_i)$ a counterexample given at the line 4 of LEARN_UNION. Assume that $S_0 = \emptyset$. In a similar way to Lemmas 6 and 7 in [10], from Lemmas 1, 3, 4 and 6, we can prove that for every $i \geq 1$, $g_i \in \mathcal{L}_\Lambda(\Gamma, \mathcal{T}_*)$, $S_{i-1} \sqsubseteq \mathcal{T}_*$ and $S_{i-1} \sqsubset S_i$ hold. Hence, we can see that the algorithm LEARN_UNION correctly outputs a set S such that $\mathcal{L}_\Lambda(\Gamma, S) = \mathcal{L}_\Lambda(\Gamma, \mathcal{T}_*)$, and that LEARN_UNION terminates in polynomial time.

Next, we consider the numbers of restricted subset queries and equivalence queries. In the loop of the lines 5-6, LEARN_UNION uses at most $|E_i|$ restricted subset queries. Moreover, the loop of the lines 7-10 uses at most $|\Gamma| \times |E_i|^2$ restricted subset queries. The while-loop from the line 2 to the line 12 is repeated at most $|\mathcal{T}_*|$ times. Therefore, LEARN_UNION uses at most $m(n+rn^2)$ restricted subset queries and at most $m + 1$ equivalence queries, where $m = |\mathcal{T}_*|$, n the maximum number of edges of counterexamples and $r = |\Gamma|$, if the number of labels of edges is infinite.   □

Let $\mathcal{FMI} = \mathcal{FGP}_{\Lambda\cup\chi}(\mathcal{MI})$, $\mathcal{FUI} = \mathcal{FGP}_{\Lambda\cup\chi}(\mathcal{UI})$, $\mathcal{FOI} = \mathcal{FGP}_{\Lambda\cup\chi}(\mathcal{OI})$ and $\mathcal{FTTSP} = \mathcal{FGP}_{\Lambda\cup\chi}(\mathcal{TTSP})$. From the definitions of $\mathcal{MI}$ and $\dashv_{\mathcal{MI}}$, we can reduce the number of restricted subset queries as the following corollary.

Corollary 1. *Any set $\mathcal{T}_* \in \mathcal{FMI}$ is exactly identified in polynomial time using at most $m + 1$ equivalence queries and at most $m(n + 3n^2)$ restricted subset queries, where $m = |\mathcal{T}_*|$ and n is the maximum number of edges of counterexamples, if the number of labels of edges is infinite.*

Moreover, since $\mathcal{GP}_{\Lambda\cup\chi}(\mathcal{UI})$, $\mathcal{GP}_{\Lambda\cup\chi}(\mathcal{OI})$ and $\mathcal{GP}_{\Lambda\cup\chi}(\mathcal{TTSP})$ are closed with respect to the binary relations $\dashv_{\mathcal{UI}}$, $\dashv_{\mathcal{OI}}$ and $\dashv_{\mathcal{TTSP}}$, respectively, we have the following corollary.

Corollary 2. *The algorithm* LEARN_UNION *in Fig. 7 exactly identifies any finite set $\mathcal{T}_*$ of either $\mathcal{FUI}$, $\mathcal{FOI}$ or $\mathcal{FTTSP}$ in polynomial time using at most $m + 1$ equivalence queries and at most $m(n + n^2)$ restricted subset queries, where $m = |\mathcal{T}_*|$ and n is the maximum number of edges of counterexamples, if the number of labels of edges is infinite.*

4   Hardness Results on the Learnability

In this section, we show the insufficiency of learning of $\mathcal{FUI}$, $\mathcal{FOI}$, $\mathcal{FMI}$ and $\mathcal{FTTSP}$ in exact learning model. For a graph pattern $g = (V_g, E_g, H_g)$, the sum of numbers of edges and variables in g is called the *size* of g, i.e., $|g| = |E_g| + |H_g|$.

Lemma 7. *(László Lovász [8]) Let W_n be the number of all rooted unordered unlabeled trees of size n. Then, $2^{n+1} < W_n < 4^{n+1}$, where $n \geq 5$.*

From the above lemma, if $|\Lambda| \geq 1$, then the number of rooted unordered (ordered, mixed) trees of size n is greater than 2^{n+1}. The following lemma is known to show the insufficiency of learning in exact learning model.

Lemma 8. *(Angluin [2]) Suppose the hypothesis space contains a class of distinct sets $L_1, \ldots, L_N$. If there exists a set $L_\cap$ in the hypothesis space such that for any pair of distinct indices i, j ($1 \leq i, j \leq N$), $L_\cap = L_i \cap L_j$, then any algorithm that exactly identifies each of the hypotheses L_i using restricted equivalence, membership, and subset queries must make at least $N - 1$ queries in the worst case.*

By Lemmas 7 and 8, we have the following Theorems 2 and 3.

Theorem 2. *Let $\mathcal{F}$ be either $\mathcal{FUI}$, $\mathcal{FOI}$ or $\mathcal{FMI}$ and $\mathcal{F}_n$ the collection of all sets in $\mathcal{F}$ each of which contains only graph patterns of size n. Then, any learning algorithm that exactly identifies all sets in $\mathcal{F}_n$ using restricted equivalence, membership and subset queries must make greater than 2^{n+1} queries in the worst case, where $|\Lambda| \geq 1$ and $n \geq 5$.*

Proof. We prove the insufficiently of learning for $\mathcal{FUI}$. In a similar way to it, we can prove the insufficiently of learning for $\mathcal{FOI}$ and $\mathcal{FMI}$. We denote by

$\mathcal{S}_n$ the class of singleton sets of rooted unordered trees of size n. The class $\mathcal{S}_n$ is a subclass of $\mathcal{FUT}$ and for any L and L' in $\mathcal{S}_n$, $L \cap L' = \emptyset$. Since the empty set $\mathcal{L}_\Lambda(\mathcal{UT}, \emptyset) = \emptyset$ is a hypothesis in $\mathcal{FUT}$, by Lemmas 7 and 8, any learning algorithm that exactly identifies all the finite sets of rooted unordered term trees of size n using restricted equivalence, membership and subset queries must make more than 2^{n+1} queries in the worst case, even when $|\Lambda| = 1$. □

Theorem 3. *Any learning algorithm that exactly identifies all sets in $\mathcal{FTTSP}$ each of which contains only graph patterns of size n, using restricted equivalence, membership and subset queries, must make greater than $2^{\frac{n}{2}}$ queries in the worst case, where $|\Lambda| \geq 1$ and $n \geq 10$.*

5 Conclusion

We have considered polynomial time learnabilities of finite unions of graph structured datasets in exact learning model of Angluin [2]. In order to represent structural features of graph structured data, we have given a linear graph pattern with structural features such as "having tree structures" and "having TTSP graph structures" by using Formal Graph System given in [15]. Then, for a simple FGS program Γ, we have shown that any set $\mathcal{T}_*$ of m linear graph patterns is exactly identified in polynomial time using at most $m+1$ equivalence queries and at most $m(n + rn^2)$ restricted subset queries, where n is the maximum number of edges of counterexamples and $r = |\Gamma|$, if the number of labels of edges is infinite. Next, as a negative result, we show that finite sets of linear graph patterns having tree structures and two-terminal series parallel graph structures are not learnable in polynomial time using restricted equivalence, membership and subset queries.

As future works, we will consider polynomial time learnabilities of finite unions of graph patterns having structural features generated by non-simple FGS programs such as planar graphs, balanced binary trees, complete graphs. We conclude by summarizing our results and remained open problems in Table 1.

Table 1. Our results and remained open problems

	polynomial time exact learning	polynomial time inductive inference from positive data
$\mathcal{FGP}_{\Lambda \cup \mathcal{X}}(\Gamma)$		
$\quad \mathcal{FTTSP}$		
$\quad \mathcal{FMT}$	*Yes*[This Work]	*Open*
$\quad \mathcal{FUT}$		
$\quad \mathcal{FOT}$	*Yes*[10]	*Yes*[6] (for 2 unions)
$\mathcal{FGP}_{\Lambda \cup \mathcal{X}}(\Delta)$	*Open*	*Open*
$\mathcal{FEXGP}_{\Lambda \cup \mathcal{X}}(\mathcal{OT})$	*Yes*[11]	*Open*
$\mathcal{FEXGP}_{\Lambda \cup \mathcal{X}}(\Delta)$	*Open*	*Open*

Here, Γ and Δ are a simple FGS program and a (non-simple) FGS program, respectively. $\mathcal{FEXGP}_{\Lambda \cup \mathcal{X}}(\mathcal{OT})$ denotes the finite sets of (non-linear) graph patterns with ordered tree structures generated by the simple FGS program $\mathcal{OT}$.

References

1. Amoth, T.R., Cull, P., Tadepalli, P.: On exact learning of unordered tree patterns. Machine Learning 44, 211–243 (2001)
2. Angluin, D.: Queries and concept learning. Machine Learning 2, 319–342 (1988)
3. Arimura, H., Sakamoto, H., Arikawa, S.: Efficient learning of semi-structured data from queries. In: Abe, N., Khardon, R., Zeugmann, T. (eds.) ALT 2001. LNCS (LNAI), vol. 2225, pp. 315–331. Springer, Heidelberg (2001)
4. Arimura, H., Shinohara, T., Otsuki, S.: Polynomial time algorithm for finding finite unions of tree pattern languages. In: Proc. NIL-91. LNCS (LNAI), vol. 659, pp. 118–131. Springer, Heidelberg (1993)
5. Duffin, R.J.: Topology of series parallel networks. J. Math. Anal. Appl. 10, 303–318 (1965)
6. Hirashima, H., Suzuki, Y., Matsumoto, S., Uchida, T., Nakamura, Y.: Polynomial time inductive inference of unions of two term tree languages. In: Proc. ILP'06, pp. 92–94 (2006) (short papers)
7. Lloyd, J.W.: Foundations of Logic Programming, 2nd edn. Springer, Heidelberg (1987)
8. Lovász, L.: Combinatorial Problems and Exercises. ch. Two classical enumeration problems in graph theory. North-Holland Publishing Company (1979)
9. Matsumoto, S., Hayashi, Y., Shoudai, T.: Polynomial time inductive inference of regular term tree languages from positive data. In: ALT 1997. LNCS (LNAI), vol. 1316, pp. 212–227. Springer, Heidelberg (1997)
10. Matsumoto, S., Shoudai, T., Miyahara, T., Uchida, T.: Learning of finite unions of tree patterns with internal structured variables from queries. In: McKay, B., Slaney, J.K. (eds.) AI 2002: Advances in Artificial Intelligence. LNCS (LNAI), vol. 2557, pp. 523–534. Springer, Heidelberg (2002)
11. Matsumoto, S., Suzuki, Y., Shoudai, T., Miyahara, T., Uchida, T.: Learning of finite unions of tree patterns with repeated internal structured variables from queries. In: Gavaldá, R., Jantke, K.P., Takimoto, E. (eds.) ALT 2003. LNCS (LNAI), vol. 2842, pp. 144–158. Springer, Heidelberg (2003)
12. Miyahara, T., Suzuki, Y., Shoudai, T., Uchida, T., Takahashi, K., Ueda, H.: Discovery of frequent tag tree patterns in semistructured web documents. In: Chen, M.-S., Yu, P.S., Liu, B. (eds.) PAKDD 2002. LNCS (LNAI), vol. 2336, pp. 341–355. Springer, Heidelberg (2002)
13. Suzuki, Y., Akanuma, R., Shoudai, T., Miyahara, T., Uchida, T.: Polynomial time inductive inference of ordered tree patterns with internal structured variables from positive data. In: Kivinen, J., Sloan, R.H. (eds.) COLT 2002. LNCS (LNAI), vol. 2375, pp. 169–184. Springer, Heidelberg (2002)
14. Takami, R., Suzuki, Y., Uchida, T., Shoudai, T., Nakamura, Y.: Polynomial time inductive inference of TTSP graph languages from positive data. In: Kramer, S., Pfahringer, B. (eds.) ILP 2005. LNCS (LNAI), vol. 3625, pp. 366–383. Springer, Heidelberg (2005)
15. Uchida, T., Shoudai, T., Miyano, S.: Parallel algorithm for refutation tree problem on formal graph systems. IEICE Transactions on Information and Systems E78-D(2), 99–112 (1995)
16. Yamasaki, H., Shoudai, T.: A polynomial time algorithm for finding linear interval graph patterns. In: Proc. TAMC-2007. LNCS, vol. 4484, pp. 67–78. Springer, Heidelberg (2007)

Polynomial Summaries of
Positive Semidefinite Kernels

Kilho Shin[1] and Tetsuji Kuboyama[2]

[1] Carnegie Mellon CyLab Japan
yshin@cmuj.jp
[2] University of Tokyo
kuboyama@ccr.u-tokyo.ac.jp

Abstract. Although polynomials have proven to be useful tools to tailor generic kernels to context of specific applications, little was known about generic rules for tuning parameters (*i.e.* coefficients) to engineer new positive semidefinite kernels. This not only may hinder intensive exploitation of the flexibility of the kernel method, but also may cause misuse of indefinite kernels. Our main theorem presents a sufficient condition on polynomials such that applying the polynomials to known positive semidefinite kernels results in positive semidefinite kernels. The condition is very simple and therefore has a wide range of applications. In addition, in the case of degree 1, it is a necessary condition as well. We also prove the effectiveness of our theorem by showing three corollaries to it: the first one is a generalization of the polynomial kernels, while the second one presents a way to extend the principal-angle kernels, the trace kernels, and the determinant kernels. The third corollary shows corrected sufficient conditions for the codon-improved kernels and the weighted-degree kernels with shifts to be positive semidefinite.

1 Introduction

To exploit the flexibility of the kernel method, it is critical that sufficiently wide latitude is allowed in selecting kernel functions. On the other hand, using polynomials has proven effective to tailor known basic kernels (we call them *underlying kernels*) to context of specific applications (*e.g.* polynomial kernels [1], principal-angle and determinant kernels [2,3], codon-improved and weighted-degree-with-shift kernels [4,5]; See Sect. 2.1).

Little, however, was known about generic methodologies on the use of polynomials for this purpose, more specifically, about conditions on polynomials which lead the resulting kernels to be *positive semidefinite*. Positive semidefiniteness of a kernel $K(x,y)$ indicates the property that arbitrary Gram matrices of $K(x,y)$ are positive semidefinite (*i.e.* the matrices include no negative eigenvalues; See also Definition 3) — for an arbitrary set of data points $\{x_1, \ldots, x_n\} \subseteq \mathcal{X}$, the corresponding Gram matrix is defined as the $n \times n$ matrix $[K(x_i, x_j)]_{i,j=1}^n$. Also, if $\mathcal{X}$ is a finite set, this property is equivalent to the property that there exists a mapping (*feature decomposition*) $\Phi : \mathcal{X} \longrightarrow \mathbb{R}^N$ such that $K(x,y) = \Phi(x)^\mathsf{T} \Phi(y)$.

M. Hutter, R.A. Servedio, and E. Takimoto (Eds.): ALT 2007, LNAI 4754, pp. 313–327, 2007.

Positive semidefinite kernels are also known as *reproducing* kernels and *Mercer's* kernels.

Lack of general rules to discriminate between *fertile polynomials* that always generate positive semidefinite kernels and the other infertile polynomials is a serious problem, since positive semidefiniteness is a fundamental premise for many kernel-based learning machines to work properly (*e.g.* SVM [1]).

The present paper addresses this problem, and presents a sufficient condition for *fertile polynomials* (Theorem 1). Moreover, the condition turns out to be a *necessary* condition in the case of degree 1. This implies that the condition is generic. As additional collateral evidences of the generality of the condition, we employ a few known positive semidefinite kernels as examples, which are all derived using polynomials from other positive semidefinite underlying kernels, and show that they are special cases of our main theorem.

2   Problem Identification and Our Contributions

2.1   A Review of Polynomial-Based Composition of Kernels

To start with, we show three examples of polynomial-based composition of kernels.

Polynomial (Poly) kernels. The *polynomial kernels* are given in the form of $(k(x,y)+c)^d$ for a positive semidefinite underlying kernel $k(x,y)$, and are positive semidefinite if the constant c is non-negative (*e.g.* [1]). Polynomial kernels have proven useful for two main reasons – (1) a separating *hypersurface*[1] in a feature space of the underlying kernel is mapped to a *hyperplane* in a higher-dimensional feature space so that learning machines (*e.g.* SVM) can discover it; (2) polynomial kernels reflect the correlation of tuples of features of the underlying kernel [4]. Polynomial kernels can be generalized to the form of $K(x,y) = \sum_{i=0}^{d} c_i k(x,y)^i$ with arbitrary $c_i \geq 0$ without harming positive semidefiniteness.

Principal-angle (PA) kernels and determinant (Det) kernels. When a data point is represented as a set of vectors, the principal angles of the linear subspaces spanned by the representing vectors of two data points have proven to be effective measures for similarity between the data points (*e.g.* [6]). Wolf *et al.* [2] showed that principal angles can be computed using the *kernel trick*, and introduced the positive semidefinite kernels defined by (1).

$$K(x,y) = \left(\det[Q_X{}^\mathsf{T} Q_Y] \right)^2 = \prod_{i=1}^{k} \cos^2 \theta_i \tag{1}$$

In (1), x and y are tuples $(x^{(1)}, \ldots, x^{(D)})$ and $(y^{(1)}, \ldots, y^{(D)})$ in $\mathcal{X}^D$, θ_i denotes the i-th principal angle between the column spaces of the matrices $X =$

---

[1] In this paper, by a hypersurface, we mean a subspace of $\mathbb{R}^N$ defined by an algebraic equation whose degree is higher than 1.

$[\Phi(x^{(1)}),\ldots,\Phi(x^{(D)})]$ and $Y = [\Phi(y^{(1)}),\ldots,\Phi(y^{(D)})]$, Q_X and Q_Y are matrices obtained by the QR decomposition of X and Y, and $\Phi : \mathcal{X} \longrightarrow \mathbb{R}^n$ is a feature decomposition of an underlying kernel $k(x,y)$ (i.e. $k(x^{(i)}, y^{(j)}) = \Phi(x^{(i)})^{\mathsf{T}}\Phi(y^{(j)})$). Also, Zhou [3] introduced the determinant kernels for matrix-type data points.

Positive semidefiniteness of the principal-angle kernels and the determinant kernels is reduced to that of $K(x,y)$ defined as follows.

$$K(x,y) = \det \begin{bmatrix} k(x^{(1)}, y^{(1)}) & \cdots & k(x^{(1)}, y^{(D)}) \\ \vdots & \ddots & \vdots \\ k(x^{(D)}, y^{(1)}) & \cdots & k(x^{(D)}, y^{(D)}) \end{bmatrix}$$

$$= \sum_{\sigma \in \mathfrak{S}_D} \mathrm{sgn}(\sigma) \prod_{i=1}^{D} k(x^{(i)}, y^{(\sigma(i))}) \tag{2}$$

$K(x,y)$ is definitely a polynomial in $k(x^{(i)}, y^{(j)})$, and its positive semidefiniteness can be proven by Binet-Cauchy Theorem [2,3,7].

Codon-improved (CI) kernels and weighted-degree-with-shift (WDwS) kernels. The codon-improved kernels [4] and their generalization, namely the weighted-degree-with-shift kernels [5], are similar to the spectrum kernels [8] in that they count matching substrings between a pair of strings, but are different in that matches are weighted according to their positional information. Although these kernels are defined using polynomials, different from the examples seen so far, their positive semidefiniteness is not proven in a straightforward manner. In fact, appropriate selection of the coefficients of the polynomials is required to maintain positive semidefiniteness, and both [4] and [5] made mistakes in this regard.

For example, the codon-improved kernels [4] are designed so as to exploit the *a priori* knowledge "a coding sequence (CDS) shifted by three nucleotides still looks like CDS." In fact, in addition to the matches of substrings starting at the same position, they count substrings starting at the positions differ exactly by 3. A precise definition is given as follows. For sequences of nucleotides x and y, we let x_p (resp. y_p) denote the nucleotide at position p in x (resp. y). Then, $k_p(x,y)$ is defined as follows.

$$k_p(x,y) = \sum_{j=-\ell}^{\ell} w_j \delta(x_{p+j}, y_{p+j}) \tag{3}$$

In (3), w_j's are non-negative weights, and $\delta(x_{p+j}, y_{p+j})$ is Kronecker's delta: $\delta(x_{p+j}, y_{p+j})$ is 1, if x_{p+j} and y_{p+j} represent the same nucleotide, and it is 0, otherwise. When T denotes the shift operator that chops off the leading 3 nucleotides, the window score $\mathrm{win}_p(x,y)$ at position p, and the codon-improved kernel $K(x,y)$ are respectively defined by (4) and (5), where $\bar{w}$ is another non-negative weight.

$$\mathrm{win}_p(x,y) = \{k_p(x,y) + \bar{w}\,(k_p(Tx,y) + k_p(x,Ty))\}^{d_1} \tag{4}$$

$$K(x,y) = \left(\sum_{p=\ell+1}^{L-\ell} \mathrm{win}_p(x,y) \right)^{d_2} \tag{5}$$

Although [4] claims that the codon-improved kernels are unconditionally positive semidefinite, the fact is that the weights should be chosen appropriately. We will illustrate this by a simplified example. Assume $w_j = 1$ $(j = -\ell, \ldots, \ell)$, $d_1 = 1$, $\ell = 3q$ and $p = 3q + 1$. When x and y are the strings of length $6q + 1$ defined as follows, $\mathrm{win}_p(x,x) = \mathrm{win}_p(y,y) = 6q + 1$ and $\mathrm{win}_p(x,y) = 4q + 4\bar{w}q$ hold.

$$x = \underbrace{\texttt{ATGCGT ATGCGT}\ldots\texttt{ATGCGT}}_{6q}\texttt{ A}$$

$$y = \underbrace{\texttt{CTGAGT CTGAGT}\ldots\texttt{CTGAGT}}_{6q}\texttt{ C}$$

Therefore, the corresponding Gram matrix is positive semidefinite, if, and only if, $1 \geq 2(2\bar{w} - 1)q$ holds. In particular, $\bar{w} \leq \frac{1}{2}$ proves necessary for $\mathrm{win}_p(x,y)$ to be positive semidefinite regardless of q.

On the other hand, [5] claims that $\bar{w} \leq \frac{1}{2}$ is a sufficient condition for the codon-improved kernels to be positive semidefinite, but the proof presented in [5] holds true only for cases of $w_j \geq w_{j-3}$ for $j = -\ell + 3, \ldots, \ell$. In Sect. 6.3, we will present $\bar{w} \leq \min\left\{ \frac{w_j}{w_j + w_{j-3}} \big| j = -\ell + 3, \ldots, \ell \right\}$ as a corrected sufficient condition.

2.2   Problem Identification

Table 1 gives the list of the polynomials used in the above examples. Of the listed polynomials, that for the polynomial kernels is in the most generic form. Nevertheless, it is still specific for two reasons: (1) it is simply univariate; (2) it cannot include negative coefficients. In fact, the examples other than the polynomial kernels use multivariate polynomials, and the polynomials for the principal-angle and determinant kernels include negative coefficients.

Thus, the polynomial kernels cannot be a generic formula to discriminate between the *fertile polynomials* that generate positive semidefinite kernels and the other infertile ones. Eventually, little was known about such formulas, and

Table 1. Polynomials in Examples

Kernels	Variables	Polynomial
Poly	ξ	$\sum_{j=1}^{d} c_j \xi^i$, $c_j \geq 0$
PA, Det	$\{\xi_{i,j}\}_{i,j=1}^{D}$	$\sum_{\sigma \in \mathfrak{S}_D} \mathrm{sgn}(\sigma) \prod_{i=1}^{D} \xi_{i,\sigma(i)}$
CI, WDwS	$\{\xi_{i,j}\}_{i,j=1}^{L+3}$	$\left[\sum_{j=-\ell}^{\ell} w_j \{ \xi_{p+j,p+j} + \bar{w}(\xi_{p+j,p+j+3} + \xi_{p+j+3,p+j}) \} \right]^{d_1}$, $\bar{w} \leq \min\left\{ \frac{w_j}{w_j+w_{j-3}} \big| j = -\ell+3, \ldots, \ell \right\}$

the lack not only would restrict the ranges of polynomials that can be used to *engineer* new kernels, but also could cause misuse of indefinite kernels.

The present paper addresses this problem, and, in fact, presents a strong sufficient condition for the *fertile polynomials*.

2.3 Our Contributions

Below, the contributions of the present paper are summarized.

- Four settings are known for deriving polynomial-based kernels from underlying kernels (Sect. 3.1). We first show that one of the types is truly the most expressive — the other types are reduced to it, but the converse does not hold. Then, we define *polynomial summaries* under the setting (Definition 1 in Sect. 3.2).
- Our main theorem (Theorem 1 in Sect. 4.2) presents a sufficient condition on polynomials whose *polynomial summaries* result in positive semidefinite kernels regardless of underlying positive semidefinite kernels. In the case of degree 1, the condition is also a necessary one, in the sense that, if a given linear polynomial p does not meet the condition, there exists an underlying positive semidefinite kernel $k(x, y)$ such that the p-summary of $k(x, y)$ is not positive semidefinite (Sect. 5.4).
- We introduce three corollaries to Theorem 1. The first two generalize the polynomial kernels and the determinant kernels (Sect. 6.1 and 6.2). The third one presents a corrected sufficient condition for the codon-improved kernels and the weighted-degree-with-shift kernels to be positive semidefinite (Sect. 6.3).

3 Polynomial Summaries

In this section, we first pursue the *most expressive* setting for polynomial-based composition of kernels, and then define *polynomial summaries* under the setting.

3.1 Relation Among the Known Settings

In the literature, we see four settings for polynomial-based kernels according to the answers to the following questions (Type A to D, and also see Table 2).

1. Are the polynomial to be used to derive $K(x, y)$ univariate or multivariate? This question is equivalent to the question whether the domain $\mathcal{X}$ of $K(x, y)$ is identical with the domain of the underlying kernel(s) or a non-trivial cartesian product of the domain(s) of the underlying kernel(s)?
2. Is the resulting kernel $K(x, y)$ derived from a single underlying kernel?

Type A. A univariate polynomial $p(\xi)$ is applied to a single underlying kernel $k : \mathcal{X} \times \mathcal{X} \to \mathbb{R}$. $K(x, y)$ is simply defined as $p(k(x, y))$ (*e.g.* the polynomial kernels [1]).

Type B. A multivariate polynomial p in the D^2 variables ξ_{ij} for $i, j = 1, \ldots, D$ is applied to a single underlying kernel $k : \mathcal{X}_* \times \mathcal{X}_* \to \mathbb{R}$. The domain $\mathcal{X}$ is defined as $\mathcal{X}_*^D$, and $K((x_1, \ldots, x_D), (y_1, \ldots, y_D))$ is obtained by substituting $k(x_i, y_j)$ for ξ_{ij}. (e.g. the principal-angle kernels [2] and the determinant kernels [7,3]).

Type C. A multivariate polynomial p in the D variables ξ_d for $d = 1, \ldots, D$ is applied to multiple underlying kernels $k_d : \mathcal{X} \times \mathcal{X} \to \mathbb{R}$. $K(x, y)$ is obtained by substituting $k_d(x, y)$ for ξ_d (e.g. [1, Proposition 3.12]).

Type D. A multivariate polynomial p in the D variables ξ_d for $d = 1, \ldots, D$ is applied to multiple underlying kernels $k'_d : \mathcal{X}_d \times \mathcal{X}_d \to \mathbb{R}$. The domain $\mathcal{X}$ is defined as $\mathcal{X}_1 \times \cdots \times \mathcal{X}_D$, and $K((x_1, \ldots, x_D), (y_1, \ldots, y_D))$ is obtained by substituting $k'_d(x_d, y_d)$ for ξ_d (e.g. Haussler's R-convolution kernels [9]).

Table 2. Types of polynomial-based kernels

Type	Domain of K	Polynomial	Underlying kernel(s)	Substitution
A	$\mathcal{X}$	$p(\xi)$	$k : \mathcal{X} \times \mathcal{X} \to \mathbb{R}$	$\xi = k(x, y)$
B	$\mathcal{X}_*^D$	$p(\xi_{11}, \ldots, \xi_{ij}, \ldots, \xi_{DD})$	$k : \mathcal{X}_* \times \mathcal{X}_* \to \mathbb{R}$	$\xi_{ij} = k(x_i, y_j)$
C	$\mathcal{X}$	$p(\xi_1, \ldots, \xi_D)$	$\{k_d : \mathcal{X} \times \mathcal{X} \to \mathbb{R}\}_{d=1,\ldots,D}$	$\xi_d = k_d(x, y)$
D	$\mathcal{X}_1 \times \cdots \times \mathcal{X}_D$	$p(\xi_1, \ldots, \xi_D)$	$\{k'_d : \mathcal{X}_d \times \mathcal{X}_d \to \mathbb{R}\}_{d=1,\ldots D}$	$\xi_d = k'_d(x_d, y_d)$

Type A is the special case of the other types where $D = 1$. Type D is the special case of Type C where $\mathcal{X} = \mathcal{X}_1 \times \cdots \times \mathcal{X}_D$ and $k_d((x_1, \ldots, x_D)) = k'_d(x_d)$. Also, Type B, when p only includes the variables ξ_{ii}, is a special case of Type D where $\mathcal{X}_1 = \cdots = \mathcal{X}_D = \mathcal{X}_*$ and $k_1 = \cdots = k_D = k$. Furthermore, Lemma 1 asserts that Type C is a special case of Type B.

Lemma 1. *For an arbitrary family of positive semidefinite kernels $\{k_d : \mathcal{X} \times \mathcal{X} \to \mathbb{R}\}_{d=1,\ldots,D}$, there exist a set $\mathcal{X}_*$, a positive semidefinite kernel $k : \mathcal{X}_* \times \mathcal{X}_* \to \mathbb{R}$ and an inclusion mapping $i : \mathcal{X} \to \mathcal{X}_*^D$ such that $k_d(x, y) = k(i_d(x), i_d(y))$ for $i(x) = (i_1(x), \ldots, i_D(x))$ and $i(y) = (i_1(y), \ldots, i_D(y))$.*

Proof. Let $\mathcal{X}_*$ be $\mathcal{X} \times \{1, 2, \ldots, D\}$. When $k : \mathcal{X}_* \times \mathcal{X}_* \to \mathbb{R}$ is defined so that (6) holds, it is obvious that k is positive semidefinite.

$$k((x, a), (y, b)) = \begin{cases} k_d(x, y) & \text{if } a = b = d, \\ 0 & \text{otherwise.} \end{cases} \tag{6}$$

We obtain the assertion by defining $i : \mathcal{X} \to \mathcal{X}_*^D$ by $i(x) = ((x, 1), \ldots, (x, D))$. □

For a polynomial $p(\xi_1, \ldots, \xi_D)$, (7) follows from Lemma 1. Therefore, if $p(k(x_1, y_1), \ldots, k(x_D, y_D))$ are positive semidefinite for arbitrary positive semidefinite k, so are $p(k_1(x, y), \ldots, k_D(x, y))$ for arbitrary positive semidefinite k_d.

$$p(k_1(x,y), \ldots, k_D(x,y)) = p(k(i_1(x), i_1(y)), \ldots, k(i_D(x), i_D(y))) \quad (7)$$

Thus, the set of *fertile* polynomials of the form $p(\xi_1, \ldots, \xi_D)$ that produce positive semidefinite $K(x,y)$ is identical regardless of the type of setting, and, therefore, the setting of Type B is truly more expressive than those of the other types, since it allows D^2-variate polynomials $p(\xi_{11}, \ldots, \xi_{ij}, \ldots, \xi_{DD})$.

3.2 Definition of Polynomial Summaries

We define polynomial summaries assuming the setting of Type B.

Definition 1. *Let $p(\xi_{11}, \xi_{12}, \ldots, \xi_{ij}, \ldots, \xi_{DD})$ be a real polynomial in the D^2 variables of $\{\xi_{ij} \mid i,j = 1, \ldots, D\}$. The p-summary of an underlying kernel $k : \mathcal{X} \times \mathcal{X} \longrightarrow \mathbb{R}$ is the kernel $p[k] : \mathcal{X}^D \times \mathcal{X}^D \longrightarrow \mathbb{R}$ defined as below.*

$$p[k]((x_1, \ldots, x_D), (y_1, \ldots, y_D))$$
$$= p(k(x_1, y_1), k(x_1, y_2), \ldots, k(x_i, y_j), \ldots, k(x_D, y_D))$$

Example 1. The kernel of (2) is a polynomial summary with respect to the polynomial p as below.

$$p(\xi_{11}, \ldots, \xi_{DD}) = \sum_{\sigma \in \mathfrak{S}_D} \mathrm{sgn}(\sigma) \prod_{i=1}^{D} \xi_{i\sigma(i)}$$

If the underlying kernel k is positive semidefinite, the p-summary for k is positive semidefinite (*e.g.* [7,3]). □

Example 2. Define f_p as follows.

$$f_p(\xi_{11}, \ldots, \xi_{DD}) = \left[\sum_{j=-\ell}^{\ell} w_j \{\xi_{p+j,p+j} + \bar{w}(\xi_{p+j,p+j+3} + \xi_{p+j+3,p+j})\} \right]^{d_1}$$

Then, the window score $\mathrm{win}_p(x,y)$ of (4) is the f_p-summary of Kronecker's delta $\delta(\xi, \eta)$ defined over the alphabet $\{\mathsf{A}, \mathsf{T}, \mathsf{G}, \mathsf{C}\}$. Although $\delta(\xi, \eta)$ is positive semidefinite, it is necessary to choose appropriate w_j and $\bar{w}$ to make the resulting f_p-summary positive semidefinite (Sect. 2.1 and 6.3). □

As seen in Example 2, even if an underlying kernel k is positive semidefinite, p-summaries of k may or may not be positive semidefinite dependent on choice of polynomials p. Thus, the following question naturally arises.

What is a condition on p for the p-summaries to be positive semidefinite?

Theorem 1 answers to this question.

4   Main Theorem and a Plot of the Proof

4.1   Coefficient Matrices of Polynomials

Assume that a polynomial p of degree d in the D^2 variables $\xi_{11}, \ldots, \xi_{ij}, \ldots, \xi_{DD}$ is given the following representation for $\Delta = \{1, \ldots, \Delta\}$.

$$p = c_{\varnothing, \varnothing} + \sum_{\delta=1}^{d} \sum_{(k_1, \ldots, k_\delta) \in \Delta^\delta} \sum_{(l_1, \ldots, l_\delta) \in \Delta^\delta} c_{(k_1, \ldots, k_\delta), (l_1, \ldots, l_\delta)} \cdot \xi_{k_1 l_1} \xi_{k_2 l_2} \cdots \xi_{k_\delta l_\delta}$$

Then, we define a $\frac{D^{d+1}-1}{D-1}$-dimensional square matrix C, and refer to it as a *coefficient matrix* of p, whose rows and columns are indexed by vectors $\boldsymbol{i} \in \Delta^{[0..d]} = \{\varnothing\} \cup \bigcup_{\delta=1}^{d} \Delta^\delta$ and whose $(\boldsymbol{i}, \boldsymbol{j})$-element is $c_{\boldsymbol{i}, \boldsymbol{j}}$, if $|\boldsymbol{i}| = |\boldsymbol{j}|$, and 0, otherwise. When $\deg p \geq 2$ and $D \geq 2$, there exist more than one coefficient matrices for the same p. For example, C_c are all coefficient matrices for the polynomial $p = \xi_{11}\xi_{22} = c\xi_{11}\xi_{22} + (1-c)\xi_{22}\xi_{11}$ for arbitrary c.

$$C_c = \begin{bmatrix} 0 & 0 & 0 & 0 & 0 & 0 & 0 \\ 0 & 0 & 0 & 0 & 0 & 0 & 0 \\ 0 & 0 & 0 & 0 & 0 & 0 & 0 \\ 0 & 0 & 0 & 0 & 0 & 0 & 0 \\ 0 & 0 & 0 & 0 & c & 0 & 0 \\ 0 & 0 & 0 & 0 & 0 & 1-c & 0 \\ 0 & 0 & 0 & 0 & 0 & 0 & 0 \end{bmatrix}$$

- The 1st row (column) corresponds to the empty vector $\varnothing$.
- The 2nd and 3rd rows (columns) respectively correspond to the vectors (1) and (2).
- The $(2i+j+1)$-th row (column) for $(i, j) \in \{1, 2\}^2$ corresponds to the vector (i, j).

4.2   Statement of the Main Theorem

Theorem 1. *Let p be a real polynomial in the D^2 variables of $\{\xi_{ij} \mid (i,j) \in \{1, \ldots, D\}^2\}$. If p is given a representation whose coefficient matrix is positive semidefinite, then the p-summary $p[k]$ of an arbitrary underlying kernel k is positive semidefinite, if k is positive semidefinite.*

4.3   Key Lemma

Before we sketch a plot of the proof of Theorem 1, we first introduce Lemma 2, which will play a key role in the proof. The proof of Lemma 2 is given in 5.2.

Let X^{ij} be m-dimensional square matrices parameterized by $(i, j) = \{1, \ldots, n\}^2$, and let X denote the derived mn-dimensional square matrix $[X^{ij}]_{i,j=1,\ldots,n}$: the $(mi + k, mj + l)$-element of X, denoted by X_{kl}^{ij}, is defined to be the (k, l)-element of X^{ij}. For m- and n-dimensional vector spaces V and W, the matrix X represents a linear endomorphism of $V \otimes W$.

Definition 2. *For an m-dimensional square matrix A, the n-dimensional square matrix $\left[\mathrm{tr}(A^\mathsf{T} X^{ij})\right]_{i,j=1,\ldots,n} = \left[\sum_{k=1}^{m} \sum_{l=1}^{m} A_{kl} X_{kl}^{ij}\right]_{i,j=1,\ldots,n}$ is called the A-linear summary matrix of X, and is denoted by $\mathrm{smry}_A(X)$.*

Lemma 2. *For an m-dimensional real matrix A, the following are equivalent to each other.*

(1) A is positive semidefinite.

(2) The linear summary matrix $\mathrm{smry}_A(X)$ is positive semidefinite for an arbitrary mn-dimensional positive semidefinite matrix X.

4.4   A Plot of the Proof of Theorem 1

Now, we are ready to sketch a plot of the proof of Theorem 1. The objective of the proof is to show that arbitrary Gram matrices G of $p[k]$ are positive semidefinite (Definition 3 and 5.3). To show a plot, we will prove the assertion of Theorem 1 under the limited situation stated below.

- $D = 2$;
- G is for two data points $x^{(1)} = (x_1^{(1)}, x_2^{(1)})$ and $x^{(2)} = (x_1^{(2)}, x_2^{(2)})$;
- p is homogeneous of degree 1 or 2.

Therefore, we will see that the Gram matrix G is positive semidefinite.

$$G = \begin{bmatrix} p[k](x^{(1)}, x^{(1)}), \, p[k](x^{(1)}, x^{(2)}) \\ p[k](x^{(2)}, x^{(1)}), \, p[k](x^{(2)}, x^{(2)}) \end{bmatrix}$$

Case deg$(p) = 1$. Let $p(\xi_{11}, \xi_{12}, \xi_{21}, \xi_{22})$, C and X be as follows.

$$p(\xi_{11}, \xi_{12}, \xi_{21}, \xi_{22}) = c_{11}\xi_{11} + c_{12}\xi_{12} + c_{21}\xi_{21} + c_{22}\xi_{22}$$

$$C = \begin{bmatrix} c_{11} & c_{12} \\ c_{21} & c_{22} \end{bmatrix}, \quad X_{kl}^{ij} = k(x_k^{(i)}, x_l^{(j)}), \quad X^{ij} = \begin{bmatrix} X_{11}^{ij} & X_{12}^{ij} \\ X_{21}^{ij} & X_{22}^{ij} \end{bmatrix}, \quad X = \begin{bmatrix} X^{11} & X^{12} \\ X^{21} & X^{22} \end{bmatrix}$$

The matrix X is positive semidefinite, since it is a Gram matrix with respect to k. Therefore, it is concluded that $G = \mathrm{smry}_C(X)$ is positive semidefinite by Lemma 2.

Case deg$(p) = 2$. Let p, C and Y be as follows.

$$p(\xi_{11}, \xi_{12}, \xi_{21}, \xi_{22}) = \sum_{k_1=1}^{2} \sum_{k_2=1}^{2} \sum_{l_1=1}^{2} \sum_{l_2=1}^{2} c_{(k_1,k_2),(l_1,l_2)} \xi_{k_1 l_1} \xi_{k_2 l_2}$$

$$C = \left[c_{(k_1,k_2),(l_1,l_2)} \right]_{k_1,k_2,l_1,l_2=1,2}$$

$$Y^{ij} = \left[X_{k_1,l_1}^{ij} X_{k_2,l_2}^{ij} \right]_{k_1,k_2,l_1,l_2=1,2}, \quad Y = \begin{bmatrix} Y^{11} & Y^{12} \\ Y^{21} & Y^{22} \end{bmatrix}$$

To apply Lemma 2 to $G = \mathrm{smry}_C(Y)$, we claim that Y is positive semidefinite. The claim proves true since Y is a submatrix of $X \otimes X$ such that the diagonal elements of Y are also diagonal elements in $X \otimes X$ (Proposition 2).

5 Proof of Theorem 1

5.1 Mathematical Preliminary

In the remainder of this paper, a matrix always means *real* matrix.

When the transpose of a matrix A is denoted by A^{T}, a symmetric A satisfies $A^{\mathsf{T}} = A$, and an orthogonal A does $A^{\mathsf{T}} = A^{-1}$.

A positive semidefinite kernel $K : \mathcal{X} \times \mathcal{X} \longrightarrow \mathbb{R}$ is defined so that, for arbitrary $x_1, \ldots, x_n \in \mathcal{X}$, the Gram matrix G as defined below is *positive semidefinite*.

$$G = \begin{bmatrix} K(x_1, x_1) & \ldots & K(x_1, x_n) \\ \vdots & & \vdots \\ K(x_n, x_1) & \ldots & K(x_1, x_n) \end{bmatrix}$$

The definition of positive semidefinite matrices is given below.

Definition 3. *A real matrix A is called* positive semidefinite, *if, and only if, it is symmetric and one of, hence all of, the conditions of Proposition 1 hold.*

Proposition 1. *For an n-dimensional symmetric real matrix A, the following are equivalent to each other.*

(1) $(c_1, \ldots, c_n)A(c_1, \ldots, c_n)^{\mathsf{T}} \geq 0$ for arbitrary $(c_1, \ldots, c_n) \in \mathbb{R}^n$.
(2) A has only non-negative real eigenvalues.
(3) There exists an n-dimensional orthogonal matrix P such that $P^{\mathsf{T}}AP$ is diagonal with non-negative elements.
(4) $A = B^{\mathsf{T}}B$ for some $m \times n$ real matrix B.

Note that, from now on in this paper, we deploy the notion that a positive semidefinite matrix is necessarily symmetric.

Proposition 2. *Let $A = [A_{ij}]$ be an n-dimensional positive semidefinite matrix. For $\alpha_1, \ldots, \alpha_m \in \{1, \ldots, n\}$ for an arbitrary m, the m-dimensional matrix $A[\alpha_1, \ldots, \alpha_m]$ whose (i, j)-element is $A_{\alpha_i \alpha_j}$ is also positive semidefinite. In particular, any diagonal element of a positive semidefinite matrix is non-negative.*

5.2 Proof of Lemma 2

The claim that A and $\mathrm{smry}_A(X)$ are symmetric follows from Lemma 3.

Lemma 3. *For an m-dimensional real matrix A, the following are equivalent to each other.*

(1) A is symmetric.
(2) The linear summary matrix $\mathrm{smry}_A(X)$ is symmetric for an arbitrary mn-dimensional positive semidefinite matrix X.

Proof. Easy to see.   □

To prove the remainder of the assertion of Lemma 2, we first prove for the case where A is diagonal, and then prove the general case.

Assume that A is a positive semidefinite diagonal matrix, *i.e.* the I-th diagonal element α_I is equal to or greater than 0 for $I = 1, \ldots, m$.

If X is positive semidefinite, there exists an mn-dimensional matrix $Y = [Y^{ij}]_{i,j=1,\ldots,n}$ such that $X = Y^\mathsf{T} Y$ by Proposition 1 (4).

$$\mathrm{tr} A^\mathsf{T} X^{ij} = \sum_{I=1}^m \alpha_I \left(\sum_{k=1}^n \sum_{l=1}^m Y_{lI}^{ki} Y_{lI}^{kj} \right) = \sum_{I=1}^m \sum_{k=1}^n \sum_{l=1}^m (\sqrt{\alpha_I} Y_{lI}^{ki})(\sqrt{\alpha_I} Y_{lI}^{kj})$$

Therefore, $\mathrm{smry}_A(X) = Z^\mathsf{T} Z$ holds for the $m^2 n \times n$ matrix Z such that

$$Z_{mn(I-1)+m(k-1)+l,i} = \sqrt{\alpha_I} Y_{lI}^{ki}.$$

This means that $\mathrm{smry}_A(X)$ is positive semidefinite.

To prove the inverse, we assume $n = 1$. Let X_I is an m-dimensional positive semidefinite X_I whose elements are 0 except for $X_{II} = 1$. Since $\mathrm{smry}_A(X_I) = \alpha_I \geq 0$ holds for arbitrary I, A turns out positive semidefinite.

Now, we claim that the assertion for the general case where A is not necessarily diagonal is reduced to that of the diagonal case. Note that A is symmetric (Lemma 3). Therefore, $P^\mathsf{T} A P$ is diagonal for some orthogonal matrix P. Our claim immediately follows from the properties shown below.

1. A is positive semidefinite, if, and only if, so is $P^\mathsf{T} A P$.
2. X is positive semidefinite, if, and only if, so is $Y = \left[P^\mathsf{T} X^{ij} P \right]_{i,j=1,\ldots n}$.
3. $\mathrm{smry}_A(X) = \mathrm{smry}_{P^\mathsf{T} A P}(Y)$, since

$$\mathrm{tr}(A^\mathsf{T} X^{ij}) = \mathrm{tr} \left(P^\mathsf{T}(A^\mathsf{T} X^{ij}) P \right) = \mathrm{tr} \left((P^\mathsf{T} A P)^\mathsf{T} (P^\mathsf{T} X^{ij} P) \right).$$

Now, we have completed the proof.

Corollary 1. *For an m-dimensional real matrix A, the following are equivalent to each other.*

(1) A is positive semidefinite.
(2) A is a symmetric matrix such that $\mathrm{tr}(A^\mathsf{T} X) \geq 0$ for an arbitrary m-dimensional positive semidefinite matrix X.

5.3   Proof of Theorem 1

The concept of linear summaries of matrices can be naturally extended to that of *polynomial summaries* of matrices, and Lemma 2 is generalized as Lemma 4, to which Theorem 1 is a direct corollary.

Let p be a polynomial of degree d in the m^2 variables $\xi_{11}, \xi_{12}, \ldots, \xi_{mm}$. The *$p$-polynomial summary matrix* of an mn-dimensional matrix $X = [X^{ij}]_{i,j=1,\ldots,n}$ is denoted by $\mathrm{smry}_p[X]$, and is defined as follows.

$$\mathrm{smry}_p[X] = \left[p(X_{11}^{ij}, X_{12}^{ij}, \ldots, X_{mm}^{ij}) \right]$$

Lemma 4. *Let p be a real polynomial in the m^2 variables $\{\xi_{ij}\}_{i,j=1,\ldots,m}$. If p is given a representation whose coefficient matrix C is positive semidefinite, $\mathrm{smry}_p[X]$ is positive semidefinite for an arbitrary mn-dimensional positive semidefinite matrix X.*

Proof. Symmetry of $\mathrm{smry}_p[X]$ follows from that of C.

Let $\bar{X}^{ij}$ be the $\frac{m^{d+1}-1}{m-1}$-dimensional matrix defined as follows.

– The rows and columns are indexed by $k \in \{1,\ldots,m\}^{[0..d]}$.
– The (k,l)-element for $k,l \in \{1,\ldots,m\}^{[0..d]}$ is defined as follows.

$$\bar{X}^{ij}_{k,l} = \begin{cases} 1 & \text{if } k = l = \emptyset; \\ \displaystyle\prod_{\alpha=1}^{\delta} X^{ij}_{k_\alpha l_\alpha} & \text{if } k,l \in \{1,\ldots,m\}^\delta \text{ for } 1 \leq \delta \leq d; \\ 0 & \text{otherwise.} \end{cases}$$

If we can prove that $\bar{X} = [\bar{X}^{ij}]_{i,j=1,\ldots,n}$ is positive semidefinite, the assertion follows from Lemma 2, since $\mathrm{smry}_p[X] = \mathrm{smry}_C(\bar{X})$.

First, we will define the m^δ-dimensional matrix $\bar{X}\langle\delta\rangle^{ij}$ for $0 \leq \delta \leq d$.

– The rows and columns are identified by $k \in \{1,\ldots,m\}^\delta$.
– For $k,l \in \{1,\ldots,m\}^\delta$, the (k,l)-element is defined as follows.

$$\bar{X}\langle\delta\rangle^{ij}_{k,l} = \bar{X}^{ij}_{k,l}$$

When $\bar{X}\langle\delta\rangle$ denotes $[\bar{X}\langle\delta\rangle^{ij}]_{i,j=1,\ldots,n}$, $\bar{X}$ is isomorphic to the direct sum $\bigoplus_{\delta=0}^d \bar{X}\langle\delta\rangle$. Therefore, proving that $\bar{X}$ is positive semidefinite is equivalent to proving that so are $\bar{X}\langle\delta\rangle$ for all $0 \leq \delta \leq d$.

The tensor product $X^{\otimes\delta}$ has the following properties.

– The rows and columns are indexed by $(i,k) \in \{1,\ldots,n\}^\delta \times \{1,\ldots,m\}^\delta$.
– The $((i,k),(j,l))$-element $\left(\bar{X}^{\otimes\delta}\right)_{(i,k),(j,l)}$ is $\prod_{\alpha=1}^\delta X^{i_\alpha j_\alpha}_{k_\alpha l_\alpha}$.

Therefore, when i^δ denotes $(i,\ldots,i)$,

$$X\langle\delta\rangle = X^{\otimes\delta}[\{1^\delta, 2^\delta, \ldots, n^\delta\} \times \{1,\ldots,m\}^\delta]$$

holds under the notation of Proposition 2. Since X is positive semidefinite, so is $X^{\otimes d}$, and therefore, so is $\bar{X}\langle\delta\rangle$ by Proposition 2. $\qquad\square$

5.4 Consideration of the Case of Degree 1

For a linear polynomials $p(\xi_{11}, \xi_{12}, \ldots, \xi_{DD})$, if its unique coefficient matrix C is not positive semidefinite, neither is the D^2-dimensional submatrix C_1 corresponding to the terms of degree 1. Therefore, Corollary 1 asserts that there exists a D^2-dimensional positive semidefinite matrix X such that $C_1^\mathsf{T} X < 0$.

When we define a positive semidefinite kernel k over $\mathcal{X}_* = \{x^{(1)}, \ldots, x^{(D)}\}$ by $k(x^{(i)}, x^{(j)}) = \gamma X_{ij}$ for a positive γ, the following holds.

$$p[k]((x^{(1)}, \ldots, x^{(D)}), (x^{(1)}, \ldots, x^{(D)})) = c_{\emptyset,\emptyset} + \gamma C^{\mathsf{T}}_1 X.$$

Therefore, $p[k]$ is not positive semidefinite for a sufficiently large γ.

Proposition 3. *Assume that $p(\xi_{11}, \xi_{12}, \ldots, \xi_{DD})$ is a linear polynomial such that its unique coefficient matrix is not positive semidefinite. Then, there exists $\mathcal{X}_*$ and $k : \mathcal{X}_* \times \mathcal{X}_* \longrightarrow \mathbb{R}$ such that $\mid \mathcal{X}_* \mid \leq D$, k is positive semidefinite, and $p[k]$ is not positive semidefinite.*

6   Applications

6.1   Generalization of Polynomial Kernels

The following direct corollary to Theorem 1 presents the multivariate version of the polynomial kernels (Sect. 2.1).

Corollary 2. *Let p be a real polynomial in the D variables of $\{\xi_{ii} \mid i = 1, \ldots, D\}$. If p includes only non-negative coefficients, then the p-summary $p[k]$ is always a kernel function for an arbitrary kernel k.*

Proof. The polynomial p can be represented by a diagonal coefficient matrix with non-negative elements, which is apparently positive semidefinite. Hence, we have the assertion by Theorem 1. $\qquad\square$

6.2   Generalization of the Determinant Kernels

Take two data points $x = (x^{(1)}, \ldots, x^{(D)})$ and $y = (y^{(1)}, \ldots, y^{(D)})$. Letting $\Phi(\xi)$ be the characteristic polynomial of a Gram matrix $G = \left[k(x^{(i)}, y^{(j)})\right]_{i,j=1,\ldots,D}$, Wolf *et al.* [2] and Zhou [3] proved that the constant part (*i.e.* $\det(G)$) of $\Phi(\xi)$ is a positive semidefinite kernel, if $k(x^{(i)}, y^{(j)})$ is positive semidefinite.

Generalizing the result, we will see that all of the coefficients of $\Phi(\xi)$ are also positive semidefinite. When $(-1)^d \Phi_{D,d}$ denotes the coefficient of the term of ξ^d of (8), $\Phi_{D,d}$ is a $(D-d)$-degree homogeneous polynomial in $\{\xi_{11}, \xi_{12}, \ldots, \xi_{DD}\}$.

$$\det \begin{bmatrix} \xi_{11} - \xi & \cdots & \xi_{1D} \\ \vdots & \ddots & \vdots \\ \xi_{D1} & \cdots & \xi_{DD} - \xi \end{bmatrix} \tag{8}$$

Thus, we have the following.

$$\Phi(\xi) = \sum_{d=1}^{D} (-1)^d \left(\Phi_{D,d}[k](x, y)\right) \xi^d$$

Corollary 3. *For an arbitrary kernel* k, $\Phi_{D,d}[k]((x^{(1)}, \ldots, x^{(D)}), (y^{(1)}, \ldots, y^{(D)}))$
are positive semidefinite kernels for $d = 0, \ldots, D$.

Proof. $\Phi_{D,d}(\xi_{11}, \xi_{12}, \ldots, \xi_{DD})$ is evaluated as follows, where $\mathfrak{S}_{D-d}$ denotes the
permutation group acting on $\{1, \ldots, D-d\}$ and $\mathrm{sgn}(\pi)$ does the sign of the
permutation $\pi \in \mathfrak{S}_{D-d}$.

$$\Phi_{D,d} = \sum_{1 \leq \alpha_1 < \cdots < \alpha_{D-d} \leq D} \det([\xi_{\alpha_i \alpha_j}]_{i,j=1,\ldots,D-d})$$

$$= \sum_{1 \leq \alpha_1 < \cdots < \alpha_{D-d} \leq D} \sum_{\pi \in \mathfrak{S}_{D-d}} \mathrm{sgn}(\pi) \prod_{i=1}^{D-d} \xi_{\alpha_i \alpha_{\pi(i)}}$$

We fix an instance of $1 \leq \alpha_1 < \cdots < \alpha_{D-d} \leq D$, and show that the coefficient
matrix $C_{\alpha_1, \ldots, \alpha_{D-d}}$ for $\sum_{\tau \in \mathfrak{S}_{D-d}} \mathrm{sgn}(\pi) \prod_{i=1}^{D-d} \xi_{\alpha_i \alpha_{\pi(i)}}$ is positive semidefinite.

$$\sum_{\pi \in \mathfrak{S}_{D-d}} \mathrm{sgn}(\pi) \prod_{i=1}^{D-d} \xi_{\alpha_i \alpha_{\pi(i)}} = \sum_{\pi \in \mathfrak{S}_{D-d}} \sum_{\sigma \in \mathfrak{S}_{D-d}} \frac{\mathrm{sgn}(\sigma)\mathrm{sgn}(\pi \circ \sigma)}{(D-d)!} \prod_{i=1}^{D-d} \xi_{\alpha_{\sigma(i)} \alpha_{\pi(\sigma(i))}}$$

$$= \sum_{\sigma \in \mathfrak{S}_{D-d}} \sum_{\tau \in \mathfrak{S}_{D-d}} \frac{\mathrm{sgn}(\sigma)}{\sqrt{(D-d)!}} \frac{\mathrm{sgn}(\tau)}{\sqrt{(D-d)!}} \prod_{i=1}^{D-d} \xi_{\alpha_{\sigma(i)} \alpha_{\tau(i)}}$$

Therefore, $C_{\alpha_1, \ldots, \alpha_{D-d}}$ is equal to $\boldsymbol{c}^{\mathsf{T}} \boldsymbol{c}$ for the row vector $\boldsymbol{c} = \left(\dfrac{\mathrm{sgn}(\sigma)}{\sqrt{(D-d)!}} \right)_{\sigma \in \mathfrak{S}_{D-d}}$,
and therefore, is positive semidefinite. $\qquad \square$

6.3   Correction to the CI and WDwS Kernels

Let x and y be strings of an alphabet Σ with length L (*i.e.* $x = x_1 x_2 \ldots x_L$
and $y = y_1 y_2 \ldots y_L$). We will inspect when the kernels K defined by (9) are
positive semidefinite, where $\{w_1, w_2, \ldots, w_{L-k+1}\}$ and $\bar{w}$ are positive weights, s
is a constant shift, $x_{i,\ell}$ denotes the contiguous substring $x_i x_{i+1} \ldots x_{i+\ell-1}$ of x,
and $k : \Sigma^\ell \times \Sigma^\ell \longrightarrow \mathbb{R}$ is a positive semidefinite kernel[2]. The codon-improved
kernels [4] corresponds to the special case of $s = 3$, and the weighted-degree-
with-shift kernels [5] are derived from sums over plural shifts s.

$$K(x, y) = \sum_{i=1}^{L-\ell+1} w_i \left\{ k(x_{i,\ell}, y_{i,\ell}) + \bar{w} \left(k(x_{i,\ell}, y_{i+s,\ell}) + k(x_{i+s,\ell}, y_{i,\ell}) \right) \right\} \quad (9)$$

Let C_L be the matrix whose (i,j)-element is defined by: $C_{L,i,j} = w_i$, if $i = j$;
$C_{L,i,j} = \bar{w} w_{\min\{i,j\}}$, if $i = j + s$ or $j = i + s$; $C_{L,i,j} = 0$, otherwise. C_L is a
coefficient matrix of the polynomial determining the kernels of (9). Further, let
q and r such that $L - \ell + 1 = sq + r$ and $r \in \{1, \ldots, s\}$. Equation (10) holds for

---

[2] $k(x_{i,\ell}, y_{j,\ell}) = 0$ if $i > L - \ell + 1$ or $i > L - \ell + 1$.

$\alpha_j^{(i)}$, which is the determinant of the $(j+1)$-dimensional submatrix of C_L whose (a,b)-element is $C_{L,s(a-1)+i,s(b-1)+i}$. Moreover, $\alpha_j^{(i)}$'s are calculated by (11).

$$\det(C_L) = \prod_{i=1}^{r} \alpha_q^{(i)} \prod_{i=r+1}^{s} \alpha_{q-1}^{(i)} \tag{10}$$

$$\alpha_{-1}^{(i)} = 1, \quad \alpha_0^{(i)} = w_i, \quad \alpha_j^{(i)} = w_{sj+i}\alpha_{j-1}^{(i)} - \bar{w}^2 w_{s(j-1)+i}^2 \alpha_{j-2}^{(i)} \tag{11}$$

C_L is positive definite (*i.e.* all the eigenvalues are *strictly* positive), if, and only if, $\alpha_j^{(i)} > 0$ hold for all (i,j) such that $sj + i \leq L - \ell$ 1. Corollary 4 follows from this property, and Corollary 5 does from Corollary 4.

Corollary 4. *The kernels K defined by (9) are positive semidefinite, if $\alpha_j^{(i)} > 0$ hold for all (i,j) such that $sj + i \leq L - \ell + 1$.*

Corollary 5. *The kernels K defined by (9) are positive semidefinite, if $w_1, \ldots, w_{L-\ell+1}$ and $\bar{w}$ satisfy the following inequality.*

$$\bar{w} \leq \min\left\{ \frac{w_a}{w_{a-s} + w_a} \mid a = s+1, \ldots, L - \ell + 1 \right\}$$

Proof. For $i \in \{1, \ldots, s\}$, $\alpha_j^{(i)} > 0$ is proved by mathematical induction on j. □

References

1. Cristianini, N., Shawe-Taylor, J.: An Introduction to Support Vector Machines and other kernel-based learning methods. Cambridge University Press, Cambridge (2000)
2. Wolf, L., Shashua, A.: Learning over sets using kernel principal angles. Journal of Machine Learning Research 4, 913–931 (2003)
3. Zhou, S.K.: Trace and determinant kernels between matrices. SCR technical report (2004)
4. Zien, A., Rätsch, G., Mika, S., Schölkopf, B., Lengauer, T., Müller, K.R.: Engineering support vector machine kernels that recognize translation initiation sites. Bioinformatics 16(9), 799–807 (2000)
5. Rätsch, G., Sonnenburg, S., Schölkopf, B.: Rase: recognition of alternatively spliced exons in C.elegans. Bioinformatics 21, i369–i377 (2005)
6. Yamaguchi, O., Fukui, K., Maeda, K.: Face recognition using temporal image sequence. In: IEEE International Conference on Automatic Face & Gesture Recognition, IEEE Computer Society Press, Los Alamitos (1998)
7. Vishwanathan, S., Smola, A., Vidal, R.: Binet-cauchy kernels on dynamical systems and its application to the analysis of dynamic scenes. International Journal of Computer Vision 73, 95–119 (2007)
8. Leslie, C., Eskin, E., Noble, W.: The spectrum kernel: a string kernel for svm protein classification. In: 7th Pacific Symposium of Biocomputing (2002)
9. Haussler, D.: Convolution kernels on discrete structures. UCSC-CRL 99-10, Dept. of Computer Science, University of California at Santa Cruz (1999)

Learning Kernel Perceptrons on Noisy Data Using Random Projections

Guillaume Stempfel and Liva Ralaivola

Laboratoire d'Informatique Fondamentale de Marseille, UMR CNRS 6166
Université de Provence, 39, rue Joliot Curie, 13013 Marseille, France
{guillaume.stempfel,liva.ralaivola}@lif.univ-mrs.fr

Abstract. In this paper, we address the issue of learning nonlinearly separable concepts with a kernel classifier in the situation where the data at hand are altered by a uniform classification noise. Our proposed approach relies on the combination of the technique of random or deterministic projections with a classification noise tolerant perceptron learning algorithm that assumes distributions defined over finite-dimensional spaces. Provided a sufficient separation margin characterizes the problem, this strategy makes it possible to envision the learning from a noisy distribution in any separable Hilbert space, regardless of its dimension; learning with any appropriate Mercer kernel is therefore possible. We prove that the required sample complexity and running time of our algorithm is polynomial in the classical PAC learning parameters. Numerical simulations on toy datasets and on data from the UCI repository support the validity of our approach.

1 Introduction

For a couple of years, it has been known that kernel methods [1] provide a set of efficient techniques and associated models for, among others, classification supported by strong theoretical results (see, e.g. [2,3]), mainly based on *margin* criteria and the fact they constitute a generalization of the well-studied class of linear separators.

Astonishingly enough however, there is, to our knowledge, very little work on the issue of learning noisy distributions with kernel classifiers, a problem which is of great interest if one aims at using kernel methods on real-world data. Assuming a *uniform classification noise* process [4], the problem of learning from noisy distributions is a key challenge in the situation where the *feature space* associated with the chosen kernel is of *infinite dimension*, knowing that approaches to learn noisy linear classifiers in finite dimension do exist [5,6,7,8].

In this work, we propose an algorithm to learn noisy distributions defined on general Hilbert spaces (not necessarily finite dimensional) from a reasonable number of data (where reasonable is specified later on); this algorithm combines the technique of random projections with a known finite-dimensional noise-tolerant linear classifier.

The paper is organized as follows. In Section 2, the problem setting is depicted together with the assumed classification noise model. Our strategy to learn kernel classifiers from noisy distributions is described in Section 3. Section 4 reports some contributions related to the questions of learning noisy perceptrons and learning kernel classifiers using projections methods. Numerical simulations carried out on synthetic

M. Hutter, R.A. Servedio, and E. Takimoto (Eds.): ALT 2007, LNAI 4754, pp. 328–342, 2007.

datasets and on benchmark datasets from the UCI repository proving the effectiveness of our approach are presented in Section 5.

2 Problem Setting and Main Result

Remark 1 (Binary classification in Hilbert spaces, zero-bias separating hyperplanes). From now on, $\mathcal{X}$ denotes the input space, assumed to be a *Hilbert space* equipped with an inner product denoted by $\cdot$. In addition, we will restrict our study to the binary classification problem and the target space $\mathcal{Y}$ will henceforth always be $\{-1, +1\}$.

We additionally make the simplifying assumption of the existence of zero-bias separating hyperplanes (i.e. hyperplanes defined as $\mathbf{w} \cdot \mathbf{x} = 0$).

2.1 Noisy Perceptrons in Finite Dimension

The Perceptron algorithm [9] (cf. Fig. 1) is a well-studied greedy strategy to derive a linear classifier from a sample $\mathcal{S} = \{(\mathbf{x}_1, y_1) \dots (\mathbf{x}_m, y_m)\}$ of m labeled pairs $(\mathbf{x}_i, y_i)$ from $\mathcal{X} \times \mathcal{Y}$ assumed to be drawn independently from an *unknown* and *fixed* distribution D over $\mathcal{X} \times \mathcal{Y}$. If there exists a separating hyperplane $\mathbf{w}^* \cdot \mathbf{x} = 0$ according to which the label y of $\mathbf{x}$

Input: $\mathcal{S} = \{(\mathbf{x}_1, y_1) \dots (\mathbf{x}_m, y_m)\}$
Output: a linear classifier $\mathbf{w}$

 $t \leftarrow 0, \mathbf{w}_0 \leftarrow \mathbf{0}$
 while there is i s.t. $y_i \mathbf{w}_t \cdot \mathbf{x}_i \leq 0$ **do**
  $\mathbf{w}_{t+1} \leftarrow \mathbf{w}_t + y_i \mathbf{x}_i / \|\mathbf{x}_i\|$
  $t \leftarrow t + 1$
 end while
 return $\mathbf{w}$

Fig. 1. Perceptron algorithm

---

is set, i.e. y is set to $+1$ if $\mathbf{w}^* \cdot \mathbf{x} \geq 0$ and -1 otherwise[1], then the Perceptron algorithm, when given access to $\mathcal{S}$, converges towards a hyperplane $\mathbf{w}$ that correctly separates $\mathcal{S}$ and might with high probability exhibit good generalization properties [10].

We are interested in the possibility of learning linearly separable distributions on which a random *uniform classification noise*, denoted as CN [4], has been applied, that is, distributions where correct labels are flipped with some given probability η. In order to tackle this problem, [5] has proposed a simple algorithmic strategy later exploited by [6]: it consists in an iterative learning process built upon the Perceptron algorithm where update vectors are computed as sample averages of training vectors fulfilling certain properties. The expectations of those update vectors guarantee the convergence of the learning process and, thanks in part to Theorem 1 stated just below, it is guaranteed with probability $1 - \delta$ ($\delta \in (0, 1)$) that whenever the dimension n of $\mathcal{X}$ is *finite* and there exists a separating hyperplane of margin $\gamma > 0$, a polynomial number of training data is sufficient for the sample averages to be close enough to their expectations; this, in turn implies a polynomial running time complexity of the algorithm together with a $1 - \delta$ guarantees for a generalization error of ε. Here, *polynomiality* is defined with respect to n, $1/\delta$, $1/\varepsilon$, $1/\gamma$ and $1/(1 - 2\eta)$. Note that despite the availability of generalization bounds for soft-margin SVM expressed in terms of margin and the values of

---

[1] We assume a deterministic labelling of the data according to the target hyperplane $\mathbf{w}^*$, i.e. $Pr(y = 1|\mathbf{x}) = 1$ or $Pr(y = 1|\mathbf{x}) = 0$; a nondeterministic setting can be handled as well.

Algorithm 1. RP-classifier

Input: • $S = \{(\mathbf{x}_1, y_1) \ldots (\mathbf{x}_m, y_m)\}$ in $\mathcal{X} \times \{-1, +1\}$
      • n, projection dimension
Output: • a random projection $\pi = \pi(S, n) : \mathcal{X} \to \mathcal{X}', \mathcal{X}' = \mathrm{span}\langle \mathbf{x}_{i_1}, \ldots, \mathbf{x}_{i_n} \rangle$
       • projection classifier $f(\mathbf{x}) = \mathbf{w} \cdot \pi(\mathbf{x}), \mathbf{w} \in \mathcal{X}'$

    learn an orthonormal random projection $\pi : \mathcal{X} \to \mathcal{X}'$
    learn a linear classifier $\mathbf{w}$ from $S = \{(\pi(\mathbf{x}_1), y_1) \ldots (\pi(\mathbf{x}_m), y_m)\}$
    return $\pi, \mathbf{w}$

---

slack variables, which account for possible classification errors, there is no result, to our knowledge, which characterizes the solution obtained by solving the quadratic program when the data is uniformly corrupted by classification noise. It is therefore not possible to control beforehand the values of the slack variables, and, hence, the non-triviality of the bounds (i.e. bounds with values lower than 1).

Theorem 1 ([11]). *If* $\mathcal{F} = \{f_\varphi(\mathbf{x}) | \varphi \in \Phi\}$ *has a pseudo-dimension of h and a range R (i.e. $|f_\varphi(\mathbf{x})| \leq R$ for any φ and $\mathbf{x}$), and if a random sample of $M \geq m_0(h, R, \delta, \varepsilon) = \frac{8R^2}{\varepsilon^2}\left(2h \ln \frac{4R}{\varepsilon} + \ln \frac{9}{\delta}\right)$ i.i.d examples are drawn from a fixed distribution, then with probability $1 - \delta$, the sample average of every indicator function $f_\varphi(\mathbf{x}) > \alpha$ is within $\frac{\varepsilon}{R}$ of its expected value, and the sample average of every f_φ is within ε of its expected value. (The pseudo-dimension of $\mathcal{F}$ is the VC dimension of $\{f_\varphi(\mathbf{x}) > \alpha | \varphi \in \Phi \wedge \alpha \in \mathbb{R}\}$.)*

2.2 Main Result: RP Classifiers and Infinite-Dimensional Spaces

h The question that naturally arises is whether it is possible to learn linear classifiers from noisy distributions defined over *infinite dimensional spaces* with similar theoretical guarantees with respect to the polynomiality of sample and running time complexities. We answer to this question positively by exhibiting a family of learning algorithm called *random projection classifiers* capable of doing so. Classifiers of this family learn from a training sample S according to Algorithm 1: given a finite projection dimension n, they first learn a projection π from $\mathcal{X}$ to a space $\mathcal{X}'$ spanned by n (randomly chosen) vectors of S dimensional space and then, learn a finite dimensional noisy perceptron from the labeled data projected according to π. An instantiation of RP-classifiers simply consists in a choice of a random projection learning algorithm and of a (noise-tolerant) linear classifier.

Let us more formally introduce some definitions and state our main result.

Remark 2 (Labeled Examples Normalization). In order to simplify the definitions and the writing of the proofs we will use the handy transformation that consists in converting every labeled example $(\mathbf{x}, y)$ to $y\mathbf{x}/\|\mathbf{x}\|$. From now on, we will therefore consider distributions and samples defined on $\mathcal{X}$ (instead of $\mathcal{X} \times \mathcal{Y}$).

Note that the transformation does not change the difficulty of the problem and that the search for a separating hyperplane between +1 and -1 classes boils down to the search for a hyperplane $\mathbf{w}$ verifying $\mathbf{w} \cdot \mathbf{x} > 0$.

Definition 1 ((γ, ε)-separable distributions $\mathcal{D}^{\gamma, \varepsilon}$). *For $\gamma > 0, \varepsilon \in [0, 1), \mathcal{D}^{\gamma, \varepsilon}$ is the set of distributions on $\mathcal{X}$ such that for any D in $\mathcal{D}^{\gamma, \varepsilon}$, there exists a unit vector $\mathbf{w}$ in $\mathcal{X}$ such that $Pr_{\mathbf{x} \sim D}[\mathbf{w} \cdot \mathbf{x} < \gamma] \leq \varepsilon$.*

Definition 2 (CN distributions $\mathcal{U}^{\gamma,\eta}$ [4]). *For $\eta \in [0, 0.5)$, let the random transformation U^η map $\mathbf{x}$ to $-\mathbf{x}$ with probability η and leave it unchanged with probability $1 - \eta$. The set of distributions $\mathcal{U}^{\gamma,\eta}$ is defined as $\mathcal{U}^{\gamma,\eta} := U^\eta(\mathcal{D}^{\gamma,0})$.*

Uniform classification noise may appear as a very limited model but learnability results int this framework can be easily extended to more general noise model [12]. We can now state our main result.

Theorem 2 (Dimension-Independent Learnability of Noisy Perceptrons). *There are an algorithm $\mathcal{A}$ and polynomials $p(\cdot, \cdot, \cdot, \cdot)$ and $q(\cdot, \cdot, \cdot, \cdot)$ such that the following holds.*
$\forall \varepsilon \in (0, 1), \forall \delta \in (0, 1), \forall \gamma > 0, \forall \eta \in [0, 0.5), \forall D \in \mathcal{D}^{\gamma,0}$, if a random sample $\mathcal{S} = \{\mathbf{x}_1, \ldots, \mathbf{x}_m\}$ with $m \geq p(\frac{1}{\varepsilon}, \frac{1}{\delta}, \frac{1}{1-2\eta}, \frac{1}{\gamma})$ is drawn from $U^\eta(D)$, then with probability at least $1 - \delta$, $\mathcal{A}$ runs in time $q(\frac{1}{\varepsilon}, \frac{1}{\delta}, \frac{1}{1-2\eta}, \frac{1}{\gamma})$ and the classifier $f := \mathcal{A}(\mathcal{S})$ output by $\mathcal{A}$ has a generalization error $Pr_{\mathbf{x} \sim D}(f(\mathbf{x}) \leq 0)$ bounded by ε.

3 Combining Random Projections and a Noise-Tolerant Algorithm

This section gives a proof of Theorem 2 by showing that an instance of RP-classifier using a linear learning algorithm based on a specific perceptron update rule, Cnoise-update, proposed by [8] and on properties of simple random projections proved by [13] is capable of efficiently learning CN distributions (see Definition 2) independently of the dimension of the input space.

The proof works in two steps. First (section 3.1) we show that Cnoise-update (Algorithm 2) in finite dimension can tolerate a small amount of *malicious noise* and still returns relevant update vectors. Then (section 3.2) thanks to properties of random projections (see [13]) we show that they can be efficiently used to transform a CN problem into one that meets the requirements of Cnoise-update (and Theorem 4 below).

3.1 Perceptron Learning with Mixed Noise

We suppose in this subsection that $\mathcal{X}$ is of finite dimension n. We make use of the following definitions.

Definition 3 (Sample and population accuracies). *Let $\mathbf{w}$ be a unit vector, D be a distribution on $\mathcal{X}$ and $\mathcal{S}$ be a sample drawn from D. We say that $\mathbf{w}$ has sample accuracy $1 - \varepsilon$ on $\mathcal{S}$ and (population) accuracy $1 - \varepsilon'$ if:*

$$Pr_{\mathbf{x} \in \mathcal{S}}[\mathbf{w} \cdot \mathbf{x} < 0] = \varepsilon, \quad and \quad Pr_{\mathbf{x} \sim D}[\mathbf{w} \cdot \mathbf{x} < 0] = \varepsilon'.$$

Definition 4 (CN-consistency). *A unit vector $\mathbf{w}^*$ is CN-consistent on $D \in \mathcal{U}^{\gamma,\eta}$ if $Pr_{\mathbf{x} \sim D}[\mathbf{w}^* \cdot \mathbf{x} < \gamma] = \eta$. It means $\mathbf{w}^*$ makes no error on the noise free version of D.*

We recall that according to the following theorem [8], Cnoise-update, depicted in Algorithm 2, when used in a perceptron-like iterative procedure, renders the learning of CN-distributions possible in finite dimension.

Algorithm 2. Cnoise-Update [8]

Input: $\mathcal{S}$: training data, $\mathbf{w}$: current weight vector, ν a nonnegative real value
Output: an update vector $\mathbf{z}$

$$\mu \leftarrow \frac{1}{|\mathcal{S}|} \sum_{\mathbf{x} \in \mathcal{S}} \mathbf{x}, \quad \mu' \leftarrow \frac{1}{|\mathcal{S}|} \sum_{\mathbf{x} \in \mathcal{S} \wedge \mathbf{w} \cdot \mathbf{x} \leq 0} \mathbf{x}$$

if $\mathbf{w} \cdot \mu \leq \nu \|\mathbf{w}\|$ then

    $\mathbf{z} \leftarrow \mu$

else

    $a \leftarrow \dfrac{\mathbf{w} \cdot \mu - \nu \|\mathbf{w}\|}{\mathbf{w} \cdot \mu - \mathbf{w} \cdot \mu'}, \quad b \leftarrow \dfrac{-\mathbf{w} \cdot \mu' + \nu \|\mathbf{w}\|}{\mathbf{w} \cdot \mu - \mathbf{w} \cdot \mu'}, \quad \mathbf{z} \leftarrow a\mu' + b\mu$

end if

if $\mathbf{w} \cdot \mathbf{z} > 0$ then

    $\mathbf{z} \leftarrow \mathbf{z} - \mathbf{w} \dfrac{\mathbf{w} \cdot \mathbf{z}}{\mathbf{w} \cdot \mathbf{w}}$ /* projection step */

end if

return $\mathbf{z}$

Theorem 3 ([8]). *Let* $\gamma \in [0, 1], \eta \in [0, 0.5), \varepsilon \in (0, 1 - 2\eta]$. *Let* $D \in \mathcal{U}^{\gamma, \eta}$. *If* $\mathbf{w}^*$ *is CN-consistent on* D, *if a random sample* $\mathcal{S}$ *of* $m \geq m_0 \left(10(n + 1), 2, \delta, \frac{\varepsilon\gamma}{4}\right)$ *examples are drawn from* D *and if the perceptron algorithm uses update vectors from* Cnoise-Update$(\mathcal{S}, \mathbf{w}_t, \frac{\varepsilon\gamma}{4})$ *for more than* $\frac{16}{(\varepsilon\gamma)^2}$ *updates on these points, then the* $\mathbf{w}_t$ *with the highest sample accuracy has accuracy at least* $1 - \eta - \varepsilon$ *with probability* $1 - \delta^2$.

The question that is of interest to us deals with a little bit more general situation than simple CN noise. We would like to show that Cnoise-update is still applicable when, in addition to being CN, the distribution on which it is called is also corrupted by *malicious noise* [14], i.e. a noise process whose statistical properties cannot be exploited in learning (this is an 'incompressible' noise). Envisioning this situation is motivated by the projection step, which may introduce some amount of *projection noise* (cf. Theorem 5), that we treat as malicious noise.

Of course, a limit on the amount of malicious noise must be enforced if some reasonable generalization error is to be achieved. Working with distributions from $\mathcal{U}^{\gamma, \eta}$ we therefore set $\theta_{\max}(\gamma, \eta) = \frac{\gamma(1 - 2\eta)}{8}$ as the maximal amount tolerated by the algorithm. For $\theta \leq \theta_{\max}$, a minimal achievable error rate $\varepsilon_{\min}(\gamma, \eta, \theta) = \frac{64\theta}{\gamma(1 - \eta)(\frac{1}{8} - \theta)}$ will be our limit[3]. Provided that the amount of malicious noise is lower than $\theta_{\max}$, we show that learning can be achieved for any error $\varepsilon \geq \varepsilon_{\min}(\gamma, \eta, \theta)$. The proof non trivially extends that of [8] and roughly follows its lines.

Definition 5 (Mixed-Noise distributions, $\mathcal{U}^{\gamma, \eta, \theta}$). *For* $\theta \in [0, 1)$, *let the random transformation* U^θ *leave an input* $\mathbf{x}$ *unchanged with probability* $1 - \theta$ *and change it to any arbitrary* $\mathbf{x}'$ *with probability* θ *(nothing can be said about* $\mathbf{x}'$*). The set of distributions* $\mathcal{U}^{\gamma, \eta, \theta}$ *is defined as* $\mathcal{U}^{\gamma, \eta, \theta} := U^\theta \left(U^\eta(\mathcal{D}^{\gamma, 0})\right)$.

---

[2] For the remaining of the paper, ε is not the usual error parameter ε' used in PAC, but $\varepsilon'(1 - 2\eta)$.

[3] Slightly larger amount of noise and smaller error rate could be theoretically targeted. But the choices we have made suffice to our purpose.

Remark 3 (CN and MN decomposition). For $\gamma > 0, \eta \in [0, 0.5), \theta \in [0, 1)$, the image distribution $D^{\gamma, \eta, \theta} := U^\theta \left(U^\eta (D^{\gamma, 0}) \right)$ of $D^{\gamma, 0} \in \mathcal{D}^{\gamma, 0}$ is therefore a mixture of two distributions: the first one, of weight $1 - \theta$, is a CN distribution with noise η and margin γ while nothing can be said about the second, of weight θ. This latter distribution will be referred to as the malicious part (MN) of $D^{\gamma, \eta, \theta}$. In order to account for the malicious noise, we introduce the random variable $\theta : \mathcal{X} \to \{0, 1\}$ such that $\theta(\mathbf{x}) = 1$ if $\mathbf{x}$ is altered by malicious noise and $\theta(\mathbf{x}) = 0$ otherwise.

From now on, we will use $E[f(\mathbf{x})]$ for $E_{\mathbf{x} \sim D}[f(\mathbf{x})]$ and $\hat{E}[f(\mathbf{x})]$ for $E_{\mathbf{x} \in S}[f(\mathbf{x})]$.

Lemma 1. *Let* $\gamma > 0$, $\eta \in [0, 0.5)$ *and* $\delta \in (0, 1)$. *Let* $\theta \in [0, \theta_{\max}(\gamma, \eta))$ *such that* $\varepsilon_{min}(\gamma, \eta, \theta) < 1$, $\varepsilon \in (\varepsilon_{min}(\gamma, \eta, \theta), 1]$ *and* $D \in \mathcal{D}^{\gamma, \eta, \theta}$. *Let* $m' > 1$. *If a sample* S *of size* $m \geq m_1(m', \gamma, \theta, \varepsilon, \delta) = m' \frac{64^2}{2(1 - \theta - \frac{\varepsilon \gamma}{64})(\varepsilon \gamma)^2} \ln \frac{2}{\delta}$ *is drawn from* D *then, with probability* $1 - \delta$:

$$1. \left| \frac{1}{m} \sum_{x \in S} \theta(\mathbf{x}) - E[\theta(\mathbf{x})] \right| \leq \frac{\varepsilon \gamma}{64} \qquad 2. |\{\mathbf{x} \in S | \theta(\mathbf{x}) = 0\}| > m'.$$

Proof. Simple Chernoff bounds arguments prove the inequalities. (It suffices to observe that $\frac{1}{m} \sum_{x \in S} \theta(\mathbf{x}) = \hat{E}[\theta(\mathbf{x})]$ and $\sum_{x \in S} \theta(\mathbf{x}) = m - |\{\mathbf{x} \in S | \theta(\mathbf{x}) = 0\}|$.) □

Definition 6 (CN-consistency on Mixed-Noise distributions). *Let* $\gamma > 0, \eta \in [0, 0.5)$, $\theta \in [0, \theta_{\max}(\gamma, \eta))$. *Let* $D \in \mathcal{U}^{\gamma, \eta, \theta}$. *Let* $\mathbf{w}^* \in \mathcal{X}$. *If* $Pr_{\mathbf{x} \sim D}[\mathbf{w}^* \cdot \mathbf{x} \leq \gamma | \theta(\mathbf{x}) = 0] = \eta$ *then* $\mathbf{w}^*$ *is said to be CN-consistent.*

The next lemma says how much the added malicious noise modify the sample averages on the CN part of a distribution.

Lemma 2. *Let* $\gamma > 0, \eta \in [0, 0.5)$ *and* $\delta \in (0, 1]$. *Let* $\theta \in [0, \theta_{\max}(\gamma, \eta))$ *such that* $\varepsilon_{min}(\gamma, \eta, \theta) < 1 - 2\eta$, *and* $\varepsilon \in (\varepsilon_{min}(\gamma, \eta, \theta), 1 - 2\eta]$. *Let* $D \in \mathcal{U}^{\gamma, \eta, \theta}$. *Let* $M(n, \gamma, \eta, \theta, \varepsilon, \delta) = m_1 \left(m_0 \left(10(n+1), 2, \frac{3\delta}{4}, \frac{\varepsilon \gamma}{16} \right), \gamma, \theta, \varepsilon, \frac{\delta}{4} \right)$ *and* $\mathbf{w}$ *be a unit vector. If* S *is a sample of size* $m > M(n, \gamma, \eta, \theta, \varepsilon, \delta)$ *drawn from* D *then, with probability* $1 - \delta$, $\forall R \in [-1, 1]$:

$$\left| \hat{E}[(\mathbf{w} \cdot \mathbf{x}) \mathbf{1}_{\leq R}(\mathbf{w} \cdot \mathbf{x})] - E[(\mathbf{w} \cdot \mathbf{x}) \mathbf{1}_{\leq R}(\mathbf{w} \cdot \mathbf{x})] \right| \leq \frac{\varepsilon \gamma}{8}$$

where $\mathbf{1}_{\leq R}(\alpha) = 1$ *if* $\alpha \leq R$ *and* 0 *otherwise.*

Proof. By Lemma 1, we know that $|\{\mathbf{x} \in S | \theta(\mathbf{x}) = 0\}| > m_0 \left(10(n+1), 2, \frac{3\delta}{4}, \frac{\varepsilon \gamma}{16} \right)$ with probability $1 - \frac{3\delta}{4}$. So, by Theorem 1, with probability $1 - \frac{3\delta}{4} - \frac{\delta}{4}, \forall R \in [-1, 1]$

$$\left| \hat{E}[(\mathbf{w} \cdot \mathbf{x}) \mathbf{1}_{\leq R}(\mathbf{w} \cdot \mathbf{x}) | \theta(\mathbf{x}) = 0] - E[(\mathbf{w} \cdot \mathbf{x}) \mathbf{1}_{\leq R}(\mathbf{w} \cdot \mathbf{x}) | \theta(\mathbf{x}) = 0] \right| \leq \frac{\varepsilon \gamma}{16} \qquad (1)$$

In addition, we have

$$\left| \hat{E}[(\mathbf{w} \cdot \mathbf{x}) 1_{\leq R}(\mathbf{w} \cdot \mathbf{x})] - E[(\mathbf{w} \cdot \mathbf{x}) 1_{\leq R}(\mathbf{w} \cdot \mathbf{x})] \right|$$

$$= \left| \hat{E}[(\mathbf{w} \cdot \mathbf{x}) 1_{\leq R}(\mathbf{w} \cdot \mathbf{x}) | \theta(\mathbf{x}) = 0] \left(Pr_{\mathbf{x} \in S}[\theta(\mathbf{x}) = 0] - Pr_{\mathbf{x} \sim D}[\theta(\mathbf{x}) = 0] \right) \right.$$

$$+ \left(\hat{E}[(\mathbf{w} \cdot \mathbf{x}) 1_{\leq R}(\mathbf{w} \cdot \mathbf{x}) | \theta(\mathbf{x}) = 0] - E[(\mathbf{w} \cdot \mathbf{x}) 1_{\leq R}(\mathbf{w} \cdot \mathbf{x}) | \theta(\mathbf{x}) = 0] \right) Pr_{\mathbf{x} \sim D}[\theta(\mathbf{x}) = 0]$$

$$+ \hat{E}[(\mathbf{w} \cdot \mathbf{x}) 1_{\leq R}(\mathbf{w} \cdot \mathbf{x}) | \theta(\mathbf{x}) = 1] \left(Pr_{\mathbf{x} \in S}[\theta(\mathbf{x}) = 1] - Pr_{\mathbf{x} \sim D}[\theta(\mathbf{x}) = 1] \right)$$

$$+ \left. \left(\hat{E}[(\mathbf{w} \cdot \mathbf{x}) 1_{\leq R}(\mathbf{w} \cdot \mathbf{x}) | \theta(\mathbf{x}) = 1] - E[(\mathbf{w} \cdot \mathbf{x}) 1_{\leq R}(\mathbf{w} \cdot \mathbf{x}) | \theta(\mathbf{x}) = 1] \right) Pr_{\mathbf{x} \sim D}[\theta(\mathbf{x}) = 1] \right|$$

$$\leq \left| \hat{E}[(\mathbf{w} \cdot \mathbf{x}) 1_{\leq R}(\mathbf{w} \cdot \mathbf{x}) | \theta(\mathbf{x}) = 0] \right| \left| Pr_{\mathbf{x} \in S}[\theta(\mathbf{x}) = 0] - Pr_{\mathbf{x} \sim D}[\theta(\mathbf{x}) = 0] \right|$$

$$\hspace{8cm} (\leq \tfrac{\varepsilon\gamma}{64} \text{ by lemma 1})$$

$$+ \left| \hat{E}[(\mathbf{w} \cdot \mathbf{x}) 1_{\leq R}(\mathbf{w} \cdot \mathbf{x}) | \theta(\mathbf{x}) = 0] - E[(\mathbf{w} \cdot \mathbf{x}) 1_{\leq R}(\mathbf{w} \cdot \mathbf{x}) | \theta(\mathbf{x}) = 0] \right| Pr_{\mathbf{x} \sim D}[\theta(\mathbf{x}) = 0]$$

$$\hspace{8cm} (\leq \tfrac{\varepsilon\gamma}{16} \text{ by equation 1})$$

$$+ \left| \hat{E}[(\mathbf{w} \cdot \mathbf{x}) 1_{\leq R}(\mathbf{w} \cdot \mathbf{x}) | \theta(\mathbf{x}) = 1] \right| \left| Pr_{\mathbf{x} \in S}[\theta(\mathbf{x}) = 1] - Pr_{\mathbf{x} \sim D}[\theta(\mathbf{x}) = 1] \right|$$

$$\hspace{8cm} (\leq \tfrac{\varepsilon\gamma}{64} \text{ by lemma 1})$$

$$+ \left| \hat{E}[(\mathbf{w} \cdot \mathbf{x}) 1_{\leq R}(\mathbf{w} \cdot \mathbf{x}) | \theta(\mathbf{x}) = 1] - E[(\mathbf{w} \cdot \mathbf{x}) 1_{\leq R}(\mathbf{w} \cdot \mathbf{x}) | \theta(\mathbf{x}) = 1] \right| Pr_{\mathbf{x} \sim D}[\theta(\mathbf{x}) = 1]$$

$$\leq 1 \times \frac{\varepsilon\gamma}{64} + \frac{\varepsilon}{16}(1 - \theta) + 1 \times \frac{\varepsilon\gamma}{64} + 2\theta \hspace{2cm} \text{(with probability } 1 - \delta)$$

$$\leq \frac{6\varepsilon}{64} + 2\theta \leq 2\varepsilon \hspace{2cm} \text{(according to the values of } \varepsilon_{\min} \text{ and } \theta_{\max})$$

$$\square$$

The following lemma shows that a CN-consistent vector $\mathbf{w}^*$ allows for a positive expectation of $\mathbf{w}^* \cdot \mathbf{x}$ over a Mixed-Noise distribution.

Lemma 3. *Let* $\gamma > 0, \eta \in [0, 0.5), \theta \in [0, \theta_{\max}(\gamma, \eta))$. *Suppose that* $D \in \mathcal{U}^{\gamma, \eta, \theta}$. *If* $\mathbf{w}^*$ *is CN-consistent on the CN-part of* D, *then* $E[\mathbf{w}^* \cdot \mathbf{x}] \geq (1 - 2\eta)(1 - \theta)\gamma - \theta$.

Proof.

$$E[\mathbf{w}^* \cdot \mathbf{x}] = E[\mathbf{w}^* \cdot \mathbf{x} | \theta(\mathbf{x}) = 0] Pr(\theta(\mathbf{x}) = 0) + E[\mathbf{w}^* \cdot \mathbf{x} | \theta(\mathbf{x}) = 1] Pr(\theta(\mathbf{x}) = 1)$$

$$= E[\mathbf{w}^* \cdot \mathbf{x} | \theta(\mathbf{x}) = 0](1 - \theta) + E[\mathbf{w}^* \cdot \mathbf{x} | \theta(\mathbf{x}) = 1]\theta$$

$$\geq E[\mathbf{w}^* \cdot \mathbf{x} | \theta(\mathbf{x}) = 0](1 - \theta) - \theta \geq (1 - 2\eta)(1 - \theta)\gamma - \theta$$

It is easy to check that the lower bound is strictly positive. $\square$

We will make use of the following lemma due to Bylander and extend it to the case of Mixed-noise distributions.

Lemma 4 ([8])
Let $\gamma > 0, \eta \in [0, 0.5), \varepsilon \in (0, 1 - 2\eta]$. *Let* $D \in \mathcal{U}^{\gamma, \eta}$. *Let* $\mathbf{w}$ *be an arbitrary weight vector. If* $\mathbf{w}^*$ *is CN-consistent on* D, *and if* $\mathbf{w}$ *has accuracy* $1 - \eta - \varepsilon$, *then:*

$$(1 - 2\eta) E[(\mathbf{w}^* \cdot \mathbf{x}) 1_{\leq 0}(\mathbf{w} \cdot \mathbf{x})] + \eta E[\mathbf{w}^* \cdot \mathbf{x}] \geq \varepsilon\gamma \tag{2}$$

$$(1 - 2\eta) E[(\mathbf{w} \cdot \mathbf{x}) 1_{\leq 0}(\mathbf{w} \cdot \mathbf{x})] + \eta E[\mathbf{w} \cdot \mathbf{x}] \leq 0 \tag{3}$$

Lemma 5. *Let* $\gamma > 0, \eta \in [0, 0.5)$ *and* $\delta \in (0, 1]$. *Let* $\theta \in [0, \theta_{\max}(\gamma, \eta))$ *such that* $\varepsilon_{\min}(\gamma, \eta, \theta) < \frac{4(1 - 2\eta)}{3}$, *and* $\varepsilon \in (\varepsilon_{\min}(\gamma, \eta, \theta), \frac{4(1 - 2\eta)}{3}]$. *Let* $D \in \mathcal{U}^{\gamma, \eta, \theta}$. *Let* $\mathbf{w}$ *be* an arbitrary weight vector and $D \in \mathcal{U}^{\gamma, \eta, \theta}$. If $\mathbf{w}^*$ is CN-consistent on the CN part of D, and if $\mathbf{w}$ has accuracy $1 - \eta - \frac{3\varepsilon}{4}$ on the CN part of D, then the following holds:

$$(1 - 2\eta)\, E\left[(\mathbf{w}^* \cdot \mathbf{x})\mathbf{1}_{\leq 0}(\mathbf{w} \cdot \mathbf{x})\right] + \eta E\left[\mathbf{w}^* \cdot \mathbf{x}\right] \geq \frac{5\varepsilon\gamma}{8} \tag{4}$$

$$(1 - 2\eta)\, E\left[(\mathbf{w} \cdot \mathbf{x})\mathbf{1}_{\leq 0}(\mathbf{w} \cdot \mathbf{x})\right] + \eta E\left[\mathbf{w} \cdot \mathbf{x}\right] \leq \eta\theta \tag{5}$$

Proof. For the first inequality, we have:

$$
\begin{aligned}
&(1 - 2\eta)\, E\left[(\mathbf{w}^* \cdot \mathbf{x})\mathbf{1}_{\leq 0}(\mathbf{w} \cdot \mathbf{x})\right] + \eta E\left[\mathbf{w}^* \cdot \mathbf{x}\right] \\
&= (1 - 2\eta)\, E\left[(\mathbf{w}^* \cdot \mathbf{x})\mathbf{1}_{\leq 0}(\mathbf{w} \cdot \mathbf{x})|\theta(\mathbf{x}) = 1\right] Pr\left[\theta(\mathbf{x}) = 1\right] \\
&\quad + \eta E\left[\mathbf{w}^* \cdot \mathbf{x}|\theta(\mathbf{x}) = 1\right] Pr\left[\theta(\mathbf{x}) = 1\right] \\
&\quad + (1 - 2\eta)\, E\left[(\mathbf{w}^* \cdot \mathbf{x})\mathbf{1}_{\leq 0}(\mathbf{w} \cdot \mathbf{x})|\theta(\mathbf{x}) = 0\right] Pr\left[\theta(\mathbf{x}) = 0\right] \\
&\quad + \eta E\left[\mathbf{w}^* \cdot \mathbf{x}|\theta(\mathbf{x}) = 0\right] Pr\left[\theta(\mathbf{x}) = 0\right] \\
&\geq (1 - \theta)\frac{3}{4}\varepsilon\gamma \qquad\qquad\qquad\qquad\qquad\qquad\text{(by lemma 4 eq. 2)} \\
&\quad + (1 - 2\eta)\, E\left[(\mathbf{w}^* \cdot \mathbf{x})\mathbf{1}_{\leq 0}(\mathbf{w} \cdot \mathbf{x})|\theta(\mathbf{x}) = 1\right] Pr\left[\theta(\mathbf{x}) = 1\right] \\
&\quad + \eta E\left[\mathbf{w}^* \cdot \mathbf{x}|\,\theta(\mathbf{x}) = 1\right] Pr\left[\theta(\mathbf{x}) = 1\right] \\
&\geq (1 - \theta)\frac{3}{4}\varepsilon\gamma - (1 - 2\eta)\,\theta - \eta\theta \\
&\geq (1 - \theta)\frac{3}{4}\varepsilon\gamma - (1 - \eta)\,\theta \geq \frac{5\varepsilon\gamma}{8} \qquad\qquad\text{(by definition of ε)}
\end{aligned}
$$

For the second inequality, we have:

$$
\begin{aligned}
&(1 - 2\eta)\, E\left[(\mathbf{w} \cdot \mathbf{x})\mathbf{1}_{\leq 0}(\mathbf{w} \cdot \mathbf{x})\right] + \eta E\left[\mathbf{w} \cdot \mathbf{x}\right] \\
&= (1 - 2\eta)\, E\left[(\mathbf{w} \cdot \mathbf{x})\mathbf{1}_{\leq 0}(\mathbf{w} \cdot \mathbf{x})|\theta(\mathbf{x}) = 1\right] Pr\left[\theta(\mathbf{x}) = 1\right] \\
&\quad + \eta E\left[\mathbf{w} \cdot \mathbf{x}|\theta(\mathbf{x}) = 1\right] Pr\left[\theta(\mathbf{x}) = 1\right] \\
&\quad + (1 - 2\eta)\, E\left[(\mathbf{w} \cdot \mathbf{x})\mathbf{1}_{\leq 0}(\mathbf{w} \cdot \mathbf{x})|\theta(\mathbf{x}) = 0\right] Pr\left[\theta(\mathbf{x}) = 0\right] \\
&\quad + \eta E\left[\mathbf{w} \cdot \mathbf{x}|\theta(\mathbf{x}) = 0\right] Pr\left[\theta(\mathbf{x}) = 0\right] \\
&\leq 0 \qquad\qquad\qquad\qquad\qquad\qquad\qquad\qquad\text{(by lemma 4 eq.3)} \\
&\quad + (1 - 2\eta)\, E\left[(\mathbf{w} \cdot \mathbf{x})\mathbf{1}_{\leq 0}(\mathbf{w} \cdot \mathbf{x})|\theta(\mathbf{x}) = 1\right] Pr\left[\theta(\mathbf{x}) = 1\right] \\
&\quad + \eta E\left[\mathbf{w} \cdot \mathbf{x}|\,\theta(\mathbf{x}) = 1\right] Pr\left[\theta(\mathbf{x}) = 1\right] \leq 0 + \eta\theta \qquad\qquad\square
\end{aligned}
$$

We now state our core lemma. It says that, with high probability, Algorithm 2 outputs a vector that can be used as an update vector in the Perceptron algorithm (cf. Fig. 1), that is a vector erroneously classified by the current classifier but correctly classified by the target hyperplane (i.e. the vector is noise free). Calling Algorithm 2 iteratively makes it possible to learn a separating hyperplane from a mixed-noise distribution.

Lemma 6. *Let $\gamma > 0, \eta \in [0, 0.5)$ and $\delta \in (0, 1)$. Let $\theta \in [0, \theta_{\max}(\gamma, \eta))$ such that $\varepsilon_{\min}(\gamma, \eta, \theta) < \frac{4}{3}(1 - \eta)$. Let $D \in \mathcal{U}^{\gamma, \eta, \theta}$ and $\mathbf{w}^*$ be the target hyperplane (CN-consistent on the CN-part of D). $\forall \varepsilon \in \left[\varepsilon_{\min}(\gamma, \eta, \theta), \frac{4}{3}(1 - \eta)\right)$, for all input samples S of size $M(n, \gamma, \eta, \theta, \delta, \varepsilon)$, with probability at least $1 - \delta$, $\forall \mathbf{w} \in \mathcal{X}$ if $\mathbf{w}$ has accuracy at most $1 - \eta - \frac{3\varepsilon}{4}$ on the CN-part of D then **Cnoise-update** (Algorithm 2), when given inputs $S, \mathbf{w}, \frac{\varepsilon\gamma}{4}$, outputs a vector $\mathbf{z}$ such that $\mathbf{w} \cdot \mathbf{z} \leq 0$ and $\mathbf{w}^* \cdot \mathbf{z} \geq \frac{\varepsilon\gamma}{4}$.*

Proof. The projection step guarantees that $\mathbf{w} \cdot \mathbf{z} \leq 0$. We focus on the second inequality.

Case 1. Suppose that $\mathbf{w} \cdot \boldsymbol{\mu} < \|\mathbf{w}\| \frac{\varepsilon\gamma}{4}$: $\mathbf{z}$ is set to $\boldsymbol{\mu}$ by the algorithm, and, if needed, is projected on the $\mathbf{w}$ hyperplane.

Every linear threshold function has accuracy at least η on the CN-part of D, so an overall accuracy at least $(1 - \theta)\eta$. $\mathbf{w}$ has accuracy on the CN-part of D of, at most, $1 - \eta - \frac{3\varepsilon}{4}$ and so an overall accuracy at most of $1 - (1 - \theta)\left(\eta + \frac{3\varepsilon}{4}\right) + \theta$.

It is easy to check that

$$1 - (1 - \theta)\left(\frac{3\varepsilon}{4} + \eta\right) + \theta \geq (1 - \theta)\eta \Leftrightarrow (1 - 2\eta)(1 - \theta)\gamma - \theta \geq (1 - \theta)\frac{3\varepsilon}{4}\gamma - (2\gamma + 1)\theta,$$

and thus, from Lemma 3, $E\left[\mathbf{w}^* \cdot \mathbf{x}\right] \geq (1 - \theta)\frac{3\varepsilon}{4}\gamma - (2\gamma + 1)\theta$. Because $\theta < \theta_{\max}(\gamma, \eta)$ and $\varepsilon > \varepsilon_{\min}(\gamma, \eta, \theta)$, we have $E\left[\mathbf{w}^* \cdot \mathbf{x}\right] \geq \frac{5\varepsilon\gamma}{8}$. Because of Lemma 2 and because $|\mathcal{S}| \geq M(n, \gamma, \eta, \theta, \delta, \varepsilon)$, we know that $\mathbf{w}^* \cdot \mathbf{z}$ is, with probability $1 - \delta$, within $\frac{\varepsilon\gamma}{8}$ of its expected value on the entire sample; hence we can conclude that $\mathbf{w}^* \cdot \boldsymbol{\mu} \geq \frac{\varepsilon\gamma}{2}$.

If $\mathbf{w} \cdot \boldsymbol{\mu} < 0$, then the lemma follows directly.

If $0 < \mathbf{w} \cdot \boldsymbol{\mu} < \|\mathbf{w}\| \frac{\varepsilon\gamma}{4}$, then $\mathbf{z}$ is set to $\boldsymbol{\mu}$ and, if needed, projected to $\mathbf{w}$. Let $\mathbf{z}_\| = \boldsymbol{\mu} - \mathbf{z}$ ($\mathbf{z}_\|$ is parallel to $\mathbf{w}$). It follows that

$$\mathbf{w}^* \cdot \boldsymbol{\mu} \geq \frac{\varepsilon\gamma}{2} \Leftrightarrow \mathbf{w}^* \cdot \mathbf{z} + \mathbf{w}^* \cdot \mathbf{z}_\| \geq \frac{\varepsilon\gamma}{2} \Rightarrow \mathbf{w}^* \cdot \mathbf{z} \geq \frac{\varepsilon\gamma}{2} - \|\mathbf{z}_\|\| \Rightarrow \mathbf{w}^* \cdot \mathbf{z} \geq \frac{\varepsilon\gamma}{2} - \|\boldsymbol{\mu}\|$$

$$\Rightarrow \mathbf{w}^* \cdot \mathbf{z} \geq \frac{\varepsilon\gamma}{4}.$$

And the lemma again follows.

Case 2. Suppose instead that $\mathbf{w} \cdot \boldsymbol{\mu} \geq \|\mathbf{w}\| \frac{\varepsilon\gamma}{4}$. Let $a \geq 0$ and $b \geq 0$ be chosen so that $a\frac{\mathbf{w}}{\|\mathbf{w}\|} \cdot \boldsymbol{\mu}' + b\frac{\mathbf{w}}{\|\mathbf{w}\|} \cdot \boldsymbol{\mu} = \frac{\varepsilon\gamma}{4}$ and $a + b = 1$. $\mathbf{w} \cdot \boldsymbol{\mu}'$ is negative and $\frac{\mathbf{w}}{\|\mathbf{w}\|} \cdot \boldsymbol{\mu} \geq \frac{\varepsilon\gamma}{4}$ in this case, so such an a and b can always be chosen. Note that in this case, Cnoise-update sets $\mathbf{z}$ to $a\boldsymbol{\mu}' + b\boldsymbol{\mu}$ and then projects $\mathbf{z}$ to the $\mathbf{w}$ hyperplane. Because $\mathbf{w} \cdot \mathbf{z} = \|\mathbf{w}\| \frac{\varepsilon\gamma}{4}$ before $\mathbf{z}$ is projected to the $\mathbf{w}$ hyperplane, then the projection will decrease $\mathbf{w}^* \cdot \mathbf{z}$ by at most $\frac{\varepsilon\gamma}{4}$ (recall that $\mathbf{w}^*$ is a unit vector).

Note that $a\frac{\mathbf{w}}{\|\mathbf{w}\|} \cdot \boldsymbol{\mu}' + b\frac{\mathbf{w}}{\|\mathbf{w}\|} \cdot \boldsymbol{\mu} = a\hat{E}\left[\left(\frac{\mathbf{w}}{\|\mathbf{w}\|} \cdot \mathbf{x}\right)\mathbf{1}_{\leq 0}(\mathbf{w} \cdot \mathbf{x})\right] + b\hat{E}\left[\frac{\mathbf{w}}{\|\mathbf{w}\|} \cdot \mathbf{x}\right]$. Because, by lemma 2, sample averages are, with probability $1 - \delta$, within $\frac{\varepsilon\gamma}{8}$ of their expected values, it follows that

$$aE\left[\left(\frac{\mathbf{w}}{\|\mathbf{w}\|} \cdot \mathbf{x}\right)\mathbf{1}_{\leq 0}(\mathbf{w} \cdot \mathbf{x})\right] + bE\left[\frac{\mathbf{w}}{\|\mathbf{w}\|} \cdot \mathbf{x}\right] \geq \frac{\varepsilon\gamma}{8}.$$

Lemma 5 implies that $a' = \frac{\eta}{1-\eta}$ and $b' = \frac{1-2\eta}{1-\eta}$ results in $a'E\left[\left(\frac{\mathbf{w}}{\|\mathbf{w}\|} \cdot \mathbf{x}\right)\mathbf{1}_{\leq 0}(\mathbf{w} \cdot \mathbf{x})\right] + b'E[\frac{\mathbf{w}}{\|\mathbf{w}\|} \cdot \mathbf{x}] \leq \frac{\eta\theta}{1-\eta}$ and so less than $\frac{\varepsilon\gamma}{8}$. So, it must be the case when $a \leq \frac{\eta}{1-\eta}$ because a larger a would result in an expected value less than $\frac{\varepsilon\gamma}{8}$ and a sample average less than $\frac{\varepsilon\gamma}{4}$.

Lemma 5 also implies that choosing $a' = \frac{\eta}{1-\eta}$ and $b' = \frac{1-2\eta}{1-\eta}$ results in $a'E[(\mathbf{w}^* \cdot \mathbf{x})\mathbf{1}_{\leq 0}(\mathbf{w} \cdot \mathbf{x})] + b'E[\mathbf{w}^* \cdot \mathbf{x}] \geq \frac{5\varepsilon\gamma}{8}$

Because $a' \geq a$ and $b' \leq b$, and because Lemma 3 implies $E\left[\mathbf{w}^* \cdot \mathbf{x}\right] \geq \frac{5\varepsilon\gamma}{8}$, it follows that $aE[(\mathbf{w}^* \cdot \mathbf{x})\mathbf{1}_{\leq 0}(\mathbf{w} \cdot \mathbf{x})] + bE[\mathbf{w}^* \cdot \mathbf{x}] \geq \frac{5\varepsilon\gamma}{8}$ and $a\mathbf{w}^* \cdot \boldsymbol{\mu}' + b\mathbf{w}^* \cdot \boldsymbol{\mu} \geq \frac{\varepsilon\gamma}{2}$.

Thus, when $\mathbf{z}$ is projected onto hyperplane $\mathbf{w}$, $\mathbf{w}^* \cdot \mathbf{z} \geq \frac{\varepsilon\gamma}{4}$ and $\mathbf{w} \cdot \mathbf{z} = 0$. Consequently a total of m examples, implies , with probability $1 - \delta$, that $\mathbf{w}^* \cdot \mathbf{z} \geq \frac{\varepsilon\gamma}{4}$ and $\mathbf{w} \cdot \mathbf{z} \leq 0$ for the $\mathbf{z}$ computes by Cnoise-update. This proves the Lemma.   □

We finally have the Theorem 4 for Mixed-Noise learnability using Cnoise-update.

Theorem 4. *Let $\gamma > 0, \eta \in [0, 0.5)$ and $\delta \in (0,1)$. Let $\theta \in [0, \theta_{\max}(\gamma, \eta))$ such that $\varepsilon_{\min}(\gamma, \eta, \theta) < 1 - 2\eta$. Let $D \in \mathcal{U}^{\gamma, \eta, \theta}$ and $\mathbf{w}^*$ be the target hyperplane (CN-consistent on the CN-part of D). $\forall \varepsilon \in (\varepsilon_{\min}(\gamma, \eta, \theta), 1 - 2\eta], \forall \mathbf{w} \in \mathcal{X}$, when given inputs $\mathcal{S}$ of size at least $M(n, \gamma, \eta, \theta, \delta, \varepsilon)$, if the Perceptron algorithm uses update vectors from CNoise update for more than $\frac{16}{\varepsilon^2 \gamma^2}$ updates, then the $\mathbf{w}_i$ with the highest sample accuracy on the CN-part has accuracy on the CN-part of D at least $1 - \eta - \varepsilon$ with probability $1 - \delta$.*

Proof. By lemma 6, with probability $1 - \delta$, whenever $\mathbf{w}_i$ has accuracy at most $1 - \eta - \frac{3\varepsilon}{4}$ on the CN-part of S then Cnoise-update$(X, \mathbf{w}_i, \frac{\varepsilon\gamma}{16})$ will return an update vector $\mathbf{z}_i$ such that $\mathbf{w}^* \cdot \mathbf{z}_i \geq \frac{\varepsilon\gamma}{4}$ and $\mathbf{w}_i \cdot \mathbf{z}_i \leq 0$. The length of a sequence $(\mathbf{z}_1, \ldots, \mathbf{z}_l)$ where each $\mathbf{z}_i$ has $\frac{\varepsilon\gamma}{4}$ separation, is at most $\frac{16}{(\varepsilon\gamma)^2}$ [15,16]. Thus, if more than $\frac{16}{(\varepsilon\gamma)^2}$ update vectors are obtained, then at least one update vector must have less than $\frac{\varepsilon\gamma}{4}$ separation, which implies at least one $\mathbf{w}$ has more than $1 - \eta - \frac{3\varepsilon\gamma}{4}$ accuracy on CN-part.

The sample accuracy of $\mathbf{w}_i$ corresponds to the sample average of an indicator function. By Theorem 1, the indicator functions are covered with probability $1 - \delta$. So, assuming that the situation is in the $1 - \delta$ region, the sample accuracy of each $\mathbf{w}_i$ on the CN-part of the distribution will be within $\frac{\varepsilon\gamma}{16}$ of its expected value. Since at least one $\mathbf{w}_i$ will have $1 - \eta - \frac{3\varepsilon}{4}$ accuracy on the CN-part, this implies that its sample accuracy on the CN-part is at least $1 - \eta - \frac{13\varepsilon}{16}$. The accuracy on the distribution is more than $1 - (1 - \theta)\left(\eta - \frac{13\varepsilon}{16}\right) - \theta < 1 - (1 - \theta)\left(\eta - \frac{13\varepsilon}{16}\right) - \frac{\varepsilon}{32}$. Any other w_i with a better sample accuracy will have accuracy of at least $1 - (1 - \theta)\left(\eta - \frac{13\varepsilon}{16}\right) - \frac{5\varepsilon}{32}$ and so an accuracy on the CN-part of at least $1 - \eta - \varepsilon$.   □

Remark 4. An interpretation of the latter result is that distributions from $\mathcal{D}^{\gamma, \varepsilon}$, for $\varepsilon > 0$ can also be learned if corrupted by classification noise. The extent to which the learning can take place of course depends on the value of ε (which would play the role of θ in the derivation made above).

In the next section, we show how random projections can help us reduce a problem of learning from a possibly infinite dimensional CN distribution to a problem of finite Mixed-Noise distribution where the parameters of the Mixed-Noise distribution can be controlled. This will directly give a proof to Theorem 2.

3.2   Random Projections and Separable Distributions

Here, we do not make the assumption that $\mathcal{X}$ is finite-dimensional.

Theorem 5 ([13]). *Let $D \in \mathcal{D}^{\gamma, 0}$. For a random sample $\mathcal{S} = \{\mathbf{x}_1, \ldots, \mathbf{x}_n\}$ from D, let $\pi(\mathcal{S}) : \mathcal{X} \to span\langle \mathcal{S} \rangle$ be the orthogonal projection on the space spanned by $\mathbf{x}_1, \ldots, \mathbf{x}_n$.*

If *a sample S of size $n \geq \frac{8}{\theta}[\frac{1}{\gamma^2} + \ln\frac{1}{\delta}]$ is drawn according to D then with probability at least $1 - \delta$, the mapping $\pi = \pi(S)$ is such that $\exists \mathbf{w} \ Pr_{\mathbf{x} \sim D}\left[\mathbf{w} \cdot \pi(\mathbf{x}) > \gamma/2\right] < \theta$ on* span$\langle S \rangle \subseteq \mathcal{X}$.

This theorem says a random projection can transform a linearly separable distribution into an *almost linearly separable* one defined in a finite dimensional space. We can therefore consider that such a transformation incurs a *projection noise*; this noise should possess some exploitable regularities for learning, but we leave the characterization of these regularities for a future work and apprehend in the sequel this projection noise as malicious. In RP-classifier, the vectors used to define π are selected randomly within the training set.

Corollary 1 (of Theorem 2). *Let $\gamma > 0, \eta \in [0, 0.5)$ and $D \in \mathcal{U}^{\gamma, \eta}$. $\forall \varepsilon \in (0, 1 - 2\eta], \forall \delta \in (0, 1]$, if a sample S of $m > M(\frac{K}{\varepsilon\gamma(1-2\eta)}\left[\frac{1}{\gamma^2} + \ln\frac{2}{\delta}\right], \frac{\gamma}{2}, \eta, \frac{\delta}{2}, \frac{\varepsilon}{2})$ examples drawn from D is input to RP-classifier, then with probability $1 - \delta$ RP-classifier outputs a classifier with accuracy at least $1 - \eta - \varepsilon$. ($K > 0$ is a universal constant.)*

Proof. Fix $\gamma, \eta, D \in \mathcal{U}^{\gamma, \eta}$ and ε. Fix $\theta = \frac{\gamma\varepsilon(1-2\eta)}{2080}$.

First, it is straightforward to check that $\theta \leq \theta_{\max}(\gamma, \eta)$, $\varepsilon_{min} \leq \min(\frac{\varepsilon}{2}, 1 - 2\eta)$ and, since $\theta \leq \varepsilon_{min}(\gamma, \eta, \theta)$, $\theta \leq \frac{\varepsilon}{2}$. (The assumptions of Theorem 4 hold true.)

By Theorem 5, choosing $n = \frac{8}{\theta}[\frac{1}{\gamma^2} + \ln\frac{2}{\delta}]$ guarantees with probability $1 - \frac{\delta}{2}$, that the projection D' of D onto a random subspace of dimension n is a distribution having a CN part of weight $1 - \theta$ and another part of weight θ corrupted by projection noise. D' can therefore be considered as an element of $\mathcal{U}^{\frac{\gamma}{2}, \eta, \theta^4}$.

By Theorem 4, using m examples (with m set as in the Theorem) allows with probability $1 - \frac{\delta}{2}$ the learning algorithm that iteratively calls Cnoise-update to return in polynomial time a classifier with accuracy at least $\frac{\varepsilon}{2}$ on the CN-part of the distribution.

Therefore, the accuracy of the classifier on the examples drawn from D is, with probability $1 - \frac{\delta}{2} - \frac{\delta}{2} = 1 - \delta$, at least $1 - (1 - \theta)\frac{\varepsilon}{2} - \theta \geq 1 - \frac{\varepsilon}{2} - \frac{\delta}{2} = 1 - \delta$. Theorem 2 now follows.   □

Remark 5. We could also learn with an initial malicious noise θ_{init} less than $\theta_{\max}$. In this case, the maximum amount of noise added by random projections must obviously be less than $\theta_{\max} - \theta_{\text{init}}$.

Remark 6. Random projections based on the Johnson-Lindenstrauss lemma could be directly combined with a CN-noise tolerant perceptron to achieve the same kind of learnability results. It however requires numerous data resamplings and the resulting sample and time complexities are very high.

4   Related Work

Learning from a noisy sample of data implies that the linear problem at hand might not necessarily be consistent, that is, some linear constraints might contradict others. In that

---

[4] The choices of θ and n give $K = 2080 \times 8$.

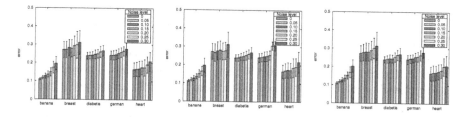

Fig. 2. Error rates on UCI datasets with random projections, KPCA and KGS projection with different amount of classification noise; 1-standard deviation error bars are shown

case, as stated before, the problem at hand boils down to that of finding an approximate solution to a linear program such that a minimal number of constraints are violated, which is know as an NP-hard problem (see, e.g., [17]).

In order to cope with this problem, and leverage the classical perceptron learning rule to render it tolerant to noise classification, one line of approaches has mainly been exploited. It relies on exploiting the statistical regularities in the studied distribution by computing various sample averages as it is presented here; this makes it possible to 'erase' the classification noise. As for Bylander's algorithms [5,8], whose analysis we have just extended, the other notable contributions are those of [6] and [7]. However, they tackle a different aspect of the problem of learning noisy distributions and are more focused on showing that, in finite dimensional spaces, the running time of their algorithms can be lowered to something that depends on $\log 1/\gamma$ instead of $1/\gamma$.

Regarding the use of kernel projections to tackle classification problems, the *Kernel Projection Machine* of [18] has to be mentioned. It is based on the use of Kernel PCA as a feature extraction step. The main points of this interesting work are a proof on the regularizing properties of KPCA and the fact that it gives a practical model selection procedure. However, the question of learning noisy distributions is not addressed.

Freund and Schapire [19] provide data-dependent bounds for the voted kernel perceptron that support some robustness against outliers. However, as for SVM, it is not clear whether this algorithm is tolerant to 'systematic' uniform classification noise.

Cesa-Bianchi and al. in [20] propose bounds for online perceptron on non-separable data. However, the authors specify that their algorithms tolerate only a low rate of non-linearly separable examples and thus are not valid for uniform classification noise.

Finally, the empirical study of [21] provides some insights on how random projections might be useful for classification. No sample and running time complexity results are given and the question of learning with noise is not addressed.

5   Numerical Simulations

UCI Datasets. We carried out numerical simulations on benchmark datasets from the UCI repository preprocessed and made available by Gunnar Rätsch[5]. For each problem (Banana, Breast Cancer, Diabetes, German, Heart), we have 100 training and 100 test

---

[5] http://ida.first.fraunhofer.de/projects/bench/benchmarks.htm

samples. All these problems contain a few hundred training examples, which is far from what the theoretical results require to get interesting accuracy and confidence.

We have tested three projection procedures: random, Kernel PCA (KPCA), Kernel Gram-Schmidt (KGS) [22]. This latter projection is sometimes referred to as a 'sparse version of Kernel PCA' (note that KPCA and KGS are deterministic projections and that RP-classifier is not a random-projection learning algorithm anymore). In order to cope with the non separability of the problems, we have used Gaussian kernels, and thus infinite-dimensional spaces, whose widths have been set to the best value for SVM classification as reported on Gunnar Rätsch's website.

In our protocol, we have corrupted the data with classification noises of rates 0.0, 0,05, 0.10, 0.15, 0.20, 0.25, 0.30. Instead of carrying out a cumbersome cross-validation procedure, we provide the algorithm RP-classifier with the actual value of η.

To determine the right projection size we resort to the same cross-validation procedure as in [23], trying subspace sizes of 2 to 200. The results obtained are summarized on Figure 2. We observe that classifiers produced on a dataset with no extra noise have an accuracy a little lower than that of the classifiers tested by Gunnar Rätsch, with a very reasonable variance. We additionally note that, when the classification noise amount artificially grows, the achieved accuracy decreases very weakly and the variance grows rather slowly. It is particularly striking since again, the sample complexities used are far from meeting the theoretical requirements; moreover, it is interesting to see that the results are good even if no separation margin exists. We can also note that when the actual values of the accuracies (not reported here for sake of space) are compared, KGS and KPCA roughly achieve the same accuracies and both are a little (not significantly though) better than random projection. Eventually, the main conclusion from the numerical simulations is that RP-classifier has a very satisfactory behavior on real data.

Toy Problems. We have carried out additional simulations on five 2-dimensional toy problems. Due to space limitations however, we only discuss and show the learning results for three of them[6] (cf. Figure 3). Here, we have used the KGS projection since due to the uniform distribution of points on $[-10; 10] \times [-10; 10]$, random projections provide exactly the same results. For each problem, we have produced 50 train sets and 50 test sets of 2000 examples each. Note that we do not impose any separation margin.

We have altered the data with 5 different amounts of noise (from 0.0 to 0.40), 12 Gaussian kernel width (from 10.0 to 0.25) and 12 projection dimensions (from 5 to 200) have been tested and for each problem and for each noise rate, we have selected the couple which minimizes the error rate of the produced classifier (proceeding as above). Figure 3 depicts the learning results obtained with a noise rate of 0.20.

The essential point showed by these simulations is that, again, RP-classifier is very effective in learning from noisy nonlinear distributions. Numerically (the numerical results are not reported here due to space limitations), we have observed that our algorithm can tolerate noise levels as high as 0.4 and still provide small error rates (around 10%). Finally, our simulations show that the algorithm is tolerant to classification noise and thus illustrate our theoretical results, while extending already existing experiments to this particular framework of learning.

---

[6] Full results are available at http://hal.archives-ouvertes.fr/hal-00137941

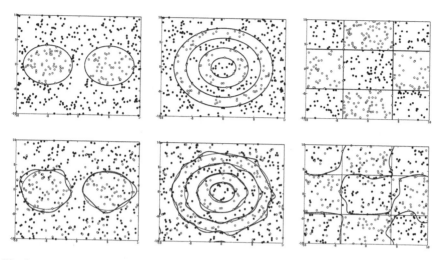

Fig. 3. Toy problems: first row show the clean concepts with black disks being of class +1 and white ones of class -1. Second row shows the concept learned by RP-classifier with a uniform classification noise rate of 0.20 and KGS projection.

6   Conclusion and Outlook

In this paper, we have given theoretical results on the learnability of kernel perceptrons when faced to classification noise. The keypoint is that this result is independent of the dimension of the kernel feature space. In fact, it is the use of finite-dimensional projections having good generalization that allows us to transform a possibly infinite dimensional problem into a finite dimension one that, in turn, we tackle with Bylander's noise tolerant perceptron algorithm. This algorithm is shown to be robust to some additional 'projection noise' provided the sample complexity are adjusted in a suitable way. A better characterization of the projection noise, more intelligent than 'malicious', could, in a future work, allow us to use projection dimensions appreciably smaller. Several simulation results support the soundness of our approach. Note that the random projection procedure using Johnson-Lindenstrauss lemma, described in [13], could be associated with RP-learn and would lead to lower sample and time complexities for the perceptron learning step.

Several questions are raised by this work. Among them, the question about the generalization properties of the Kernel Gram-Schmidt projector: we think tight generalization bounds can be exhibited in the framework of PAC Bayesian bounds, by exploiting, in particular, the sparseness of this projection. Resorting again to the PAC Bayesian framework it might be interesting to work on generalization bound on noisy projection classifiers, which would potentially provide a way to automatically estimate a reasonable projection dimension *and* noise level. Finally, we have been recently working on the harder problem of learning optimal separating hyperplane from noisy distributions.

References

1. Schölkopf, B., Smola, A.J.: Learning with Kernels, Support Vector Machines, Regularization, Optimization and Beyond. MIT University Press, Cambridge (2002)
2. Vapnik, V.: The nature of statistical learning theory. Springer, New York (1995)
3. Cristianini, N., Shawe-Taylor, J.: An Introduction to Support Vector Machines and other Kernel-Based Learning Methods. Cambridge University Press, Cambridge (2000)
4. Angluin, D., Laird, P.: Learning from Noisy Examples. Machine Learning 2 (1988)
5. Bylander, T.: Learning Linear Threshold Functions in the Presence of Classification Noise. In: Proc. of 7th Ann. Work. on Computat. Learning Theory., pp. 340–347 (1994)
6. Blum, A., Frieze, A.M., Kannan, R., Vempala, S.: A Polynomial-Time Algorithm for Learning Noisy Linear Threshold Functions. In: Proc. of 37th IEEE Symposium on Foundations of Computer Science, pp. 330–338. IEEE Computer Society Press, Los Alamitos (1996)
7. Cohen, E.: Learning Noisy Perceptrons by a Perceptron in Polynomial Time. In: Proc. of 38th IEEE Symposium on Foundations of Computer Science, pp. 514–523. IEEE Computer Society Press, Los Alamitos (1997)
8. Bylander, T.: Learning Noisy Linear Threshold Functions (1998) (submitted to journal)
9. Rosenblatt, F.: The Perceptron: A probabilistic model for information storage and organization in the brain. Psychological Review 65, 386–407 (1958)
10. Graepel, T., Herbrich, R., Williamson, R.C.: From Margin to Sparsity. In: Adv. in Neural Information Processing Systems, vol. 13, pp. 210–216 (2001)
11. Vapnik, V.: Statistical Learning Theory. John Wiley and Sons, inc., West Sussex, England (1998)
12. Ralaivola, L., Denis, F., Magnan, C.N.: CN=CPCN. In: Proc. of the 23rd Int. Conf. on Machine Learning (2006)
13. Balcan, M.F., Blum, A., Vempala, S.: Kernels as Features: on Kernels, Margins, and Low-dimensional Mappings. In: Ben-David, S., Case, J., Maruoka, A. (eds.) ALT 2004. LNCS (LNAI), vol. 3244, Springer, Heidelberg (2004)
14. Kearns, M., Li, M.: Learning in the presence of malicious errors. SIAM Journal on Computing 22(4), 807–837 (1993)
15. Block, H.D.: The perceptron: A model for brain functioning. Reviews of Modern Physics 34, 123–135 (1962)
16. Novikoff, A.B.J.: On convergence proofs on perceptrons. In: Proc. of the Symp. on the Mathematical Theory of Automata, pp. 615–622 (1962)
17. Amaldi, E., Kann, V.: On the approximability of some NP-hard minimization problems for linear systems. Electronic Colloquium on Computational Complexity (ECCC) 3(015) (1996)
18. Zwald, L., Vert, R., Blanchard, G., Massart, P.: Kernel projection machine: a new tool for pattern recognition. In: Adv. in Neural Information Processing Systems, vol. 17 (2004)
19. Freund, Y., Schapire, R.E.: Large Margin Classification Using the Perceptron Algorithm. Machine Learning 37(3), 277–296 (1999)
20. Cesa-Bianchi, N., Conconi, A., Gentile, C.: On the generalization ability of online learning algorithms. IEEE Transactions on Information Theory 50(9), 2050–2057 (2004)
21. Fradkin, D., Madigan, D.: Experiments with random projections for machine learning. In: Proc. of the 9th ACM SIGKDD int. conf. on Knowledge discovery and data mining, ACM Press, New York (2003)
22. Shawe-Taylor, J., Cristianini, N.: Kernel Methods for Pattern Analysis. Cambridge University Press, Cambridge (2004)
23. Rätsch, G., Onoda, T., Müller, K.R.: Soft Margins for AdaBoost. Machine Learning 42, 287–320 (2001)

Continuity of Performance Metrics for Thin Feature Maps

Adam Kowalczyk

National ICT Australia and
Department of Electrical & Electronic Engineering,
The University of Melbourne,
Parkville, Vic. 3010, Australia
adam.kowalczyk@nicta.com.au

Abstract. We study the class of hypothesis composed of linear functionals superimposed with smooth feature maps. We show that for "typical" smooth feature map, the pointwise convergence of hypothesis implies the convergence of some standard metrics such as error rate or area under ROC curve with probability 1 in selection of the test sample from a (Lebesgue measurable) probability density. Proofs use transversality theory. The crux is to show that for every "typical", sufficiently smooth feature map into a finite dimensional vector space, the counter-image of every affine hyperplane has Lebesgue measure 0.

The results extend to every real analytic, in particular polynomial, feature map if its domain is connected and the limit hypothesis is non-constant. In the process we give an elementary proof of the fundamental lemma that locus of zeros of a real analytic function on a connected domain either fills the whole space or forms a subset of measure 0.

1   Introduction

The issue of approximation of classifiers (or hypothesis, in the language of machine learning theorists) and of convergence of sequences of classifiers to limits is encountered very often in the theoretical research and practical applications of supervised machine learning. This is explicit in some algorithms such as boosting or online learning which stop after predefined number of iterations and implicit in some practical solvers such as LIBSVM's implementation of SMO algorithm [1] which is in fact an iterative procedure with a stopping criteria depending on a predefined precision constant.

The performance of classifiers is typically measured by metrics, in particular, the *error rate* (ERR) or the *area under receiver operating characteristic* (AROC). Such metrics are fully determined by the values of a classifier on the test set. However, this dependence could be non-continuous, as it is exemplified by the above two metrics. With what confidence and under what conditions we can claim that given a metric and a sequence of classifiers converging (pointwise) to a limit classifier, the sequence of values of a metric evaluated for these classifiers also converges to the metric value for the limit? It is easy to see that

M. Hutter, R.A. Servedio, and E. Takimoto (Eds.): ALT 2007, LNAI 4754, pp. 343–357, 2007.

potential problems exist. Indeed, given a converging, non-constant, sequence of linear classifiers we can select a malicious test set (e.g. fully positioned on the zero-hyperplane of the limit) such that test error rates for the classifiers do not converge to the error rate evaluated for the limit. It is somewhat harder to show that such malicious test sets can arise 'spontaneously', say by random sampling from a Lebesgue measurable density on the input space for some non-linear hypothesis classes, where the hypothesis is generated by an algorithm from randomly sampled training data. We show such examples in Section 2 mainly as an illustration of technical issues to be dealt with by transversality theory in the main Section 3. Accepting that the discontinuity of metrics can occur, at least theoretically, the question arises as to what extent one should be concerned with this issue. Actually, there are two questions hidden here:

(i) To what extent it impacts on everyday practice and
(ii) To what extent this is an issue for theorists producing rigorous theorems.

Concisely, our results say that in some popular applications such as kernel machines with a polynomial or radial basis kernel, the discontinuities does not occur with probability 1. However, the rigorous proofs should observe some special assumptions and restriction, e.g. analyticity or conditions assuring "thinness" of kernels, implicit in Corollary 3.

The starting point for discussion of these issues is an observation that the problem is practically non existent in the linear case, i.e. for linear classifiers

$$f : \mathbb{R}^N \to \mathbb{R}, \quad z \mapsto z \cdot w + b,$$

where $w \in \mathbb{R}^N$ and $b \in \mathbb{R}$, when the test data is sampled from an N-dimensional Lebesgue measurable probability density. This is attributed to the fact that any level set $f(z) = $ const has measure 0, being an $(N-1)$-dimensional hyperplane. However, this picture changes if kernel machines are used, with associated feature maps $\Phi : \mathbb{R}^n \to \mathbb{R}^N$. Although we can still conceptualize a classifier as a linear function f on the feature space $\mathbb{R}^N$, the data resides on a subset $\Phi(\mathbb{R}^n) \subset \mathbb{R}^N$ of measure 0. The relevant probability measure in this case is n but not N-dimensional, originally residing on $\mathbb{R}^n$ and then 'pushed up' to $\Phi(\mathbb{R}^n) \subset \mathbb{R}^N$ by the mapping Φ. The level (sub) sets of the composed classifier, defined as solutions of the equation

$$f \circ \Phi(x) = w \cdot \Phi(x) + b = \text{const} \tag{1}$$

for $x \in \mathbb{R}^n$, may have n-dimensional Lebesgue measure $\neq 0$; indeed, for every Lebesgue measurable probability density such C^∞-smooth feature map can be easily constructed. Moreover, in Section 2 we show examples that this can result in discontinuity of standard performance metrics for naturally converging sequences of kernel machines.

Now we have arrived to the technical crux of the paper which is as follows. We specify conditions for the feature map Φ under which the level subsets (1) have always Lebesgue measure 0. In other words, we specify the conditions for the counter-image $\Phi^{-1}(H) \subset \mathbb{R}^n$ to be a subset of measure 0 for any affine hyperplane $H \subset \mathbb{R}^N$, see Figure 1. Furthermore, we show that feature mappings which fail these conditions are 'exceptional', so that the 'typical' majority of feature mappings has the desired property of $\Phi^{-1}(H)$ always having measure 0, resulting in continuity of performance metrics. The way we have casted the problem so far lends itself to methodology of transversality theory [2,3,4]. The main link is to observe that the equations for the level subsets (1) having measure 0 can be formulated as conditions for derivatives of the feature map Φ, see Lemma 1, which in turn leads to conditions for transversality of submanifolds in the appropriate 'jet' spaces (Theorem 4). Then the 'typical majority' of transformations Φ with desired continuity property is specified in terms of a residual dense subset in a Whitney topology on the space of feature maps. In the process we prove that for a real analytic[1] transformation Φ, the level subsets (1) have the desired property of measure 0 (Proposition 1). In this case it is natural to restate the problem in terms of the so-called kernel function [5,6,7]. This allows covering simultaneously the case of infinite dimensional feature space, which occurs, for instance, for the popular radial basis kernel. The conclusion here is that the popular polynomial or radial basis kernels are 'well behaving', so discontinuity of the performance metrics does not occur with probability 1 in the selection of a test set.

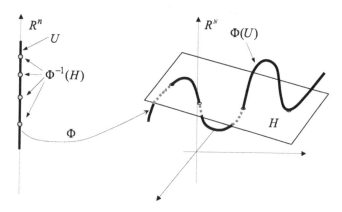

Fig. 1. An illustration for the technical crux of the paper: for any 'typical', sufficiently smooth feature map $\Phi : U \to \mathbb{R}^N$, $U \subset \mathbb{R}^n$, $n < N$, the counter-image $\Phi^{-1}(H) \subset \mathbb{R}^n$ of any affine hyperplane $H \subset \mathbb{R}^N$ has measure 0. In other words, $\Phi(U)$ is very 'wiggly' and cannot contain 'thick flat parts'. This also holds for any real analytic, in particular polynomial, feature map Φ, if U is connected and $\Phi(U) \not\subset H$.

2 Convergence of Metrics for Thin Limit Hypothesis

Throughout this section we assume that $X \subset \mathbb{R}^n$ is a subset of positive n-dimensional Lebesgue measure [8] and that there is chosen a Lebesgue measurable probability density on $X \times \{\pm 1\} \subset \mathbb{R}^n \times \{\pm 1\}$ with both subset $X \times \{-1\}$ and $X \times \{+1\}$ having positive probability. We assume that the finite data subsets,

---

[1] A transformation is called *analytic* if it is locally equal to its Taylor sequence expansion about every point of its domain.

the *training* set $\mathbb{X} = \{(\boldsymbol{x}_i, y_i)\}_{i=1,...,m}$ and the *test* set $\mathbb{X}' = \{(\boldsymbol{x}_i, y_i)\}_{i=1,...,m'}$, are randomly drawn from $X \times \{\pm 1\}$ with this probability density and contain data samples from both labels.

As it has been indicated already we are concerned here with the classes of functions $f : X \to \mathbb{R}$, synonymously called *hypothesis*. The hypothesis of particular interest are of the form

$$\boldsymbol{x} \mapsto \boldsymbol{w} \cdot \Phi(\boldsymbol{x}) + b,$$

where $\Phi : X \to \mathbb{R}^N$ is called *feature map* and $\mathbb{R}^N$ is called a *feature space*. Another class of hypothesis are *kernel* machines of the form

$$f(\boldsymbol{x}) := \sum_{i=1}^{m} \beta_i k(\boldsymbol{x}_i, \boldsymbol{x}) + b, \tag{2}$$

where $\boldsymbol{x}_i \in X$ is a training sample, $\beta_i \in \mathbb{R}$, for $i = 1, ..., m$ and $k : X \times X \to \mathbb{R}$ is a function called a *kernel*. Note that for our considerations we do not require, unless explicitly stated, that k is symmetric and positive definite, which are the typical additional assumptions on kernel in the machine learning literature [5,6,7]. These additional conditions characterise a special class of kernel functions having the form

$$k(\boldsymbol{x}, \boldsymbol{x}') = \Phi(\boldsymbol{x}) \cdot \Phi(\boldsymbol{x}')$$

for a suitably chosen feature map $\Phi : \mathbb{R}^n \to H$ into a Hilbert space $(H, \cdot)$.

2.1   Thin Hypothesis

We say that a subset A of $\mathbb{R}^n$ is *negligible* if it has n-dimensional Lebesgue measure 0 [8, 2]. We recall that this means that for every $\varepsilon > 0$ there exists a sequence $K_1, K_2, ...$ of hyper-cubes with

$$A \subset \bigcup_i K_i \ \& \ \sum_i vol(K_i) < \varepsilon.$$

Any union of a countable number of negligible subsets is negligible.

Thin hypothesis. We say that the function (hypothesis) $f : X \mapsto \mathbb{R}$ is *thin* if its level subsets $f^{-1}(v) \subset \mathbb{R}^n$ are negligible for $v \in \mathbb{R}$.

Thin feature map. A feature map $\Phi : X \to \mathbb{R}^m$ is called *thin* if every hypothesis $f(\boldsymbol{x}) := \boldsymbol{w} \cdot \Phi(\boldsymbol{x}) + b \not\equiv const$ is thin, for $\boldsymbol{w} \in \mathbb{R}^m$ and $b \in \mathbb{R}$.

Thin kernel. A kernel function $k : X \times X \to \mathbb{R}$ is called *thin* if every kernel machine (2) which is $\not\equiv const$ is a thin hypothesis, for every $\boldsymbol{x}_i \in X$ and $\beta_i, b \in \mathbb{R}$, $i = 1, ..., m$ and every $m = 1, 2, ...$

2.2   Performance Metrics

Now we introduce formally two performance metrics which will be used in our formal results. We consider a hypothesis $f : \mathbb{R}^n \to \mathbb{R}$ and a non-void test subset

$$\mathbb{X}' = \{(\boldsymbol{x}_i, y_i)\}_{i=1,...,m'} \subset \mathbb{R}^n \times \{\pm 1\}$$

containing samples from both labels. By $\mathbb{X}'_-$ and $\mathbb{X}'_+$, we denote the subsets of examples with negative $(y_i = -1)$ and positive $(y_i = +1)$ labels, respectively. The first metric is the typical *Error Rate*:

$$\text{ERR}(f, \mathbb{X}') := \frac{\#\{i \; ; \; y_i f(\boldsymbol{x}_i) \le 0 \text{ for } (\boldsymbol{x}_i, y_i) \in \mathbb{X}'\}}{\#\mathbb{X}'}.$$

The second metric is *the Area under the Receiver Operating Characteristic curve (*AROC*)*[2], i.e. the area under the plot of true positive vs. false positive rate while decision threshold is varied over the whole range of possible values [9]. Following [10] we use an order statistics formulation:

$$\text{AROC}(f, \mathbb{X}') = \frac{\#\{(i,j) \mid f(\boldsymbol{x}_i) < f(\boldsymbol{x}_j) \; \& \; -y_i = y_j = 1\}}{\#\mathbb{X}'_- \times \#\mathbb{X}'_+} + \frac{1}{2}\mathbb{P}_=[f, \mathbb{X}'],$$

where

$$\mathbb{P}_=[f, \mathbb{X}'] := \frac{\#\{(i,j) \mid f(\boldsymbol{x}_i) = f(\boldsymbol{x}_j) \; \& \; -y_i = y_j = 1\}}{\#\mathbb{X}'_- \times \#\mathbb{X}'_+} \tag{3}$$

is the probability that two test points with different labels obtain the same score.

The expected value of $AROC(f, \mathbb{X}')$ for the trivial uniformly random predictor is 0.5. This is also the value for this metric for the trivial constant classifier mapping all $\mathbb{X}'$ to a constant value, i.e. $\equiv y_o \in \mathbb{R}$.

2.3   Limits for Metrics

Theorem 1. *Assume that a sequence $f_1, f_2, \ldots$ of hypothesis is pointwise converging to a limit f, which is a thin hypothesis on X. For either of the two metrics, $\mu = $ AROC or $\mu = $ ERR, the following equality holds*

$$\lim_{n \to \infty} \mu(f_n, \mathbb{X}') = \mu(f, \mathbb{X}'), \tag{4}$$

with probability 1 (in selection of the test set $\mathbb{X}'$).

Proof. The classifiers f_n are pointwise convergent to the limit f, hence we have the upper bounds:

$$\left| \lim_{n \to \infty} \text{AROC}(f_n, \mathbb{X}') - \text{AROC}(f, \mathbb{X}') \right| \le \frac{1}{2}\mathbb{P}_=[f, \mathbb{X}'], \tag{5}$$

$$\left| \lim_{n \to \infty} \text{ERR}(f_n, \mathbb{X}') - \text{ERR}(f, \mathbb{X}') \right| \le \mathbb{P}_0[f, \mathbb{X}'], \tag{6}$$

where $\mathbb{P}_0[f, \mathbb{X}'] := \frac{\#(f^{-1}(0) \cap \mathbb{X}')}{\#\mathbb{X}'}$. The hypothesis f is thin hence the event that $\mathbb{X}'$ contains two data points which are mapped by f to the same value has probability 0. Thus, with probability 1, $\mathbb{P}_=[f, \mathbb{X}'] = \mathbb{P}_0[f, \mathbb{X}'] = 0$. □

---

[2] Also known as *the area under the curve, AUC*; it is essentially the well known order statistics U.

2.4   A Counterexample

The following result shows that there exist non-thin smooth kernels. Moreover, we shall show that there exist sequences of kernel machines which pointwise converge but are such that the limits do not extend to the corresponding limits for either performance metric AROC or ERR. We shall introduce the sequences of converging machines first.

Given a constant $\lambda > 0$, a training set $\mathbb{X} = \{(\boldsymbol{x}_i, y_i)\}_{i=1,...,m}$ and a kernel function k such that the matrix $[k(\boldsymbol{x}_i', \boldsymbol{x}_j')]_{1 \leq i,j \leq m'}$ is positive semidefinite for any selection of samples $\boldsymbol{x}_i' \in X$, $i = 1, ..., m'$ and every $m' = 1, 2, ...$ In such a case there exists a map $\Phi : X \to \mathbb{R}^m$ such that $k(\boldsymbol{x}_i, \boldsymbol{x}) = \Phi(\boldsymbol{x}_i)^T \cdot \Phi(\boldsymbol{x})$ for every $i = 1, ..., m$ and every $\boldsymbol{x} \in X$. We consider a homogeneous kernel machine

$$f_\lambda(\boldsymbol{x}) := \sum_{i=1}^m \beta_i k(\boldsymbol{x}_i, \boldsymbol{x}) = \sum_{i=1}^m \beta_i \Phi(\boldsymbol{x}_i) \cdot \Phi(\boldsymbol{x}),$$

where coefficients (β_i) are defined as minimizers of the following functional (regularised risk):

$$(\beta_i) = \arg\min_{\beta_i} \lambda \sum_{i,j=1}^m \beta_i \beta_j k(\boldsymbol{x}_i, \boldsymbol{x}_j) + \sum_{i=1}^m \left(y_i - \sum_{j=1}^m \beta_j k(\boldsymbol{x}_j, \boldsymbol{x}_i) \right)^2. \tag{7}$$

This is the well known ridge regression solution which can be also defined by the following closed form formula for sufficiently small $\lambda > 0$ [6]:

$$f_\lambda(\boldsymbol{x}) = \boldsymbol{y}\big(\lambda I + \Phi(\mathbb{X})^T \Phi(\mathbb{X})\big)^{-1} \Phi(\mathbb{X})^T \Phi(\boldsymbol{x}), \tag{8}$$

where $\Phi(\mathbb{X}) := [\Phi(\boldsymbol{x}_1), ..., \Phi(\boldsymbol{x}_m)]$ is the $m \times m$ matrix and $\boldsymbol{y} = [y_1, .., y_m]^T$.

It is known that the following pointwise limits exist

$$\lim_{\lambda \to \infty} \lambda f_\lambda(\boldsymbol{x}) = f_{centr}(\boldsymbol{x}) := \sum_{i=1}^m \frac{y_i}{m(y_i)} k(\boldsymbol{x}_i, \boldsymbol{x}), \tag{9}$$

$$\lim_{\lambda \to 0+} f_\lambda(\boldsymbol{x}) = \boldsymbol{y}\Phi(\mathbb{X})^\dagger \Phi(\boldsymbol{x}) = f_{regr}(\boldsymbol{x}), \tag{10}$$

for any $\boldsymbol{x} \in X$, where $f_{regr}(\boldsymbol{x})$ is the ordinary regression, i.e. the solution of (7) for $\lambda = 0$. In Eqn. 10, '$\dagger$' denotes the Moore-Penrose pseudoinverse [11]

$$A^\dagger := \lim_{\lambda \to 0+} (A^T A + \lambda I)^{-1} A^T = \lim_{\lambda \to 0+} A^T (AA^T + \lambda I)^{-1}. \tag{11}$$

The equation (9) is a special case of [12, Theorem 2], while the justification of (10) follows easily from (8).

Theorem 2. *There exists a C^∞-smooth kernel $k : \mathbb{R}^n \times \mathbb{R}^n \to \mathbb{R}$ which is not thin. Moreover, for either metric $\mu = $ AROC or $\mu = $ ERR, the following inequalities hold with probability > 0 in selection of the training and test subset:*

$$\lim_{\lambda \to \infty} \mu\Big(f_\lambda, \mathbb{X}'\Big) = \lim_{\lambda \to \infty} \mu\Big(\lambda f_\lambda, \mathbb{X}'\Big) \neq \mu\Big(\lim_{\lambda \to \infty} \lambda f_\lambda, \mathbb{X}'\Big) = \mu\big(f_{centr}, \mathbb{X}'\big),$$

$$\lim_{\lambda \to 0+} \mu\Big(f_\lambda, \mathbb{X}'\Big) \neq \mu\Big(\lim_{\lambda \to 0+} f_\lambda, \mathbb{X}'\Big) = \mu\big(f_{regr}, \mathbb{X}'\big).$$

We emphasize that the following result holds for every Lebesgue measurable probability density on $X \times \{\pm 1\} \subset \mathbb{R}^n \times \{\pm 1\}$.

Proof outline. The proof consists in a construction of an example such that the right-hand-side of the upper bounds (5 - 6) is non-zero and actually equal to the discrepancy for the corresponding limits.

The idea is sketched in Figure 2. First we construct a seven-point dataset $((\boldsymbol{z}_i, y_i)) \subset (\mathbb{R}^2 \times \{\pm 1\})^7$, four points for training (blue and red circles) and three points for a 'malicious' testset $\mathbb{X}'$ (blue and red squares). It is insured that for the (linear) ridge regression solutions $\boldsymbol{w}_\lambda$, the homogeneous classifiers $f_\lambda(\boldsymbol{z}) := \boldsymbol{w}_\lambda \cdot \boldsymbol{z}$, $\boldsymbol{z} \in \mathbb{R}^2$ classify correctly the testdata, hence $\text{AROC}(f_\lambda, \mathbb{X}') = 1$ and $\text{ERR}(f_\lambda, \mathbb{X}') = 0$ for any $\lambda > 0$. The limit $\lim_{\lambda \to \infty} \lambda \boldsymbol{w}_\lambda$ exists and is precisely the directional vector of the centroid of the training data [12]

$$\boldsymbol{w}_{centr} := \frac{1}{2} \sum_{i, y_i = +1} y_i \boldsymbol{z}_i,$$

which a vector of the form $[C, 0]^T \in \mathbb{R}^2$, $C > 0$. However, the centroid $f_{centr}(\boldsymbol{z}) := \boldsymbol{w}_{centr} \cdot \boldsymbol{z}$ maps two of the three test samples to 0. Consequently, $\text{AROC}(f_{centr}, \mathbb{X}') = 0.75$ and $\text{ERR}(f_{centr}, \mathbb{X}') = 1/3$. Hence, the test set values of the performance metric for any linear hypothesis λf_λ differs from that for their limit as $\lambda \to \infty$.

The extension to the general case relies on an observation that for any Lebesgue measurable probability density on $X \times \{\pm 1\} \subset \mathbb{R}^n \times \{\pm 1\}$ there exist a feature mapping $\Phi : X \to \mathbb{R}^2$ which maps data onto the points of the example with positive probability (in Figure 2 we illustrate this for $n = 1$). Indeed, even more generally, using a smooth partition of unity it can be easily shown [2] that for any values $\boldsymbol{v}_1, ..., \boldsymbol{v}_s \in \mathbb{R}^2$ and any s disjoint closed subsets of $V_1, ..., V_s \subset \mathbb{R}^n$ there exists C^∞ feature mapping $\Phi : \mathbb{R}^n \to \mathbb{R}^2$ such that $V_i \subset \Phi^{-1}(\boldsymbol{v}_i)$ for $i = 1, ..., s$. Hence it is sufficient to select the subsets V_i such that $V_i \times \{-1\}$ or $V_i \times \{+1\}$ has positive measure according to the probability measure on $X \times \{\pm 1\}$, respectively. Finally, the kernel is defined as $k(\boldsymbol{x}, \boldsymbol{x}') := \Phi(\boldsymbol{x}) \cdot \Phi(\boldsymbol{x}')$. □

3   Theorems on Abundance of Thin Hypothesis

Let $r \geq 0$ be an integer. Let $P^r(\mathbb{R}^n; \mathbb{R}^N)$ denote the vector space of polynomial maps of degree $\leq r$ from $\mathbb{R}^n$ to $\mathbb{R}^N$. It has dimension $m \times d_n^r$, where

$$d_n^r := \sum_{i=0}^r \binom{n+i-1}{i}$$

is the dimension of the vector space of polynomials of degree $\leq r$ in n-variables. In particular, $d_n^0 = 1$, $d_n^1 = n + 1$ and $d_n^2 = (n+1)(n+2)/2$.

Let us consider an open non-empty subset U of $\mathbb{R}^n$ and a transformation $f = (f_1, ..., f_N) : U \to \mathbb{R}^N$. We say that f is C^r-smooth, and write $f \in C^r(U, \mathbb{R}^N)$,

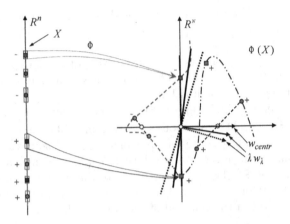

Fig. 2. Illustration to the proof of Theorem 2. The right-hand figure shows an idea of a four point training set ("+" and "-" circles) and a malicious three point test set ("+" and "-" squares) in $\mathbb{R}^2$ with a properties as follows. For the (trained) ridge regression classifiers $f_\lambda(z) := w_\lambda \cdot z$ the limit $\lim_{\lambda \to \infty} \lambda f_\lambda(x) = f_{centr}(x)$ exists but it does not extend to the level of performance metrics. Indeed, $\mathrm{ERR}(\lambda f_\lambda(x), \mathbb{X}') = 0 \neq 1/3 = \mathrm{ERR}(f_{centr}, \mathbb{X}')$; similarly, $\mathrm{AROC}(\lambda f_\lambda(x), \mathbb{X}') = 1 \neq 0.75 = \mathrm{AROC}(f_{centr}, \mathbb{X}')$ for $\lambda > 0$. This is then shown to occur with probability > 0 for a smooth kernel k on $\mathbb{R}$ and with training and test set samples from a continuous probability density on $\mathbb{R}^2$. We construct such a kernel $k(x, x') := \Phi(x) \cdot \Phi(x')$ on $\mathbb{R}$, by choosing $\Phi : \mathbb{R} \to \mathbb{R}^2$ which maps multiple input samples onto the seven points in $\mathbb{R}^2$ as indicated, e.g. with a positive probability for any Gaussian distribution on $\mathbb{R}$.

where $r \in \{1, 2, ...\}$, if it has all partial derivatives of order $\leq r$ and they are all continuous functions on U. Additionally, we say f is C^∞-smooth if f is C^r-smooth for any finite $r = 1, 2, ...$; we write $f \in C^\omega(\mathbb{R}^n, \mathbb{R}^N)$ and call f a *real analytic function* if it is additionally locally equal to its Taylor series expansion about every point of its domain. It is well known that $C^\omega \neq C^\infty$.

3.1   Main Results for Analytic Hypothesis

Theorem 3. *Any analytic hypothesis $\not\equiv$ const is thin.*

The formal proof is shifted to Section 4, where we show a slightly reformulated result, Proposition 1.

Corollary 1. *All analytic kernels $k \in C^\omega(\mathbb{R}^n \times \mathbb{R}^n, \mathbb{R})$ and all feature maps $\Phi \in C^\omega(\mathbb{R}^n, \mathbb{R}^N)$ are thin, for $n, N = 1, 2, ...$*

Proof. This is an instantaneous conclusion from Theorem 3 as for such k the kernel machine (2) is an analytic function on $\mathbb{R}^n$. Similarly, the hypothesis $x \mapsto w \cdot \Phi(x) + b$ is analytic for every $(w, b) \in \mathbb{R}^N \times \mathbb{R}$. $\qquad\square$

As every polynomial is an analytic function, we have:

Corollary 2. *All polynomial hypothesis $\not\equiv$ const, all polynomial kernels and all polynomial feature maps are thin.*

3.2   Main Results for Non-analytic Hypothesis

For the formulation of the result in this section we need a constructive way of saying that a property is "almost always true" or "typical". Ideally, this should be equivalent to stating that it holds for an open and dense subset in some "natural" topology. This condition is often a little bit too strong in practice. Fortunately, in an important class of Baire spaces, it can be replaced by a concept of a residual subset. The formal definitions follow.

A subset of a topological space which is an intersection of a countable family of open and dense subsets is called *residual*. A *Baire space* is a topological space in which every residual subset is dense in it. Note that an intersection of a countable family of residual subsets is a residual subset. The Euclidean space $\mathbb{R}^m$ with the natural topology is an example of a Baire space [13, 2, 4].

Theorem 4. *Let $\emptyset \neq U \subset \mathbb{R}^n$ be an open connected subset, $\Phi \in C^r(\mathbb{R}^n, \mathbb{R}^N)$ and an integer $r \geq 1$ be such that $n + d_n^r > N$. There exists a polynomial $p \in P^r(\mathbb{R}^n, \mathbb{R}^N)$ such that the feature mapping $\Phi + p$ is thin. Moreover, the set of such polynomials contains a dense residual subset of $P^r(\mathbb{R}^n, \mathbb{R}^N)$ with negligible complement.*

The proof is given in Section 4.3.

For the following result we need to introduce a topology on the space of feature maps $C^r(\mathbb{R}^n, \mathbb{R}^N)$, where $r = 1, 2, ..., \infty$. Let us define first the operator of mixed partial derivatives $f \mapsto \partial_\alpha f$ for a multi-index $\alpha = (\alpha_1, \alpha_2, \cdots \alpha_n)$ of integers $\alpha_i \geq 0$:

$$\partial_\alpha f(x) := \frac{\partial^{|\alpha|} f}{\partial x_1^{\alpha_1} \cdots \partial x_n^{\alpha_n}}(x)$$

where $|\alpha| := \alpha_1 + \alpha_2 + \cdots + \alpha_n$, $x \in \mathbb{R}^n$. In our case it is very straightforward to define the C^r-*Whitney topology* via specification of the neighborhood basis [13, 2]. This is defined as the family of all sets of the form

$$\{g \in C^r(\mathbb{R}^n, \mathbb{R}^N) \,;\, \|\partial_\alpha f(x) - \partial_\alpha g(x)\| < \delta(x) \,\forall x \in \mathbb{R}^n\},$$

for all $f \in C^r(\mathbb{R}^n, \mathbb{R}^N)$, all α such that $|\alpha| \leq r$ and all $\delta : \mathbb{R} \to \mathbb{R}^+$. It is well known that $C^r(\mathbb{R}^n, \mathbb{R}^N)$ with C^r-Whitney topology is a Baire space [2, 3, 4].

The C^r-Whitney topology, sometimes called fine topology [4], allows for a fine controll of perturbations at "infinity", which is not possible with coarser topologies such as the topology of uniform C^r convergence on compact set, used in [4] or the natural topology on the polynomial perturbations $P^r(\mathbb{R}^n, \mathbb{R}^N)$ used in Theorem 4.

Corollary 3. *Let $r \geq 1$ and $n + d_n^r > N$ or $r = \infty$. The subset of all thin feature maps in $C^r(\mathbb{R}^n, \mathbb{R}^N)$ contains a dense residual subset of $C^r(\mathbb{R}^n, \mathbb{R}^N)$ with C^r-Whitney topology.*

The proof uses the smooth partitions of unity and the fact (Weierstrass' Theorem) that polynomials are dense in $C^r(K, \mathbb{R}^N)$ with the topology induced from the C^r-Whitney topology on $C^r(\mathbb{R}^n, \mathbb{R}^N)$ by restrictions to a compact set $K \subset \mathbb{R}^n$. Details are omitted.

4   Key Proofs

This section contains main technical result: an application of Thom transversality theorem to characterisation of zeros of compositions of linear functions with smooth transformations. We derive also a characterisation of zeros of analytic functions. We start with introduction of basic definitions and notation.

4.1   The Space of r-Jets

Let $r \geq 0$ be an integer. For every open set $U \subset \mathbb{R}^n$ we write

$$J^r(U; \mathbb{R}^N) = U \times P^r(\mathbb{R}^n; \mathbb{R}^N)$$

and call this *the space of jets of order r* of maps from U to $\mathbb{R}^N$.

Let $f : U \to \mathbb{R}^N$ be a C^r-map, where $r = 1, 2, \ldots$ and let $a \in U$. Write the Taylor expansion for f in the form

$$f(a + h) = (j_a^r f)(h) + (R_a^r f)(h) \tag{12}$$

$$= f(a) + \sum_\alpha \frac{\partial f}{\partial x_\alpha}(a) h_\alpha + \cdots + (R_a^r f)(h) \tag{13}$$

$$= \sum_{\substack{\alpha_1, \cdots \alpha_n \geq 0 \\ 0 \leq |\alpha| \leq r}} \frac{1}{\alpha_1! \cdots \alpha_n!} \partial_\alpha f(a) h_1^{\alpha_1} \cdots h_n^{\alpha_n} + (R_a^r f)(h), \tag{14}$$

with $j_a^r f \in P^r(\mathbb{R}^n; \mathbb{R}^N)$ and $(R_a^r f)(h) = o(\|h\|^r)$. We say that $j_a^r f$ is *the jet of order r* or *r-jet* of f at a. The maps

$$j_*^r f : U \to P^r(U, \mathbb{R}^N), \quad (j_*^r f)(a) = j_a^r f,$$
$$j^r f : U \to J^r(U, \mathbb{R}^N), \quad (j^r f)(a) = (a, j_*^r f) = (a, j_a^r f)$$

are called the *r-jet* of f. In particular we have $j^0 f(a) = (a, f(a))$,

$$j^1 f(a) = \left(a, f(a), df(a) \right)$$

$$= \left(a, (f_i(a))_{1 \leq i \leq N}, \left(\frac{\partial f_i}{\partial x_j}(a) \right)_{\substack{1 \leq i \leq n \\ 1 \leq i \leq N}} \right) \in \mathbb{R}^n \times \mathbb{R}^N \times \mathbb{R}^{nN}.$$

4.2   Locus of Zeros of Smooth Functions

For completeness we recall now a simplified concept of submanifold and transversality [2, 3, 4]. We say that $W \subset \mathbb{R}^m$ is a C^r-submanifold for $r \in \{1, 2, \ldots, \infty, \omega\}$

of dimension d, $1 \leq d \leq m$ if for every $z_o \in W$ there exists and open neighbourhood $z_o \in U_{z_o} \subset \mathbb{R}^m$, an open subset $V \subset \mathbb{R}^m$ and a C^r-smooth diffeomorphism $\phi = (\phi_i) : U_{z_o} \to V$, called a *local co-ordinate map*, such that

$$\phi(W \cap U_{z_o}) = \{(z_i) \in V \; ; \; z_{d+1} = z_{d+2} = \cdots = z_m = 0\}.$$

The difference $m - d$ is called co-dimension of W and is denoted codim$W$. If codim$W > 0$, then $W \subset \mathbb{R}^m$ is a subset of measure 0 [8, 2, 4].

We say that the transformation $f : \mathbb{R}^n \to \mathbb{R}^m$ is *transversal* to the submanifold W if for every $z_o \in f(\mathbb{R}^n) \cap W$ and every $x_o \in f^{-1}(z_o)$ we have

$$\text{rank}[\partial_j (\phi_i \circ f)(x_o)]_{\substack{d < i \leq m \\ 1 \leq j \leq n}} = m - d = \text{codim}W.$$

Note that $\text{rank}[\partial_j (\phi_i \circ f)(x_o)] \leq n$, hence if codim$W > n$, the transversality of f to W means that that $W \cap f(\mathbb{R}^n) = \emptyset$.

Lemma 1. *Let $r \geq 1$ be an integer and $f \in C^r(U, \mathbb{R})$, where $U \subset \mathbb{R}^n$ is an open subset. Then the subset $f^{-1}(0) \backslash (j_*^r f)^{-1}(0)$ of $\mathbb{R}^n$ is negligible.*

Proof. Proof is by induction on r.

Case $r = 1$. Let $x_0 \in f^{-1}(0) \backslash (j_*^1 f)^{-1}(0)$ and i be an index such that $\frac{\partial f}{\partial x_i}(x_0) \neq 0$. By implicit function theorem [4] there exists an open subset $U_{x_0} \subset U$ such that $f^{-1}(0) \cap U_{x_0}$ is an $(n-1)$-dimensional submanifold (since the equation $f(x) = 0$ can be solved for the ith variable). Thus $f^{-1}(0) \cap U_{x_0}$ has measure 0 [4, 2]. Now note that $f^{-1}(0) \backslash (j_*^1 f)^{-1}(0)$ can be covered by countably many such subsets U_{x_0}, hence it is negligible.

Case $r > 1$. Assume that the lemma holds for $r \gets 1, 2, ..., r-1$. We will show that in such a case $f^{-1}(0) \backslash (j_*^r f)^{-1}(0)$ is negligible. Each function $\partial_\alpha f$ for a multi-index $\alpha = (\alpha_1, \alpha_2, \cdots \alpha_n)$, $|\alpha| = r - 1$, is C^1-smooth on U. We can write

$$f^{-1}(0) = \left[f^{-1}(0) \backslash (j_*^{r-1} f)^{-1}(0) \right] \cup (j_*^{r-1} f)^{-1}(0)$$

$$= \left[f^{-1}(0) \backslash (j_*^{r-1} f)^{-1}(0) \right] \cup \left[\bigcup_\alpha \partial_\alpha f(0) \backslash (j_*^1 \partial_\alpha f)^{-1}(0) \right] \cup (j_*^r f)^{-1}(0),$$

where the union $\bigcup_\alpha$ is over all multi-indices $\alpha = (\alpha_1, \alpha_2, \cdots \alpha_n)$ such that $|\alpha| = r-1$. By the inductive assumption the term in the first square bracket is negligible and the second square bracket term is negligible as the finite union of sets which are negligible by the proof of Case $r = 1$. Thus $f^{-1}(0) \backslash (j_*^r f)^{-1}(0)$ is negligible being contained in the union of these two negligible terms.   □

Proposition 1. *Let U be a connected non-empty open subset of $\mathbb{R}^n$. If $f \in C^\omega(U; R)$, then either $f \equiv 0$ or $f^{-1}(0) \subset \mathbb{R}^n$ is negligible.*

In the case of complex analytic functions (holomorphic) the analog of this result can be deduced relatively easily from the so called Weierstrass Preparatory Theorem [2], though some non-trivial work will be required if justification is to be made rigorous and that require more effort than the following direct proof[3].

---

[3] In a nutshell, the Weierstrass theorem implies that the locus of zeros of a holomorphic extension to $\mathbb{C}^n \approx \mathbb{R}^{2n}$ is negligible in $2n$-dimensional Lebesgue measure, but we need to deduce that the locus of the original real analytic function is negligible in n-dimensional Lebesgue measure on $\mathbb{R}^n$.

Proof. Let $A \subset U$ be a compact connected subset. The set U is a countable union of such sets A, hence it is sufficient to prove that either $f|A \equiv$ const or the set $A \cap f^{-1}(0)$ is negligible.

Consider the descending sequence of closed subsets (they are compact if $\neq \emptyset$):

$$A \supset A \cap (j_*^1 f)^{-1}(0) \supset A \cap (j_*^2 f)^{-1}(0) \supset \cdots \supset A \cap (j_*^r f)^{-1}(0) \supset \cdots .$$

There are two possibilities.

(i) The sequence does not terminate. In such a case there exists

$$x_0 \in \bigcap_{r=1}^{\infty} (j_*^r f)^{-1}(0) \cap A \neq \emptyset$$

by virtue of Riesz condition for compact spaces [13]. The Taylor expansion of the analytic function f around x_0 is identically 0, hence $f \equiv 0$ near x_0, hence on the whole connected domain U.

(ii) The above sequence terminates, i.e. there exists r_0 such that $(j_*^r f)^{-1}(0) \cap A = \emptyset$ for $r \geq r_0$. In such a case

$$A \cap f^{-1}(0) = A \cap f^{-1}(0) \setminus (j_*^{r_0} f)^{-1}(0) \subset f^{-1}(0) \setminus (j_*^{r_0} f)^{-1}(0)$$

is negligible by Lemma 1. □

The assumption of analyticity in the above lemma is essential as the lemma does not hold for some C^∞-functions. The function $f \in C^\infty(\mathbb{R})$, defined as $\exp(-x^{-2})$ for $x > 0$ and 0 otherwise is a counter example here. Note that for any closed subset $V \subset \mathbb{R}^n$ there exists $f \in C^\infty(\mathbb{R}^n, \mathbb{R})$ such that $V = f^{-1}(0)$ [2].

4.3 Characterisation of Typical C^r-Smooth Feature Maps

In this section we show the key lemma facilitating application of the Thom transversality theorem to the proof of our main result, Theorem 4.

Let us consider the subset

$$W := \left\{ (b, x_1, ..., x_m) \in \mathbb{R}^s \times (\mathbb{R}^d)^N ; \ \exists_{0 \neq (w_i) \in \mathbb{R}^N} \sum_{i=1}^{N} w_i x_i = 0 \right\}. \tag{15}$$

Lemma 2. *If $d > N$, then we have a decomposition $W = \bigcup_{q=1}^{N-1} W_q$, where $W_q \subset \mathbb{R}^s \times (\mathbb{R}^d)^N$ is an analytic submanifold of codimension*

$$\mathrm{codim}(W_q) = (d - q)(N - q) \geq d - N + 1 \tag{16}$$

for $q = 1, 2, ..., N - 1$.

Proof. For any $(b, x_1, ..., x_N) \in \mathbb{R}^s \times (\mathbb{R}^d)^N$ the condition

$$\exists_{0 \neq (w_i) \in \mathbb{R}^N} \sum_{i=1}^{N} w_i x_i = 0$$

is equivalent to

$$\text{rank } [\boldsymbol{x}_1, ..., \boldsymbol{x}_N] < N,$$

where $[\boldsymbol{x}_1, ..., \boldsymbol{x}_N]$ denotes $d \times N$-matrix with column vectors $\boldsymbol{x}_i \in \mathbb{R}^d$. Let

$$W_q := \left\{ (\boldsymbol{b}, \boldsymbol{x}_1, ..., \boldsymbol{x}_N) \in \mathbb{R}^s \times (\mathbb{R}^d)^N \; ; \; \text{rank } [\boldsymbol{x}_1, ..., \boldsymbol{x}_N] = q \right\},$$

for $q = 1, ..., N - 1$. Obviously $W = \bigcup_{q=1}^{N-1} W_q$. It remains to show that W_q is a submanifold of the codimension $(d - q)(N - q)$. To that end, in [14, Chapter 2.2] we find an explicit proof that the subset of $d \times N$ matrices of rank q forms an analytic, hence C^∞ smooth, sub-manifold of $\mathbb{R}^{dN}$ of dimension $q(d + N - q)$. This implies immediately

$$\text{codim}(W_q) = s + dN - (s + q(d + N - q)) = (d - q)(N - q) \geq d - N + 1$$

for every $q = 1, 2, ..., N - 1$. $\qquad\square$

Proof of Theorem 4. Let the assumptions of the theorem hold. Let $\Phi = (\phi_i) \in C^r(\mathbb{R}^n, \mathbb{R}^N)$, $p = (p_i) \in P^r(\mathbb{R}^n, \mathbb{R}^N)$ and $(\boldsymbol{w}, b) \in \mathbb{R}^N \times \mathbb{R}$. We define

$$f_{\Phi+p,\boldsymbol{w},b}(\boldsymbol{x}) := \boldsymbol{w} \cdot \big((\Phi + p)(\boldsymbol{x})\big) + b$$

for $\boldsymbol{x} \in \mathbb{R}^n$. By Lemma 1 the subset

$$f_{\Phi+p,\boldsymbol{w},b}^{-1}(0) \backslash (j_*^r f_{\Phi+p,\boldsymbol{w},b})^{-1}(0) \subset \mathbb{R}^n$$

is always negligible. Thus in order to show that $f_{\Phi+p,\boldsymbol{w},b}$ is thin, i.e. that $f_{\Phi+p,\boldsymbol{w},b}^{-1}(v)$ is negligible for all $v \in \mathbb{R}$, it is sufficient to show that

$$(j_*^r f_{\Phi+p,\boldsymbol{w},b})^{-1}(v) = \emptyset$$

or, equivalently, after absorbing v into the bias b, that for a sufficiently large r,

$$\sum_{i=1}^m w_i \, j_*^r (\Phi_i + p_i)(\boldsymbol{x}) + b \neq 0 \in J^r(\mathbb{R}^n, \mathbb{R}) \tag{17}$$

for all $0 \neq \boldsymbol{w} = (w_i) \in \mathbb{R}^N$, $b \in \mathbb{R}$.

For completion of the proof we shall show that the set of polynomials p satisfying (17) contains a subset S which is a dense residual subset of $P^r(\mathbb{R}^n, \mathbb{R}^N)$ with negligible complement.

Let $P_H^r(\mathbb{R}^n, \mathbb{R}^N)$ denote the subset of all homogeneous polynomials in $P^r(\mathbb{R}^n, \mathbb{R}^N)$ and for $f \in C^r(\mathbb{R}^n, \mathbb{R}^N)$ let $j_H^r f \in P_H^r(\mathbb{R}^n, \mathbb{R}^N)$ denote the homogeneous part of the jet $j^r f$, i.e. all polynomial terms of order 1 or higher. The vector space $P_H^r(\mathbb{R}^n, \mathbb{R}^N)$ has dimension $N(d_n^r - 1)$. We have isomorphisms $P^r(\mathbb{R}^n, \mathbb{R}) = \mathbb{R} \times P_H^r(\mathbb{R}^n, \mathbb{R}) = \mathbb{R} \times \mathbb{R}^{d_n^r - 1}$, thus

$$J^r(\mathbb{R}^n, \mathbb{R}^N) = \mathbb{R}^n \times P^r(\mathbb{R}^n, \mathbb{R}^N) = \mathbb{R}^{n+N} \times \left(\mathbb{R}^{d_n^r - 1}\right)^N.$$

Let W and W_q be defined as in (15) and Lemma 2 (with the substitutions $s \leftarrow n + N$ and $d \leftarrow d_n^r - 1$). Hence

$$W = \bigcup_{q=1}^{N-1} W_q \subset R^{n+N} \times (\mathbb{R}^{d_n^r - 1})^m = J^r(\mathbb{R}^n, \mathbb{R}^N),$$

where each $W_q \subset J^r(\mathbb{R}^n, \mathbb{R}^N)$ is a submanifold of codimension (see Eqn. 16):

$$\mathrm{codim}(W_q) = (d_n^r - 1 - q)(N - q) > d_n^r - N > n, \tag{18}$$

where the last inequality is the theorem's assumption. The definition of W means that for $\Psi = (\Psi_i) \in C^r(\mathbb{R}^n, \mathbb{R}^N)$ we have

$$j^r \Psi(\boldsymbol{x}) \notin W \quad \Leftrightarrow \quad \forall_{(w_i) \neq 0} \sum_{i=1}^{N} w_i \, j_H^r \Psi_i(\boldsymbol{x}) \neq 0 \in P_H^r(\mathbb{R}^N, \mathbb{R}). \tag{19}$$

According to the 'concrete' version of the Thom transversality theorem [4, Proposition 3.9.1] the set S of those polynomials $p \in P^r(\mathbb{R}^n, \mathbb{R}^N)$, where the map $j^r(\Phi + p)$ is transversal to every W_q, is dense residual with negligible complement. The inequality (18) implies that transversality in our case means that $W_q \cap j^q(\Phi + p)(\mathbb{R}^n) = \emptyset$ for every $q = 1, ..., N - 1$ [4]. Hence

$$j^r(\Phi + p)(\boldsymbol{x}) \notin W$$

for every $p \in S$ and $\boldsymbol{x} \in \mathbb{R}^n$. Now, an application of (19) for $\Psi = \Phi + p$ implies $\sum_i w_i \, j_H^r(\Phi + p)(\boldsymbol{x}) \neq 0$ for all $0 \neq (w_i) \in \mathbb{R}^N$ proving that (17) holds for every $p \in S$. $\qquad\square$

5   Discussion

Note that in order to make use of Thom transversality theorem it is necessary to move from the feature space $\mathbb{R}^N$ to the space of higher order jets, $J^r(U, \mathbb{R}^N)$, $r \geq 1$. More specifically, for any differentiable feature mapping $\Phi : \mathbb{R}^n \to \mathbb{R}^N$ there exists a hyperplane $H \subset \mathbb{R}^n$, which are non-transversal to Φ at some points $\boldsymbol{x} \in U$. This will be in particular a case of hyperplane H tangent to $\Phi(U)$ (see Fig. 1). Thus of necessity, we have reformulated our original problem (of characterisation of the thin feature map) into a set of transversality conditions in $J^{r'}(U, \mathbb{R}^N)$, $r' = 1, ..., r - 1$: they are implicit in the key Lemma 1 and equivalent to $j^{r'} f(\boldsymbol{x}) \neq 0$.

Theorem 2 shows that there exist non-thin C^r-smooth feature maps. Theorem 4 and Corollary 3 show that such feature maps are exceptional and disappear under suitable perturbations, in particular under polynomial perturbations from a dense residual set.

Acknowledgements

Many thanks to Justin Bedo for help in preparation of this paper.

National ICT Australia (NICTA) is funded by the Australian Government's *Backing Australia's Ability* initiative, in part through the Australian Research Council.

This work was supported in part by the IST Programme of the European Community, under the PASCAL Network of Excellence, IST-2002-506778. This publication only reflects the authors' views.

References

1. Platt, J.: Fast training of support vector machines using sequential minimal optimization. In: Schölkopf, B., Burges, C., Smola, A. (eds.) Advances in kernel methods: support vector learning, pp. 185–208. MIT Press, Cambridge, MA (1998)
2. Golubitsky, M., Guillemin, V.: Stable Mapping and Their Singularities. Springer, New York (1973)
3. Arnold, V., Gussein-Zade, S., Varchenko, A.: Singularities of Differentiable Maps. Birkhauser, Boston (1985)
4. Demazure, M.: Bifurcations and Catastrophes. Springer, New York (2000)
5. Vapnik, V.: Statistical Learning Theory. John Wiley and Sons, New York (1998)
6. Cristianini, N., Shawe-Taylor, J.: An Introduction to Support Vector Machines. Cambridge University Press, Cambridge, UK (2000)
7. Schölkopf, B., Smola, A.: Learning with Kernels. MIT Press, Cambridge, MA (2002)
8. Bartle, R.: The Elements of Integration and Lebesgue Measure. Wiley, Chichester (1995)
9. Provost, F., Fawcett, T.: Robust classification for imprecise environments. Machine Learning 42(3), 203–231 (2001)
10. Bamber, D.: The area above the ordinal dominance graph and the area below the receiver operating characteristic graph. J. Math. Psych. 12, 387–415 (1975)
11. Albert, A.: Regression and the Moore-Penrose Pseudoinverse. Academic Press, New York (1972)
12. Bedo, J., Sanderson, C., Kowalczyk, A.: An efficient alternative to svm based recursive feature elimination with applications in natural language processing and bioinformatics. In: Sattar, A., Kang, B.-H. (eds.) AI 2006. LNCS (LNAI), vol. 4304, pp. 170–180. Springer, Heidelberg (2006)
13. Kuratowski, K.: Introduction to Set Theory and Topology. PWN, Warszawa (1962)
14. Sternberg, S.: Lectures on Differential Geometry. Prentice-Hall, N.J. (1964)

Multiclass Boosting Algorithms for Shrinkage Estimators of Class Probability

Takafumi Kanamori

Nagoya University, Furocho, Chikusaku, Nagoya 464-8603, Japan
kanamori@is.nagoya-u.ac.jp

Abstract. Our purpose is to estimate conditional probabilities of output labels in multiclass classification problems. Adaboost provides highly accurate classifiers and has potential to estimate conditional probabilities. However, the conditional probability estimated by Adaboost tends to overfit to training samples. We propose loss functions for boosting that provide shrinkage estimator. The effect of regularization is realized by shrinkage of probabilities toward the uniform distribution. Numerical experiments indicate that boosting algorithms based on proposed loss functions show significantly better results than existing boosting algorithms for estimation of conditional probabilities.

1   Introduction

Over the past two decades, statistical learning methods have been highly developed. Boosting [1] is one of the most significant achievements in machine learning. By applying boosting to a so-called weak learner such as decision trees, one will obtain accurate classifiers or decision functions. That is, boosting is regarded as a meta-learning algorithm. Friedman *et al.* [2] pointed out that boosting algorithms are derived from coordinate descent methods for loss functions. Friedman *et al.*'s work also clarified that boosting has the potential to estimate conditional probabilities. Indeed, there is the correspondence between decision functions and conditional probabilities. Thus, the correspondence provides an estimator of probabilities based on estimated decision functions given by boosting algorithm. In practice, however, common boosting algorithms, such as Adaboost [1] or Logitboost [2], do not provide reliable estimator of conditional probabilities.

In this paper, we propose loss functions for multiclass boosting algorithms that provide more accurate estimate of conditional probabilities than Adaboost or Logitboost. The estimation accuracy is measured by the cross-entropy, or Kullback-Leibler divergence [12], on the test samples. Loss functions inducing a shrinkage estimator have a significant role in our methods.

Key ideas of boosting algorithm for conditional probability estimate are summarized as follows: (i) by using appropriate loss function, one has a shrinkage estimator of conditional probability that will reduce variance of estimation, (ii) by slowing down learning process of boosting, one can look closely into estimates in learning process.

We introduce boosting algorithm from the viewpoint of optimization according to Friedman *et al.* [2]. First, some notations are defined. Let $\mathcal{D} =$

M. Hutter, R.A. Servedio, and E. Takimoto (Eds.): ALT 2007, LNAI 4754, pp. 358–372, 2007.
© Springer-Verlag Berlin Heidelberg 2007

Boosting based on Loss Function L:
Input: Training samples, $\mathcal{D}$. Initialize $f^{(0)} = 0 \in \mathcal{F}$.
For $m = 1, \ldots, M$:

- Find a weak hypothesis such as

$$h^{(m)} = \arg\min_{h \in \mathcal{H}} \frac{\partial}{\partial \alpha} L(\mathcal{D}, f^{(m-1)} + \alpha h)\big|_{\alpha=0}.$$

- Find a coefficient $\alpha^{(m)} \geq 0$ attaining the minimum value of $L(\mathcal{D}, f^{(m-1)} + \alpha h^{(m)})$.
- Update the decision function:
    $f^{(m)} = f^{(m-1)} + \alpha^{(m)} h^{(m)}$.
Output: $f^{(M)}$.

Fig. 1. Boosting based on Loss Function L

$\{(x_1, y_1), \ldots, (x_n, y_n)\}$ be training samples, where x_i is input vector in $\mathcal{X}$ and y_i is output label in $\mathcal{Y} = \{1, \ldots, K\}$. When $K = 2$, the problem is called binary classification, and if $K > 2$, it is called multiclass classification. Training samples are independently and identically distributed from a probability $\mu(x)p(y|x)$, where μ is a marginal distribution on $\mathcal{X}$ and $p(y|x)$ is a conditional probability of output labels such as $\sum_{y \in \mathcal{Y}} p(y|x) = 1$ for any $x \in \mathcal{X}$. Let us define a countable set of functions, $\mathcal{H} = \{h_t : \mathcal{X} \times \mathcal{Y} \to [-1, 1] \mid t \in Z\}$, where Z is a countable index set. Weak learner outputs an element of $\mathcal{H}$ for given training samples. Functions in $\mathcal{H}$ are referred to as weak hypothesis. The set of decision functions defined from $\mathcal{H}$ is given as $\mathcal{F} = \{\sum_{t \in Z} \alpha_t h_t \mid h_t \in \mathcal{H}, \alpha_t \in \mathbb{R}, \sum_{t \in Z} |\alpha_t| < \infty\}$. For a decision function $f \in \mathcal{F}$, the label of input x is predicted by $\hat{y} := \arg\max_{y' \in \mathcal{Y}} f(x, y')$.

The loss function $L(\mathcal{D}; f)$ on $\mathcal{F}$ is used to estimate decision function. For example, in Adaboost [1], $L(\mathcal{D}; f) = \frac{1}{n} \sum_{i=1}^{n} \sum_{y \in \mathcal{Y}} e^{f(x_i, y) - f(x_i, y_i)}$, and in Logitboost, $L(\mathcal{D}; f) = \frac{1}{n} \sum_{i=1}^{n} \log \left(\sum_{y \in \mathcal{Y}} e^{f(x_i, y) - f(x_i, y_i)} \right)$. The intuition behind these loss functions is that the resulting optimization formulation favors small values $f(x_i, y) - f(x_i, y_i)$ for all $y \neq y_i$. Therefore, it favors a decision function such that $f(x_i, y_i) = \arg\max_{y \in \mathcal{Y}} f(x_i, y)$. Loss functions of common boosting algorithms are shown in [2].

Boosting algorithm shown in Fig 1 searches for an approximate minimum solution of L over $\mathcal{F}$. Common learning algoriths such as decision trees are avairable to find weak hypothesis $h^{(m)}$. In Adaboost algorithm for multiclass classification problems, the minimization of $\frac{\partial}{\partial \alpha} L(\mathcal{D}, f + \alpha h)\big|_{\alpha=0}$ with respect to $h \in \mathcal{H}$ is done by the learning algorithms that minimize pseudo-loss function [1]. Easier implementation of finding hypothesis is error correcting output coding (ECOC) [3,4], which will be briefly introduced in the next section.

2   Estimation of Conditional Probabilities

For probability estimation, we propose statistical models and loss functions. Those loss functions are compared with existing ones.

For a decision function $f \in \mathcal{F}$, let us define the conditional probability q_f as $q_f(y|x) = \frac{e^{f(x,y)}}{\sum_{y' \in \mathcal{Y}} e^{f(x,y')}}$, and let $\ell(y|x; f)$ be the negative log-likelihood of q_f, i.e., $\ell(y|x; f) = -\log q_f(y|x) = -f(x,y) + \log \sum_{y' \in \mathcal{Y}} e^{f(x,y')}$.

For a decision function $f \in \mathcal{F}$ and a set of samples $\mathcal{D} = \{(x_i, y_i) \mid i = 1, \ldots, n\}$, let us define the cross entropy $\mathcal{S}(\mathcal{D}; f)$ as $\mathcal{S}(\mathcal{D}; f) = -\frac{1}{n} \sum_{i=1}^{n} \log q_f(y_i|x_i)$. The accuracy of estimated probability q_f is measured by the cross entropy on the set of test samples.

The class of loss functions we consider is defined as

$$L_\phi(\mathcal{D}; f) = \frac{1}{n} \sum_{i=1}^{n} \phi(\ell(y_i|x_i; f)), \tag{1}$$

where $\phi : [0, \infty) \to [0, \infty)$ is an increasing convex function. It is easy to see that L_ϕ is convex in f. In Adaboost, $\phi(z) = e^z$, and in Logitboost, $\phi(z) = z$ is respectively applied.

For binary problems, the loss function L_ϕ is reduced to well-known margin-based loss function [5]. When $\mathcal{Y} = \{1, 2\}$ and $f(x, 1) + f(x, 2) = 0$ holds for all $f \in \mathcal{F}$ and all $x \in \mathcal{X}$, the loss function L_ϕ is represented as $L_\phi(\mathcal{D}; f) = \frac{1}{n} \sum_{i=1}^{n} U(f(x_i, y_i))$, where $U(z) = \phi(\log(1 + e^{-2z}))$. The loss function L_ϕ for multiclass problems is regarded as a direct extension of margin-based loss functions. For binary classification problems, margin-based loss functions have been deeply investigated [5,7,8]. In the present paper, we focus on multiclass problems.

Example 1 (L_ϕ for binary classification). For $\phi(z) = z$, one has $U(z) = \log(1 + e^{-2z})$ that is the loss function of Logitboost for binary problems. For $\phi(z) = e^z - 1$, $U(z)$ is equal to e^{-2z} that is the loss function of binary Adaboost [1,2]. The loss function $U(z) = e^{-z}$ is also used, and it is obtained by $\phi(z) = \sqrt{e^z - 1}$ that is not convex in z. The convexity of $\phi(z)$ in z is a sufficient condition for the convexity of L_ϕ in $f \in \mathcal{F}$, and one can apply non-convex ϕ if L_ϕ becomes convex in f. Applying $\phi(z) = \max\{0, 1 + \frac{1}{2}\log(e^z - 1)\}$, one has $U(z) = \max\{0, 1 - z\}$ that is the hinge loss function for the support vector machine [9,10].

We apply L_ϕ to boosting algorithm in Fig. 1. When ϕ is differentiable, one has

$$\frac{\partial}{\partial \alpha} L_\phi(\mathcal{D}; f + \alpha h)\Big|_{\alpha=0} = \sum_{i=1}^{n} \sum_{y \in \mathcal{Y}} w(i, y)(h(x_i, y) - h(x_i, y_i)), \tag{2}$$

where $w(i, y)$ denotes a weight function depending on ϕ. In Adaboost, the weight function is equal to $w(i, y) = e^{f(x_i, y) - f(x_i, y_i)}/n$, and in Logitboost, $w(i, y) = q_f(y|x_i)/n$.

Weak hypothesis $h \in \mathcal{H}$ minimizing the right-hand of (2) can be sought by the learning technique called error correcting output coding (ECOC) [3]. ECOC is a learning method that uses binary classification algorithms for multiclass classification problems. In the numerical experiments in our paper, we apply boosting with ECOC proposed by Schapire [4] to solve the minimization problem (2) with respect to $h \in \mathcal{H}$.

3   Boosting with L_ϕ

In this section, we apply the loss function L_ϕ to the boosting algorithm. The probability estimator with L_ϕ has shrinkage effect. That is, the deformation of log-likelihood has same role with the regularization that is commonly used in machine learning.

3.1   Learning Algorithm

We propose to apply the loss function L_ϕ for the boosting algorithm. Determining the number of the boosting step is significant to achieve accurate estimation. We apply validation technique for estimation of the number of boosting steps. First, one divides observed samples into training set $\mathcal{D}_{\mathrm{tr}}$ and validation set $\mathcal{D}_{\mathrm{val}}$, and estimate decision function by applying boosting algorithm based on L_ϕ over $\mathcal{D}_{\mathrm{tr}}$. During the boosting process, one computes values of $S(\mathcal{D}_{\mathrm{val}}; f^{(m)})$, $m = 1, \ldots, M$, and determine the number of the iteration, $\hat{m}$, attaining the minimum value of $S(\mathcal{D}_{\mathrm{val}}; f^{(m)})$. Finally, the conditional probability is estimated by $q_{f^{(\hat{m})}}$. When our concern is to estimate accurate classifier, the cross entropy is replaced by the error rate on the validation set. Cross validation can be also applied to estimate the number of boosting steps. Here, we use only one validation set in order to reduce the computational cost.

Additionally, one can apply Platt's scaling method [15] to improve the estimator of conditional probability. Estimated decision function $f^{(\hat{m})}$ is modified to $\hat{c}_0 f^{(\hat{m})} + \hat{c}_1$ so as to attain the minimum value of the cross entropy of $\hat{c}_0 f^{(\hat{m})} + \hat{c}_1$ on $\mathcal{D}_{\mathrm{val}}$. Then, the estimator of conditional probability is given as $q_{\hat{c}_0 f^{(\hat{m})} + \hat{c}_1}$. Note that in Platt's scaling method, one may use slightly modified output labels to avoid overfitting. In numerical experiments of section 5, unmodified validation set is used to estimate c_0 and c_1, because there is no promising criteria for the modification of output labels.

3.2   Shrinkage Effect of L_ϕ

Under some conditions on ϕ, probability estimator given by L_ϕ is regarded as shrinkage estimator that has been intensively studied in statistics.

First, we derive the correspondence between decision functions and conditional probabilities under the loss function L_ϕ. The differential of ϕ is denoted as ϕ', ϕ'', and so forth. When the population distribution is $p(y|x)$, the conditional expectation of the loss function for the conditional probability $q(y|x)$ is given as $\sum_{y \in \mathcal{Y}} \phi(-\log q(y|x)) p(y|x)$. Applying Lagrange multiplier method, one finds that the minimum solution with respect to $q(y|x)$ subject to $\sum_{y \in \mathcal{Y}} q(y|x) = 1$ satisfies

$$p(y|x) \propto \frac{q(y|x)}{\phi'(-\log q(y|x))} \qquad (3)$$

at each $x \in \mathcal{X}$. The proportional relation is defined as the function of y. For Logitboost derived from $\phi(z) = z$, the above correspondence is reduced to

$p(y|x) = q(y|x)$, and for Adaboost given by $\phi(z) = e^z$, the correspondence is given as $p(y|x) = q(y|x)^2 / \sum_{y' \in \mathcal{Y}} q(y'|x)^2$.

We assume that there exists a decision function $f \in \mathcal{F}$ such that q_f satisfies (3) with $q = q_f$. Note that the correspondence between f and q_f is not necessarily biunivoque. To recover the one-to-one correspondence, one can impose a constraint on $\mathcal{F}$ such as $\sum_{y \in \mathcal{Y}} f(x, y) = 0$ for any $f \in \mathcal{F}$ and $x \in \mathcal{X}$. Here, we focus on the relation between the population distribution p and the probability estimator q. Thus, it does not matter whether one-to-one correspondence between f and q_f holds or not, as long as there exists f such that q_f enjoys (3).

We show a proposition on the relation between p and q in (3). The proposition indicates that q is regarded as a shrinkage estimator of p. Remember that the Kullback-Leibler divergence is defined as $\mathrm{KL}(p, q) = \sum_{y \in \mathcal{Y}} p_y \log(p_y/q_y)$. for probability distributions p, q on $\mathcal{Y}$.

Proposition 1. *Let u be the uniform distribution on $\mathcal{Y}$, i.e. $u_y = 1/K$ for all $y \in \mathcal{Y}$. Let g be a function $g : (0, 1) \to (0, \infty)$, and suppose that there exists a positive constant C such that*

$$0 < z \le \frac{1}{K} \Longrightarrow g(z) \le Cz, \quad \frac{1}{K} \le z < 1 \Longrightarrow g(z) \ge Cz. \tag{4}$$

Assume that two probability distributions, p and q, on $\mathcal{Y}$ take positive probabilities on all labels. When $p_y \propto g(q_y)$ holds as the function of y, we have $\mathrm{KL}(u, p) \ge \mathrm{KL}(u, q)$.

Proof. Eq. (4) leads $g(q_y)\left(1 - \frac{1}{Kq_y}\right) \ge C\left(q_y - \frac{1}{K}\right)$ for all $y \in \mathcal{Y}$. Summing over $y \in \mathcal{Y}$, one has inequality, $\sum_{y \in \mathcal{Y}} g(q_y)\left(1 - \frac{1}{Kq_y}\right) \ge 0$. Thus, the positivity of $g(q_y)$ gives $0 < \frac{1}{K \sum_{y' \in \mathcal{Y}} g(q_{y'})} \sum_{y \in \mathcal{Y}} \frac{g(q_y)}{q_y} \le 1$. On the other hand, one has $\mathrm{KL}(u, p) - \mathrm{KL}(u, q) = -\sum_{y \in \mathcal{Y}} \frac{1}{K} \log \frac{g(q_y)}{q_y \sum_{y' \in \mathcal{Y}} g(q_{y'})} \ge -\log \frac{1}{K \sum_{y' \in \mathcal{Y}} g(q_{y'})} \sum_{y \in \mathcal{Y}} \frac{g(q_y)}{q_y}$, due to convexity of the negative-log function. The claim follows these inequalities.

Remark 1. Let g be a convex, strictly increasing function on $(0, 1)$ satisfying $g(+0) = 0$. Then, Eq. (4) holds for $C = Kg(1/K)$, and the ranking of probability is preserved under the transformation of g. Thus, the decision boundary is unchanged.

Corollary 1. *The lower bound in the proof of proposition 1 is denoted as $\beta(g; q) = -\log \frac{1}{K \sum_{y' \in \mathcal{Y}} g(q_{y'})} \sum_{y \in \mathcal{Y}} \frac{g(q_y)}{q_y}$. Let g_1 and g_2 be functions satisfying $0 < z \le \frac{1}{K} \Longrightarrow 0 < g_1(z) \le g_2(z) \le Cz$, $\frac{1}{K} \le z < 1 \Longrightarrow g_1(z) = g_2(z) \ge Cz$. Then, $\beta(g_1; q) \ge \beta(g_2; q)$ holds.*

Applying the similar argument of proposition 1 with the inequality $\sum_{y \in \mathcal{Y}} g_1(q_y) \le \sum_{y \in \mathcal{Y}} g_2(q_y)$, one can prove the corollary. Corollary 1 qualitatively explains the difference of shrinkage effect induced by g.

Proposition 1 provides a sufficient condition such that p is shrunk toward the uniform distribution. Applying boosting algorithm based on the loss function

L_ϕ, one can estimate the conditional probability q_f. If the function $z/\phi'(-\log z)$ in Eq. (3) satisfies the condition in Proposition 1, then, the estimator of q_f is regarded as a shrinkage estimator of p. When the decision function $\hat{f}$ is estimated, the probability which is proportional to $q_{\hat{f}}/\phi'(-\log q_{\hat{f}})$ is an asymptotically unbiased estimator of conditional probability under mild conditions. On the other hand, we adopt $q_{\hat{f}}$ as an estimator of p.

In the statistical decision theory, shrinkage estimators such as Stein's estimator have been deeply investigated [11]. By using estimators shrunk toward the uniform distribution, one will attain variance reduction, and will be able to avoid overfitting to training samples. Thus, shrinkage estimator is regarded as a variant of regularization that is commonly used in machine learning.

3.3 Regularization Introduced by Deformation of Log-Likelihood Function

In the context of machine learning, regularization is commonly incorporated into loss functions as additive form. But, generally additive form does not suit to boosting algorithms. For the loss function with additive regularization term, the differential does not have the form of Eq. (2) in general. Hence, we cannot directly apply ECOC technique to search the weak hypothesis minimizing the differential. Applying our loss function to boosting algorithm, the effect of regularization is easily incorporated without any specific modification of weak learner.

There are some works on regularized boosting. Schapire et al. [6] have introduced additive regularization term for Logitboost. This is an interesting example in which additive regularization term works with boosting. Their algorithm needs some modification of weak learner, since additional training samples are necessary to realize the weighted loss function in Eq. (2). Rätsch *et al.* [7] proposed regularized boosting in which the regularization term is built in loss functions as multiplicative form. On the other hand, in our methods, regularization is incorporated as deformation of log-loss functions, and the deformation leads shrinkage of estimator toward the uniform distribution.

Here, we study two kind of minimization problems to investigate the relation between additive regularization and deformation of loss function.

First one is the minimization of negative log-likelihood function with an additive regularization term such as

$$\min_q -\sum_{y\in\mathcal{Y}} p_y \log q_y + \Lambda(q), \quad \text{s.t.} \quad q \in \Delta_K^\circ \tag{5}$$

The additive term $\Lambda(q)$ is introduced for regularization, and Δ_K° denotes the interior of the probability simplex in $\mathbb{R}^K$. We assume that Λ is defined on an open set including Δ_K°. Applying the Lagrange multiplier methods to (5), we find that the optimal solution $q \in \Delta_K^\circ$ satisfies

$$p_y = q_y \left(\frac{\partial \Lambda}{\partial q_y}(q) + \lambda \right), \quad y \in \mathcal{Y}, \tag{6}$$

where λ is the Lagrange multiplier given by $\lambda = 1 - \sum_{y\in\mathcal{Y}} q_y \frac{\partial\Lambda}{\partial q_y}(q)$.

The second problem is the minimization of L_ϕ under the probability distribution p,

$$\min_q \sum_{y\in\mathcal{Y}} p_y \phi(-\log q_y), \quad \text{s.t.} \quad q \in \Delta_K^\circ \tag{7}$$

As shown in Eq. (3), the optimal solution $q \in \Delta_K^\circ$ satisfies

$$p_y = \mu \frac{q_y}{\phi'(-\log q_y)}, \quad y \in \mathcal{Y}, \tag{8}$$

where μ is the Lagrange multiplier defined as $\mu = (\sum_{y\in\mathcal{Y}} q_y/\phi'(-\log q_y))^{-1}$. Let $g_\phi(z)$ be $z/\phi'(-\log z)$, then $p_y \propto g_\phi(q_y)$ holds.

If the equality

$$q_y\left(\frac{\partial\Lambda}{\partial q_y}(q) + \lambda\right) = \mu \frac{q_y}{\phi'(-\log q_y)}, \quad y \in \mathcal{Y} \tag{9}$$

is satisfied for any $q \in \Delta_K^\circ$, the optimality conditions (6) and (8) holds simultaneously for common distribution p. This denotes that the optimal solution of (5) is identical to that of (7). We show some examples of Λ and ϕ satisfying (9).

Example 2. Let $\phi_s(z)$ be $(e^{sz} - 1)/s$ for $s \geq 0$, where $\phi_0(z) = z$. In Adaboost $\phi_1(z)$ is used. We define $\Lambda_s(q)$ as $\Lambda_s(q) = \log\|q\|_{s+1}$, where $\|q\|_{s+1}$ denotes $(s + 1)$-norm of $q \in \mathbb{R}^K$. For $q \in \Delta_K$, $\Lambda_0(q)$ is equal to zero, and thus, the regularization term Λ_0 does not lead shrinkage effect to the estimator. For any $s \geq 0$, $\phi = \phi_s$ and $\Lambda = \Lambda_s$ satisfy (9). Since $g_{\phi_s}(z) = z/\phi_s'(-\log z) = z^{s+1}$, the relation between p and q is given as $q_y \propto \exp\left\{\frac{1}{1+s}\log p_y + (1 - \frac{1}{1+s})\log u_y\right\}$. Hence, the estimator q is given by shifting p toward the uniform distribution u along the e-geodesic [12] on the set of multinomial distributions.

Example 3. Let $\phi_c(z)$ be $-\log(e^{-z} - \frac{1}{K(1+1/c)})$, and $\Lambda_c(q)$ be $c\,\mathrm{KL}(u, q)$ for $c \geq 0$, where u is the uniform distribution on $\mathcal{Y}$. The regularization term has exactly same effect as the Dirichlet prior with parameters taking a common value. Note that the Dirichlet prior is the conjugate prior of the multinomial distribution. We regard $\mathrm{KL}(u, q)$ as the function on $\{(q_1, \ldots, q_K) \mid q_y > 0, y \in \mathcal{Y}\}$. We define $\phi_0(z)$ as z. For any $c \geq 0$, $\phi = \phi_c$ and $\Lambda = \Lambda_c$ satisfy (9), and the proportional relation, $p_y \propto g_\phi(q_y)$, holds, where $g_{\phi_c}(z) = z/\phi_c'(-\log z) = z - \frac{1}{K(1+1/c)}$. Although the domain of g_{ϕ_c} is $(1/K(1 + 1/c), 1)$, proposition 1 is applicable with some modification. We find that the minimizer of (7) is given as $q_y = \frac{1}{c+1}p_y + (1 - \frac{1}{c+1})u_y$, and hence, q is given by shifting p toward the uniform distribution u along the m-geodesic [12].

In general, the intensity of regularization is controlled by the regularization parameter such as c in example 3. In our methods, one can control the intensity of regularization according to corollary 1. If there is simple correspondence between ϕ and Λ as shown in example 2 or 3, one can apply the loss function L_ϕ

to boosting algorithm in order to impose the regularization effect same as Λ. From example 2 and empirical studies in section 4 and 5, the order of $g_\phi(\varepsilon)$ for infinitesimal ε seems to be main factor to determine the intensity of regularization.

We can apply the deformation of negative log-likelihood not only to regularization but also to incorporation of the prior knowledge. Schapire et al. [6] incorporated prior knowledge into boosting algorithm by adding a modified regularization term to log-likelihood function. Here, we introduce a different approach from [6]. Applying the deformation function ϕ depending on the class label, one will obtain the estimator that is shifted to any specified class probability.

We show some examples in which the function ϕ depends on class labels. Let $r \in \Delta_K^\circ$ be a probability distribution on $\mathcal{Y}$. If we use the function $\phi_s(z, y) = ((r_y e^z)^s - 1)/s$ for the deformation of the negative log-likelihood, the minimizer of the loss function $\sum_{y \in \mathcal{Y}} p_y \phi_s(-\log q(y|x), y)$ with respect to $q \in \Delta_K^\circ$ satisfies $q_y \propto \exp\left\{\frac{1}{1+s} \log p_y + (1 - \frac{1}{1+s}) \log r_y\right\}$. Hence, the estimator q is regarded as a shrinkage estimator toward the distribution r. The corresponding regularization term is given as $\Lambda_s(q) = \frac{1}{s+1} \log \sum_{y \in \mathcal{Y}} q_y^{s+1}/r_y^s$. The second example is a modification of example 3. Let $\phi_c(z, y)$ be $-\log(e^{-z} - \frac{r_y}{1+1/c})$. The minimizer of the loss function $\sum_{y \in \mathcal{Y}} p_y \phi_c(-\log q(y|x), y)$ satisfies $q_y = \frac{1}{1+c} p_y + (1 - \frac{1}{1+c}) r_y$. The corresponding regularization term is given as $\Lambda_c(q) = c \, \mathrm{KL}(r, q)$. The regularization term $\Lambda_c(q)$ has same role as the Dirichlet prior with the parameter proportional to r. When the functions such as ϕ_s or ϕ_c are applied to boosting algorithm, the regularization parameter, s or c, may depend on class label $y \in \mathcal{Y}$ and input variable $x \in \mathcal{X}$ in order to incorporate more detailed prior knowledge.

4   Overfitting in Boosting Process

As shown in section 2, one can transform the estimated decision function f to the estimator of conditional probability q_f. Thus, Adaboost or Logitboost has potential to estimate conditional probabilities via decision functions. In practice, however, probability estimation by Adaboost or Logitboost tends to overfit to training samples, even though estimated decision boundary provides highly accurate classification rule [13]. We intuitively explain the reason that estimated probability overfits to training samples, and propose ϕ that tends to avoid overfitting.

In the boosting process, training error rapidly decreases to zero, and as the result, most of $\ell(y_i|x_i; \hat{f})$ take nearly zero, where $\hat{f}$ is an estimated decision function. To analyze the behavior of boosting algorithm, we introduce a naive assumption such that all of negative-likelihoods on training samples, $\ell(y_i|x_i; \hat{f})$, take common infinitesimal value, ε. Note that this assumption is not totally correct, because in some numerical experiments the mean value of $\ell(y_i|x_i; \hat{f})$ on training samples has almost same order of its deviation. However, the assumption is very useful for rough understanding of boosting process. In the last part of this section, we study the learning process without the naive assumption.

In the line search of boosting algorithm, the coefficient $\alpha \in \mathbb{R}$ minimizing $\sum_{i=1}^{n} \phi(\ell(y_i|x_i; \hat{f} + \alpha h))$ is sought, where h is a weak hypothesis. Let ℓ_i, $\partial \ell_i$, and $\partial^2 \ell_i$ be $\ell(y_i|x_i; \hat{f})$, $\frac{\partial}{\partial \alpha} \ell(y_i|x_i; \hat{f} + \alpha h)\big|_{\alpha=0}$, and $\frac{\partial^2}{\partial \alpha^2} \ell(y_i|x_i; \hat{f} + \alpha h)\big|_{\alpha=0}$, respectively. We approximate the loss function by quadratic function, and obtain an approximate solution of line search as

$$\alpha = -\frac{\sum_{i=1}^{n} \phi'(\ell_i)\partial \ell_i}{\sum_{i=1}^{n} \phi''(\ell_i)(\partial \ell_i)^2 + \sum_{i=1}^{n} \phi'(\ell_i)\partial^2 \ell_i}. \tag{10}$$

From the assumption of $q_{\hat{f}}(y_i|x_i) = e^{-\varepsilon} = 1 - \varepsilon + o(\varepsilon)$, one has $\partial \ell_i = \sum_{y \in \mathcal{Y}} q_{\hat{f}}(y|x_i)h(x_i, y) - h(x_i, y_i) = O(\varepsilon)$, and in the same way, one also has $\partial^2 \ell_i = O(\varepsilon)$. Thus, derivatives are represented as $\partial \ell_i = a_i \varepsilon + o(\varepsilon)$, $\partial^2 \ell_i = b_i \varepsilon + o(\varepsilon)$ by some constants, a_i and b_i. As the result, the solution of the line search is given as

$$\alpha = \frac{C_0 + o(1)}{C_1 + \phi''(\varepsilon)\varepsilon/\phi'(\varepsilon)}, \tag{11}$$

where C_0 and C_1 are constants depending on $a_1, \ldots, a_n$ and $b_1, \ldots, b_n$.

The derivative value of ϕ in a vicinity of zero affects the order of α. For Adaboost with $\phi(z) = e^z$, one has $\phi''(\varepsilon)\varepsilon/\phi'(\varepsilon) = \varepsilon$, and $\alpha = C_0/C_1 + o(1)$. This order is same as that of Logitboost with $\phi(z) = z$. For the polynomial function $\phi(z) = z^\gamma$, one has $\alpha = C_0/(C_1 + \gamma - 1) + o(1)$. Although the order is $O(1)$ in common with Adaboost and Logitboost, polynomial functions with high degree will reduce α to some extent. If $\phi(z)$ is defined as $e^{-1/z}$ in a vicinity of zero, one has $\alpha = C_0 \varepsilon + o(\varepsilon)$ that is of the order less than $O(1)$.

Numerical experiments on several loss functions are illustrated in Figure 2. The figure shows $\alpha^{(m)}$ and the cross entropy on test set in boosting process. Here, $\xi(z)$ is the function defined as $e^{-10/z}$ for $0 < z \leq 5$ and $az + b$ for $5 \leq z$, where a and b are constants such that $\xi(z)$ is differentiable at $z = 5$. Training samples are generated from a logistic model with input space of dimension two, and three output labels. Rpart [14] with max depth two, i.e. decision stumps, is used as weak learner. In the process of Adaboost, $\alpha^{(m)}$ takes almost constant values. The same thing is said of boosting based on $\phi(z) = z^4$, but values are smaller than those of Adaboost. For $\phi(z) = \xi(z)$, $\alpha^{(m)}$ decreases as boosting algorithm proceeds. These observations are consistent with the analysis given above.

We have rough understanding about the behavior of boosting. Applying the function ϕ such that $\lim_{\varepsilon \searrow 0} \phi''(\varepsilon)\varepsilon/\phi'(\varepsilon)$ takes large value or diverges to infinity, one obtains a learning algorithm with short step size in line search. From the viewpoint of estimation, short step size in the learning process is helpful to determine appropriate boosting step, m, in boosting algorithm. In other words, estimators given by Adaboost or Logitboost soon pass by the appropriate conditional probability in the learning process. This tendency is illustrated in right panel of Figure 2. The cross entropy on the test set is plotted to the number of boosting step.

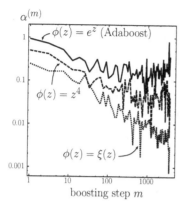
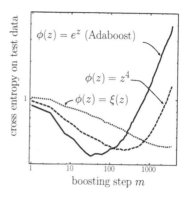

Fig. 2. Left figure: Double logarithmic plot of $\alpha^{(m)}$ in boosting process for $\phi(z) = z$, z^4, $\xi(z)$, respectively. Right figure: Double logarithmic plot of cross entropy on test set (risk) in boosting process for each loss function.

Under more general assumption, we derive an upper bound of $|\alpha|$. Here, we do not assume that all ℓ_i, $i = 1, \ldots, n$ take common value. Let ℓ_1 be the maximum value of ℓ_i, *i.e.* $\ell_1 = \max_i \ell_i$. Note that all ℓ_i's are non-negative, and that $\partial^2 \ell_i$ takes positive value in general. We assume that ϕ is an increasing and strictly convex function. Thus, ϕ' and ϕ'' takes positive value, and ϕ' is an increasing function. As the result, Eq. (10) provides the inequality, $|\alpha| \leq \frac{n \max_i |\partial \ell_i| \, \phi'(\ell_1)}{\phi''(\ell_1)(\partial \ell_1)^2 + \phi'(\ell_1)\partial^2 \ell_1} = \frac{n \max_i |\partial \ell_i|}{\frac{\phi''(\ell_1)}{\phi'(\ell_1)}(\partial \ell_1)^2 + \partial^2 \ell_1}$. That is, the upper bound is governed by the order of ϕ''/ϕ' in a vicinity of zero. When the order of $\partial \ell_i$ and $\partial^2 \ell_i$ is $O(\varepsilon)$, one has a formula like (11) as an upper bound of $|\alpha|$.

5   Numerical Results

We examine some loss functions on benchmark data. We use data sets in the "mlbench" of the R library [16]. Multiclass data sets in mlbench are Glass, LetterRecognition, Satellite, Vehicle, Vowel, DNA, Shuttle, and Soybean. We use first five data sets. The input dimension of DNA is 180, and thus, some dimension reduction techniques will be needed. In the data set of Shuttle, the number of training samples in each label varies greatly, and Soybean includes lots of missing data. These are main reason that we omit those data sets from our numerical experiments. The data set named Synthetic is generated from a logistic model. Each data set in mlbench is split up into training set, validation set and test samples. LetterRecognition and Satellite contain a large amount of data, and we use a part of them. For each data set, we repeat numerical experiments 20-50 times with different random splits of the data in order to evaluate the generalization performance of each learning algorithm. The properties of each data sets are shown in Table 1, where "dim", "class", "tr.", "val.", "test" and "rep." denote the input dimension, the number of labels, the size of

Table 1. Properties of each data set

name	dim	class	tr.	val.	test	rep.
Glass	9	6	121	53	40	50
Letter	16	26	500	200	200	20
Satellite	36	6	700	200	100	20
Vehicle	18	4	487	209	150	50
Vowel	10	11	553	237	200	30
Synthetic	2	3	300	100	1000	50

training set, the size of validation set, the size of test set, and the number of replication of learning, respectively.

Table 2 shows the results of comparison among loss functions. The function $\xi(z)$ is defined in Section 4. For all $\phi(z)$ in experiments, the function $g_\phi(z) = z/\phi'(-\log z)$ satisfies the condition of Remark 1. In numerical experiments, the decision tree algorithm called "Rpart" for binary problems [14] with error correcting output coding [4] is applied as the weak learner. Besides boosting algorithms, we examine support vector machines (svm) [9], and linear-logistic regression (logit) [17]. As the kernel function for svm, radial basis function (rbf) and linear function are used. To estimate conditional probability by svm, we apply pairwise coupling technique with svm [18]. The svm with pairwise coupling is implemented in the library of kernlab [19], and we use kernlab for the numerical experiments. In kernlab library, the hyper-parameter of the rbf kernel, *i.e.* bandwidth σ, is estimated by the mean value of 10% and 90% quantile of $\|x_i - x_j\|, i, j = 1, \ldots, n$, and the value of C in C-SVM is set to 1 as default.

The values of "err" in the table 2 denotes the test error (%) of estimated decision functions. The cross entropy on the test set for the estimated conditional probability without (with) Platt's scaling is denoted by "risk" ("risk$_p$"). Significant difference of test error or cross entropy is decided by the p-values of the corrected resampled t-test [20] against Adaboost. Here, the statistic proposed in [20] takes into account the correlation of resampling methods, and depresses the type I error of the statistical test. Test error, risk, and risk$_p$ of each loss function is compared with those of Adaboost.

When our purpose is to construct accurate classifier, there is no dominating loss function. On some data sets, the test error of $\xi(z)$ is larger than that of the other loss functions .

On the other hand, on estimation of conditional probabilities, polynomial functions and $\xi(z)$ work better than the others. Especially, $\phi(z) = z^6$ performs significantly better than $\phi(z) = e^z$ in present numerical experiments. The Logitboost with $\phi(z) = z$ does not perform well, even though the loss function leads log-likelihood estimator for parametric statistical models. In practical situations, the dimension of $\mathcal{F}$ is extremely high and thus, estimators given by $\phi(z) = z$ will soon overfit to training samples without appropriate regularization. We see that Platt's scaling method reduces the risk value. But, polynomial $\phi(z)$ and $\xi(z)$ still have advantage in comparison to Adaboost or Logitboost. Due to shrinkage

Table 2. Comparison of loss functions, L_ϕ. We examine $\phi(z) = e^z, z, z^2, z^4, z^6$, and $\xi(z)$. Test errors are shown in %. Significant difference is decided by the p-values of 0.01 under the corrected resampled paired t-test. Dots • (crosses ×) on the left of the numbers denote that the performance of estimator given by the loss function is 1% significantly better (worse) than that given by Adaboost.

$\phi(z)$		Glass	Letter.	Sate.	Vehicle	Vowel	Synth.
e^z	err.	27.6	24.8	12.5	23.2	9.65	6.40
	risk	1.10	1.65	0.58	0.65	0.81	0.29
	risk$_p$	0.98	1.46	0.52	0.64	0.52	0.23
z	err.	26.9	24.9	11.3	22.7	9.90	6.56
	risk	1.41	×2.26	×0.84	×0.79	0.94	×0.35
	risk$_p$	1.11	1.56	×0.70	0.73	0.50	×0.26
z^2	err.	28.8	23.4	12.4	23.6	10.2	6.40
	risk	1.16	1.70	0.61	0.67	0.68	0.24
	risk$_p$	1.01	1.28	0.53	0.63	0.45	0.22
z^4	err.	29.5	22.5	11.9	24.4	9.45	6.52
	risk	0.97	•1.34	0.49	•0.57	•0.48	•0.21
	risk$_p$	0.90	•1.12	0.45	•0.56	0.37	•0.20
z^6	err.	27.8	22.6	11.6	24.5	9.17	6.42
	risk	0.88	•1.17	•0.44	•0.56	•0.41	•0.20
	risk$_p$	0.85	•1.04	•0.42	•0.55	•0.34	•0.19
$\xi(z)$	err.	30.3	×31.3	13.6	25.9	×16.4	6.61
	risk	0.86	•1.26	•0.38	•0.55	•0.49	•0.19
	risk$_p$	0.86	1.23	•0.37	•0.54	0.49	•0.19
svm	err.	32.6	×30.7	14.4	25.3	×16.0	•4.47
(rbf)	risk	×1.91	×2.67	×0.76	×1.16	×2.26	•0.11
svm	err.	×37.6	23.6	14.3	19.5	×30.3	×21.0
(linear)	risk	×1.86	×2.54	×1.19	×0.95	×4.37	×0.73
logit	err.	37.2	×32.5	×19.7	19.8	×42.1	×21.0
	risk	2.34	2.30	×2.24	0.52	×1.17	×0.42

of probability distribution toward the uniform distribution, one will be able to stabilize estimators.

As a whole, polynomial loss functions provide better estimate of conditional probabilities than Adaboost or Logitboost, and the test error is competitive to existing boosting algorithms. Especially, $\phi(z) = z^6$ provides accurate estimation of conditional probabilities in the present experiments. The loss function $\xi(z)$ also provides good estimate of conditional probabilities, while the classification error rate is larger than the other methods on some data sets. This is because the number of boosting step M in the algorithm is not enough to achieve low test error rate, when we apply $\xi(z)$ as the loss function.

The table 3 and 4 show the number of boosting steps, $\widehat{m}$, that is estimated by the validation set. Table 3 indicates the average number of $\widehat{m}$ with standard deviation for the estimation of decision boundary, and table 4 does that of conditional probability. For the estimation of decision boundary, the error rate

Table 3. The number of boosting steps for the estimate of decision boundary

$\phi(z)$		Glass	Letter.	Sate.	Vehicle	Vowel	Synth.
e^z	ave.	125.9	1112.5	371.2	125.9	1093.8	42.4
	s.d.	99.5	644.8	284.6	99.5	587.5	54.1
z	ave.	115.8	1315.3	371.2	115.8	1046.8	32.3
	s.d.	87.6	641.2	257.9	87.6	550.5	42.4
z^2	ave.	121.2	430.0	237.7	121.2	375.7	32.7
	s.d.	97.1	185.4	212.8	97.1	239.6	41.9
z^4	ave.	128.4	597.2	298.7	128.4	330.2	31.1
	s.d.	113.3	321.6	180.0	113.3	120.2	45.2
z^6	ave.	124.0	744.3	280.0	124.0	534.7	31.5
	s.d.	107.7	140.1	193.9	107.7	169.9	50.0
$\xi(z)$	ave.	134.1	1654.9	522.3	134.1	1257.6	24.1
	s.d.	106.2	293.0	305.9	106.2	278.1	40.8

Table 4. The number of boosting steps for the estimate of conditional probability

$\phi(z)$		Glass	Letter.	Sate.	Vehicle	Vowel	Synth.
e^z	ave.	14.0	47.3	12.0	7.4	67.9	11.1
	s.d.	5.7	11.2	5.1	2.7	33.2	16.6
z	ave.	10.4	49.2	7.1	4.6	63.3	8.5
	s.d.	7.8	24.5	4.4	3.1	29.0	7.6
z^2	ave.	18.4	85.8	14.7	12.1	92.0	11.1
	s.d.	23.6	20.5	6.4	6.4	33.7	4.8
z^4	ave.	27.5	142.7	28.1	23.9	143.4	21.5
	s.d.	14.2	45.3	7.2	6.4	49.3	9.6
z^6	ave.	41.4	163.7	39.7	40.2	181.6	35.0
	s.d.	19.3	44.9	11.2	12.6	53.4	15.6
$\xi(z)$	ave.	180.5	1998.6	626.9	198.5	1935.4	258.4
	s.d.	120.6	2.3	258.1	135.0	61.3	61.1

over the validation set $\mathcal{D}_{\text{val}}$ is applied to determine $\widehat{m}$, and for the estimation of conditional probability, the cross entropy over $\mathcal{D}_{\text{val}}$ is used.

In table 3, boosting with $\xi(z)$ needs more number of boosting steps than the other estimators. Because, boosting with $\xi(z)$ takes small step size $\alpha^{(m)}$, and the convergence speed is slow as illustrated in the figure 2. Adaboost ($\phi(z) = e^z$) and Logitboost ($\phi(z) = z$) also need relatively large number of boosting steps, especially in LetterRecognition and Vowel. Some works have pointed out that Adaboost is hard to overfit to training samples for the estimation of decision boundary [21], and that large number of boosting steps does not considerably degrade the prediction accuracy of output labels. We think that this is a reason that Adaboost takes relatively large number of boosting steps in Table 3. Logitboost indicates similar tendency with Adaboost.

In table 4, as a whole, the number of boosting steps is much smaller than that for the estimation of decision boundary. As illustrated in section 4, Adaboost takes small number of $\hat{m}$, while boosting with polynomial function or $\xi(z)$ needs larger $\hat{m}$. The maximum number of boosting steps is limited to $M = 2000$, and for $\xi(z)$ the boosting step reaches almost maximum value on LetterRecognition and Vowel. In Adaboost or Logitboost, $\alpha^{(m)}$ takes larger value than the other boosting algorithms examined in the numerical experiments. As the result, detailed adjustment of the probability is not attained while the number of boosting steps is small. In general, probability estimate is more difficult than prediction of output labels. Thus, careful adjustment of estimator in learning process will be necessity to achieve highly accurate estimation of probability distribution.

6 Conclusion

We proposed loss functions that provide shrinkage estimators of conditional probabilities. In numerical experiments, proposed methods work better than existing boosting algorithms for conditional probability estimation. An important future work is to study the relation between estimation of classifiers and that of conditional probabilities. Empirical choice of $\phi(z)$ is also an important future work. The correspondence between the deformation of log-likelihood and additive regularization term will provide an appropriate way of determining the function ϕ.

References

1. Freund, Y., Schapire, R.E.: A decision-theoretic generalization of on-line learning and an application to boosting. Journal of Computer and System Sciences 55(1), 119–139 (1997)
2. Friedman, J.H., Hastie, T., Tibshirani, R.: Additive logistic regression: A statistical view of boosting. Annals of Statistics 28, 337–407 (2000)
3. Dietterich, T.G., Bakiri, G.: Solving multiclass learning problems via error-correcting output codes. Journal of Artificial Intelligence Research 2, 263–286 (1995)
4. Schapire, R.E.: Using output codes to boost multiclass learning problems. In: Proc. 14th International Conference on Machine Learning, pp. 313–321. Morgan Kaufmann, San Francisco (1997)
5. Bartlett, P.L., Jordan, M.I., McAuliffe, J.D.: Convexity, classification, and risk bounds. Technical Report 638, Statistics Department, University of California, Berkeley (November 2003)
6. Schapire, R.E., Rochery, M., Rahim, M., Guputa, N.: Incorporating Prior Knowledge into Boosting. In: Proc. 19th International Conference on Machine Learning, pp. 538–545. Morgan Kaufmann, San Francisco (2002)
7. Rätsch, G., Onoda, T., Müller, K.R.: Soft margins for adaboost. Machine Learning 42(3), 287–320 (2001)
8. Zhang, T.: Statistical behavior and consistency of classification methods based on convex risk minimization. Annals of Statistics 32(1), 56–85 (2004)
9. Cortes, C., Vapnik, V.: Support-vector networks. Machine Learning 20, 273–297 (1995)
10. Vapnik, V.: Statistical Learning Theory. Wiley, Chichester, UK (1998)

11. Gruber, M.: Improving Efficiency by Shrinkage. Marcel Dekker (1998)
12. Amari, S., Nagaoka, H.: Methods of Information Geometry. Oxford University Press, Oxford (2000)
13. Niculescu-Mizil, A., Caruana, R.: Obtaining calibrated probabilities from boosting. In: UAI '05. Proc. 21st Conference on Uncertainty in Artificial Intelligence, AUAI Press (2005)
14. Therneau, T.M., Atkinson, B.: Rpart: Recursive Partitioning, R package version 3, pp. 1-23 (2005)
15. Platt, J.: Probabilistic outputs for support vector machines and comparison to regularized likelihood methods. Advance in Large Margin Classifiers, 61–74 (2000)
16. Leisch, F., Dimitriadou, E.: mlbench: Machine Learning Benchmark Problems, Original data sets from various sources., R package version 0.5-8 (2003)
17. MacCullagh, P.A., Nelder, J.: Generalized Linear Models. Chapman and Hall, London (1989)
18. Wu, T.F., Lin, C.J., Weng, R.C.: Probability Estimates for Multi-class Classification by Pairwise Coupling. Journal of Machine Learning Research 5, 975–1005 (2004)
19. Karatzoglou, A., Smola, A., Hornik, K., Zeileis, A.: kernlab: Kernel Methods Lab., R package version 0.4-2 (2004)
20. Nadeau, C., Bengio, Y.: Inference for the generalization error. Machine Learning 52(3), 239–281 (2003)
21. Drucker, H., Cortes, C.: Boosting decision trees. Advances in Neural Information Processing Systems 8, 479–485 (1996)

Pseudometrics for State Aggregation in Average Reward Markov Decision Processes

Ronald Ortner

University of Leoben, A-8700 Leoben, Austria
ronald.ortner@unileoben.ac.at

Abstract. We consider how state similarity in average reward Markov decision processes (MDPs) may be described by pseudometrics. Introducing the notion of *adequate* pseudometrics which are well adapted to the structure of the MDP, we show how these may be used for state aggregation. Upper bounds on the loss that may be caused by working on the aggregated instead of the original MDP are given and compared to the bounds that have been achieved for discounted reward MDPs.

1 Introduction

Most work done in hierarchical reinforcement learning, relational reinforcement learning, function approximation, factorization and state aggregation ultimately addresses the problem of how to deal with large state spaces in Markov decision processes (MDPs). Here we are concerned with *state aggregation* (for references see [1]), which tries to convert the idea that similar states (with respect to rewards and transition probabilities) may be aggregated to meta-states, and calculation of the optimal policy may then be conducted on the meta-MDP.

For discounted reward MDPs, upper bounds on the loss that may be caused by aggregation have been obtained by Even-Dar and Mansour [2] and more recently by Ferns et al. [3]. We are particularly interested in the latter work, as it has introduced the idea that state similarity may be described by pseudometrics. Here we try to extend this approach, first by giving a general definition of *adequate* metrics which are useful for state aggregation, and secondly by generalizing the results of [3] and [2] to average reward MDPs.

The paper is organized as follows. After preliminary definitions in Sect. 2, we show in Sect. 3 how to conduct state aggregation with respect to a given metric. We consider a very simple distance function d_v and give an upper bound on the loss by state aggregation with respect to d_v. Then in Sect. 4, we generally define *adequate* distance functions and generalize the results accordingly. In Sect. 5, we compare our bounds to those obtained in the discounted case and show why the loss by aggregation may be significantly larger for average reward MDPs. In the final section, we consider basic questions on the possibility of online aggregation and other open problems for future research.

M. Hutter, R.A. Servedio, and E. Takimoto (Eds.): ALT 2007, LNAI 4754, pp. 373–387, 2007.

2   Preliminaries

Definition 1. *A Markov decision process* (MDP) $\mathcal{M} = \langle S, A, \mu_0, p, r \rangle$ *consists of **(i)** a finite set of states S with **(ii)** a finite set of actions A available in each state $\in S$, **(iii)** an initial distribution μ_0 over S, **(iv)** the transition probabilities $p_a(s, s')$ which give the probability of reaching state s' when choosing action a in state s, and **(v)** the payoff distributions with mean $r_a(s)$ and support in $[0, 1]$ that specify the random reward obtained for choosing action a in state s.*

A *policy* on an MDP $\mathcal{M}$ is a mapping $\pi : S \to A$. Note that each policy π induces a Markov chain $\mathcal{M}_\pi$ on $\mathcal{M}$. We will only consider *ergodic* MDPs, where all policies induce ergodic Markov chains (in which states are reachable from each other after a finite number of steps). For a policy π let μ_π be the *stationary distribution* of $\mathcal{M}_\pi$. Remember that for ergodic Markov chains with probability matrix P this is the unique distribution μ with $\mu P = \mu$ (cf. e.g. [4]). The *average reward of* π then may be defined as

$$\rho_\pi(\mathcal{M}) := \sum_{s \in S} \mu_\pi(s) \, r_{\pi(s)}(s).$$

A policy π^* is *optimal on* $\mathcal{M}$, if $\rho_\pi(\mathcal{M}) \leq \rho_{\pi^*}(\mathcal{M}) =: \rho^*$ for all policies π. As ρ_π is independent of the initial distribution μ_0, in the following we ignore μ_0 and write MDPs as tuples $\mathcal{M} = \langle S, A, p, r \rangle$.

Definition 2. *Given a set X and a nonnegative function $d : X \times X \to \mathbb{R}$, we call (X, d) a pseudometric space with pseudometric d, if for all $x, y, z \in X$,*

$$\begin{aligned} &(i) \quad d(x, x) = 0, \\ &(ii) \quad d(x, y) = d(y, x), \\ &(iii) \quad d(x, y) + d(y, z) \geq d(x, z). \end{aligned}$$

In general, for d being a *metric* on X it is additionally demanded that $d(x, y) = 0$ implies $x = y$. As we will consider pseudometrics on state spaces of MDPs, this is obviously not a desired property (i.e., we want to include the possibility of having distinct states with equal properties).

Definition 3. *Given a Markov chain $\mathcal{C}$ with state space S and stationary distribution μ, its* mixing time *with respect to state s is defined as*

$$\kappa_s := \sum_{s' \in S} m_{ss'} \mu(s'),$$

where $m_{ss'}$ is the mean first passage time from s to s' if $s \neq s'$, while m_{ss} is the mean return time to s. It can be shown that κ_s is independent of s (see [5]), so that we may speak of the the mixing time *of $\mathcal{C}$, denoted by $\kappa_\mathcal{C}$.*

3   A Simple Pseudometric for State Similarity

3.1   Block MDPs

Definition 4. *An MDP* $\mathcal{M} = \langle S, A, p, r \rangle$ *is a* block MDP *with blocks* $S_1, \ldots, S_k$, *if the block set* $\{S_1, \ldots, S_k\}$ *is a partition of* S, *and for all* $a \in A$, *all* $s'' \in S$, *and all* s, s' *in the same block* S_i,

$$r_a(s) = r_a(s'), \quad \text{and} \quad p_a(s, s'') = p_a(s', s'').$$

A policy π *on a block MDP is called* uniform, *if* $\pi(s) = \pi(s')$ *for* s, s' *in the same block.*

Obviously, block MDPs are predestined to be aggregated. However, the following definition is also applicable to arbitrary MDPs.

Definition 5. *Given an MDP* $\mathcal{M} = \langle S, A, p, r \rangle$ *and a partition* $\widehat{S} = \{S_1, \ldots, S_k\}$ *of its state space* S, *the* aggregated MDP *with respect to* $\widehat{S}$ *is defined as* $\widehat{\mathcal{M}} := \langle \widehat{S}, A, \widehat{p}, \widehat{r} \rangle$, *where*

$$\widehat{r}_a(S_i) := \frac{1}{|S_i|} \sum_{s \in S_i} r_a(s), \quad \text{and} \quad \widehat{p}_a(S_i, S_j) := \frac{1}{|S_i|} \sum_{s \in S_i} \sum_{s' \in S_j} p_a(s, s').$$

It is easy to check that $\widehat{p}_a(S_i, \cdot)$ is a probability distribution for each $S_i \in \widehat{S}$.

Any policy π on an aggregated MDP $\widehat{\mathcal{M}}$ with state space $\widehat{S} = \{S_1, \ldots, S_k\}$ can be naturally extended to a policy π^e on the original MDP $\mathcal{M}$ by

$$\pi^e(s) := a, \text{ if } s \in S_j \text{ and } \pi(S_j) = a.$$

We continue with some considerations on block MDPs, the first one being trivial if the stationary distribution μ in state s is interpreted as probability of being in s after an infinite number of steps. However, we give a proof which refers only to the properties of stationary distributions mentioned in Sect. 2.

Lemma 1. *Let* $\mathcal{M} = \langle S, A, p, r \rangle$ *be a block MDP with block set* $\widehat{S} = \{S_1, \ldots, S_k\}$ *and respective aggregated MDP* $\widehat{\mathcal{M}} = \langle \widehat{S}, A, \widehat{p}, \widehat{r} \rangle$. *Given a policy* π *on* $\widehat{\mathcal{M}}$ *and its extended counterpart* π^e *on* $\mathcal{M}$ *with stationary distributions* $\widehat{\mu}_\pi$ *and* μ_{π^e}, *respectively, for all* $S_i \in \widehat{S}$,

$$\widehat{\mu}_\pi(S_i) = \sum_{s \in S_i} \mu_{\pi^e}(s).$$

Proof. First, note that since $\mathcal{M}$ is a block MDP, for all $s \in S_j$ and $a \in A$,

$$\widehat{p}_a(S_j, S_i) = \sum_{s' \in S_i} p_a(s, s'). \tag{1}$$

As $\mu P = \mu$ for the stationary distribution μ of a transition matrix P, we have[1] for all $s' \in S$,

$$\sum_{s \in S} \mu_{\pi^e}(s)\, p(s, s') = \mu_{\pi^e}(s'). \tag{2}$$

Let $\widehat{P}$ be the transition matrix of $\widehat{\mathcal{M}}$ under π. We set $\mu'(S_j) := \sum_{s \in S_j} \mu_{\pi^e}(s)$ for $S_j \in \widehat{S}$, and have by (2) and (1) for each $S_i \in \widehat{S}$,

$$
\begin{aligned}
(\mu'\widehat{P})_{S_i} &= \sum_{S_j \in \widehat{S}} \mu'(S_j)\, \widehat{p}(S_j, S_i) = \sum_{S_j \in \widehat{S}} \sum_{s \in S_j} \mu_{\pi^e}(s)\, \widehat{p}(S_j, S_i) \\
&= \sum_{S_j \in \widehat{S}} \sum_{s \in S_j} \mu_{\pi^e}(s) \sum_{s' \in S_i} p(s, s') = \sum_{s' \in S_i} \sum_{S_j \in \widehat{S}} \sum_{s \in S_j} \mu_{\pi^e}(s)\, p(s, s') \\
&= \sum_{s' \in S_i} \sum_{s \in S} \mu_{\pi^e}(s)\, p(s, s') = \sum_{s' \in S_i} \mu_{\pi^e}(s') = \mu'(S_i).
\end{aligned}
$$

Consequently, by the uniqueness of the stationary distribution we have $\widehat{\mu}_\pi = \mu'$, which proves the lemma. □

Theorem 1. *Each block MDP has an optimal policy which is uniform.*

In the proof of Theorem 1 we will make use of a minor result about optimal policies on ergodic MDPs.

Definition 6. *Given policies $\pi_1, \ldots, \pi_\ell$ on an MDP with state space S, a policy π is called a* combination *of $\pi_1, \ldots, \pi_\ell$, if for each $s \in S$ there is an $i \in \{1, \ldots, \ell\}$ such that $\pi(s) = \pi_i(s)$.*

The following proposition can be derived from the Bellman equations, which may also be used to prove Theorem 1 directly (cf. the proof of the more general Theorem 4 in Sect. 4 below). As a corollary to a more general result Proposition 1 has been proved in [6].

Proposition 1. *On ergodic MDPs, any combination of optimal policies is optimal.*

Proof of Theorem 1. Consider an arbitrary non-uniform, optimal policy π^* on a block MDP $\mathcal{M}$ with blocks $S_1, \ldots, S_k$. Take some block $S_j = \{s_1, \ldots, s_m\}$ on which π^* is not uniform. As $\mathcal{M}$ is a block MDP, all states in S_j have the same rewards and transition probabilities under each action $a \in A$. Hence, a policy π is optimal, if it coincides with π^* on $S \setminus S_j$ and swaps the actions in S_j according to some permutation $\sigma : S_j \to S_j$, that is, $\pi(s_i) = \pi^*(\sigma(s_i))$ for $i = 1, \ldots, m$.

Thus in particular, for each $i \in \{1, \ldots, m\}$ there is an optimal policy π such that $\pi(s_i) = \pi^*(s_1)$. It follows from Proposition 1 that there is an optimal policy which is uniform on S_j. This argument can be repeated for each single block to yield the theorem. □

---

[1] In the following, we usually skip indices for actions when the policy is fixed.

3.2   A Simple Pseudometric, ε-Aggregation, and an Upper Bound

Given an MDP $\mathcal{M} = \langle S, A, p, r \rangle$ and positive constants c_r, c_p, we set for $s, s' \in S$,

$$d_v(s, s') := \max_{a \in A} \left\{ c_r \left| r_a(s) - r_a(s') \right| + c_p \sum_{s'' \in S} \left| p_a(s, s'') - p_a(s', s'') \right| \right\}.$$

It is easy to check that d_v is a pseudometric on S. However, d_v is not a metric. If $d_v(s, s') = 0$, then all rewards and transition probabilities coincide in states s and s', which however does not entail that $s = s'$. The pseudometric d_v is basically the *bisimulation metric* induced by the *total variation probability metric*, which has been introduced for discounted MDPs in [3]. Ferns et al. consider also other probability metrics that measure the distance between two transition probability distributions $p_a(s, \cdot)$ and $p_a(s', \cdot)$.

Definition 7. *For fixed $\varepsilon > 0$, an ε-partition of the state space S with respect to a pseudometric d on S is a minimal partition of S into aggregated states (or blocks) $S_1, \ldots, S_k$ such that for $s, s' \in S_i$ one has $d(s, s') < \varepsilon$. Minimality here means that one cannot aggregate any S_i, S_j to $S_i \cup S_j$, that is, for distinct S_i, S_j there are $s \in S_i$, $s' \in S_j$ with $d(s, s') \geq \varepsilon$.*

When aggregating an MDP $\mathcal{M}$ with respect to an ε-partition we speak of an ε-aggregation of $\mathcal{M}$.

Theorem 2. *Let $\mathcal{M} = \langle S, A, p, r \rangle$ be an MDP and $\widehat{\mathcal{M}} = \langle \widehat{S}, A, \widehat{p}, \widehat{r} \rangle$ an ε-aggregation of $\mathcal{M}$ with respect to d_v. Then for each policy π on $\widehat{\mathcal{M}}$ and its respective extended policy π^e on $\mathcal{M}$,*

$$\left| \rho_{\pi^e}(\mathcal{M}) - \rho_\pi(\widehat{\mathcal{M}}) \right| < \left(\frac{1}{c_r} + \frac{\kappa_{\mathcal{M}_\pi} - 1}{c_p} \right) \varepsilon,$$

where $\kappa_{\mathcal{M}_\pi}$ is the mixing time of the Markov chain induced by π on $\mathcal{M}$.

For the proof of Theorem 2 we will need the following result of [5] on perturbations of Markov chains.

Theorem 3 (Hunter[5]). *Let $\mathcal{C}, \widetilde{\mathcal{C}}$ be two ergodic Markov chains on the same state space S with transition probabilities $p(\cdot, \cdot)$, $\widetilde{p}(\cdot, \cdot)$ and stationary distributions $\mu, \widetilde{\mu}$. Then*

$$\left\| \mu - \widetilde{\mu} \right\|_1 \leq (\kappa_{\mathcal{C}} - 1) \max_{s \in S} \sum_{s' \in S} \left| p(s, s') - \widetilde{p}(s, s') \right|,$$

where $\kappa_{\mathcal{C}}$ is the mixing time of $\mathcal{C}$.

Proof of Theorem 2. Let us first modify the original MDP $\mathcal{M}$ by redefining the rewards in each state $s \in S_j$ ($1 \leq j \leq k := |\widehat{S}|$) and each $a \in A$ as

$$\widetilde{r}_a(s) := \frac{1}{|S_j|} \sum_{s' \in S_j} r_a(s').$$

Then using the assumption that two states s, s' in the same block S_j have distance $d_v(s, s') < \varepsilon$, the difference in the average rewards of the original and the thus modified MDP $\mathcal{M}_{\widetilde{r}} := \langle S, A, p, \widetilde{r} \rangle$ under some fixed policy π can be upper bounded by

$$
\begin{aligned}
\left| \rho_\pi(\mathcal{M}) - \rho_\pi(\mathcal{M}_{\widetilde{r}}) \right| &= \\
= \left| \sum_{s \in S} \mu_\pi(s)\, r(s) - \sum_{s \in S} \mu_\pi(s)\, \widetilde{r}(s) \right| &= \left| \sum_{j=1}^{k} \sum_{s \in S_j} \mu_\pi(s) \left(r(s) - \frac{1}{|S_j|} \sum_{s' \in S_j} r(s') \right) \right| \\
= \left| \sum_{j=1}^{k} \sum_{s \in S_j} \mu_\pi(s) \left(\frac{1}{|S_j|} \sum_{s' \in S_j} (r(s) - r(s')) \right) \right| &< \sum_{j=1}^{k} \sum_{s \in S_j} \mu_\pi(s) \left(\frac{1}{|S_j|} \sum_{s' \in S_j} \frac{\varepsilon}{c_r} \right) \\
= \sum_{s \in S} \mu_\pi(s) \frac{\varepsilon}{c_r} &= \frac{\varepsilon}{c_r}.
\end{aligned}
\tag{3}
$$

Now we also redefine the transition probabilities for $s \in S_j$ and $a \in A$ to be

$$
\widetilde{p}_a(s, s') := \frac{1}{|S_j|} \sum_{s'' \in S_j} p_a(s'', s').
$$

It is easily checked that the $p_a(s, \cdot)$ are indeed probability distributions for all $s \in S$. Considering any policy π, for each $s \in S_j$,

$$
\begin{aligned}
\sum_{s' \in S} |p(s, s') - \widetilde{p}(s, s')| &= \sum_{s' \in S} \left| p(s, s') - \frac{1}{|S_j|} \sum_{s'' \in S_j} p(s'', s') \right| \\
= \sum_{s' \in S} \left| \frac{1}{|S_j|} \sum_{s'' \in S_j} (p(s, s') - p(s'', s')) \right| &\leq \sum_{s' \in S} \frac{1}{|S_j|} \sum_{s'' \in S_j} |p(s, s') - p(s'', s')| \\
= \frac{1}{|S_j|} \sum_{s'' \in S_j} \sum_{s' \in S} |p(s, s') - p(s'', s')| &< \frac{1}{|S_j|} \sum_{s'' \in S_j} \frac{\varepsilon}{c_p} = \frac{\varepsilon}{c_p},
\end{aligned}
\tag{4}
$$

again using that states s, s'' in the same block have distance $d_v(s, s'') < \varepsilon$. As rewards are upper bounded by 1, Theorem 3 and (4) give for the difference of the average rewards of $\mathcal{M}_{\widetilde{r}}$ and $\widetilde{\mathcal{M}} := \langle S, A, \widetilde{p}, \widetilde{r} \rangle$ under policy π (with respective stationary distributions μ_π and $\widetilde{\mu}_\pi$),

$$
\begin{aligned}
\left| \rho_\pi(\mathcal{M}_{\widetilde{r}}) - \rho_\pi(\widetilde{\mathcal{M}}) \right| &= \left| \sum_{s \in S} \mu_\pi(s)\, \widetilde{r}(s) - \sum_{s \in S} \widetilde{\mu}_\pi(s)\, \widetilde{r}(s) \right| = \\
&= \left| \sum_{s \in S} (\mu_\pi(s) - \widetilde{\mu}_\pi(s))\, \widetilde{r}(s) \right| \leq \sum_{s \in S} |\mu_\pi(s) - \widetilde{\mu}_\pi(s)|\, \widetilde{r}(s) \\
&\leq \sum_{s \in S} |\mu_\pi(s) - \widetilde{\mu}_\pi(s)| = \|\mu_\pi - \widetilde{\mu}_\pi\|_1 < (\kappa_{\mathcal{M}_\pi} - 1) \frac{\varepsilon}{c_p}.
\end{aligned}
$$

Combining this with (3) yields

$$
\begin{aligned}
\left|\rho_\pi(\mathcal{M}) - \rho_\pi(\widetilde{\mathcal{M}})\right| &\leq \left|\rho_\pi(\mathcal{M}) - \rho_\pi(\mathcal{M}_{\tilde{r}})\right| + \left|\rho_\pi(\mathcal{M}_{\tilde{r}}) - \rho_\pi(\widetilde{\mathcal{M}})\right| \\
&< \frac{\varepsilon}{c_r} + (\kappa_{\mathcal{M}_\pi} - 1)\frac{\varepsilon}{c_p}.
\end{aligned}
\tag{5}
$$

So far, π has been an arbitrary policy on $\mathcal{M}$. Now we fix π to be a policy on $\widehat{\mathcal{M}}$ and claim that $\rho_{\pi^e}(\widetilde{\mathcal{M}}) = \rho_\pi(\widehat{\mathcal{M}})$ for the extension π^e of π. It is easy to see that by definition of the rewards and transition probabilities, $\widetilde{\mathcal{M}}$ is a block MDP with block set $\widehat{S}$ and respective aggregated MDP $\widehat{\mathcal{M}}$. In particular, $\widehat{r}_a(S_j) = \widetilde{r}_a(s)$ for all $a \in A$ and $s \in S_j$, so that by Lemma 1

$$
\rho_\pi(\widehat{\mathcal{M}}) = \sum_{S_j \in \widehat{S}} \widehat{\mu}_\pi(S_j)\,\widehat{r}(S_j) = \sum_{S_j \in \widehat{S}} \sum_{s \in S_j} \widetilde{\mu}_{\pi^e}(s)\,\widetilde{r}(s) = \sum_{s \in S} \widetilde{\mu}_{\pi^e}(s)\,\widetilde{r}(s) = \rho_{\pi^e}(\widetilde{\mathcal{M}}),
$$

which together with (5) proves the theorem.   □

Corollary 1. *Let π^* be an optimal policy on an MDP $\mathcal{M}$ with optimal average reward $\rho^* := \rho_{\pi^*}(\mathcal{M})$, and let $\widehat{\pi}^*$ be an optimal policy with optimal average reward $\widehat{\rho}^* := \rho_{\widehat{\pi}^*}(\widehat{\mathcal{M}})$ on an ε-aggregation $\widehat{\mathcal{M}}$ of $\mathcal{M}$ with respect to d_v. Then*

$$
\text{(i)} \quad |\rho^* - \widehat{\rho}^*| < \left(\frac{1}{c_r} + \frac{\kappa_{\mathcal{M}} - 1}{c_p}\right)\varepsilon,
$$

$$
\text{(ii)} \quad \rho^* < \rho_{\widehat{\pi}^{*e}}(\mathcal{M}) + \left(\frac{2}{c_r} + \frac{2(\kappa_{\mathcal{M}} - 1)}{c_p}\right)\varepsilon,
$$

where $\kappa_{\mathcal{M}} := \max_\pi \kappa_{\mathcal{M}_\pi}$.

Proof. First note that the extension $\widehat{\pi}^{*e}$ of $\widehat{\pi}^*$ to the block MDP $\widetilde{\mathcal{M}}$ (as defined in the proof of Theorem 2) is optimal on $\widetilde{\mathcal{M}}$ with reward $\widehat{\rho}^*$. This follows from Theorem 1 and the fact that $\rho_{\pi^e}(\widetilde{\mathcal{M}}) = \rho_\pi(\widehat{\mathcal{M}})$ (cf. proof of Theorem 2). Now if $\rho^* > \widehat{\rho}^*$, then by optimality of $\widehat{\pi}^{*e}$ on $\widetilde{\mathcal{M}}$,

$$
\rho_{\pi^*}(\mathcal{M}) = \rho^* > \widehat{\rho}^* = \rho_{\widehat{\pi}^{*e}}(\widetilde{\mathcal{M}}) \geq \rho_{\pi^*}(\widetilde{\mathcal{M}}),
$$

so that by (5),

$$
|\rho^* - \widehat{\rho}^*| \leq \left|\rho_{\pi^*}(\mathcal{M}) - \rho_{\pi^*}(\widetilde{\mathcal{M}})\right| < \frac{\varepsilon}{c_r} + (\kappa_{\pi^*} - 1)\frac{\varepsilon}{c_p}.
\tag{6}
$$

On the other hand, if $\rho^* \leq \widehat{\rho}^*$, then by optimality of π^* on $\mathcal{M}$,

$$
\rho_{\widehat{\pi}^{*e}}(\widetilde{\mathcal{M}}) = \widehat{\rho}^* \geq \rho^* = \rho_{\pi^*}(\mathcal{M}) \geq \rho_{\widehat{\pi}^{*e}}(\mathcal{M}),
$$

and it follows again from (5) that

$$
|\widehat{\rho}^* - \rho^*| \leq \left|\rho_{\widehat{\pi}^{*e}}(\widetilde{\mathcal{M}}) - \rho_{\widehat{\pi}^{*e}}(\mathcal{M})\right| < \frac{\varepsilon}{c_r} + (\kappa_{\widehat{\pi}^{*e}} - 1)\frac{\varepsilon}{c_p},
$$

which together with (6) finishes the proof of (i).

Concerning (ii), note that by optimality of $\widehat{\pi}^{*e}$ on $\widetilde{\mathcal{M}}$ it follows from (5) that

$$
\begin{aligned}
\rho^* - \rho_{\widehat{\pi}^{*e}}(\mathcal{M}) \ &\leq\ \rho^* - \rho_{\widehat{\pi}^{*e}}(\mathcal{M}) + \left(\widehat{\rho}^* - \rho_{\pi^*}(\widetilde{\mathcal{M}})\right) \\
&=\ \rho_{\pi^*}(\mathcal{M}) - \rho_{\pi^*}(\widetilde{\mathcal{M}}) + \rho_{\widehat{\pi}^{*e}}(\widetilde{\mathcal{M}}) - \rho_{\widehat{\pi}^{*e}}(\mathcal{M}) \\
&\leq\ \left|\rho_{\pi^*}(\mathcal{M}) - \rho_{\pi^*}(\widetilde{\mathcal{M}})\right| + \left|\rho_{\widehat{\pi}^{*e}}(\widetilde{\mathcal{M}}) - \rho_{\widehat{\pi}^{*e}}(\mathcal{M})\right| \\
&<\ 2\left(\frac{\varepsilon}{c_r} + (\kappa_{\mathcal{M}} - 1)\frac{\varepsilon}{c_p}\right). \qquad\qquad \square
\end{aligned}
$$

Theorem 2 and Corollary 1 (i) can be seen as generalizations of the bounds for discounted reward MDPs obtained in Theorem 5.2 of [3].

4   Adequate Similarity Metrics

Obviously, ε-aggregation with respect to d_v is a rather restricted model which will be applicable only to very special problems. In this section, we want to develop a more general view on similarity metrics on an MDP's state space.

4.1   Generalized Block MDPs

Definition 8. *An MDP* $\mathcal{M} = \langle S, A, p, r\rangle$ *is a* generalized block MDP *with blocks* $S_1, \ldots, S_k$, *if the block set* $\{S_1, \ldots, S_k\}$ *is a partition of* S, *and for all* s, s' *in the same block* S_i, *all* $a \in A$, *and all blocks* S_j *there is an* $a' \in A$ *such that*

$$
r_a(s) = r_{a'}(s'), \quad \text{and} \quad \sum_{s'' \in S_j} p_a(s, s'') = \sum_{s'' \in S_j} p_{a'}(s', s''). \tag{7}
$$

With this definition, we could also consider MDPs in which each state has an individual set of possible actions at its disposal. All results presented easily generalize to this setting. However, for the sake of simplicity, we assume in the following without loss of generality that within a block S_i the actions in A are labelled uniformly, such that for states $s, s' \in S_i$, (7) holds for $a' = a$. Consequently, we may define *uniform policies* as we have done before.

Generalized block MDPs (yet with discounted rewards) have already been considered by Givan et al. [1] under the name of *stochastic bisimulation*, which is the equivalence relation that corresponds to the partition $\{S_1, \ldots, S_k\}$ in Definition 8 (cf. also the discussion in [3]).

Note that block MDPs are also generalized block MDPs, so that most results in this section can be considered as generalizations of the results in the previous section.

Lemma 2. *Let* $\mathcal{M} = \langle S, A, p, r\rangle$ *be a generalized block MDP with block set* $\widehat{S} = \{S_1, \ldots, S_k\}$ *and respective aggregated MDP* $\widehat{\mathcal{M}} = \langle \widehat{S}, A, \widehat{p}, \widehat{r}\rangle$. *Given a policy* π

on $\widehat{\mathcal{M}}$ and its extended counterpart π^e on $\mathcal{M}$ with stationary distributions $\widehat{\mu}_\pi$ and μ_{π^e}, respectively, one has for all $S_j \in \widehat{S}$,

$$\widehat{\mu}_\pi(S_j) = \sum_{s \in S_j} \mu_{\pi^e}(s).$$

Proof. As proof of Lemma 1. □

Theorem 4. *On generalized block MDPs there is always a uniform policy which gives optimal average return.*

Proof. Let $\mathcal{M} = \langle S, A, p, r \rangle$ be a generalized block MDP with block set $\widehat{S} = \{S_1, \dots, S_k\}$. It is a well-known fact (cf. e.g. [7]) that a policy on an ergodic MDP is optimal if it solves the Bellman equations, that is, if there is ρ^* and a value function $v : S \to \mathbb{R}$ such that for all $s \in S$,

$$v(s) + \rho^* = \max_{a \in A} \left(r_a(s) + \sum_{s' \in S} p_a(s, s')\, v(s') \right). \qquad (8)$$

Thus, an optimal policy $\widehat{\pi}^*$ on the aggregated MDP $\widehat{\mathcal{M}} = \langle \widehat{S}, A, \widehat{p}, \widehat{r} \rangle$ solves for all $S_i \in \widehat{S}$,

$$
\begin{aligned}
\widehat{v}(S_i) + \widehat{\rho}^* &= \max_{a \in A} \left(\widehat{r}_a(S_i) + \sum_{S_j \in \widehat{S}} \widehat{p}_a(S_i, S_j)\, \widehat{v}(S_j) \right) \\
&= \widehat{r}_{\widehat{\pi}^*(S_i)}(S_i) + \sum_{S_j \in \widehat{S}} \widehat{p}_{\widehat{\pi}^*(S_i)}(S_i, S_j)\, \widehat{v}(S_j).
\end{aligned}
\qquad (9)
$$

However, setting $v(s) := v(S_j)$ for $s \in S_j$, it follows from (9) that the Bellman equations (8) hold for the extension $\widehat{\pi}^{*e}$ of $\widehat{\pi}^*$ to $\mathcal{M}$ for all $s \in S$, which means that $\widehat{\pi}^{*e}$ is optimal on $\mathcal{M}$. □

4.2   Adequate Similarity Metrics

The key idea an *adequate* similarity metric shall grasp is that in similar states there should be equivalent actions available which lead to similar states again. Such a metric may then be used to partition the state space. As similarity in general is not a transitive relation, not any partition will work (for more about the problem of obtaining adequate partitions from similarity relations see e.g. [8]). Thus before formalizing our basic idea, we start with a condition for the utility of a given partition induced by a distance metric.

Definition 9. *Given $\varepsilon > 0$ and a pseudometric space (S, d), we say that $S' \subseteq S$ is ε-maximal, if (i) for all $s, s' \in S'$, $d(s, s') < \varepsilon$, and (ii) for all $s'' \in S \setminus S'$ there is $s \in S'$ with $d(s, s'') \geq \varepsilon$.*

An ε-partition $\widehat{S} = \{S_1, \dots, S_k\}$ of S with respect to a metric d is called consistent, if each $S_i \in \widehat{S}$ is ε-maximal.

Unfortunately, existence of consistent ε-aggregations of the state space cannot be guaranteed for each $\varepsilon > 0$.

Example 1. Let $S = \{s_1, s_2, s_3\}$ with $d(s_1, s_2), d(s_2, s_3) < \varepsilon$ and $d(s_1, s_3) \geq \varepsilon$. Then neither of the two possible ε-partitions $\widehat{S}_1 = \{\{s_1, s_2\}, \{s_3\}\}$ and $\widehat{S}_2 = \{\{s_1\}, \{s_2, s_3\}\}$ is consistent, because the singletons $\{s_3\}$ and $\{s_1\}$ are not ε-maximal.

Sometimes, things are easier if (S, d) can be embedded into some larger metric space (X, d), e.g. if $S \subset \mathbb{R}^n$ and d coincides on S with some arbitrary metric d on $\mathbb{R}^n$. In this case one may relax the condition for ε-maximality as follows:

A set $S' \subseteq S$ is ε-*maximal*, if $S' = S \cap U_\varepsilon(x)$ for some ε-ball $U_\varepsilon(x) := \{y \in X : d(x, y) < \varepsilon\}$ with center $x \in X$. Then an ε-partition $\widehat{S} = \{S_1, \ldots, S_k\}$ is *consistent* if it can be represented by non-intersecting ε-balls, that is, if

(i) there are $x_1, \ldots, x_k \in \mathbb{R}^n$ such that $S_i = S \cap U_\varepsilon(x_i)$ for $i = 1, \ldots, k$,
(ii) $U_\varepsilon(x_i) \cap U_\varepsilon(x_j) = \varnothing$ for $i \neq j$.

However, such an embedding may fail to give consistency either.

Example 2. Let $S = \{s_1, s_2, s_3\}$ consist of three points s_1, s_2, s_3 equidistantly distributed on a circle $C := \{y \in \mathbb{R}^2 : \|x - y\|_2 = r\}$ with center x and radius r. Considering the metric space (C, d) with $d(y, z) := \|y - z\|_2$ for $y, z \in C$, it is easy to see that for $\varepsilon = \sqrt{2}r$ (so that for $y \in C$, $U_\varepsilon(y)$ contains one half of C), there is no consistent ε-partition of S. This example can easily be extended to arbitrary n-dimensional spheres.

Also, $\mathbb{R}^n$ with Euclidean distance may not be favorable anyway, as it is impossible to cover $\mathbb{R}^n$ with non-intersecting ε-balls with respect to Euclidean distance. Thus, the metric with respect to $\| \cdot \|_\infty$, which evidently guarantees a consistent ε-partition in $\mathbb{R}^n$ for each $\varepsilon > 0$, will be preferred.

Definition 10. *Given an MDP* $\mathcal{M} = \langle S, A, p, r \rangle$, *we say that a pseudometric d on S is* adequate *to* $\mathcal{M}$, *if* $d(s, s') < \varepsilon$ *implies that for all* $a \in A$ *there is an $a' \in A$ such that*

(i) $c_r \left| r_a(s) - r_{a'}(s') \right| < \varepsilon$,

(ii) $c_p \left| \sum_{s'' \in S'} p_a(s, s'') - \sum_{s'' \in S'} p_{a'}(s', s'') \right| < \varepsilon$ *for all ε-maximal $S' \subseteq S$.*

As in the case of generalized block MDPs we assume without loss of generality that for states s, s' in the same block, actions are labelled uniformly so that $a' = a$ in the definition above.

Of course, one may as well define a particular partition $\widehat{S} = \{S_1, \ldots, S_k\}$ of the state space to be ε-*adequate*, if for all s, s' in the same block S_j,

(i) $c_r \left| r_a(s) - r_a(s') \right| < \varepsilon$,

(ii') $c_p \left| \sum_{s'' \in S_i} p_a(s, s'') - \sum_{s'' \in S_i} p_a(s', s'') \right| < \varepsilon$ for all $S_i \in \widehat{S}$.

This modified definition is similar to the definition of ε-*homogeneous* partitions for discounted reward MDPs in [2]. The only difference is that in condition (ii'), Even-Dar and Mansour consider arbitrary norms and sum up over all aggregated states.

Further, one still may work with the metric d_v defined in the previous section. Even though the kind of state similarity which may be grasped by d_v is rather restricted, aggregating states with respect to d_v for given $\varepsilon > 0$ evidently gives ε-adequate partitions of the state space. By definition, d_v is also an adequate metric.

4.3   A General Upper Bound on the Loss by Aggregation

By the remarks at the end of the previous section, the following theorem can be seen as a generalization of Lemma 3 of [2] to average reward MDPs.

Theorem 5. *Given an MDP* $\mathcal{M} = \langle S, A, p, r \rangle$ *and a consistent ε-aggregation* $\widehat{\mathcal{M}} = \langle \widehat{S}, A, \widehat{p}, \widehat{r} \rangle$ *of $\mathcal{M}$ with respect to an adequate pseudometric d, for each policy π on $\widehat{\mathcal{M}}$ and its respective extended policy π^e on $\mathcal{M}$,*

$$\left| \rho_{\pi^e}(\mathcal{M}) - \rho_\pi(\widehat{\mathcal{M}}) \right| < \left(\frac{1}{c_r} + \frac{(\kappa_{\mathcal{M}_\pi} - 1)|\widehat{S}|}{c_p} \right) \varepsilon.$$

Proof. As in the proof of Theorem 2, we start by modifying the rewards in $\mathcal{M}$ slightly to be

$$\widetilde{r}_a(s) := \frac{1}{|S_j|} \sum_{s' \in S_j} r_a(s') \tag{10}$$

for $s \in S_j$ and $a \in A$. Then the same argument can be repeated to see that for the modified MDP $\mathcal{M}_{\widetilde{r}} = \langle S, A, p, \widetilde{r} \rangle$,

$$\left| \rho_\pi(\mathcal{M}) - \rho_\pi(\mathcal{M}_{\widetilde{r}}) \right| < \frac{\varepsilon}{c_r} \tag{11}$$

for each policy π. In the next step we want to modify the transition probabilities in $\mathcal{M}_{\widetilde{r}}$ so that for s, s' in the same block and for all blocks $S_i \in \widehat{S}$,

$$\sum_{s'' \in S_i} \widetilde{p}_a(s, s'') = \sum_{s'' \in S_i} \widetilde{p}_a(s', s''). \tag{12}$$

In order to attain this, we set for all $s \in S_j$, $s' \in S_i$, and all $a \in A$,

$$\widetilde{p}_a(s, s') := p_a(s, s') + \frac{1}{|S_i|} \left(\frac{1}{|S_j|} \sum_{\bar{s} \in S_j} \sum_{s'' \in S_i} p_a(\bar{s}, s'') - \sum_{s'' \in S_i} p_a(s, s'') \right)$$

(note that the $\widetilde{p}_a(s, \cdot)$ are indeed probability distributions for all $s \in S$), so that for s in any block S_j,

$$\sum_{s' \in S_i} \widetilde{p}_a(s, s') = \sum_{s' \in S_i} p_a(s, s') + \frac{1}{|S_j|} \sum_{\bar{s} \in S_j} \sum_{s'' \in S_i} p_a(\bar{s}, s'') - \sum_{s'' \in S_i} p_a(s, s'')$$

$$= \frac{1}{|S_j|} \sum_{\bar{s} \in S_j} \sum_{s'' \in S_i} p_a(\bar{s}, s''), \tag{13}$$

independently of s, which entails (12). As $\widehat{S}$ is assumed to be a consistent ε-aggregation with respect to an adequate metric, we have by definition of $\widetilde{p}$ for transition probabilities $p(\cdot, \cdot)$, $\widetilde{p}(\cdot, \cdot)$ under any policy π and for $s \in S_j$, $s' \in S_i$,

$$|\widetilde{p}(s, s') - p(s, s')| = \frac{1}{|S_i|} \cdot \left| \frac{1}{|S_j|} \sum_{\bar{s} \in S_j} \sum_{s'' \in S_i} p(\bar{s}, s'') - \sum_{s'' \in S_i} p(s, s'') \right|$$

$$\leq \frac{1}{|S_i|} \cdot \frac{1}{|S_j|} \sum_{\bar{s} \in S_j} \sum_{s'' \in S_i} |p(\bar{s}, s'') - p(s, s'')| < \frac{\varepsilon}{c_p |S_i|},$$

so that for all $s \in S$,

$$\sum_{s' \in S} |\widetilde{p}(s, s') - p(s, s')| = \sum_{i=1}^{k} \sum_{s' \in S_i} |\widetilde{p}(s, s') - p(s, s')| < \sum_{i=1}^{k} \frac{\varepsilon}{c_p} = \frac{|\widehat{S}|}{c_p} \varepsilon.$$

Thus, by Theorem 3 we have for the difference of the average rewards of $\mathcal{M}_{\widetilde{r}}$ and $\widetilde{\mathcal{M}} := \langle S, A, \widetilde{p}, \widetilde{r} \rangle$ under some policy π (with respective stationary distributions μ_π and $\widetilde{\mu}_\pi$),

$$\left| \rho_\pi(\mathcal{M}_{\widetilde{r}}) - \rho_\pi(\widetilde{\mathcal{M}}) \right| = \left| \sum_{s \in S} \left(\mu_\pi(s) - \widetilde{\mu}_\pi(s) \right) \widetilde{r}(s) \right| \leq \sum_{s \in S} \left| \mu_\pi(s) - \widetilde{\mu}_\pi(s) \right| \widetilde{r}(s)$$

$$\leq \sum_{s \in S} \left| \mu_\pi(s) - \widetilde{\mu}_\pi(s) \right| = \left\| \mu_\pi - \widetilde{\mu}_\pi \right\|_1 < (\kappa_{\mathcal{M}_\pi} - 1) \frac{|\widehat{S}|}{c_p} \varepsilon. \qquad (14)$$

Now $\widetilde{\mathcal{M}}$ is a generalized block MDP with block set $\widehat{S}$, and by (10) and (13), its respective aggregated MDP is precisely $\widehat{\mathcal{M}}$. Analogously to the proof of Theorem 2, it follows from Lemma 2 that $\rho_\pi(\widetilde{\mathcal{M}}) = \rho_{\pi^e}(\widehat{\mathcal{M}})$ for all policies π on $\widetilde{\mathcal{M}}$. Thus (11) and (14) yield

$$\left| \rho_{\pi^e}(\mathcal{M}) - \rho_\pi(\widehat{\mathcal{M}}) \right| = \left| \rho_{\pi^e}(\mathcal{M}) - \rho_{\pi^e}(\widehat{\mathcal{M}}) \right|$$

$$\leq \left| \rho_{\pi^e}(\mathcal{M}) - \rho_{\pi^e}(\mathcal{M}_{\widetilde{r}}) \right| + \left| \rho_{\pi^e}(\mathcal{M}_{\widetilde{r}}) - \rho_{\pi^e}(\widehat{\mathcal{M}}) \right| < \frac{\varepsilon}{c_r} + (\kappa_{\mathcal{M}_\pi} - 1) \frac{|\widehat{S}|}{c_p} \varepsilon. \quad \square$$

Corollary 2. *Let π^* be an optimal policy on an MDP $\mathcal{M}$ with optimal average reward ρ^*, and let $\widehat{\pi}^*$ be an optimal policy with optimal average reward $\widehat{\rho}^*$ on a consistent ε-aggregation $\widehat{\mathcal{M}}$ of $\mathcal{M}$ with respect to an adequate metric. Then for $\kappa_{\mathcal{M}} := \max_\pi \kappa_{\mathcal{M}_\pi}$,*

$$(i) \quad |\rho^* - \widehat{\rho}^*| \leq \left(\frac{1}{c_r} + \frac{(\kappa_{\mathcal{M}} - 1)|\widehat{S}|}{c_p} \right) \varepsilon,$$

$$(ii) \quad \rho^* \leq \rho_{\widehat{\pi}^{*e}}(\mathcal{M}) + \left(\frac{2}{c_r} + \frac{2(\kappa_{\mathcal{M}} - 1)|\widehat{S}|}{c_p} \right) \varepsilon.$$

Proof. Analogously to the proof of Corollary 1. $\qquad \square$

Corollary 2 can be seen as a generalization of Lemma 4 of [2] to average reward MDPs.

5 Dependence on the Mixing Time

5.1 Why Bounds Are Worse in the Average Reward Case

The bounds obtained for ε-aggregation in the discounted case [3,2] are basically of the form $\frac{\varepsilon}{1-\gamma}$, i.e., in average one loses only ε reward in each step. This may give the impression that the mixing time parameter in the obtained bounds for average reward MDPs is redundant and should be eliminated, or at least replaced with something substantially smaller. However, it turns out that aggregation may go terribly wrong if mixing times are large.

Theorem 6. *For each $\varepsilon > 0$ and each $\delta \in (0, \varepsilon)$ there is an MDP $\mathcal{M}$ and an ε-aggregation $\widehat{\mathcal{M}}$ of $\mathcal{M}$ with respect to d_v, such that for some policy π on $\widehat{\mathcal{M}}$,*

$$|\rho_{\pi^e}(\mathcal{M}) - \rho_\pi(\widehat{\mathcal{M}})| \geq 1 - \delta.$$

Proof. Fix some $\varepsilon > 0$ and consider for $\delta \in (0, \varepsilon)$ the Markov chain $\mathcal{C}$ with $S = \{s_1, s_2, s_3\}$ and the following nonzero transition probabilities $p_{ij} := p(s_i, s_j)$,

$$p_{12} = 1 - \delta, \quad p_{13} = \delta, \quad p_{21} = p_{31} = \delta/n, \quad p_{22} = p_{33} = 1 - \delta/n,$$

where $n \in \mathbb{N}$. Then we may ε-aggregate states s_1 and s_2 with respect to d_v and obtain a Markov chain $\widehat{\mathcal{C}}$ with states $S_1 = \{s_1, s_2\}$, $S_2 = \{s_3\}$ and transition probabilities

$$\widehat{p}(S_1, S_2) = \delta/2, \quad \widehat{p}(S_2, S_1) = \delta/n, \quad \widehat{p}(S_1, S_1) = 1 - \delta/2, \quad \widehat{p}(S_2, S_2) = 1 - \delta/n.$$

The original chain $\mathcal{C}$ has stationary distribution $\mu = \left(\frac{\delta}{n+\delta}, \frac{n-\delta n}{n+\delta}, \frac{\delta n}{n+\delta}\right)$, while the stationary distribution of $\widehat{\mathcal{C}}$ is $\widehat{\mu} = \left(\frac{2}{n+2}, \frac{n}{n+2}\right)$. Thus, for $n \to \infty$ one has $\widehat{\mu} \to (0, 1)$, while $\mu \to (0, 1 - \delta, \delta)$. Thus any MDP whose induced Markov chain under some policy π is $\mathcal{C}$ satisfies the claim of the theorem, provided that π gives reward 1 in s_3 and reward 0 in s_1, s_2 (which is in accordance with ε-aggregation in respect to d_v). $\qquad\square$

Thus the results for discounted MDPs are not transferable to the average reward case. Indeed, as shown in [9], the average reward ρ_π may be expressed via the discounted rewards $\rho_\pi^\gamma(s)$ as $\rho_\pi = (1 - \gamma) \sum_s \mu_\pi(s) \rho_\pi^\gamma(s)$. This means that the stationary distribution μ_π under π plays an important role. The loss by aggregation remains small (just as in the discounted case) as long as $\widehat{\mu}$ approximates μ well, that is, $\widehat{\mu}(S_i) \approx \sum_{s \in S_i} \mu(s)$. The quality of approximation however can be estimated using the mixing time as Theorem 3 shows. Note that the mixing time in the example of Theorem 6 becomes arbitrarily large.

5.2 Alternative Perturbation Bounds

The perturbation bound for stationary distributions of Markov chains of Theorem 3, which we used in the proofs of Theorems 2 and 5, may be replaced with an arbitrary alternative perturbation bound of the form

$$\|\mu - \widetilde{\mu}\|_q \leq \lambda \|P - \widetilde{P}\|_\infty.$$

There are several such bounds in the literature (for an overview see [10]). These differ from each other in at most two aspects, namely (i) the used norm q (which is either 1 or ∞) and (ii) the *conditioning number* λ. Obviously, bounds which hold for the ∞-norm instead of the 1-norm are impractical, as they would amount to an additional factor $|S|$ in the bounds of Theorems 2 and 5. Among the 1-norm bounds the conditioning number in terms of the mixing time used by Hunter has the advantage of being rather intuitive. However, there is little general knowledge about the size of the mixing time (cf. [5] for results in some special cases and also some comparison to other 1-norm conditioning numbers, which complements the overview given in [10]). Moreover, Seneta's *ergodicity coefficient* [11], which among the 1-norm conditioning numbers considered in [10] is the smallest, is in general also not larger than Hunter's mixing time parameter (see [12]), so that one may want to replace Theorem 3 with Seneta's perturbation bound [11]. Of course, this basically gives the same aggregation bounds, only that Hunter's mixing time parameter is replaced with Seneta's ergodicity coefficient.

6   Online Aggregation and Other Open Problems

Online Aggregation. Consider an agent who starts in an MDP unknown to her and tries to aggregate states while still collecting information about the MDP. Obviously, if she is given access to an adequate distance function, the aggregation may be done online. For given $\varepsilon > 0$ the most straightforward way to do this is to assign each newly visited state s to an existing block S_i if possible (i.e., if all states s' in S_i have distance $< \varepsilon$ to s), or otherwise create a new block $S_j \ni s$. This is an obvious sequential clustering algorithm (called e.g. BSAS in [13]). Also, Ferns et al. [3] suggest a similar approach for offline aggregation.

Unfortunately, even if the existence of a consistent ε-partition is guaranteed (which, as we have seen, need not be the case), in general this online aggregation algorithm will give inconsistent ε-partitions. It is an interesting question whether there are more prospective algorithms for online aggregation.

More generally, a related open question is whether any online regret bounds are achievable for a combination of a suitable online aggregation algorithm with an online reinforcement learning algorithm (such as e.g. UCRL [14]). As it is of course hard to choose an appropriate ε in advance without having any information about the MDP at hand, one would need a mechanism which adapts the aggregation parameter ε to the MDP.

It may be relevant that generally, *optimal* aggregation is hard even if the MDP is known (cf. [2]). Although Even-Dar and Mansour consider discounted MDPs, their results hold generally, as the question is to find for given $\varepsilon > 0$ a minimal ε-adequate aggregation (see the modification of Definition 10).

Similarity of Actions. We have concentrated on MDPs with large state spaces. It is an interesting question whether an analogous approach will work for a similarity metric on actions, and in particular how the two approaches may be combined.

Relaxing Similarity. In many real-world problems one would want to relax the given similarity conditions. In particular, the idea that similar states shall lead to similar states under equivalent actions may not mean that states s, s' with $d(s, s') < \varepsilon$ will lead to states whose distance is $< \varepsilon$ as well. Rather one may e.g. demand that for some constant $c > 1$ the distance will be $< c\varepsilon$. Of course, under this generalized assumption no aggregation in the sense of a strict partition of the state space is possible anymore. Thus in order to deal with this setting, new methods will have to be developed.

Acknowledgements. The author would like to thank the anonymous reviewers for their helpful comments and suggestions. This work was supported in part by the the the Austrian Science Fund FWF (S9104-N04 SP4) and the IST Programme of the European Community, under the PASCAL Network of Excellence, IST-2002-506778. We also acknowledge support by the PASCAL pump priming projects "Sequential Forecasting" and "Online Performance of Reinforcement Learning with Internal Reward Functions". This publication only reflects the authors' views.

References

1. Givan, R., Dean, T., Greig, M.: Equivalence notions and model minimization in Markov decision processes. Artif. Intell. 147(1-2), 163–223 (2003)
2. Even-Dar, E., Mansour, Y.: Approximate equivalence of Markov decision processes. In: Proc. 16th COLT, pp. 581–594. Springer, Heidelberg (2003)
3. Ferns, N., Panangaden, P., Precup, D.: Metrics for finite Markov decision processes. In: Proc. 20th UAI, pp. 162–169. AUAI Press (2004)
4. Kemeny, J., Snell, J., Knapp, A.: Denumerable Markov Chains. Springer, Heidelberg (1976)
5. Hunter, J.J.: Mixing times with applications to perturbed Markov chains. Linear Algebra Appl. 417, 108–123 (2006)
6. Ortner, R.: Linear dependence of stationary distributions in ergodic Markov decision processes. Oper. Res. Lett. (in press 2007), doi:10.1016/j.orl.2006.12.001
7. Puterman, M.L.: Markov Decision Processes. Wiley, Chichester (1994)
8. Leitgeb, H.: A new analysis of quasianalysis. J. Philos. Logic 36(2), 181–226 (2007)
9. Singh, S.P., Jaakkola, T., Jordan, M.I.: Learning without state-estimation in partially observable Markovian decision processes. In: Proc. 11th ICML, pp. 284–292. Morgan Kaufmann, San Francisco (1994)
10. Cho, G.E., Meyer, C.D.: Comparison of perturbation bounds for the stationary distribution of a Markov chain. Linear Algebra Appl. 335, 137–150 (2001)
11. Seneta, E.: Sensitivity analysis, ergodicity coefficients, and rank-one updates for finite Markov chains. In: Numerical solution of Markov chains, pp. 121–129. Dekker, New York (1991)
12. Seneta, E.: Markov and the creation of Markov chains. In: MAM 2006: Markov Anniversary Meeting, pp. 1–20. Boson Books, Raleigh (2006)
13. Theodoridis, S., Koutroumbas, K.: Pattern Recognition, 3rd edn. Academic Press, San Diego (2006)
14. Auer, P., Ortner, R.: Logarithmic online regret bounds for reinforcement learning. In: Proc. 19th NIPS, pp. 49–56. MIT Press, Cambridge (2006)

On Calibration Error of Randomized Forecasting Algorithms

Vladimir V. V'yugin

Institute for Information Transmission Problems, Russian Academy of Sciences,
Bol'shoi Karetnyi per. 19, Moscow GSP-4, 127994, Russia
vyugin@iitp.ru

Abstract. Recently, it was shown that calibration with an error less than $\delta > 0$ is almost surely guaranteed with a randomized forecasting algorithm, where forecasts are chosen using randomized rounding up to δ of deterministic forecasts. We show that this error can not be improved for a large majority of sequences generated by a probabilistic algorithm: we prove that combining outcomes of coin-tossing and a transducer algorithm, it is possible to effectively generate with probability close to one a sequence "resistant" to any randomized rounding forecasting with an error much smaller than δ.

1 Introduction

A minimal requirement for testing of any prediction algorithm is that it should be calibrated (see Dawid [1]). An informal explanation of calibration would go something like this. Let a binary sequence $\omega_1, \omega_2, \ldots, \omega_{n-1}$ of outcomes is observed by a forecaster whose task is to give a probability p_n of a future event $\omega_n = 1$. A typical example is that p_n is interpreted as a probability that it will rain. Forecaster is said to be well-calibrated if it rains as often as he leads us to expect. It should rain about 80% of the days for which $p_n = 0.8$, and so on. So, for simplicity we consider binary sequences, i.e. $\omega_n \in \{0, 1\}$ for all n. We give a rigorous definition of calibration later.

We suppose that the forecasts p_n are computed by some algorithm. If the weather acts adversarially, then Oakes [6] and Dawid [2] show that a deterministic forecasting algorithm will not be always be calibrated. V'yugin [9] proved that this result holds for a large majority of sequences generated by a probabilistic algorithm.

Foster and Vohra [4] show that calibration is almost surely guaranteed with a randomizing forecasting rule, i.e., where the forecasts are chosen using private randomization and the forecasts are hidden from the weather until weather makes its decision to rain or not. Kakade and Foster [5] obtained an analogous positive result for deterministic forecasting systems and for the case where the class of "selection rules" is restricted to "continuous selection rules". This approach was further developed in Vovk et al. [8].

In Section 2 we give the definition of calibration and randomized rounding. Main result of this paper - Theorem 1, is presented in Section 3, the proof of the

M. Hutter, R.A. Servedio, and E. Takimoto (Eds.): ALT 2007, LNAI 4754, pp. 388–402, 2007.

main result is given in Section 4. This theorem shows that combining outcomes of coin-tossing and a transducer algorithm, it is possible to effectively generate with probability close to one a sequence "resistant" to randomized rounding forecasting with error much smaller than the precision of rounding. Theorems 2 and 3 show that the calibration error may be much bigger if we check calibration using "deterministic selection rules".

2 Background

Let Ω be the set of all infinite binary sequences, Ξ be the set of all finite binary sequences, and λ be the empty sequence. For any finite or an infinite sequence $\omega = \omega_1 \ldots \omega_n \ldots$ we write $\omega^n = \omega_1 \ldots \omega_n$ (we put $\omega_0 = \omega^0 = \lambda$). Also, $l(\omega^n) = n$ denotes the length of the sequence ω^n. If x is a finite sequence and ω is a finite or infinite sequence then $x\omega$ denotes the concatenation of these sequences, $x \sqsubseteq \omega$ means that $x = \omega^n$ for some n.

A deterministic *forecasting system* f is a real-valued function $f : \Xi \to [0,1]$. We consider computable forecasting systems; there is an algorithm, which given a finite sequence $\omega_1 \ldots \omega_{n-1} \in \Xi$ and an arbitrary positive rational number κ, when halts, outputs a rational approximation of $f(\omega_1 \ldots \omega_{n-1})$ up to κ. A forecasting system f is called *total* if it is defined on each finite sequence $\omega_1 \ldots \omega_{n-1}$. Any total forecasting system defines the corresponding overall probability distribution P on the set of all sequences such that its conditional probabilities satisfy

$$p_n = P(\omega_n = 1 | \omega_1, \omega_2, \ldots, \omega_{n-1}),$$

where $p_n = f(\omega_1 \ldots \omega_{n-1})$. In the following we consider only total forecasting systems.

The evaluation of probability forecasts is based on a method called *calibration* (see Dawid [1], [2]). Let f be some forecasting system and $I(p)$ be a characteristic function of some subinterval $I \subseteq [0,1]$, i.e., $I(p) = 1$ if $p \in I$, and $I(p) = 0$, otherwise. Let $\omega = \omega_1 \omega_2 \ldots$ be an infinite binary sequence.

A forecasting system f is well-calibrated for an infinite sequence $\omega_1 \omega_2 \ldots$ if for the characteristic function $I(p)$ of any subinterval of $[0,1]$ *the calibration error* tends to zero, i.e.,

$$\frac{\sum_{i=1}^n I(p_i)(\omega_i - p_i)}{\sum_{i=1}^n I(p_i)} \longrightarrow 0 \tag{1}$$

as the denominator of the relation (1) tends to infinity; we denote $p_i = f(\omega^{i-1})$. Here, $I(p_i)$ determines some "selection rule" which define moments of time where we compute the deviation between forecasts p_i and outcomes ω_i.

Oakes [6] proposed arguments (see Dawid [3] for different proof) that no deterministic forecasting system can be well-calibrated for all possible sequences: any total forecasting system f is not calibrated for the sequence $\omega = \omega_1 \omega_2 \ldots$, where

$$\omega_i = \begin{cases} 1 \text{ if } p_i < 0.5 \\ 0 \text{ otherwise} \end{cases}$$

and $p_i = f(\omega_1 \ldots \omega_{i-1})$, $i = 1, 2, \ldots$.

A *randomized* forecasting system $f(\omega^{n-1})$ is a random variable with range in $[0, 1]$ defined on some probability space supplied with a probability distribution Pr_n, where $\omega^{n-1} \in \Xi$ is a parameter of this variable. For any n, the predictor chooses the forecast p_n of the event $\omega_n = 1$ randomly using probability distribution Pr_n of the variable $f(\omega^{n-1})$. In this case, for any given ω we can consider the probability Pr of the event (1), where Pr is the overall probability distribution generated by probability distributions Pr_n, $n = 1, 2, \ldots$.

In the following we suppose that for any ω^{n-1} the range of the random variable $f(\omega^{n-1})$ is finite, say, $\{p_{n,1}, \ldots, p_{n,m_n}\}$. The number

$$\delta_n = \inf\{|p_{n,i} - p_{n,j}| : i \neq j\}$$

is called *the level of discreteness of f on ω^{n-1}*. We also consider $\delta = \inf_n \delta_n$ - the level of discretness of f on ω.

A typical example is the uniform rounding: for any n the rational points $p_{n,i}$ divide the unit interval into equal parts of size $0 < \delta < 1$; then the level of discreteness is constant and equals δ.

Kakade and Foster [5] presented "an almost deterministic" *randomized rounding* total forecasting algorithm f: an observer can only randomly round with the precision of rounding (level of discreteness) δ the deterministic forecast in order to calibrate. Then for any infinite sequence $\omega = \omega_1 \omega_2 \ldots$ the overall probability Pr of the event

$$\left| \frac{1}{n} \sum_{i=1}^{n} I(p_i)(\omega_i - p_i) \right| \leq \delta$$

tends to one as $n \to \infty$, where p_i is the random variable $f(\omega^{n-1})$, $I(p)$ is the characteristic function of any subinterval of $[0, 1]$.[1] This algorithm randomly rounds a forecast computed by some deterministic algorithm (constructed in [5]): for example, the forecast 0.8512 can be rounded up to second digit to 0.86 with probability 0.12, and to 0.85 with probability 0.88, at the next moment of time, the forecast 0.2588 can be rounded up to second digit to 0.26 with probability 0.88, and to 0.25 with probability 0.12. Here we have in mind some algorithm defining the direction of rounding.

---

[1] In fact, more accurate calculations show that this inequality can be replaced on

$$\left| \frac{1}{\alpha(n)\sqrt{n}} \sum_{i=1}^{n} I(p_i)(\omega_i - p_i) \right| \leq \delta,$$

where $\alpha(n)$ is any unbounded nondecreasing function; then (1) holds for Kakade and Foster's algorithm if $\lim_{n \to \infty} \frac{1}{\sqrt{n}} \sum_{i=1}^{n} I(p_i) = \infty$. We do not go into details, since in this paper we prove results in the opposite direction (see [5], [8]).

3   Main Results

We need some computability concepts. Let $\mathcal{R}$ be the set of all real numbers extended by adding the infinities $-\infty$ and $+\infty$, A is some set of finite objects; the elements of A can be effectively enumerated by positive integer numbers (see Rogers [7]). In particular, we will identify a computer program and its number. We fix some effective one-to-one enumeration of all pairs (triples, and so on) of nonnegative integer numbers. We identify any pair (t, s) and its number $\langle t, s \rangle$.

A function $\phi \colon A \to \mathcal{R}$ is called (lower) semicomputable if $\{(r, x) : r < \phi(x)\}$ (r is a rational number) is a recursively enumerable set. This means that there is an algorithm which when fed with a rational number r and a finite object x eventually stops if $r < \phi(x)$ and never stops, otherwise. In other words, the semicomputability of f means that if $\phi(x) > r$ this fact will sooner or later be learned, whereas if $f(x) \le r$ we may be for ever uncertain. A function ϕ is upper semicomputable if $-\phi$ is lower semicomputable.

Standard argument based on the recursion theory shows that there exist the lower and upper semicomputable real functions $\phi^-(j, x)$ and $\phi^+(k, x)$ universal for all lower semicomputable and upper semicomputable functions from $x \in \Xi$.[2] As follows from the definition, for every computable real function $\phi(x)$ there exist a pair $\langle j, k \rangle$ such that

$$\phi(x) = \phi^-(j, x) = \phi^+(k, x)$$

for all x. Let $\phi^-_s(j, x)$ be equal to the maximal rational number r such that the triple (r, j, x) is enumerated in s steps in the process of enumerating of the set

$$\{(r, j, x) : r < \phi(j, x), \ r \text{ is rational}\}$$

and equals $-\infty$, otherwise. Any such function $\phi^-_s(j, x)$ takes only finite number of rational values distinct from $-\infty$. By definition, $\phi^-_s(j, x) \le \phi^-_{s+1}(j, x)$ for all j, s, x, and

$$\phi^-(j, x) = \lim_{s \to \infty} \phi^-_s(j, x).$$

An analogous non-increasing sequence of functions $\phi^+_s(k, x)$ exists for any upper semicomputable function.

Let $i = \langle t, k \rangle$. We say that the function $\phi_i(x)$ is *defined on* x if given any degree of precision - positive rational number $\kappa > 0$, it holds

$$|\phi^+_s(t, x) - \phi^-_s(k, x)| \le \kappa$$

for some s; $\phi_i(x)$ undefined, otherwise. If any such s exists then for minimal such s, $\phi_{i,\kappa}(x) = \phi^-_s(k, x)$ is called the rational approximation (from below) of $\phi_i(x)$ up to κ; $\phi_{i,\kappa}(x)$ undefined, otherwise.

Any measure P on Ω can be defined as follows. Let us consider intervals

$$\Gamma_z = \{\omega \in \Omega : z \sqsubseteq \omega\},$$

---

[2] This means that each lower semicomputable function $\phi(x)$ can be represented as $\phi(x) = \phi^-(j, x)$ for some j. The same holds for upper semicomputability.

where $z \in \Xi$. We denote $P(z) = P(\Gamma_z)$ for $z \in \Xi$ and extend this function on all Borel subsets of Ω in a standard way.

A measure P is computable if there exists an algorithm which given $z \in \Xi$ and a degree of precision κ computes the number $P(z)$ up to κ.

We use also a concept of *computable operation* on $\Xi \bigcup \Omega$ [10,11]. Let $\hat{F}$ be a recursively enumerable set of ordered pairs of finite sequences satisfying the following properties:

- (i) $(x, \lambda) \in \hat{F}$ for each x;
- (ii) if $(x, y) \in \hat{F}$, $(x', y') \in \hat{F}$ and $x \sqsubseteq x'$ then $y \sqsubseteq y'$ or $y' \sqsubseteq y$ for all finite binary sequences x, x', y, y'.

A computable operation F is defined as follows

$$F(\omega) = \sup\{y \mid x \sqsubseteq \omega \text{ and } (x, y) \in \hat{F} \text{ for some } x\},$$

where $\omega \in \Omega \bigcup \Xi$ and sup is in the sense of the partial order $\sqsubseteq$ on Ξ.

Informally, the computable operation F is defined by some algorithm; this algorithm when fed with an infinite or a finite sequence ω takes it sequentially bit by bit, processes it, and produces an output sequence also sequentially bit by bit.

A *probabilistic algorithm* is a pair (P, F), where P is a computable measure on the set of all binary sequences and F is a computable operation. For any probabilistic algorithm (P, F) and a set $A \subseteq \Omega$, we consider the probability

$$P\{\omega : F(\omega) \in A\}$$

of generating by means of F a sequence from A given a sequence ω distributed according to the computable probability distribution P. In the following $P = L$, where $L(x) = L(\Gamma_x) = 2^{-l(x)}$ is the uniform measure on Ω.

A natural definition of computable randomized forecasting system f would be the following: a random variable f is computable if its probability distribution function

$$\phi(\alpha; \omega^{n-1}) = Pr_n\{f(\omega^{n-1}) < \alpha\}$$

is "a computable real function" from arguments $\alpha \in [0, 1]$ and $\omega^{n-1} \in \Xi$. The precise definition requires some technicalities. In fact, in the construction below, we compute ϕ only at one point $\alpha = 0.5$; so, we will use the following definition. A randomized forecasting system f is *weakly computable* if its *weak probability distribution function*

$$\varphi_n(\omega^{n-1}) = Pr_n\{f(\omega^{n-1}) < 0.5\}$$

is a computable function from ω^{n-1}.

Let $I_0 = I_0(p)$ be the characteristic function of the interval $(0, \frac{1}{2})$ and $I_1 = I_1(p)$ be the characteristic function of the interval $[\frac{1}{2}, 1)$. The following theorem is the main result of this paper.

Theorem 1. *For any $\epsilon > 0$ a probabilistic algorithm (L, F) can be constructed, which with probability $\geq 1 - \epsilon$ outputs an infinite binary sequence $\omega = \omega_1 \omega_2 \ldots$ such that for every weakly computable randomized forecasting system f with level of discreteness δ on ω, for some $\nu = 0$ or $\nu = 1$, the overall probability of the event*

$$\limsup_{n \to \infty} \left| \frac{1}{n} \sum_{i=1}^{n} I_\nu(p_i)(\omega_i - p_i) \right| \geq 0.25\delta \qquad (2)$$

equals one, where the overall probability is associated with f and $p_i = f(\omega^{i-1})$, $i = 1, 2, \ldots$, is a random variable.

The following deterministic analogue of Theorem 1 was obtained in V'yugin [9].

Theorem 2. *For any $\epsilon > 0$ a probabilistic algorithm (L, F) can be constructed, which with probability $\geq 1 - \epsilon$ outputs an infinite binary sequence $\omega = \omega_1 \omega_2 \ldots$ such that for every deterministic forecasting algorithm f, for some $\nu = 0$ or $\nu = 1$,*

$$\limsup_{n \to \infty} \left| \frac{1}{n} \sum_{i=1}^{n} I_\nu(p_i)(\omega_i - p_i) \right| \geq 0.5,$$

where $p_i = f(\omega^{i-1})$, $i = 1, 2, \ldots$.

Theorem 1 uses randomized "selection rules" - $I_\nu(p_i)$, $\nu = 0, 1$. In case of some natural deterministic "selection rule", we obtain the following theorem.

Let $E(f(\omega^{i-1}))$ be the mean value of the forecasts produced by a randomized forecasting system f given an input sequence ω^{i-1}.

In the following theorem we use more strong definition of computability of randomized forecasting systems - we consider randomized forecasting systems with computable mathematical expectations: for any such system f its mathematical expectation $E(f(\omega^{i-1}))$ is a computable real function from ω^{n-1}.

Theorem 3. *For any $\epsilon > 0$ a probabilistic algorithm (L, F) can be constructed, which with probability $\geq 1 - \epsilon$ outputs an infinite binary sequence $\omega = \omega_1 \omega_2 \ldots$ such that for every randomized forecasting system f with computable mathematical expectation, for some $\nu = 0$ or $\nu = 1$, the overall probability of the event*

$$\limsup_{n \to \infty} \left| \frac{1}{n} \sum_{i=1}^{n} I_\nu(E(p_i))(\omega_i - p_i) \right| \geq 0.5 \qquad (3)$$

equals one, where $p_i = f(\omega^{i-1})$, $i = 1, 2, \ldots$.

4 Proofs of Theorems 1-3

For any probabilistic algorithm (P, F), we consider the function

$$Q(x) = P\{\omega : x \sqsubseteq F(\omega)\}. \qquad (4)$$

It is easy to verify that this function is lower semicomputable and satisfies:

$$Q(\lambda) \leq 1;$$
$$Q(x0) + Q(x1) \leq Q(x)$$

for all x. Any function satisfying these properties is called semicomputable semimeasure. For any semicomputable semimeasure Q a probabilistic algorithm (L, F) exists such that (4) holds, where $P = L$ (for the proof see [10,11]).

Though the semimeasure Q is not a measure, we consider the corresponding measure on the set Ω

$$\bar{Q}(\Gamma_x) = \inf_n \sum_{l(y)=n, x \sqsubseteq y} Q(y).$$

This function can be extended on all Borel subsets A of Ω (see [11]).

We will construct a semicomputable semimeasure Q as a some sort of network flow. We define an infinite network on the base of the infinite binary tree. Any $x \in \Xi$ defines two edges $(x, x0)$ and $(x, x1)$ of length one. In the construction below we will mount to the network extra edges (x, y) of length > 1, where $x, y \in \Xi$, $x \sqsubseteq y$ and $y \neq x0, x1$. By the length of the edge (x, y) we mean the number $l(y) - l(x)$. For any edge $\sigma = (x, y)$ we denote by $\sigma_1 = x$ its starting vertex and by $\sigma_2 = y$ its terminal vertex. A computable function $q(\sigma)$ defined on all edges of length one and on all extra edges and taking rational values is called a *network* if for all $x \in \Xi$

$$\sum_{\sigma:\sigma_1=x} q(\sigma) \leq 1.$$

Let G be the set of all extra edges of the network q (it is a part of the domain of q). By *q-flow* we mean the minimal semimeasure P such that $P \geq R$, where the function R is defined by the following recursive equations

$$R(\lambda) = 1;$$
$$R(y) = \sum_{\sigma:\sigma_2=y} q(\sigma)R(\sigma_1) \tag{5}$$

for $y \neq \lambda$. We can define the semimeasure P using R as follows. A set D is prefix-free if $x \not\sqsubseteq y$ for all $x, y \in D$, $x \neq y$. Let π_x be the set of all prefix-free sets D such that $x \sqsubseteq y$ for all $y \in D$. Then it holds

$$P(x) = \sup_{D \in \pi_x} \sum_{x::x \in D} R(x).$$

A network q is called *elementary* if the set of extra edges is finite and $q(\sigma) = 1/2$ for almost all edges of unit length. For any network q, we define the *network flow delay* function (*q-delay* function)

$$d(x) = 1 - q(x, x0) - q(x, x1).$$

The construction below works with all computable real functions $\phi_t(x)$, $x \in \Xi$, $t = 1, 2, \ldots$; any $i = \langle t, s \rangle$ is considered as a program for computing the rational approximation $\phi_{t,\kappa_s}(\omega^{n-1})$ of ϕ_t from below up to $\kappa_s = 1/s$.[3] In the proof (see Lemma 6) we use a special class of these functions, namely, functions of the type

$$\phi(\omega^{n-1}) = Pr_n\{f(\omega^{n-1}) \geq 0.5\} = 1 - \varphi_n(\omega^{n-1}), \tag{6}$$

where $\varphi_n(\omega^{n-1})$ is a weak probability distribution function for some weakly computable randomized forecasting system f. We have $\phi = \phi_t$ for some t, and by the construction below we visit any function ϕ_t on infinitely many steps n. To do this, we define some function $p(n)$ such that for any positive integer number i we have $p(n) = i$ for infinitely many n. For example, we can define $p(\langle i, s \rangle) = i$ for all i and s.

For any program $i = \langle t, s \rangle$, any finite binary sequences x and y, any elementary network q, and for any integer number n, let $B(i, x, y, q, n)$ be $true$ if the following conditions hold

- (i) $l(y) = n$, $x \sqsubseteq y$,
- (ii) $d(y^k) < 1$ for all k, $1 \leq k \leq n$, where d is the q-delay function and $y^k = y_1 \ldots y_k$;
- (iii) for all k such that $l(x) \leq k < il(x)$ the values $\phi_{t,\kappa_s}(y^k)$ are defined in $\leq n$ steps and

$$y_{k+1} = \begin{cases} 0 \text{ if } \phi_{t,\kappa_s}(y^k) \geq 0.5 \\ 1 \text{ otherwise.} \end{cases}$$

Let $B(i, x, y, q, n)$ be $false$, otherwise. Define

$$\beta(x, q, n) = \min\{y : p(l(y)) = p(l(x)), B(p(l(x)), x, y, q, n)\}$$

Here min is considered for lexicographical ordering of strings; we suppose that $\min \emptyset$ is undefined.

Lemma 1. For any total function ϕ_t, $\beta(x, q, n)$ is defined for all $x \in \Xi$ and for all sufficiently large n.

Proof. The needed sequence y can be easily defined for all sufficiently large n sequentially bit-by-bit, since $\phi_{t,\kappa_s}(z)$ is defined for all z. ☐

The goal of the construction below is the following. Any extra edge σ will be assigned to some task number i such that $p(l(\sigma_1)) = p(l(\sigma_2)) = i$. The goal of the task i is to define a finite set of extra edges σ such that for any infinite binary sequence ω one of the following conditions hold: either ω contains some extra edge as a subword, or the network flow delay function d equals 1 on some initial fragment of ω. For any extra edge σ mounted to the network q, $B(i, \sigma_1, \sigma_2, q^{n-1}, n)$ is true; it is false, otherwise. Lemma 5 shows that $\bar{Q}(E_Q) > 1 - \frac{1}{2}\epsilon$, where Q is

---

[3] Recall that $t = \langle j, k \rangle$ for some j, k; we use the lower and upper semicomputable real functions $\phi^-(j, x)$ and $\phi^+(k, x)$ universal for all lower semicomputable and upper semicomputable functions from $x \in \Xi$ to compute values $\phi_t(x)$.

the q-flow and E_Q is defined by (7) below. Lemma 6 shows that for each $w \in E_Q$ the event (2) holds with the overall probability one.

Construction. Let $\rho(n) = (n + n_0)^2$ for some sufficiently large n_0 (the value n_0 will be specified below in the proof of Lemma 5).

Using the mathematical induction by n, we define a sequence q^n of elementary networks. Put $q^0(\sigma) = 1/2$ for all edges σ of length one.

Let $n > 0$ and a network q^{n-1} is defined. Let d^{n-1} be the q^{n-1}-delay function and let G^{n-1} be the set of all extra edges. We suppose also that $l(\sigma_2) < n$ for all $\sigma \in G^{n-1}$.

Let us define a network q^n. At first, we define a network flow delay function d^n and a set G^n. The construction can be split up into two cases.

Let $w(i, q^{n-1})$ be equal to the minimal m such that $p(m) = i$ and $m > (i + 1)l(\sigma_2)$ for each extra edge $\sigma \in G^{n-1}$ such that $p(l(\sigma_1))) < i$.

The inequality $w(i, q^m) \neq w(i, q^{m-1})$ can be induced by some task $j < i$ that mounts an extra edge $\sigma = (x, y)$ such that $l(y) > w(i, q^{m-1})$ and $p(l(x)) = p(l(y)) = j$. Lemma 2 (below) will show that this can happen only at finitely many steps of the construction.

Case 1. $w(p(n), q^{n-1}) = n$ (the goal of this part is to start a new task $i = p(n)$ or to restart the existing task $i = p(n)$ if it was destroyed by some task $j < i$ at some preceding step).

Put $d^n(y) = 1/\rho(n)$ for $l(y) = n$ and define $d^n(y) = d^{n-1}(y)$ for all other y. Put also $G^n = G^{n-1}$.

Case 2. $w(p(n), q^{n-1}) < n$ (the goal of this part is to process the task $i = p(n)$).

Let C_n be the set of all x such that $w(i, q^{n-1}) \leq l(x) < n$, $0 < d^{n-1}(x) < 1$, the function $\beta(x, q^{n-1}, n)$ is defined[4] and there is no extra edge $\sigma \in G^{n-1}$ such that $\sigma_1 = x$.

In this case for each $x \in C_n$ define $d^n(\beta(x, q^{n-1}, n)) = 0$, and for all other y of length n such that $x \sqsubset y$ define

$$d^n(y) = \frac{d^{n-1}(x)}{1 - d^{n-1}(x)}.$$

Define $d^n(y) = d^{n-1}(y)$ for all other y. We add an extra edge to G^{n-1}, namely, define

$$G^n = G^{n-1} \cup \{(x, \beta(x, q^{n-1}, n)) : x \in C_n\}.$$

We say that the task $i = p(n)$ *mounts* the extra edge $(x, \beta(x, q^{n-1}, n))$ to the network and that all existing tasks $j > i$ are destroyed by the task i.

After Case 1 and Case 2, define for any edge σ of unit length

$$q^n(\sigma) = \frac{1}{2}(1 - d^n(\sigma_1))$$

and $q^n(\sigma) = d^n(\sigma_1)$ for each extra edge $\sigma \in G^n$.

---

[4] In particular, $p(l(x)) = i$ and $l(\beta(x, q^{n-1}, n)) = n$.

Case 3. Cases 1 and 2 do not hold.

Define $d^n = d^{n-1}$, $q^n = q^{n-1}$, $G^n = G^{n-1}$.

As the result of the construction we define the network $q = \lim_{n \to \infty} q^n$, the network flow delay function $d = \lim_{n \to \infty} d^n$ and the set of extra edges $G = \cup_n G^n$.

The functions q and d are computable and the set G is recursive by their definitions. Let Q denotes the q-flow.

The following lemma shows that any task can mount new extra edges only at finite number of steps. Let $G(i)$ be the set of all extra edges mounted by the task i, $w(i, q) = \lim_{n \to \infty} w(i, q^n)$.

Lemma 2. *The set $G(i)$ is finite, $w(i, q)$ exists and $w(i, q) < \infty$ for all i.*

Proof. Note that if $G(j)$ is finite for all $j < i$, then $w(i, q) < \infty$. Hence, we must prove that the set $G(i)$ is finite for any i. Suppose that the opposite assertion holds. Let i be the minimal such that $G(i)$ is infinite. By choice of i the sets $G(j)$ for all $j < i$ are finite. Then $w(i, q) < \infty$.

For any x such that $l(x) \geq w(i, q)$, consider the maximal m such that for some initial fragment $x^m \sqsubseteq x$ there exists an extra edge $\sigma = (x^m, y) \in G(i)$. If no such extra edge exists define $m = w(i, q)$. By definition, if $d(x^m) \neq 0$ then $1/d(x^m)$ is an integer number. Define

$$
u(x) = \begin{cases} 1/d(x^m) \text{ if } d(x^m) \neq 0, l(x) \geq w(i, q) \\ \rho(w(i, q)) \text{ if } l(x) < w(i, q) \\ 0 \text{ otherwise} \end{cases}
$$

By construction the integer valued function $u(x)$ has the property: $u(x) \geq u(y)$ if $x \sqsubseteq y$. Besides, if $u(x) > u(y)$ then $u(x) > u(z)$ for all z such that $x \sqsubseteq z$ and $l(z) = l(y)$. Then the function

$$
\hat{u}(\omega) = \min\{n : u(\omega^i) = u(\omega^n) \text{ for all } i \geq n\}
$$

is defined for all $\omega \in \Omega$. It is easy to see that this function is continuous. Since Ω is compact space in the topology generated by intervals Γ_x, this function is bounded by some number m. Then $u(x) = u(x^m)$ for all $l(x) \geq m$. By the construction, if any extra edge of ith type was mounted to $G(i)$ at some step then $u(y) < u(x)$ holds for some new pair (x, y) such that $x \sqsubseteq y$. This is contradiction with the existence of the number m.   □

An infinite sequence $\alpha \in \Omega$ is called an *i-extension* of a finite sequence x if $x \sqsubseteq \alpha$ and $B(i, x, \alpha^n, n)$ is true for almost all n.

A sequence $\alpha \in \Omega$ is called *i-closed* if $d(\alpha^n) = 1$ for some n such that $p(n) = i$, where d is the q-delay function. Note that if $\sigma \in G(i)$ is some extra edge then $B(i, \sigma_1, \sigma_2, n)$ is true, where $n = l(\sigma_2)$.

Lemma 3. *Let for any initial fragment ω^n of an infinite sequence ω some i-extension exists. Then either the sequence ω will be i-closed in the process of the construction or ω contains an extra edge of ith type (i.e. $\sigma_2 \sqsubseteq \omega$ for some $\sigma \in G(i)$).*

Proof. Let a sequence ω is not i-closed. By Lemma 2 the maximal m exists such that $p(m) = i$ and $d(\omega^m) > 0$. Since the sequence ω^m has an i-extension and $d(\omega^k) < 1$ for all k, by Case 2 of the construction a new extra edge (ω^m, y) of ith type must be mounted to the binary tree. By the construction $d(y) = 0$ and $d(z) \neq 0$ for all z such that $\omega^m \sqsubseteq z$, $l(z) = l(y)$, and $z \neq y$. By the choice of m we have $y \sqsubseteq \omega$. $\qquad\square$

Lemma 4. *It holds $Q(y) = 0$ if and only if $q(\sigma) = 0$ for some edge σ of unit length located on y (this edge satisfies $\sigma_2 \sqsubseteq y$).*

Proof. The necessary condition is obvious. To prove that this condition is sufficient, let us suppose that $q(y^n, y^{n+1}) = 0$ for some $n < l(y)$ but $Q(y) \neq 0$. Then by definition $d(y^n) = 1$. Since $Q(y) \neq 0$ an extra edge $(x, z) \in G$ exists such that $x \sqsubseteq y^n$ and $y^{n+1} \sqsubseteq z$. But, by the construction, this extra edge can not be mounted to the network $q^{l(z)-1}$ since $d(z^n) = 1$. This contradiction proves the lemma. $\qquad\square$

For any semimeasure P define

$$E_P = \{\omega \in \Omega : \forall n (P(\omega^n) \neq 0)\}$$

the support set of P. It is easy to see that E_P is a closed subset of Ω and $\bar{P}(E_P) = \bar{P}(\Omega)$. By Lemma 4, the relation $Q(y) = 0$ is recursive and

$$E_Q = \Omega \setminus \cup_{d(x)=1} \Gamma_x. \tag{7}$$

Lemma 5. *It holds $\bar{Q}(E_Q) > 1 - \frac{1}{2}\epsilon$.*

Proof. We bound $\bar{Q}(\Omega)$ from below. Let R be defined by (5). By definition of the network flow delay function, we have

$$\sum_{u:l(u)=n+1} R(u) = \sum_{u:l(u)=n} (1 - d(u))R(u) + \sum_{\sigma:\sigma\in G, l(\sigma_2)=n+1} q(\sigma)R(\sigma_1). \tag{8}$$

Define an auxiliary sequence

$$S_n = \sum_{u:l(u)=n} R(u) - \sum_{\sigma:\sigma\in G, l(\sigma_2)=n} q(\sigma)R(\sigma_1).$$

At first, we consider the case $w(p(n), q^{n-1}) < n$. If there is no edge $\sigma \in G$ such that $l(\sigma_2) = n$ then $S_{n+1} \geq S_n$. Suppose that some such edge exists. Define

$$P(u, \sigma) \iff l(u) = l(\sigma_2) \& \sigma_1 \sqsubseteq u \& u \neq \sigma_2 \& \sigma \in G.$$

By definition of the network flow delay function, we have

$$\sum_{u:l(u)=n} d(u)R(u) = \sum_{\sigma:\sigma\in G, l(\sigma_2)=n} d(\sigma_2) \sum_{u:P(u,\sigma)} R(u) =$$

$$= \sum_{\sigma:\sigma\in G, l(\sigma_2)=n} \frac{d(\sigma_1)}{1-d(\sigma_1)} \sum_{u:P(u,\sigma)} R(u) \leq \sum_{\sigma:\sigma\in G, l(\sigma_2)=n} d(\sigma_1)R(\sigma_1) =$$

$$= \sum_{\sigma:\sigma\in G, l(\sigma_2)=n} q(\sigma)R(\sigma_1). \tag{9}$$

Here we used the inequality

$$\sum_{u:P(u,\sigma)} R(u) \le R(\sigma_1) - d(\sigma_1)R(\sigma_1)$$

for all $\sigma \in G$ such that $l(\sigma_2) = n$. Combining this bound with (8) we obtain $S_{n+1} \ge S_n$.

Let us consider the case $w(p(n), q^{n-1}) = n$. Then

$$\sum_{u:l(u)=n} d(u)R(u) \le \rho(n) = \frac{1}{(n+n_0)^2}.$$

Combining (8) and (9) we obtain

$$S_{n+1} \ge S_n - \frac{1}{(n+n_0)^2}$$

for all n. Since $S_0 = 1$, this implies

$$S_n \ge 1 - \sum_{i=1}^{\infty} \frac{1}{(i+n_0)^2} \ge 1 - \frac{1}{2}\epsilon$$

for some sufficiently large constant n_0. Since $Q \ge R$, it holds

$$\bar{Q}(\Omega) = \inf_n \sum_{l(u)=n} Q(u) \ge \inf_n S_n \ge 1 - \frac{1}{2}\epsilon.$$

Lemma is proved. $\qquad\square$

Lemma 6. *For each weakly computable randomized forecasting system f and and for each sequence $w \in E_Q$, the event (2) holds with the overall probability one.*

Proof. Let w be an infinite sequence and let f be a weakly computable randomized forecasting system, i.e., the corresponding $\phi_t(w^{n-1})$ (defined by (6)) is defined for all n. Let $i = \langle t, s \rangle$ be a program for computing the rational approximation ϕ_{t,κ_s} from below up to $\kappa_s = 1/s$. Since in the construction we visit ϕ_t on infinitely many steps n such that $p(n) = i = \langle t, s \rangle$, where $s = 1, 2, \ldots$, in the proof we will consider only sufficiently large i.

By definition $d(w^n) < 1$ for all n if $w \in E_Q$. Since w is an i-extension of w^n for each n, by Lemma 3 there exists an extra edge $\sigma \in G(i)$ such that $\sigma_2 \sqsubseteq w$. In the following, let $k = l(\sigma_1)$ and $n = ik$.

Denote $p_j^- = \max\{p_{j,s} : p_{j,s} < 0.5\}$ and $p_j^+ = \min\{p_{j,s} : p_{j,s} \ge 0.5\}$, where $\{p_{j,1}, \ldots, p_{j,m_j}\}$ is the range of the random variable $f(w^{j-1})$.[5] By definition of precision of rounding $p_j^+ - p_j^- \ge \delta$ for all j.

---

[5] For technical reason, if necessary we add 0 and 1 to values of $f(w^{n-1})$ and set their probabilities be 0.

Denote $p_j = f(\omega^{j-1})$, $j = 1, 2, \ldots$. By definition p_j is a random variable. In the following we use the inequality

$$\phi_{t,\kappa_s}(\omega^{j-1}) \leq Pr\{p_j \geq 0.5\} \leq \phi_{t,\kappa_s}(\omega^{j-1}) + \kappa_s.$$

Consider two random variables

$$\vartheta_{n,1} = \sum_{j=k+1}^{n} \xi(p_j \geq 0.5)(\omega_j - p_j), \tag{10}$$

$$\vartheta_{n,2} = \sum_{j=k+1}^{n} \xi(p_j < 0.5)(\omega_j - p_j), \tag{11}$$

where $\xi(true) = 1$, and $\xi(false) = 0$.

We compute the bounds of mathematical expectations of these variables. These expectations are taken with respect to the overall probability distribution Pr generated by probability distributions Pr_j of random variables p_j, $j = 1, 2, \ldots$ (ω is fixed). Using the definition of the subword $\sigma \in G(i)$ of the sequence ω, we obtain ($k < j \leq n$)

$$E(\vartheta_{n,1}) \leq \sum_{\omega_j=0} Pr\{p_j \geq 0.5\}(-p_j^+) + \sum_{\omega_j=1} Pr\{p_j \geq 0.5\}(1 - p_j^+) \leq \tag{12}$$

$$-0.5 \sum_{j=k+1}^{n} \xi(\omega_j = 0)p_j^+ + (0.5 + \kappa_s) \sum_{j=k+1}^{n} \xi(\omega_j = 1)(1 - p_j^+).$$

$$E(\vartheta_{n,2}) \geq \sum_{\omega_j=0} Pr\{p_j < 0.5\}(-p_j^-) + \sum_{\omega_j=1} Pr\{p_j < 0.5\}(1 - p_j^-) \geq \tag{13}$$

$$-0.5 \sum_{j=k+1}^{n} \xi(\omega_j = 0)p_j^- + (0.5 - \kappa_s) \sum_{j=k+1}^{n} \xi(\omega_j = 1)(1 - p_j^-).$$

Subtracting (12) from (13) we obtain

$$E(\vartheta_{n,2}) - E(\vartheta_{n,1}) \geq 0.5 \sum_{j=k+1}^{n} \xi(\omega_j = 0)(p_j^+ - p_j^-) +$$

$$0.5 \sum_{j=k+1}^{n} \xi(\omega_j = 1)(p_j^+ - p_j^-) - \kappa_s \sum_{j=k+1}^{n} \xi(\omega_j = 1)(2 - p_j^- - p_j^+) \geq$$

$$\geq 0.5\delta(n - k) - 2\kappa_s(n - k) = (0.5\delta - 2\kappa_s)(n - k). \tag{14}$$

Then

$$E(\vartheta_{n,1}) \leq (-0.25\delta - \kappa_s)(n - k)$$

or

$$E(\vartheta_{n,2}) \geq (0.25\delta - \kappa_s)(n - k)$$

for infinitely many n, k. Since for any fixed f_t the ratio $k/n = i^{-1}$ and the number $\kappa_s = 1/s$ become arbitrary small for large i such that $i = \langle t, s \rangle$ for some s, we have

$$\liminf_{n \to \infty} \frac{1}{n} E(\vartheta_{n,1}) \leq -0.25\delta$$

or

$$\limsup_{n \to \infty} \frac{1}{n} E(\vartheta_{n,2}) \geq 0.25\delta.$$

The martingale strong law of large numbers: for $\nu = 1, 2$, with Pr-probability one

$$\frac{1}{n} \sum_{j=1}^{n} I_\nu(p_j)(\omega_j - p_j) - \frac{1}{n} E(\vartheta_{n,\nu}) \to 0$$

as $n \to \infty$, implies that for $\nu = 0$ or for $\nu = 1$ the overall probability of the event (2) equals one. Lemma 6 and Theorem 1 are proved. $\qquad\square$

Note, that inequalities (14) show that condition (2) of Theorem 1 can be replaced on

$$\limsup_{n \to \infty} \left| \frac{1}{n} \sum_{j=1}^{n} I_\nu(p_j)(\omega_j - p_j) \right| \geq 0.25\delta - \kappa$$

if in the construction of our algorithm the function (6) is computed up to a fixed precision κ.

The proof of Theorem 2 is in the line of the proof of Theorem 1, where $\phi(\omega^{n-1})$ denote a deterministic forecasting system. We have in the proof of Lemma 6, for some $\nu = 0$ or $\nu = 1$,

$$\sum_{j=k+1}^{n} I_\nu(p_j)(\omega_i - p_j) \geq (0.5 - 2\kappa_s)(n - k) \tag{15}$$

for infinitely many $k, n = ik$, where $p_j = f_i(\omega^{j-1})$, $j = 1, 2, \ldots$.

To prove (3) of Theorem 3 we define in (6) $\phi(\omega^{j-1}) = E(f(\omega^{j-1}))$ - the mathematical expectation of a random variable $f(\omega^{j-1})$. Then in the proof of Lemma 6, for some $\nu = 0$ or $\nu = 1$, the inequality (15), where p_j is replaced on $E(f(\omega^{j-1}))$, holds for infinitely many n. By the martingale strong law of large numbers we obtain that for $\nu = 0$ and for $\nu = 1$ with the overall probability one

$$\frac{1}{n} \sum_{j=1}^{n} I_\nu(E(p_j))(p_j - E(p_j)) \to 0 \tag{16}$$

as $n \to \infty$. Combining (16) with (15) modified as above, we obtain (3).

Acknowledgements

This research was partially supported by Russian foundation for fundamental research: 03-01-00475-a; 06-01-00122-a. A part of this work was done while the author was in Poncelet Laboratoire LIF CNRS, Marseille, France.

References

1. Dawid, A.P.: The well-calibrated Bayesian [with discussion]. J. Am. Statist. Assoc. 77, 605–613 (1982)
2. Dawid, A.P.: Calibration-based empirical probability [with discussion]. Ann. Statist. 13, 1251–1285 (1985)
3. Dawid, A.P.: The impossibility of inductive inference. J. Am. Statist. Assoc. 80, 340–341 (1985)
4. Foster, D.P., Vohra, R.: Asymptotic calibration. Biometrika 85, 379–390 (1998)
5. Kakade, S.M., Foster, D.P.: Deterministic calibration and Nash equilibrium. In: Shawe-Taylor, J., Singer, Y. (eds.) COLT 2004. LNCS (LNAI), vol. 3120, pp. 33–48. Springer, Heidelberg (2004)
6. Oakes, D.: Self-calibrating priors do not exists [with discussion]. J. Am. Statist. Assoc. 80, 339–342 (1985)
7. Rogers, H.: Theory of recursive functions and effective computability. McGraw-Hill, New York (1967)
8. Vovk, Vladimir, Takemura, Akimichi, Shafer, Glenn: Defensive Forecasting. In: Proceedings of the Tenth International Workshop on Artificial Intelligence and Statistics. pp. 365–372 (2005), http://arxiv.org/abs/cs/0505083
9. V'yugin, V.V.: Non-stochastic infinite and finite sequences. Theor. Comp. Science. 207, 363–382 (1998)
10. Uspensky, V.A., Semenov, A.L., Shen, A.Kh.: Can an individual sequence of zeros and ones be random. Russian Math. Surveys 45(1), 121–189 (1990)
11. Zvonkin, A.K., Levin, L.A.: The complexity of finite objects and the algorithmic concepts of information and randomness. Russ. Math. Surv. 25, 83–124 (1970)

Author Index

Lecture Notes in Artificial Intelligence (LNAI)

Vol. 4562: D. Harris (Ed.), Engineering Psychology and Cognitive Ergonomics. XXIII, 879 pages. 2007.

Vol. 4548: N. Olivetti (Ed.), Automated Reasoning with Analytic Tableaux and Related Methods. X, 245 pages. 2007.

Vol. 4539: N.H. Bshouty, C. Gentile (Eds.), Learning Theory. XII, 634 pages. 2007.

Vol. 4529: P. Melin, O. Castillo, L.T. Aguilar, J. Kacprzyk, W. Pedrycz (Eds.), Foundations of Fuzzy Logic and Soft Computing. XIX, 830 pages. 2007.

Vol. 4520: M.V. Butz, O. Sigaud, G. Pezzulo, G. Baldassarre (Eds.), Anticipatory Behavior in Adaptive Learning Systems. X, 379 pages. 2007.

Vol. 4511: C. Conati, K. McCoy, G. Paliouras (Eds.), User Modeling 2007. XVI, 487 pages. 2007.

Vol. 4509: Z. Kobti, D. Wu (Eds.), Advances in Artificial Intelligence. XII, 552 pages. 2007.

Vol. 4496: N.T. Nguyen, A. Grzech, R.J. Howlett, L.C. Jain (Eds.), Agent and Multi-Agent Systems: Technologies and Applications. XXI, 1046 pages. 2007.

Vol. 4483: C. Baral, G. Brewka, J. Schlipf (Eds.), Logic Programming and Nonmonotonic Reasoning. IX, 327 pages. 2007.

Vol. 4482: A. An, J. Stefanowski, S. Ramanna, C.J. Butz, W. Pedrycz, G. Wang (Eds.), Rough Sets, Fuzzy Sets, Data Mining and Granular Computing. XIV, 585 pages. 2007.

Vol. 4481: J. Yao, P. Lingras, W.-Z. Wu, M. Szczuka, N.J. Cercone, D. Ślęzak (Eds.), Rough Sets and Knowledge Technology. XIV, 576 pages. 2007.

Vol. 4476: V. Gorodetsky, C. Zhang, V.A. Skormin, L. Cao (Eds.), Autonomous Intelligent Systems: Multi-Agents and Data Mining. XIII, 323 pages. 2007.

Vol. 4456: Y. Wang, Y.-m. Cheung, H. Liu (Eds.), Computational Intelligence and Security. XXIII, 1118 pages. 2007.

Vol. 4455: S. Muggleton, R. Otero, A. Tamaddoni-Nezhad (Eds.), Inductive Logic Programming. XII, 456 pages. 2007.

Vol. 4452: M. Fasli, O. Shehory (Eds.), Agent-Mediated Electronic Commerce. VIII, 249 pages. 2007.

Vol. 4451: T.S. Huang, A. Nijholt, M. Pantic, A. Pentland (Eds.), Artifical Intelligence for Human Computing. XVI, 359 pages. 2007.

Vol. 4441: C. Müller (Ed.), Speaker Classification. X, 309 pages. 2007.

Vol. 4438: L. Maicher, A. Sigel, L.M. Garshol (Eds.), Leveraging the Semantics of Topic Maps. X, 257 pages. 2007.

Vol. 4434: G. Lakemeyer, E. Sklar, D.G. Sorrenti, T. Takahashi (Eds.), RoboCup 2006: Robot Soccer World Cup X. XIII, 566 pages. 2007.

Vol. 4429: R. Lu, J.H. Siekmann, C. Ullrich (Eds.), Cognitive Systems. X, 161 pages. 2007.

Vol. 4428: S. Edelkamp, A. Lomuscio (Eds.), Model Checking and Artificial Intelligence. IX, 185 pages. 2007.

Vol. 4426: Z.-H. Zhou, H. Li, Q. Yang (Eds.), Advances in Knowledge Discovery and Data Mining. XXV, 1161 pages. 2007.

Vol. 4411: R.H. Bordini, M. Dastani, J. Dix, A.E.F. Seghrouchni (Eds.), Programming Multi-Agent Systems. XIV, 249 pages. 2007.

Vol. 4410: A. Branco (Ed.), Anaphora: Analysis, Algorithms and Applications. X, 191 pages. 2007.

Vol. 4399: T. Kovacs, X. Llorà, K. Takadama, P.L. Lanzi, W. Stolzmann, S.W. Wilson (Eds.), Learning Classifier Systems. XII, 345 pages. 2007.

Vol. 4390: S.O. Kuznetsov, S. Schmidt (Eds.), Formal Concept Analysis. X, 329 pages. 2007.

Vol. 4389: D. Weyns, H. Van Dyke Parunak, F. Michel (Eds.), Environments for Multi-Agent Systems III. X, 273 pages. 2007.

Vol. 4386: P. Noriega, J. Vázquez-Salceda, G. Boella, O. Boissier, V. Dignum, N. Fornara, E. Matson (Eds.), Coordination, Organizations, Institutions, and Norms in Agent Systems II. XI, 373 pages. 2007.

Vol. 4384: T. Washio, K. Satoh, H. Takeda, A. Inokuchi (Eds.), New Frontiers in Artificial Intelligence. IX, 401 pages. 2007.

Vol. 4371: K. Inoue, K. Satoh, F. Toni (Eds.), Computational Logic in Multi-Agent Systems. X, 315 pages. 2007.

Vol. 4369: M. Umeda, A. Wolf, O. Bartenstein, U. Geske, D. Seipel, O. Takata (Eds.), Declarative Programming for Knowledge Management. X, 229 pages. 2006.

Vol. 4343: C. Müller (Ed.), Speaker Classification I. X, 355 pages. 2007.

Vol. 4342: H. de Swart, E. Orłowska, G. Schmidt, M. Roubens (Eds.), Theory and Applications of Relational Structures as Knowledge Instruments II. X, 373 pages. 2006.

Vol. 4335: S.A. Brueckner, S. Hassas, M. Jelasity, D. Yamins (Eds.), Engineering Self-Organising Systems. XII, 212 pages. 2007.

Vol. 4334: B. Beckert, R. Hähnle, P.H. Schmitt (Eds.), Verification of Object-Oriented Software. XXIX, 658 pages. 2007.

Vol. 4333: U. Reimer, D. Karagiannis (Eds.), Practical Aspects of Knowledge Management. XII, 338 pages. 2006.

Vol. 4327: M. Baldoni, U. Endriss (Eds.), Declarative Agent Languages and Technologies IV. VIII, 257 pages. 2006.

Vol. 4314: C. Freksa, M. Kohlhase, K. Schill (Eds.), KI 2006: Advances in Artificial Intelligence. XII, 458 pages. 2007.

Vol. 4304: A. Sattar, B.-h. Kang (Eds.), AI 2006: Advances in Artificial Intelligence. XXVII, 1303 pages. 2006.

Vol. 4303: A. Hoffmann, B.-h. Kang, D. Richards, S. Tsumoto (Eds.), Advances in Knowledge Acquisition and Management. XI, 259 pages. 2006.

Vol. 4293: A. Gelbukh, C.A. Reyes-Garcia (Eds.), MICAI 2006: Advances in Artificial Intelligence. XXVIII, 1232 pages. 2006.